Praise for

*Worlds at War*

*New York Times Book Review* Editor's Choice

"Having set the stage with great deliberation, Mr. Pagden takes a majestic stroll through the centuries, covering broad swaths of very familiar history fluently, gracefully, and always entertainingly."
—*The New York Times*

"Anthony Pagden reminds us [that] the conflict between Western Europe (and the United States) and the Muslim world still simmers and boils. . . . In *Worlds at War*, Pagden provides a sweeping, searching and somber survey—from Alexander to al-Qaida—of 2,500 years of 'perpetual enmity,' filled with crusades and jihads."
—Baltimore *Sun*

"This grand panorama of world history is an imaginative structuring of the human past in a way that might have seemed odd one hundred years ago. In 1900, Western power seemed invincible, and the East appeared prostrate. The long view observed here has reasserted itself."
—*St. Louis Post-Dispatch*

"If you are going to read only one book on the Manichaean struggle between East and West, this is the book."
—Efraim Karsh, author of *Islamic Imperialism: A History*

# WORLDS
## AT WAR

# ANTHONY PAGDEN

 RANDOM HOUSE TRADE PAPERBACKS  NEW YORK

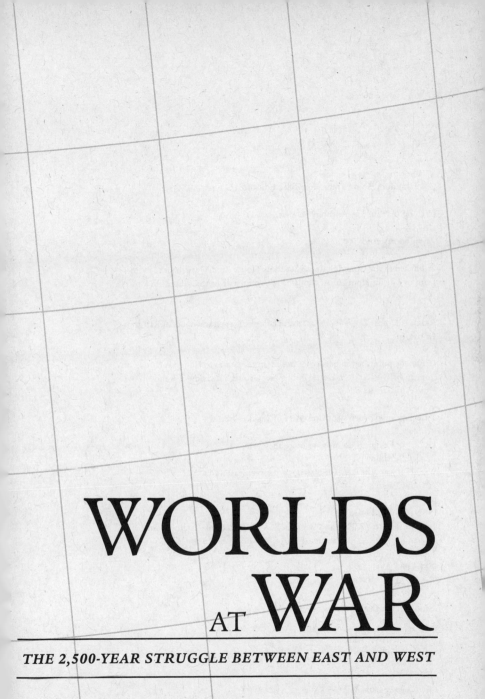

# WORLDS
## AT WAR

**THE 2,500-YEAR STRUGGLE BETWEEN EAST AND WEST**

2009 Random House Trade Paperback Edition

Published in the United States by Random House Trade Paperbacks, an imprint of The Random House Publishing Group, a division of Random House, Inc., New York.

RANDOM HOUSE TRADE PAPERBACKS and colophon are trademarks of Random House, Inc.

Originally published in hardcover in the United States by Random House, an imprint of The Random House Publishing Group, a division of Random House, Inc., in 2008.

Title page map courtesy of University of Texas Libraries

Library of Congress Cataloging-in-Publication Data
Pagden, Anthony.
Worlds at war : the 2,500-year struggle between east and west /
Anthony Pagden.
p.   cm.
ISBN 978-0-8129-6890-3
1. World history. 2. East and West. 3. Europe—Relations—Islamic countries.
4. Islamic countries—Relations—Europe. 5. Muslims. 6. Islam.
7. Balance of power. I. Title.
D21.3.P33 2008
909—dc22      2007012878

Printed in the United States of America

www.atrandom.com

9 8 7 6 5 4 3 2

Book design by Carol Malcolm Russo

Maps by Virginia Norey

*FOR GIULIA*

# CONTENTS

PREFACE                                      ix

1   PERPETUAL ENMITY                          3

2   IN THE SHADOW OF ALEXANDER               41

3   A WORLD OF CITIZENS                       69

4   THE CHURCH TRIUMPHANT                    126

5   THE COMING OF ISLAM                      157

6   HOUSES OF WAR                            224

7   THE PRESENT TERROR OF THE WORLD          251

8   SCIENCE ASCENDANT                        294

9   ENLIGHTENED ORIENTALISM                  326

10  THE MUHAMMAD OF THE WEST                 361

11  THE EASTWARD COURSE OF EMPIRE            420

12  EPILOGUE                                 512

ACKNOWLEDGMENTS                              539

NOTES                                        541

BIBLIOGRAPHY                                 583

INDEX                                        603

WE LIVE IN an increasingly united world. The boundaries that once existed between peoples are steadily dissolving; ancient divisions between tribes and families, villages and parishes, even between nations, are everywhere disintegrating. The nation-state, with which most of the peoples of the Western world have lived since the seventeenth century, may yet have a long time to live. But it is becoming increasingly hard to see it as the political order of the future. For thousands of years, few people went more than thirty miles from their place of birth. (This, it has been calculated from the places mentioned in the Gospels, is roughly the farthest Jesus Christ ever traveled from his home, and, in this respect, at least, he was not exceptional.) Today places that less than a century ago were remote, inaccessible, and dangerous have become little more than tourist sites. Today most of us in the Western world will travel hundreds, often thousands, of miles in our immensely prolonged lives. And in the process we will, inevitably, bump up against different peoples with different beliefs, wearing different clothes and holding different views. Some three hundred years ago, when the process we now label "globalization" was just beginning, it was hoped that this bumping into others, this forced recognition of all the differences that exist in the world, would smooth away the rough edges most humans acquire early in life, making them, in the process, more "polished" and "polite"—as it was called in the eighteenth century—more familiar with the preferences of others, more tolerant of their beliefs and delusions, and thus better able to live in harmony with one another.

In part this has happened. The slow withering of national boundaries and national sentiments over the past half century has brought substantial changes and some real benefits. The ancient antagonisms that tore Europe apart twice in the twentieth century (and countless other times in the preceding centuries) are no more and, we can only hope, will never be resuscitated. The virulent racism that dominated so many of the ways other peoples were seen in the West during the nineteenth century may not have vanished, but it has certainly withered. The older forms of imperialism are no more, even if many of the wounds they left behind have still not healed. Nationalism is, in most places, something of a dirty word. Anti-Semitism, alas, is still with us, but there are few places where it is as casually accepted as it was less than a century ago. Religion has not quietly died, as many, in Europe at least, hoped and believed until recently that it would. But it is certainly no longer the cause of the bitter confessional battles it once was. (Even in Northern Ireland, the last outpost of the great religious wars of the sixteenth and seventeenth centuries, the quarrel is slowly being resolved and has always been more about local politics and national identity than about faith.)

Some of the old fault lines that have divided peoples over the centuries are, however, still very much with us. One of these is the division—and the antagonism—between what was originally thought of as Europe and Asia and then, as these words began to lose their geographical significance, between "East" and "West."

The division, often illusory, always metaphorical, yet still immensely powerful, is an ancient one. The terms "East and West" are, of course, "Western," but it was probably an Eastern people, the ancient Assyrians, sometime in the second millennium B.C.E., who first made a distinction between what they called *ereb* or *irib*— "lands of the setting sun"—and Asia, *Asu*—"lands of the rising sun." For them, however, there was no natural frontier between the two, and they accorded no particular significance to the distinction. The awareness that East and West were not only different regions of the world but also regions filled with different peoples, with different cultures, worshipping different gods and, most crucially, holding different views on how best to live their lives, we owe not to an Asian but to a Western people: the Greeks. It was a Greek historian, Herodotus, writing in the fifth century B.C.E., who first stopped to ask what it was that divided Europe from Asia and why two peoples who were, in many respects, quite similar should have conceived such enduring hatreds for each other.

This East as Herodotus knew it, the lands that lay between the European peninsula and the Ganges, was inhabited by a large number of varied peoples, on whose strange peculiarities he dwelt lovingly and at length. Yet, for all their size and variety, they all seemed to have something in common, something that set them apart from the peoples of Europe, of the West. Their lands were fertile, their cities opulent. They themselves were wealthy—far wealthier than the impoverished Greeks—and they could be immensely refined. They were also fierce and savage, formidable opponents on the battlefield, something all Greeks admired. Yet for all this they were, above all else, slavish and servile. They lived in awe of their rulers, whom they looked upon not as mere men like themselves, but as gods.

For the Greeks, the West was, as it was for the Assyrians, the outer rim of the world, where, in mythology, the daughters of Hesperides lived by the shores of the Ocean, guardians of a tree of golden apples given by the goddess Earth as a wedding present to Hera, the wife of Zeus, father of all the gods. The peoples who inhabited this region were also varied and frequently divided, but they, too, shared something in common: they loved freedom above life, and they lived under the rule of laws, not men, much less gods.

Over time, the peoples of Europe and their settler populations overseas—those, that is, who live in what is now commonly understood by the term "West"—have come to see themselves as possessing some kind of common identity. What that is, and how it is to be understood, has changed radically from antiquity to the present. It is also obvious that, however strong this common heritage and shared history might be, it has not prevented bloody and calamitous conflicts among the peoples who benefitted from them. These conflicts may have abated since 1945 and, like the most recent dispute over the justice of the American-led invasion of Iraq, are now more often conducted without recourse to violence, but they have not entirely disappeared. If anything, as the ancient antagonisms of Europe have healed, a new rift between a united Europe and the United States has begun to emerge.

The term "East" was, and still often is, used to describe the territories of Asia west of the Himalayas. Obviously no one in Asia before the occupation of much of the continent by the European powers in the nineteenth and twentieth centuries gave much thought to the idea that all the nations of the region might share very much, if anything, in common. East and West, like all geographical markers, are obviously relative. If you live in Tehran, your West may be Baghdad. The current, conventional division of all of Asia into Near, Middle, and Far East is a nineteenth-century usage whose focal point was British India. What was Near or Middle lay between Europe and India, what was Far lay beyond.[1] For the inhabitants of the region, however, this classification clearly had no meaning whatsoever.

In the eighteenth century, a relatively new term, "Orient," came into use to describe everywhere from the shores of the eastern Mediterranean to the China Sea. This, too, was given, by many Westerners, a shared if not single identity. When I was studying Persian and Arabic at Oxford in the 1970s, I did so in a building called the Institute of Oriental Studies, where Persian, Sanskrit,

Turkish, Hebrew, Korean, and Chinese (not to mention Hindi, Tibetan, Armenian, and Coptic) were all studied under the same roof. Two streets away (to the east), all the languages of Europe were also studied under one roof, in an imposing neoclassical building called the Taylor Institution. They were and are called "modern languages," which firmly identified them as the true successors to the languages of the ancient world, Greek and Latin.

None of the great civilizations of what is now generally called the "Far East" belongs to my story. The Chinese may have been seen by many Westerners as sharing the same lethargic, immobile, backward-looking character as the other peoples of Asia. But there was not, nor had there ever been, any conflict between them and the West, at least before the Western powers began their own attempt to seize control of Chinese trade in the later nineteenth century. Far from presenting a challenge to the cultural assumptions of the West, China, and to some degree Japan, were for long believed to share them.

The division between Europe and Asia began as an exclusively cultural one. The Persians and the Parthians—the two great Asian and "barbarian" races of the ancient world, clearly had what would later be called "national characters." But in their origins they were very much like the Greeks and, with certain reservations, the Romans, who in giving themselves a mythic ancestry in Troy had also made themselves into an originally Asian people. Later, however, when Christianity and with it the search for the sources of human history in the Bible took hold of most of Europe, it became a commonplace to explain the origins of human diversity as the consequence of the repopulation of the world after the Flood. The sons of Noah had come down from Mount Ararat and then traveled to each of the three continents, and by "these were the nations divided in the earth after the Flood." (The subsequent discovery of two further continents—America and Australia—posed a serious threat to this story. But as with all biblical exegesis, ingenious interpretations were provided to overcome this.) Shem, it was believed, had gone to Asia (hence the subsequent classification of Jews and Arabs as "Semitic peoples"), Japhet to Europe, and Ham to Africa.

This account of human prehistory was still being taken seriously in some quarters well into the nineteenth century, largely because of its obvious racial potential. But although it seemed to provide a sound (at least from the Christian point of view) explanation as to why the peoples of Europe were so very different from those of Asia (not to mention Africa), it was always far less significant than the argument that what divided the two continents was to be found not in human origins, much less of race, but in the differences in how the worlds of men and gods were conceived.

In reality, Europe was not even a separate continent but a peninsula of Asia. The great eighteenth-century French poet, playwright, historian, and philosopher François-Marie Arouet, better known by his sobriquet, Voltaire, once remarked that if you were to situate yourself imaginatively somewhere near the Sea of Azov, just east of the Crimea, it would be impossible to tell where Europe left off and Asia began. It might, therefore, he concluded, be better to abandon both terms.[2] The now-current word "Eurasia"—which is an attempt if not to abandon, then to merge them—captures not only an obvious geographical truth but also a broadly cultural one. In Greek myth the peoples of Europe owed their very origins to an Asian princess. Greek (and subsequently all Western) science, as the Greeks were well aware, had its origins in Asia. Pagan religious beliefs were an amalgam of European—or, as we would now say, Indo-European—and Asian features. This was precisely the source of Herodotus's bewilderment. He devised, as we shall see, an explanation—one that has had a long and powerful afterlife. But the fact that such terrible wars should have been fought for so long between peoples who, at least until the seventeenth century, were divided by so little may be attributed to Sigmund Freud's famous observation that the bitterest of all human conflicts spring from what he called the "narcissism of small differences": we hate and fear those whom we most resemble, far more than those from whom we are alien and remote.

The East-West distinction is also geographically unstable. For the Assyrians, the "West" meant little more than simply "the lands over there," or what the Greeks for good mythological reasons of their own called "Europe"—a word originally applied only to central Greece, then to the Greek mainland, and finally, by the time Herodotus was writing, to the entire landmass behind it. It, too, however, was a vague region, a small and for a long time relatively insignificant peninsula of the vast Asian landmass, with no obvious western limits save for Oceanus, the massive ocean-river believed to encircle all three of the continents. The English word "West" was originally an adverb of direction. It meant, in effect, "farther down, farther away." By the Middle Ages, it was already being used by Europeans to describe Europe, and by the late sixteenth century, it had become associated with forward movement, with youth and vigor, and ultimately, as Europe expanded—westward—with "civilization."[3] Ever since the eighteenth century, the word has been applied not only to Europe but also to Europe's settlers overseas, to the wider European world.

European geographers have, ever since antiquity, spent a great deal of imagination and ingenuity in trying to establish meaningful frontiers between Europe and Asia. One was the Hellespont—the modern Dardanelles—the narrow stretch of water at the final exit of the Black Sea, which was looked upon by the ancient peoples on both sides as a divinity whose purpose was to keep the two continents apart. There it has very largely remained, something on which those who object to the attempts by modern Turkey to define itself as a European state have frequently insisted. Farther north, however, the frontier grows shadowy and uncertain. At first it was drawn at the River Don, which had the effect of placing most of modern Russia squarely in the "East." By the end of the fifteenth century, however, it had advanced to the banks of the Volga; by the late sixteenth century, it had reached the Ob; by the nineteenth the Ural and the Ural Mountains; until in the twentieth it finally came to rest on the banks of the Rivers Emba and Kerch.

When today we speak of the West or the East, however, we, like the ancients, mean something larger than mere geography. We mean the cultural peculiarities, the goals and ambitions of widely heterogeneous groups of people—and, of course, those often quoted, rarely discussed "Western values," a rough checklist of which would now include human rights, democracy, toleration, diversity, individual freedom, respect for the rule of law, and a fundamental secularism. When, in September 2006, following a somewhat undiplomatic remark by Pope Benedict XVI, the terrorist organization al-Qaeda responded by swearing to continue the "holy war," the jihad, until "the final destruction of the West," it was not so much a place they had in mind, as all those places where these values are more or less respected.[4] Their "West," therefore, now has to include a good part of the traditional East: Japan, India, even Turkey.

The beginnings of the conflicts between East and West, of which al-Qaeda's war with the West is the latest manifestation, are so old that they belong to myth. They began with probably the most famous war in history, fought between the Achaeans, Greeks from the northeast of the Peloponnese, and a quasi-mythical people of Asia Minor called the Trojans, over the slighted honor of the Spartan king, Menelaus, whose wife, Helen, had been abducted by a foppish Trojan playboy named Paris Alexander.

For the Greeks of Herodotus's generation the Trojan War, or rather Homer's account of it in the *Iliad*, celebrated the birth of Hellas, and later of Europe, and its triumph over Asia. That is not how Homer saw it. His Greeks and Trojans share the same values and apparently speak the same language. They also venerate the same gods, who take sides in the combat according to their own particular whims and even appear on the battlefield. The war was caused by irate and uncontrollable humans; but it came about because Earth complained to Zeus, the father of the gods, that there were too many humans for her to bear.

For later generations, however, who organized their own identities and their cultural longings around Homer's poem, the fall of Troy became the beginning of the history of a struggle for supremacy between two peoples whose differences grew ever more marked as time went by. When Alexander the Great invaded the mighty Persian Empire in 334 B.C.E., he did so in a precisely dramatized reenactment of the Greek assault upon Troy, with himself in the role of the greatest of the Greek heroes, Achilles. For the ancients and their heirs, the division between the peoples of the two continents became, thereafter, an immutable fact of nature. "All the natural world," the Roman scholar Varro declared bluntly in the first century B.C.E., "is divided into earth and air, as all the earth is divided into Asia and Europe."[5]

Troy, Alexander, and Rome were, however, only the beginning. In the centuries following the extinction of the Roman Empire, the cultural, political, and religious geography of Europe and Asia changed as new peoples, bearing new identities, swept through both regions: nomadic Germanic tribes in the West; Mongol, Turkic, and Arab peoples in the East. But each successive wave, as it came to rest, reassumed the ancient struggle between an ever-shifting West and an equally amorphous East. A flame had been lit at Troy that would burn steadily down the centuries, as the Trojans were succeeded by the Persians, the Persians by the Phoenicians, the Phoenicians by the Parthians, the Parthians by the Sassanids, the Sassanids by the Arabs, and the Arabs by Ottoman Turks.

The Ottoman sultan Mehmed II, the conqueror of Constantinople, capital of the Greek Byzantine Empire, in 1453, was very conscious of this history. When in 1462 he paid a visit to the supposed site of the Trojan War, he stood on the shore where the Greek invaders had beached their ships and declared that, through his efforts, the heirs of those same Greeks had been made to pay, "the right penalty, after a long period of years, for their injustice to us Asiatics at that time and so often in subsequent times." Nearly half a millennium later, in 1918, British and Italian troops entered Istanbul. The Allies stayed for less than five years; but many at the time hailed the occupation as the culmination of a centuries-long conflict, the day on which the West had finally repossessed the "second Rome" and brought what Herodotus had called the "perpetual enmity" between Europe and Asia to a close.

The struggle between the various civilizations of Europe and Asia has been a long and enduring one. But it has been neither continuous nor uninterrupted. An uneasy peace existed along the frontiers of the Byzantine and Ottoman worlds, even as Greco-Roman culture and the Christian religion steadily vanished from the Middle East. The so-called Moors, Berbers and Arabs from North Africa, who occupied much of the Iberian Peninsula in the eighth century, lived an unsteadily cooperative existence— famously and imprecisely described as *convivencia,* or "living together"—for centuries with their Christian subjects, even while a formal state of war existed between many of them. In the late sixteenth century, a fragile cooperation prevailed among the Ottomans, the Spanish, the Venetians, and the Genoese in the eastern Mediterranean, which, on more than one occasion, saw Ottoman ships involved in conflicts among Christians. Both the Valois kings of France and the Spanish Hapsburgs sought Ottoman and Safavid assistance in their own seemingly perpetual struggle with each other.

But these arrangements were always uncertain, always temporary. The old antagonisms, the conflicting visions of what nature or God had intended for man, and the memories of ancient hostilities, carefully nurtured by generation after generation of historians, poets, and preachers on both sides of the divide, were always there to justify a return to a struggle in which, as the Persian Achaemenid emperor Xerxes toward the end of the fifth century B.C.E. had seen, "there can be no middle course."[6]

The battle lines have shifted over time, and the identities of the antagonists have changed. But both sides' broader under-standing of what it is that separates them has remained, drawing, as do all such perceptions, on accumulated historical memories, some reasonably accurate, some entirely false. This book is an at-tempt to chart those histories, both true and fictive, and to ex-plain how they came to be the way they are. Although I make no pretense to be merely telling a story, nor have I made any attempt to hide my preference for an enlightened, liberal secular society over any other, nor to disguise the fact that I believe that the myths perpetrated by all monotheistic religions—all religions in-deed—have caused more lasting harm to the human race than any other single set of beliefs, this is not yet another history of how the West came to dominate the East and with it most of the known world. If Christianity seems to fare slightly better than Islam in this story, that is only because, as I explain in Chapter 8, Christianity was less well equipped to resist the destructive forces unleashed by its own internal inconsistencies and was thus un-able to resist the several forms of secularization that, by the end of the eighteenth century, had all but eliminated its presence in the civil and political life of the West. Many Europeans and, still more Americans, continue, of course, to call themselves—and some evi-dently are—Christian. Few would also deny that Christianity con-tinues to be one of the shaping cultural moments in the history of the West. But, as successive popes, patriarchs, and bishops have bitterly lamented, no matter what the personal, religious beliefs of their peoples might be, for the past three hundred years or more, the civil and political trajectory of the nations of the West has continued on its way very much as if no religion of any kind had ever existed.

"We need history," the great German philosopher Friedrich Nietzsche once said, "for the sake of life and action.... We want to serve history only to the extent that history serves life."[7] I hope that in its own way, this book of history will also serve life, by showing, if only fleetingly, that the tragic conflicts now arising from the attempts by some of the Western powers to reorder a

substantial part of the traditional "East" in their own image belong to a history far older, and potentially far more calamitous,
than most of them are even dimly aware.

ALL BOOKS BEGIN by chance. One morning, over breakfast, my
wife, the classical scholar Giulia Sissa, was looking at a picture in
*The New York Times* of a group of Iranians prostrate in prayer.
"How ironic," she remarked. "It was the habit of prostration
which most puzzled the Greeks about the ancient Persians. Perhaps," she added, "you should think about writing a book on
what Herodotus calls the 'perpetual enmity' between Europe and
Asia." So I have. To her it owes its inspiration, its shape, and most
of its chapter titles.

All chances, however, have their own prehistories. In the late
1960s, when I was unemployed, waiting to go to university, and
making a precarious living as a freelance translator, I went one
summer to stay with my sister, whose husband was then attached
to the British High Commission in Cyprus. I spent my time, when
not working on a translation of a rather dull biography of Paul
Cézanne, visiting archaeological sites, trailing behind my sister
and brother-in-law to embassy parties, and wandering through
the Turkish quarter of Nicosia, fascinated by a culture I had never
had any contact with before.

Cyprus—the legendary birthplace of Venus, to which, so myth
had it, some of the Greek heroes from the Trojan War had come to
settle and which had been Egyptian and Persian, Macedonian and
Roman, before becoming the refuge of the Crusader king Guy de
Lusignan, and then Venetian and Ottoman, and finally British—
lay squarely across the fault line that, ever since antiquity, has divided Europe from Asia. In 1878, the island was ceded by the
Ottoman sultan to the British. In 1960, after a bitter struggle for
independence, it became the Republic of Cyprus, with a mixed
Greek-Turkish parliament. Three years later, however, the government of President Archbishop Makarios collapsed, the Turkish

members of Parliament were effectively driven out of office, and the island was divided into Turkish and Greek zones along a line that separated the northern part of the island from the south. The Greek zone was prosperous and European and, in effect, constituted all that the rest of the world recognized as the Republic of Cyprus. The Turkish zone was poor, embattled, and a self-governing enclave unacknowledged then (and now) by any state other than Turkey itself.

When I was there, the frontier between the two enclaves zigzagged arbitrarily across the land, separating villages and towns that, under their previous imperial rulers, British and Ottoman, had lived, or been compelled to live, together in relative harmony. Now on one side of the line were the Greeks, who described themselves as the heirs to the oldest civilization of the West—absurdly, I thought at the time, since these people bore no obvious resemblance to Pericles or Plato (despite often being named after them). On the other were the Turks, who carried the burden of another kind of history. Their imperial past, like that of the British, was a relatively recent memory. For many, their Ottoman ancestry was a source of pride. For others, it was an embarrassment, an obstacle to their wish to become a modern European nation. I must confess that at the time I had more sympathy and liking for the Turks than I had for the Greeks.

The capital, Nicosia, was also split, as Berlin was before 1989, into two sectors separated by a narrow strip of land, known as the "Green Line," patrolled, at that time, by a largely inactive U.N. peacekeeping force. No one was prevented from crossing this line, and many Turks traveled regularly to the Greek sector to shop, some even to work. The Greeks, however, rarely ventured into the Turkish sector, swearing that they would never return if they did. From time to time I would sit and drink sweet Turkish tea with a man named Kemal Rustam, who owned a small shop that sold books and looted antiquities (a brisk business among the Turks, for whom the island's Greco-Roman past meant nothing) and who acted as an informal liaison officer between the Turkish and

Greek governments. From him I learned a great deal, anecdotal and indirect, about what it meant to live on a frontier, and on this frontier in particular. He was one of the Turks who crossed regularly into the Greek zone, and when my niece was christened, he attended the service, a smiling, ironic, unbelieving Muslim among a group of largely unbelieving Christians. On that simple day-to-day basis, the terrible religious and ethnic forces that had divided the island since independence, and even then had begun to divide the whole of the Middle East, seemed remote, grotesque, and absurd.

Cyprus introduced me to Ottoman history, and to Islam. It also showed me just how enduring the ancient divisions between Europe and Asia could be and left me wanting to know something more about how they had shaped the histories of both.

The following year, I went up to Oxford to study Persian and Arabic, having formed a vague plan to write a dissertation on the relationship between the Safavid rulers of Iran and Portugal in the seventeenth century. In the end this came to nothing, and I turned my attention elsewhere, to Spain and the Spanish Empire in America. But if Persia dropped—at least partially—out of sight, no one who studies any aspect of Spanish history, even that of the farthest westward of all its possessions, can remain for long unaware of the presence of Islam, or of the role it has played in the creation of modern Europe.

The Turks, too, never quite ceased to exercise a hold on my imagination. In the mid-1970s, and largely on a whim, I went to eastern Turkey, to what is loosely labeled "Kurdistan," the area that lies between Lake Van and the borders with Turkey, Iraq, and Iran. Then as now, the Kurds were pressing their distant masters in Istanbul for an independent homeland, and although the region was open to foreigners, it had very recently been under martial law and all the indications were—as indeed turned out to be the case—that it would soon be so again. I had a friend who had served in the British Embassy in Ankara, had useful connections among both Turks and Kurds, had always wanted to visit the East, and needed a traveling companion. It seemed too good an opportunity to miss.

I did not get very far, after contracting paratyphoid in Van. But I made it to the foot of Mount Ararat, shot vainly at indifferent eagles with a police chief just outside Tatvan, dynamited fish in a shallow river near Mus, and talked to the straggling remnants of Mustafa Barzani's Kurdish militia as it limped across the border from Iraq. I also slept under the chill Anatolian sky in the company of transhumant shepherds, where I learned something at first hand about ancient hospitality, and something, too, of the horrors, in particular for women, of those "traditional" ways of life so mourned by sentimental Westerners who have never had to experience them.

Those images, and that of the unfailing courtesy and generosity with which I, a foreign infidel with no good reason to be there and only the slenderest of connections, was treated have never faded. But the impression that is, perhaps, still most vivid in my mind is of something that occurred toward the very end of my journey. One morning, I found myself standing on a hill outside the modern town of Van, shabby and ramshackle, and looking down on the ruins of a city. It had been built almost entirely of sunbaked mud bricks, which, in the years since the place had been abandoned, had slowly dissolved in the winter rains until all that was left was the first two or three feet of the outer walls. It was an unforgettable sight. All that could be seen was street after street of the traces of former houses, shops, squares, and marketplaces with, here and there, a higher, more complete ruin of stone. At first glance it looked not unlike the pictures of the German city of Dresden after the bombing raids in February 1945. But it was not bombs but neglect and the weather that had razed this city almost to the ground. The Turk who had driven me there explained that this place was very ancient and had been abandoned for centuries. I asked him who had once lived here. "Ancient peoples," he answered, which implied that they had been, at least, pre-Islamic, "very ancient peoples." Did they have a name? No, he replied, their names had been lost. This much, he added, he had been told at school. For the rest, the place was a mystery·and, like all other ruins, of interest only to foreigners. He seemed entirely certain and perfectly sincere.

But I knew that the ghostly place we were staring at had been part of the old Armenian capital of Van, and far from being abandoned "long ago," its population had in fact been slaughtered in the Armenian massacres of June 1915—which just about everyone now, except the Turkish government, refers to as the Armenian Genocide. Between 1894 and 1896, Ottoman troops had systematically destroyed and looted Armenian villages, killing, by most accounts, as many as two hundred thousand people, which *The New York Times,* in what may be the first use of the term, described as "another Armenian Holocaust." The Armenians had been killed largely because they had been looked upon with suspicion as a Christian fifth column, plotting with the enemies of the Ottoman Empire to create a separate homeland. At the outbreak of the First World War they turned for help to Russia, the empire's most intractable foe, and in May 1915, with Russian assistance, they created an independent Armenian state. It lasted barely more than a month. In the aftermath, and, it must be said, amid rumors of the slaughter of Turks and Kurds by the victorious Armenian-Russian army, the authorities in Istanbul retaliated by deporting the entire Armenian population of the region to southeast Anatolia. In the process, thousands were massacred or systematically tortured, their homes and their belongings destroyed or seized, their churches desecrated and their ancient capital left empty, and at last, to which my Turkish guide was an eloquent, if unwitting, witness, all memory of their very existence was finally expunged.[8] What I saw that day showed me something about the ferocity of ethnic conflict and of the still yawning divide between East and West, which no one who lives a comfortable, secure life in the West could, at least until September 11, 2001, have ever imagined.

When finally I sat down to write this book, it was these two images I had in mind: the prostrate Persians and the devastated city—two moments in a history that has no obvious beginning and as yet no foreseeable ending.

Los Angeles—Paris—Venice, 2006

# WORLDS AT WAR

# 1

## *PERPETUAL ENMITY*

**I**

It all began with an abduction. The girl's name was Europa, and she was the daughter of Agenor, king of the city of Tyre on the coast of Sidon, in what is now Lebanon. She was sitting by the water, pale and golden-haired, weaving flowers—hyacinths, violets, roses, thyme—into a garland, when Zeus, father of the gods, came out of the sea in the shape of a white bull, "his breath scented with saffron." Her attendants fled. But she remained, and, in the Roman poet Ovid's telling of the story,

> *Gradually she lost her fear, and he*
> *Offered his breast for her virgin caresses*
> *His horns for her to wind with chains of flowers*[1]

Cupid, who had materialized fluttering beside her, lifted her gently onto the creature's back. Zeus carried her out to sea and across the strait that separated two worlds, to Crete, where in the meadows of Gortyn they made love under a huge, shady plane tree.[2]

There she would bear three sons, Minos, Rhadamanthys, and Sarpedon, and give her name to a continent. Then, having tired of

her, as he did of all his human consorts, Zeus married her to Asterius, king of Crete, who adopted her half-divine sons as his own.

This is the myth of the Rape of Europa. For centuries it has
been the story of the founding of the European peoples and the
source of what has ever since been called the West. But as Europa's
home had been in Asia, it meant that this "West" had been born
out of the "East." "What then," asked the great twentieth-century
French poet Paul Valéry, "is this Europe? A sort of promontory of
the Old Continent, an appendix of Asia?" Yes, he answered, but
one "which looks naturally to the West."[3]

As with all myths, however, there is another, more mundane
version. This was first suggested by the Greek historian Herodotus and later seized upon by the third-century Christian theologian Lactantius, eager to debunk and demystify all the unsettling
erotic fantasies that had reached him from the pagan world. In
this version, which Herodotus attributed to the Persians, the Rape
of Europa was an act of revenge for the seizure by Phoenician
sailors of Io, the daughter of Inachus, king of Argos. Later on,
wrote Herodotus, "some Greeks, whose name the Persians failed
to record"—they were, in fact, Cretans, known for their savagery as
"the boars of Ida"—put into the Phoenician port of Tyre "and carried off the king's daughter Europa, thus giving them tit for tat."
Lactantius provides these Cretans with a bull-shaped ship so as to
explain away Zeus and says that Europa was intended as a present
for their king, Asterius, a version taken up centuries later by the
Italian poet Boccaccio, who added his own twist to an already
overcomplicated story by renaming the Cretan ruler "Jove."[4]

The mythic abductions continued. The Cretans were what
Herodotus called "Europeans," and as Europa was an Asian
woman, her rape was taken by all Asians to be an affront. Later,
Jason, another European, would sail into the Black Sea and
abduct Medea, the daughter of Aeëtes, king of Colchian Aia, along
with the Golden Fleece she had helped him to steal. Later still, the
Trojans, a people from what we now call Asia Minor, would seize
a not wholly unwilling Helen, wife of the Spartan king Menelaus,

in revenge and carry her off to Troy. In turn, Menelaus's brother, Agamemnon, would raise an army, cross the sea, and lay siege to the great city of Troy for ten long years.

Herodotus, the "Father of History," as the great Roman jurist Marcus Tullius Cicero called him, was searching for a resolution to the question that all these stories claimed to answer but none in fact did: Why had "these two peoples—Greeks and Persian— fought with each other?" Herodotus had grown up with the enmity between Greeks and Persians and lived through its conse- quences. He had been born in Halicarnassus, modern Bodrum, on what is now the coast of Turkey, sometime around 490 B.C.E., the year the Persian "Great King" Darius I had launched the first full- scale attempt by an Asian power to subdue the whole of Europe. Halicarnassus was a Greek city, but at the time of Herodotus's birth and during all the time he spent there as a young man, it was under Persian rule. He had lived between two worlds, then appar- ently at peace if not always at ease with each other, and he wanted to know how relative amity between two peoples could be trans- formed into long years of bitter hatred. In search of his answer, he dedicated his later life and all his creative energy to telling the story of the momentous struggle between Asia and Europe that has come to be called the Persian Wars—a series of conflicts that lasted off and on from 490 until 479 B.C.E.

The tales of Io and Europa and of the Trojan War were, he knew, merely pretexts. These quasi-mythic struggles had been fought out in the shadow of the gods, in times when humans were rarely seen as exercising any will of their own. Herodotus, how- ever, was one of the very first writers to see human beings as re- sponsible for their own actions. The gods are still there, but they are shadowy creatures. They still speak, through signs, auguries, and the unreliable and devious voices of the oracles. But they make nothing happen. It is now humankind that dominates and controls the world.

Like the Trojan War, the Persian Wars were titanic struggles between Europe and Asia. But they were historical conflicts this time rather than mythological ones, and they had precise origins and no less precise consequences. Herodotus had himself spoken with many of the combatants in the course of gathering the material for what he called simply his "inquiries"—his *Histories*—and he evidently had a great measure of sympathy for all of them, no matter what their origins. Although he never learned Persian (he seems for instance to believe that all Persian names end with *s*), he claimed to be in possession of information that could only have come from Persian sources, and his view of the Persians, although it sometimes conformed to what would later become a hardened Greek stereotype of the "Oriental," was more nuanced than that of most later writers.

For all that, his vision was necessarily a Greek one, and because his is the only detailed account we have of the wars, it was his vision that for centuries dominated our understanding of what happened and why. Modern archaeology has now given us a sometimes very different account of the rise of the Achaemenids— the ruling dynasty of what is now routinely called the Persian Empire—and of the kind of society over which they ruled. From this new perspective it would indeed seem that Herodotus was not only the father of history but also, as he was once called by the Greco-Roman philosopher and biographer Plutarch, the "father of lies."[5] The literal accuracy of his story, however, is not to the point. For the *Histories* are not only an attempt to recount a succession of wars; they are also a representation of the origins— cultural, political, and to some degree psychological—of the Greek world. Although Herodotus grumbled that he could not understand "why three names, and women's names at that [Europa, Asia, and Africa], should have been given to a tract which is in reality one,"[6] he had a clear understanding of the distinction between Europe and Asia. He also had a clear sense of what "Greekness" was, and a term—*to hellenikon*—to describe it. It was, he said, "shared blood, shared language, shared religion, shared customs." Here is the origin of the European, "Western" sense of distinctiveness. But it is also the recognition that distinctiveness of any kind is never complete and that Greece—Europe—owed an immense debt to what was for centuries its most enduring enemy.

"Greekness" may have been common to all. Herodotus was, nevertheless, well aware that the cities of ancient Greece, although alike in many ways, were, in fact, also very different kinds of societies; and if they shared the same gods and versions of the same language and even, more dubiously, the same blood, their customs could often be very different. When Herodotus described what it was that separated the Greeks from their Asian adversaries, it was usually Athenian, and in particular democratic values, he had in mind.

The Greeks of Herodotus's day lived in small cities, which were also, save for those under Persian rule, self-governing political communities—city-states we call them today—scattered along the shores of the Mediterranean from Sicily to Cyprus and the Aegean coast of Asia Minor. Although the peoples of this world were all "Hellenes," their story was not one of peaceful cohabitation. Until August 338 B.C.E., when, at the battle of Chaeronea, Philip of Macedon brought it finally to an end, the ancient Greek world was, in reality, always one of constantly shifting alliances. As the Persians in Herodotus's pages frequently observed, the Greeks had always had a hard time making a common cause against any enemy. The frontier between Europe and Asia was also, in reality, highly porous. Greek cities flourished under Persian rule, and influential Greeks fleeing the anger of their own people frequently sought asylum at the Persian court.

These are not facts that Herodotus either ignored or glossed over. What he was concerned to show was that what divided the Persians from the Greeks or the Asians from the Europeans was something more profound than petty political differences. It was a view of the world, an understanding of what it was to be and to live like a human being. And while the cities of Greece, and of "Europe" more widely, were possessed of very different personalities and had created sometimes very different kinds of societies, and were all too happy to betray one another if it suited them, they nevertheless all shared the common elements of that view. They could all distinguish freedom from slavery, and they were all committed broadly to what we today would identify as an individualistic view of humanity.

The great Athenian dramatist Aeschylus knew this well. He had fought at the fabled battle of Salamis in the autumn of 480 B.C.E., the first great naval conflict in European history and a Greek victory that would seal the future fate not only of Greece but of the whole of Europe.[7] In Aeschylus's play *The Persians*—the oldest piece of drama in any language—Atossa, the widow of the Persian Great King Darius and mother of his successor, Xerxes, whose ships were destroyed at Salamis, has a dream. There are a lot of dreams in this story. Cyrus, Darius's predecessor, has dreamed that Darius appeared to him with a pair of wings on his shoulders, one of which cast a shadow over Europe, the other over Asia, and during his own campaign, Xerxes will dream of his own downfall as he attempts to fulfill his ancestor's prophecy.[8]

Atossa dreams of her son's defeat. She also has a vision of the historical origins of the conflict that brought it about. "Never has a vision showed more clear," she says,

> *Than what I saw last night*
> *In the kind-hearted dark.*
> *I'll tell you:*
>     *It seemed to me*
> *Two well-dressed women—*
> *One robed with Persian luxury*
> *In other in plain Greek tunic—*
> *Came into view, both*
> *Taller far than any woman now living,*
> *And flawless in beauty,*
> *And sisters from the one same parentage,*
> *And for a fatherland, a home,*
> *One was allotted Greek soil,*
> *The other the barbarous earth beyond.*

In this dream, Greece and Persia, Europe and Asia, are sisters. Like all sisters they are different, and in their case the differences, the opulence of the one as opposed to the simplicity of the other, will be one of the defining marks of the enduring images of the two peoples. The sisters soon begin to quarrel, and Xerxes in an attempt to "curb and gentle them" harnesses both to his chariot. One, Asia,

> *towered herself*
> *Proud in this harness*
> *And she kept her mouth*
> *Well-governed by the reins*

But the other, Greece:

> *bucked stubborn,*
> *And with both hands*
> *She wrenches the harness from the chariot fittings*
> *And drags it by sheer force,*
> *Bridle flung off, and she*
> *shatters the yoke mid-span*
> *and he falls,*
> > *My son falls*[9]

Greece—Europe—the dream says, will be subdued by no one. Anyone who attempts to "bridle" her will only engineer his own destruction, and this—as the audience, many of whom, like the playwright, had been present at Salamis, knows—is exactly what Xerxes will do.

*The Persians* gives us what is probably the only eyewitness account we have of the battle of Salamis. But, for all that, it is a drama, a fiction; like Herodotus's *Histories,* it is about what makes the Greeks who they are and makes them so unlike the Persians. Like Herodotus, Aeschylus was also aware that these qualities had been created and sustained out of conflict, above all the conflict between Europe and Asia, of which Salamis was the final devastating scene.

## II

From the mid-sixth century, until Alexander the Great burned the Persian capital, Persepolis, in 330 B.C.E., Greek history was played out in the shadow of the Persian Empire.[10] It was the largest, most powerful state in the ancient world. By ancient standards, however, its rise was relatively swift and its duration relatively brief. The people whom we refer to as the Persians were originally a small tribe settled in a territory between the Persian Gulf and the central deserts of Iran called in antiquity Persis (modern Fars), from which they took their name. For a long time they formed a part of the empire of the Medes, to whom they paid tribute. Sometime around 550 B.C.E. the Persian leader Cyrus, chief of the Achaemenid clan of the Pasargadae tribe, rose in revolt against the Medes, successfully defeated their king, Astyages, and then welded the two peoples together into one kingdom. For some years prior to Cyrus's rebellion, the Medes had been expanding eastward into Anatolia, where they came into conflict with the kingdom of Lydia with its capital at Sardis, a city that would play a significant role in subsequent Persian history. In 585 B.C.E., a peace was made between the two kingdoms that established the frontier at the River Halys. In 547 B.C.E., however, the famously wealthy Lydian king Croesus—"as rich as Croesus," as the saying goes—crossed the Halys and invaded Cappadocia. After a series of indecisive battles, and with the onset of winter, Croesus retreated to Sardis, expecting to resume hostilities in the spring with the help of Egyptian and Ionian allies. The winter in Anatolia can be fierce, and few commanders would have attempted a siege in freezing winds and heavy snow. Cyrus, however, pursued Croesus's retreating army all the way back to Sardis. After fourteen days he stormed the city and captured it and its monarch.

Once he had control of Lydia, Cyrus sent two Median generals, Mazares and Harpagus, to conquer the cities of Ionia, Caria,

Lycia, and Phrygia, while he himself prepared to attack Babylon, the Bactrians, Sakas, and finally Egypt. In October 539 B.C.E., he entered Babylon, apparently welcomed by its apprehensive inhabitants, who spread green twigs before him, much as they would two centuries later before Alexander the Great, and "a state of peace was imposed on the city." Cyrus had become the heir to the kings of Babylon, the focus of allegiance of the peoples of the Fertile Crescent all the way to the borders of Egypt.[11] In 538 B.C.E., he issued a decree allowing the Jews exiled in Babylon by Nebuchadnezzar to return to their homeland. For this he was hailed by the prophet Isaiah as "the Lord's anointed." Like so many rulers since, he now laid claim to the sovereignty of the entire world as he knew it. "I am Cyrus, king of the world," one inscription reads, "great king, legitimate king, king of Babylon, king of Sumer and Akkad, king of the four rims of the earth." "King of the religions of the world."[12] Subsequent Persian rulers would assume similar titles. All styled themselves "great king" or "king of kings," a title that their Median predecessors had probably already taken from the Assyrians and that in its modern Persian form, *shâhânshâh,* would be assumed by successive dynasties of Iranian rulers until 1979.

In Greek eyes, these inflated claims were a sign of what they feared and despised most in the Persians: their imperialism. For the Greeks of the fifth century B.C.E., the political horizon was bounded by the city-state, and no Greek before Alexander the Great would ever make a claim to be the legitimate ruler of any people other than his own. Universalism, which would later become a central feature of European expansionism, was, like so much else in European culture, a creation of Eastern origin.

Like every would-be world ruler, however, Cyrus had constant problems on his empire's outermost rims. Nomadic or seminomadic tribes pressed down upon his frontiers, as they would on the frontiers of most successive empires. The age-old struggle between pastoralists and agriculturalists, between the descendants of Cain and those of Abel, would eventually lead to the overthrow of Seljuq Baghdad and Chinese Peking, as it would to the fall of

Rome and Byzantium. Cyrus's nemesis was the Massagetae, a sun-worshiping, horse-sacrificing people on the eastern shores of the Caspian Sea. In the summer of 530 B.C.E., Cyrus led an army east. Here he was met by the Massagetae queen, Tomyris. Having first unsuccessfully attempted to seduce her, he tried to bridge the River Araxes, which separated his army from hers, by a series of ferries and bridges, an operation his descendants would attempt to use twice to cross the Hellespont. While he was doing this, Tomyris sent him a message that would be repeated in one way or another by many critics of overextended empires. "King of the Medes," it ran, "I advise you to abandon this enterprise, for you cannot know if in the end it will do you any good. Rule your own people and try to bear the sight of me ruling mine." Cyrus, as she had anticipated, would have none of this and advanced to meet her. The battle that followed was, in Herodotus's opinion, "more violent than any other fought between foreign nations." When it was over, most of the Persian army had been destroyed and Cyrus was dead.[13] Legend has it that Tomyris cut off his head and flung it into a dish filled with his blood. "You," she said, "who have lived so long on the blood of others, now drink your fill."

Cyrus was succeeded by his son Cambyses, who five years later invaded Egypt and was accepted as a new pharaoh of the twenty-seventh dynasty. Herodotus, who like other Greek sources paints a very black picture of his reign, claims that he ruled with "maniacal savagery" after having secretly assassinated his brother Smerdis and then apparently gone insane.[14] Only madness, in Herodotus's view, could have driven a foreign ruler to "make mock of holy rites and long-established customs." Like most Greeks and most Persians, Herodotus was convinced that all religions and all local customs should be respected. One of the signs, or possibly the cause, of Cambyses' madness was his attempt to kill a sacred bull calf—instead he only stabbed himself in the thigh—which the Egyptians believed to be the living presence of the god Apis.[15] In 522 B.C.E., he again stabbed himself by accident in the same place. This time he died of gangrene. The Persian throne was then seized by two Magi, members of a priestly cast and a tribal group of Median descent, from whose name the word "magician" derives and who played a major role in the dissemination of Zoroastrianism throughout the Achaemenid Empire. (Centuries later three of their distant descendants, "wise men from the east," traveled to Jerusalem to pay homage to the newborn Jesus Christ.) The reign of the Magi, however, was brief. The following year seven prominent Persians assassinated the two Magi and then set about murdering every Magus they could find so that "if darkness had not put an end to the slaughter the tribe would have been exterminated."[16]

Five days later, when the excitement had died down, the conspirators met to decide what to do next. Who among them should now become king? There then took place a remarkable, and in all probability apocryphal, debate among three of them over the best form of government. It has come to be known as the "constitutional debate," and, as with all such political debates in the ancient world, the three types of government under discussion were democracy, oligarchy, and monarchy. Herodotus, who, once again, is our only source for this, claimed that "some of the Greeks refused to believe they [the speeches] were actually made at all, nevertheless they were made." Herodotus insisted, furthermore, that one of the conspirators, Otanes, put forward a proposal that was entirely at odds with the prevailing image of Persian monarchy and of Persian society and culture in general, the image that had always been at the heart of the Greek representation of their ancient enemy and, by contrast, also of themselves.

In all likelihood, Herodotus's Greek skeptics were right. Discussion about the best form of government was a feature of Greek political life, a tradition that they handed on to the Romans and the Romans in their turn bequeathed to Renaissance—and, subsequently, modern—Europe.[17] For the Greeks, Persia was not a place where any discussion over the way a people might be ruled—something that implied that they had a choice and might have an opinion—could take place. It was not a world in which discussion of any kind played any part whatsoever. Asia was, in Aeschylus's telling image of the two sisters, she who had always "kept her mouth well-governed by the reins" of despotic government.[18]

Certainly there is no independent historical record that the Persians possessed anything resembling a tradition of political dispute, still less that they ever questioned the legitimacy or desirability of monarchy. Darius himself, who was one of the disputants and who finally emerged victorious, left something resembling an autobiography, which, perhaps unsurprisingly, makes no mention of any dispute over the succession. "From long ago we were princely," it reads, "from long ago our family was royal. Eight of my family were formerly kings, I am the ninth: nine are we in two lines."[19] Herodotus's set piece is probably fiction, one of the reasons why he insisted so much on its truthfulness. Its importance, however, lies not in what it tells us about ancient Persian political life but in what it says about how the Greeks saw the Persians and themselves.

Otanes, who significantly had instigated the revolt against the Magi, speaks first, and it is he who suggests that the Persians should abandon their tradition of monarchy in favor of popular government, "which enjoys the finest of names, *isonomia*—'the order of political equality.' "[20] Although the conflict with Greece itself was still some time in the future, what Herodotus's readers would clearly have understood Otanes to be saying was that Persians would do well to adopt the form of government that had made the Greeks unique in the ancient world. "I think," Otanes begins, "that the time has passed for any one man among us to have absolute power. Monarchy is neither pleasant nor good." Otanes's reasons for saying this are recognizably Greek ones. Monarchs are unaccountable to anyone but themselves. "How can you fit monarchy into a sound system of ethics," he asks, "when it allows a man to do whatever he likes without any responsibility or control?" Monarchs are, after all, only human, and like all humans they suffer from envy and pride. Unlike other humans, however, monarchs are usually persuaded by these vices that they are greater than other men, and from that delusion inevitably follow "acts of savage and unnatural violence" of the kind, Otanes reminds his listeners, both Cambyses and the Magi perpetrated. Kings are capricious, and they like to be revered; yet at the same time, they despise those who revere them as "toadies." They are jealous of their best subjects and take pleasure only in the company of the worst. They have ever-ready ears for all kinds of flattering lies. Worst of all, they "break up the structure of ancient tradition and law, force women to serve their pleasure, and put men to death without trial."

Contrast this, he demands, to the rule of the people. "First it has the finest of all names to describe it—equality before the law." Magistrates are chosen by lot—this was the practice in Athens—"and all questions are put up for open debate." Let us therefore, Otanes concludes, "do away with monarchy, and raise the people to power, for the people are all in all."[21] Here, then, are the guiding principles of Athenian democracy: openness, accountability, and the rule of law. In all the vast Greek literature on the nature of politics there are very many critiques of democracy, yet very few clear manifestoes in its favor. There is some irony, then, that one of the best known should have been placed in the mouth of a Persian.

Alone of all the Persians, Otanes supposes that government is about more than simple power. He alone makes the argument, which falls on deaf Persian ears, that politics is also about justice and about what the Greeks called "the good life," that politics cannot, in the end, be detached from ethics. These—justice and the pursuit of an ethical life—were the essentially Greek virtues the Persians would encounter again and again, usually, as Herodotus told it, with incredulity. And in the end it was these, not numbers or force of arms or even simple courage, that would save Greece from becoming merely another province of the mighty Persian empire.

Otanes is followed by Megabyzus, who proposes the traditional middle course: government not by the one nor by the many but by the few. Certainly, he says, agreeing with Otanes, monarchy should be abolished. But it is not a good idea to hand over power to the people, for "the masses are a feckless lot—nowhere would you find more ignorance or irresponsibility or violence." Creatures such as they are incapable of reflection; they "rush blindly into politics like a river in flood." It would be intolerable to "escape the murderous caprice of a king only to be caught by the equally wanton brutality of the mob." No, far better to entrust the affairs of the state to a select few, the "best men" (the *aristoi* in Greek, hence the term "aristocracy")—for it is only natural that the "best men will produce the best policy."

Finally comes Darius, who argues for monarchy. It is, of course, Darius who wins the day, and he does so largely by raising the specter of the one threat that haunted the political imagination of the ancient Greek world more than any other: civil war. Both democracy and oligarchy, Darius argues, must inevitably result in personal conflicts. And in time, these will inevitably "lead to civil wars, and then to bloodshed: and from that state of affairs the only way out is return to monarchy—a clear proof that monarchy is the best." Monarchy, then, is inevitable, as all peoples discover sooner or later. Monarchy is also—and this is presented as Darius's winning argument—the traditional Persian form of government, and "we should refrain from changing ancient ways, which have served us well in the past. To do so would not profit us."

When at last all of the conspirators vote to retain the monarchy, Otanes's response is to withdraw from the competition. "I have no wish to rule," he says, "and no wish to be *ruled* either." All he asks is that he and his descendants not be forced to submit to the rule of any monarch. The others agree to this condition, and, said Herodotus, "the family of Otanes continues to be the only free family in Persia."

At this moment in their history, the Persians could have followed the path the Athenians had taken at the end of the sixth century, when Cleisthenes introduced the rule of the majority into Athens. Herodotus, who is writing what amounts to a chronicle of the power of democracy to triumph over monarchy, has given them their chance, but they have turned it down. There are many reasons why they chose to do this. Otanes's appeal to democracy is hardly very compelling, based as it is largely on the undesirable properties of monarchy. He offers no striking vision of what a future world ruled by the many, what freedom and self-determination might bring with it. But without a tradition of political reflection and political debate, how could he? Otanes's failure to persuade, however, has less to do with the quality of his oratory than it does with the nature, the image of the Persians, of all the peoples of Asia, that Herodotus offered his readers—an image most Greeks shared—of a people who were intensely suspicious of any kind of personal or individual action. The Persians, for instance, were said to be a devout people, but no worshipper was "allowed to pray for any personal or private blessing but only for the king and the general good of the community of which he is himself a part."[22] Again and again in the course of his *Histories,* Herodotus scattered his main narrative with anecdotes and asides intended to reveal the Persians, for all their courage and ferocity, to be craven, slavish, reverential, and parochial, incapable of individual initiative, a horde rather than a people.[23]

When, for instance, the Spartans sent a messenger to Cyrus the Great warning him that if he were to harm any Greek city in Ionia they would make sure that he lived to regret it, Cyrus asked, "who the Spartans were, and what were their numbers that they dared send him such a command." When he was told, Cyrus replied, "I have never yet been afraid of men who have a special meeting place in the center of their city, where they swear this and that and cheat each other. Such people, if I have anything to do with it, will not have the troubles of Ionia to worry about but their own."[24] Time and again in the course of the Persian Wars, Cyrus's successors would make the same mistake of confusing sheer numbers with military power and debate for weakness, the same mistake of assuming that discord, even on occasions internecine warfare, could prevent a people from uniting against a common enemy.

Shortly before the battle of Plataea, at the very end of the war, when the Persian forces under their general, Mardonius, were attempting to extricate what was left of the Persian army from Greece, the Thebans, who supported the Persians at that stage of the conflict, invited the Persian high command to a banquet. During the course of the dinner one of the Persian generals told his Greek companion, a man whom Herodotus called Thersander, that he knew full well that within a short time, most of the Persians dining with him and the remnants of the Persian army, which was camped across the river, would be dead. When this Persian was asked why he did not attempt to do something about this, he replied, "My friend . . . What God has ordained no man can by any means prevent. Many of us know that what I have said is true; yet we are constrained by necessity, we continue to take orders from our commander."[25] This conversation, which he claimed to have heard from Thersander himself, greatly impressed Herodotus. Here was Persian passivity in the face of the inevitable, here was Persian refusal to confront authority when need demanded, to present king or commander with an undesirable truth and the need to act. It was, as we shall see, to become an enduring image not only of the Persians but, after them, of the whole of Asia.[26]

IN THE "CONSTITUTIONAL debate," monarchy wins the day not because of a desire or need for justice or because of its intrinsic merits as a form of government. Monarchy triumphs because it was, as Darius points out, a monarch who had created the Persian Empire, and monarchy was therefore one of the "ancient ways" of the Persians. It wins because the Persians, although they might be happy adopting "foreign ways," if all this meant was dressing up as Medes and Egyptians or even accepting "pederasty, which they learned from the Greeks," are always loath to abandon their traditions. They are always, in their own estimation, the best of all peoples. It is this ethnocentrism that in Herodotus's eyes makes them so vulnerable. Unable to accept any kind of outside criticism, unable to perceive the weaknesses in their own traditions, their own customs, they are also unable to change or adopt to circumstances, and this will eventually be their undoing.

The Greeks chose *isonomia,* which all later generations would identify with democracy. They did so because only they, for reasons Herodotus did not go into, were capable of making such a choice. The values associated with the respect for the rule of law, the willingness to challenge authority when it seems in error, together with the corresponding devotion to one's city, one's family, and one's gods, are beliefs that, although they are human and intelligible to all, only Greeks were able to embrace. The Persians reject Otanes because, as the story of Thersander makes clear, they can do nothing else. For the Persians, to have embraced democracy would have meant becoming, in effect, another people. It would have meant becoming, in effect, Greeks. Herodotus's debate, whether it or something like it ever took place, was the first time in history that an Asian people had been confronted by one of their own with the possibility of becoming "Western," and rejected it. It would not, however, be the last.

WITH THE DEBATE over and Otanes gone, it remained to decide which of the conspirators would become the great king. There then followed an account of one of the most absurd means of deciding who was best suited to be king in recorded history. One can only assume that, whatever his sources, Herodotus hoped that at this point his Athenian audience, wearied perhaps by his political theorizing, would laugh at the obvious absurdity by which these Persians, who, despite their solemn reflection on the best possible form of government, could allow themselves to be so easily duped.

This is what happened.

The six conspirators agreed that they would all mount their horses on the outskirts of the city and the one whose horse was the first to neigh after sunup would have the throne. Darius won this contest by a clever ruse, thus showing that he was, if nothing else, the craftiest of the six. His groom, Oebares, took a mare from the stables and tethered her on the outskirts of the city. He then led Darius's stallion around and around the tethered mare for a while until the poor animal was snorting with frustration, when he finally allowed him to mount her. The next morning the six men rode out through the city gate. As soon as Darius's horse reached the spot where the mare had been tethered the night before, he started forward and neighed. "At the same instant, though the sky was clear, there was a flash of lightning and a clap of thunder, as if it were a sign from heaven; the election of Darius was assured, and the other five leapt from their horses and bowed to the ground at his feet."[27]

Darius I, "Darius the Great," came to power by a ruse, appropriate for a man whom Herodotus says the Persians described as a "huckster . . . being only out for profit wherever he could get it"—and he was chosen by a horse.[28]

But huckster or not, Darius also became a great empire builder and a great lawmaker. "By the favor of Ahura Mazda," he wrote, "these lands walked according to my law; as was to them by me commanded, so they did." He may have been responsible, as this inscription suggests, for establishing Zoroastrianism as the dominant religion among the Persian elite. Certainly he claimed to have restored Zoroastrian sanctuaries destroyed in previous wars. Today Zoroastrianism is associated largely with the Parsees of India and thought of as an essentially pacific religion. Certainly it has few of the murderous possibilities inherent in the world's great monotheisms. But it is also a deeply Manichean creed, the world being equally divided between the principle, or god, of light—Ahura Mazda—and that of darkness—Ahriman—between whom there exists a condition of incessant war. Although there is little suggestion that the Achaemenids ever waged the kind of ideological warfare later monotheists would make against one another, there was nothing inherent in Zoroastrianism to deter them from doing so.[29]

Darius also gave to the empire a capital worthy of its size and power. The Achaemenids had ruled from Susa. But Susa was an alien city of Elamite origin with a long history of occupation by other peoples. Once the revolts that had inevitably followed his accession were quelled, he set about creating a new city worthy of his own and his dynasty's achievements. He settled it in the plain of the Median River at the foot of the Mountain of Mercy and called it Parsa. The Greeks referred to it first as "Persai," literally "the Persians." Later it came to be called, and has been called ever since, Persepolis—"city of the Persians," which Aeschylus mistranslated as "Perseptolis," "destroyer of cities."[30] With Darius, Achaemenid power had come of age.

## III

Darius's accession was also the beginning of the conflict with Greece, the beginning of the desire to unite Europe and Asia under an Asian tyranny that, in Herodotus's account, constituted the overriding arrogance of the Persian monarchs. In Herodotus's *Histories,* the Persian Wars are described as two acts in a drama in which the infinitely more powerful Asian people pits itself against a weak, but infinitely more able, more virtuous Western one and loses. Both acts begin with a crossing from Asia into Europe, which means conveying an army across the Hellespont, an act both sides look upon as at best foolhardy and possibly even unnatural. Both acts end with battles in which the Greeks, against all odds, defeat overwhelmingly superior Persian forces, first at Marathon, then at Salamis.

The struggle between Greeks and Persians first began in about 499 B.C.E., when the Ionians, Asiatic Greeks who had been overrun by Cyrus after his defeat of Croesus, together with Greeks from Cyprus and Carians from Anatolia, rose in revolt against their Persian overlords. The following year, the Athenians, persuaded by

Arsitagoras of Miletus that there were any number of "good things to be found in Asia" and that the Persians would be easy to defeat since—or so he imagined—they had neither shields nor spears, decided to intervene and sent a fleet of twenty ships to Ionia. This, as Herodotus said, was "the beginning of evils for the Greeks and the barbarians."[31]

The Athenian fleet, reinforced by five triremes from Eretria, sailed for Ephesus. Here the Greeks disembarked and marched inland to Croesus's former capital, Sardis, captured the town, and burned it to the ground. In the process they destroyed a temple to Cybele, "a goddess worshipped by the natives." The destruction of a city was one of the hazards of war. The destruction of a sacred site, however, violated a code by which both the Greeks and the Persians lived. It was an offense against a divinity and could not be forgotten or go unpunished. When Darius heard of this act of impiety—the first, as it would turn out, of many—he vowed revenge on the Greeks. He shot an arrow into the air, at the same time intoning, "Grant, O God, that I may punish the Athenians," and then ordered one of his servants to repeat to him three times every day before he sat down to dinner, "Master, remember the Athenians."

After the destruction of Sardis, the Athenians, having failed to secure any of the goods they supposed to be found in Asia, retreated and refused to take any further part in the Ionian rebellion. Darius, however, had neither forgotten nor forgiven them. From the Greek, which usually means the Athenian, point of view, the expedition had been an attempt to free and avenge the subordination of fellow Greeks by a barbarian power. From the Persian perspective, things must have looked rather different. The Asian Greek cities were the most wealthy, most populous, and most cultured part of the Greek world. For the Athenians to describe them as in need of liberation was derisory. Indeed, for Darius it must have seemed only a matter of time before the small, independent, impoverished, and faction-ridden Greek communities across the strait in Europe also passed under Persian rule.

Darius now began to gather a vast army from all the peoples who owed him allegiance. He also adopted a new strategy toward the Greek cities under his rule. Once the Ionian revolt had been suppressed, Darius dismissed the tyrants by whom they had been governed and reorganized them in accordance with the relatively new and primitive Athenian mode of government: democracy. Herodotus regarded this sudden act of liberation as a refutation of "those Greeks who cannot believe that Otanes declared to the seven conspirators that Persia should have a democratic government."[32] But there was probably a more immediate reason for Darius's move. Democracy was a very recent creation. After the expulsion from Athens in 510 B.C.E. of the tyrant Hippias (who would play an important role in the subsequent encounter between Athenians and Persians) and the collapse of an attempt to establish an oligarchy with Spartan aid, the Athenian politician Cleisthenes had reformed the Athenian constitution in such a way as to break the power of the rich families whose feuding had plagued the city for centuries. He divided the peoples of Attica into ten new tribes, organized along entirely artificial lines, thus substantially weakening the older tribal, familial allegiances. A body known as the Council of Five Hundred was then established; its function was to prepare business for the larger assembly and to manage financial and foreign affairs. Each year, each tribe appointed fifteen members, who were chosen by lot. Around 500 B.C.E. a meeting place for the assembly, the *ekklesia,* was carved out of the rock on a hill known as the Pnyx, overlooking the city. From then on, this body met regularly to frame the policies of the state. Its meetings were open to all free male citizens and were attended by upwards of six thousand people. Simple and still subject to corruption and manipulation though it was, the government that Cleisthenes created was the basis for all subsequent democratic constitutions.

By installing similar governments in Ionia, Darius might have hoped not only to secure his rear as he advanced into Europe but also to offer a number of European Greeks the prospect of freedom from a local tyrannical rule, under the aegis of a benign foreign empire. Once again we can only infer from Herodotus's account, but Darius would seem to be making a move that so many later European empire builders would emulate: he was offering not only to provide his subject peoples security from external foes but also to guarantee the preservation of their particular ways of life. In exchange he demanded allegiance and support in time of war.

If that was indeed Darius's intent, it had some initial success. In 491 B.C.E., heralds were sent to the Greek mainland with requests that the cities provide earth and water—the traditional token of submission—to the king of kings. In Athens and Sparta the unfortunate emissaries were thrown into a well and told to fetch their own earth and water. But most of the other states submitted.

In 490 B.C.E. Darius's forces landed on the shores of Attica, and their commander, Datis, declared his master's intention to conquer Eretria and Athens. Eretria was the first to fall; divided against itself, the city was finally betrayed by two democrats, Euphorbus and Philagrus, after six days of resistance. The Persians proceeded to strip and burn the temples in revenge for the Athenian sacrilege at Sardis and carried off all the inhabitants into slavery.[33] "Flushed with victory and confident that they could treat Athens in the same way," the grand army sailed for Attica. Hippias, the deposed tyrant, who had joined the Persians in the hope of regaining power, guided them to the plain at Marathon because, as Herodotus tells us, this was the "best place for the cavalry to maneuver." The Athenians hurried to meet them and sent a runner, named Pheidippides, to ask the Spartans for their help, with the plea that not only Athens but the whole of Greece was in danger. The Spartans received him courteously but answered that although they would have been pleased to assist, they were unable to do so just then because it was the ninth day of the month and custom prevented them from taking the field before the moon was full.[34] After running day and night for nearly 145 miles, poor

Pheidippides is said to have dropped dead once he had delivered his message. His name is now largely forgotten, but it is in memory of his astonishing run that the modern "marathon" is named.

The Athenian army, with the support of only a small force from Plataea, was stretched to its limits, strong enough on the wings but only a few men deep in the center. Despite their numerical weakness, they did not hesitate, wrote the orator Andocides a century later, "to place themselves before all the Greeks like a rampart, to march out against the barbarians . . . considering that their valor alone was capable of confronting the enemy multitudes."[35] Once a preliminary sacrifice, which seemed to promise success, had been made, the Athenians advanced at a run, hoping to get under the range of the formidable Persian archers before they had time to prepare themselves. The Persians, who seem to have believed that so small a force without either archers or cavalry would not dare to oppose them—an error they would make more than once—were taken off guard. The fighting lasted all day, until finally the Persians scattered and retreated to their ships as best they could. When it was all over, some 6,400 Persians had been killed, but a mere 162 Athenians.[36] Later, after the moon was full, two thousand Spartans turned up to admire the corpses and to wonder at the slaughter they had missed. As would happen again ten years later at Salamis, it was democratic Athens that had saved the entire Greek world from slavery.

Marathon marked the end of the first Persian War. It brought down the curtain on the first act of Herodotus's great tragic drama of the struggle between Europe and Asia, between Greeks and barbarians. Ever since that day, Marathon has come to be accepted as a turning point in the history of Greece and subsequently of the whole of Europe. It was the moment when the resilience of an unusual political form—democracy—and Greeks' confidence in their peculiar notion of freedom was put to the test and had emerged triumphant. It was, the English liberal philosopher John Stuart Mill said in the nineteenth century, a more important event in *English* history than the battle of Hastings. For without it the history of Europe, and with it of Britain, might have taken a very different course.


FOR DARIUS, HOWEVER, Marathon was by no means the end of the conflict. He seems, in fact, to have regarded it as little more than a temporary setback in a much larger project to control not only mainland Greece but all the islands of the Aegean.[37] The wholly unexpected Athenian victory merely increased the great king's anger and made him even more determined to punish the Greeks for what had now become a double insult. He therefore began to prepare for another invasion. Before he could complete his plans, however, he died in November 486 B.C.E.[38] He left a territory that reached from the shores of the eastern Mediterranean and the Balkans to beyond the Indus valley, from the Black Sea to the Caspian, to the Nile and the Arabian peninsula. Lordship over all these vast and varied lands now passed to his son Xerxes— Xsayarsa, in Old Persian, "he who rules over heroes"—an ironic sobriquet in light of his subsequent career.

At first, Xerxes seems to have been less interested in pursuing his father's policy toward Greece than in subduing Egypt, which had for some time been in revolt against its Achaemenid rulers. In 485 B.C.E., Xerxes sent an army into Egypt, crushed the revolt, and, in Herodotus's words, "reduced the country to a condition of worse servitude than it had ever been in the previous reign." Xerxes was now persuaded by his ambitious cousin Mardonius to turn his attention toward Europe. "The Athenians have done us great injury," Mardonius reminded Xerxes. "Destroy them," he continued, "and your name will be held in honor all over the world." Besides, he added slyly, reversing the usual image of the differences between Europe and Asia, Europe was "a very beautiful place that produced every kind of garden tree. The land was everything that land should be. It was, in short, too good for any mortal except the Persian king."

Before committing himself to another European adventure

and amassing the vast armies it would require, Xerxes held a conference. In all their history, he declared, the Persians "have never yet remained inactive," and he had no desire to "fall short of the kings who have sat on this throne before me." Now he had found a way to increase the size of the empire greatly by adding to it a country that was as large and rich as his own, indeed larger and richer, and at the same time to "get satisfaction and revenge." "I will bridge the Hellespont," he declared, "and march an army through Europe into Greece and punish the Athenians for the outrage they committed upon my father and upon us." We will, he continued, "so extend the power of Persia that its boundaries will be God's own sky, so that the sun will not look down upon any land beyond the boundaries of what is ours. With your help I shall pass through Europe from end to end and make it all one country."[39]

Xerxes would unite what the mythical journeys of Io, Europa, and Helen had undone. Under Persian rule the enmity, whose origin could be traced back to the beginning of time, would be healed, and the world, or all that he knew of the world, would be one. It would also, of course, be one in which the entire population of the globe, as he conceived it, would be enslaved to this king of kings. More than a century later, a Greek, Alexander of Macedon, would gather an even greater army than Xerxes' and travel in the opposite direction, from west to east, in the hope of securing this selfsame objective.

Mardonius then got up and gave his opinion of the Greeks. They were, he said, pugnacious enough and ever ready to start a fight on the spur of the moment without sense or judgment. But, despite the fact that they all spoke the same language, they were a divided people, who could never find a better way of settling their differences—"by negotiation, for instance, or an exchange of views"—than by fighting one another. Such a people could never resist the united power of a Persian monarch. Once again the Persians are represented as incredulous at the mere thought that any people who fought and squabbled among themselves would be able to come together when they needed to. Herodotus knew, however, as do his readers, that it was their freedom to dispute and their equality before the law that made the Greeks such good warriors. Cleisthenes had set them free, and it was as a consequence of their freedom that they had, as Herodotus phrased it, gone from strength to strength and proved, if proof were needed,

> how noble a thing is equality before the law [*isonomia*], not in one respect only but in all; for while they were oppressed under tyrants they had no better success in war than any of their neighbors, yet, once the yoke was flung off, they proved the finest fighters in the world.

Slaves will always "shirk their duty in the field" because they are only ever fighting against their will and their interest and on behalf of someone else. Freemen, by contrast, even when they are fighting for their city, fight only for themselves.[40] Liberty in the West has always in this way grown because it served the interests of power.[41]

Mardonius, however, puts his faith, as would Xerxes, in sheer force of numbers and in the despotic nature of Persian government. "Well then, my lord," he asks, "who is likely to resist you when you march against them with the millions of Asia at your back and the whole Persian fleet?" When Artabanus, Xerxes' venerable uncle, opposes the planned invasion because, as he says, sounding all too much like a Greek, "without a debate in which both sides of a question are expressed, it is not possible to choose the better course," and reminds Xerxes that the Athenians did, after all, destroy the Persian army at Marathon against overwhelming odds, Xerxes rises in indignation. "You are my father's brother, and that alone saves you from paying the price your empty and ridiculous speech deserves." Artabanus is spared death but condemned to remain behind in Persia with the women. Once again reasoned disputation is suppressed by precisely what Otanes had identified as one of monarchy's greatest weaknesses:

the inability of the monarch to listen to the voices of any but those who tell him what he wishes to hear.

But for all his despotic rage, Xerxes has another more cogent point to make. It was, he reminds his uncle, after all, the Greeks who began this present conflict by marching into Asia and burning the Persian capital at Sardis. Now either the Persians must invade to revenge the destruction of Darius's army or wait patiently to be invaded. "All we possess will pass to the Greeks," says Xerxes, "or all they possess will pass to the Persians. That is the choice before us; for in the enmity between us there can be no middle course."[42]

Xerxes now set about gathering an army able to conquer the whole of Greece, an army, said Herodotus, far larger than any yet seen, larger even than the mythic armies Agamemnon and Menelaus had assembled to besiege Troy. Its sheer size would remain a thing of wonder for centuries to come. In the middle of the second century C.E., a young Greek rhetorician, Aelius Aristides (whom we shall meet again) would stand before a Roman audience and compare his attempt to capture the glories of Rome with "some effort to describe the marvelous size of an army such as Xerxes'."[43]

This expedition, at least as it was represented by the Greeks, who would ultimately be the victors, was to be no mere act of revenge for a desecrated shrine, much less a punitive expedition against the allies of rebellious subjects. It was to be a westward march to overawe the journey eastward of the semidivine heroes who had brought about the destruction of the first Asian city.

In the spring of 480 B.C.E., Xerxes left Sardis and moved down the Valley of the Caicus into Mysia, and then on through the plain of Thebe, until finally he reached the fabled site of Ilium. Xerxes now mounted the citadel of Priam, the last ruler of Troy, and sacrificed a thousand cattle to the Trojan Athena. Xerxes had come from Asia to resume the Trojan War. The cycle of enmity between Europe and Asia was about to begin again.[44]

Xerxes, however, needed more than an army. He also needed

to get that army across the Hellespont and into Europe. To do this he built a bridge of boats across the narrow neck of water that separates the two continents. No sooner was it complete, however, than the whole thing was swept away in a sudden storm. In his anger at the setback, Xerxes ordered the engineers who had built the bridge to be beheaded, as if the weather were their responsibility. Unsatisfied even by this, Xerxes then proceeded to commit an act that, for the Greeks, epitomized not only his own personal tyranny but also the despotism of the kind of world over which he ruled. Furious, he turned his anger against the Hellespont itself. Like most peoples in antiquity, both the Greeks and the Persians believed that rivers and other expanses of water were divine beings, deities who could, when angry, be placated and appeased. Only Xerxes, however, seems to have believed that they could also be punished for failing to comply with human wishes. He ordered a pair of fetters to be thrown into the Hellespont and its waters to be lashed with whips. While Xerxes' men were performing this fruitless and absurd task, they were ordered to "utter . . . the barbarous and presumptuous words 'You salt and bitter stream, your master lays this punishment upon you for injuring him who never injured you. But Xerxes the king will cross you, with or without your permission."[45] To us this may sound merely petulant; to Herodotus's Greek readers, however, it was further proof of Xerxes' arrogance, his failure to respect the gods in much the same way as he failed to respect his own people.

Two new bridges were built in place of the previous one, by lashing together more than six hundred vessels of one kind or another. The bridges were strewn with myrtle, and incense was burned on them. Xerxes, seated on a white throne on a nearby hill, poured libations from a golden cup and prayed with his face toward the rising sun. The crossing now began. It lasted seven days and nights without a break, the troops, as Herodotus puts it "coming over under the whips." According to a contemporary inscription at Thermopylae, there were three million of them—an exaggeration to be sure. But the Athenian orator Lysias's observation that Xerxes had "made a road across the sea and sailed his fleet through the land" later became a rhetorical commonplace for megalomania.[46]

When they were across and a fleet of 1,207 triremes had been assembled, Xerxes held a review of his forces.[47] For all their barbaric splendor, for all that their armor glinted and glittered in the morning sun, for all that they made the earth tremble as they marched, this massive force was, in Greek eyes, a sign of the great king's ultimate weakness. The Greeks were few and frequently divided, but ultimately they all shared in the same "Greekness," which could make of them, however briefly, a people, an *ethnos*. Xerxes' army was a horde, held together by a sense neither of identity nor of purpose but only by a common fear, the fear of Xerxes' wrath and Xerxes' lash. In Greek eyes these men were not warriors but slaves. And in one literal sense at least they were right: every subject of the great king had no legal status as other than a *bandaka* or slave, a conception that derived from the Babylonian definition of the subject as an *ardu* or chattel of the monarch. How could, asked the great Athenian orator Isocrates a little more than a century later, "there be either an able general or a good soldier produced amid such ways of life as theirs? Most of their populations are a mob . . . more effectively trained for servitude than are the slaves in our country."[48]


XERXES NOW SENT for the exiled Spartan Demaratus, who had joined the Persian horde in the hope of revenging himself on his countrymen. This was the occasion for Herodotus to stage one of those encounters between Greek and Persian, between Asian and European, that serves to remind us just what it is that separates the two and why in the end they have no choice but to fight each other until one of them finally destroys the other. Standing before his teeming army, Xerxes demands to know of Demaratus if the

Greeks will dare make a stand against him. "My own belief," he says, "is that all the Greeks and all the other Western peoples gathered together would be insufficient to withstand the attack of my army."

Demaratus replies that although poverty has been Greece's "inheritance of old," she had won wisdom and the strength of law for herself and that this was what kept both poverty and despotism at bay. Asiatic wealth versus European poverty would be a constant theme in the ancient conception of the enduring enmity between the two continents. For later writers, like Isocrates, it was a source of envy, a just cause for Philip of Macedon and his son Alexander to turn the tables on the mighty Achaemenid Empire and invade Persia itself. In the mouths of Herodotus's Greeks, however, it is, if anything, a source of pride. The Greeks may lack the economic power of the Persians, but they possess in abundance what the Persians, for all their wealth, so markedly do not have: the strength and the courage that come from being free. Even if the Spartans—and it is about them that Demaratus is now talking—were reduced to a mere thousand men, that thousand would stand up to the massive Persian army. At this Xerxes laughs.

> Demaratus, what an extraordinary thing to say! Let me put my point as reasonably as I can—how is it possible that a thousand men, or ten thousand or fifty thousand, should stand up to an army as big as mine—especially if they were not under a single master?

The Persians can be whipped into battle or forced to fight from fear of what their commander may do to them if they do not. But the Greeks? They are perfectly free to do as they please. Why, then, should they fight against such overwhelming odds?

In reply Demaratus sets out the true nature of Greek strength. The Greeks, says Demaratus, are free, yes, "but not entirely free. They have a master, and that master is Law. This they fear far more than Xerxes' subjects fear him." Xerxes bursts out laughing and good-humoredly lets Demaratus go.[49]

At the time, Xerxes had good reason for confidence. Not only was his army larger than any other ruler had ever amassed, but his enemies were having considerable difficulty in creating a coalition against it. The Greeks may have feared the law, but, as Xerxes had rightly observed, their independence made them quarrelsome and factious, reluctant to forgive past grievances even in their own collective interests. With very little to oppose him, Xerxes overran the small Greek cities in his path, one by one, until at last all that remained were a few recalcitrant states of the Peloponnese: Sparta, Corinth, and Athens itself.

In August, an attempt was made to turn the Persian army back at Thermopylae—the "Hot Gates"—by the Spartan king Leonidas at the head of an army of about seven thousand men, a tiny force when compared with the mass of the Persians. They succeeded in keeping Xerxes' great army pinned down for several days until a Greek traitor revealed the existence of a secret pass over the mountains. Leonidas was surprised from the rear by a contingent of Xerxes' crack troops, known as the "Immortals." The Greeks fought fiercely and succeeded not only in dispatching a large number of the Immortals but also two of Xerxes' own brothers. Finally, however, the small force was overwhelmed. Leonidas, dying and riddled with spears, managed to snatch off Xerxes' crown before he fell. When he was dead, the barbarian king cut out his heart and found it was covered with hair.[50]

The battle of Thermopylae failed to check Xerxes' advance for more than a few days. For later generations, however, it became a symbol of the courage of lost causes, used to glorify the defense of the Alamo against the forces of the Mexican general Santa Anna in 1836, and to encourage underage recruits to join the Wehrmacht as Adolf Hitler's Third Reich began to disintegrate toward the end of the Second World War.[51] (Most recently the battle has been made into a gruesome film with comic-book Greek heroes, wearing cloaks and codpieces, and pierced and tattooed Persians.

The film may or may not contain some remote allusion to the American forces in Iraq—certainly many in Iran believed so—or it may simply be a celebration of video-game violence.)

THE PERSIAN VICTORY at Thermopylae opened the way to central Greece, and four months after crossing the Hellespont, Xerxes entered Attica. He found Athens deserted except for a few "temple stewards and the needy" who attempted to barricade the Acropolis against the invaders. Their resistance did not last long. The Persians climbed up the sheer wall in front of the Acropolis, burst open the gates to the temple of Athena, slaughtered everyone within it, stripped the temple of all its treasures, and then burned the building to the ground. Xerxes had now become a desecrator of temples, something both he and his successor, Darius III, would later come to regret. He had also, if only briefly, made himself absolute master of Athens.[52]

The rest of the population had fled to the island of Salamis, where the combined allied fleet was waiting to evacuate them. Some 380 ships were compressed into the narrows between Salamis and the port harbor of what is today Piraeus. In the open waters beyond the strait the far larger Persian fleet lay in wait. Immediate flight seemed the only possible hope of survival. At this point in the story, the Athenian commander, Themistocles, argued that the Greeks should stay and fight in the narrows rather than risk destruction on the open sea, and with it the certain loss of both Salamis itself and the still unconquered cities of the Peloponnese. By threat, bribery, and subterfuge, Themistocles won the argument and the fleet prepared to fight. It was, however, effectively bottled in. So far as we know—and we know remarkably little about them—ancient triremes were, by modern standards, small vessels, little more than a hundred feet long and no more than fifteen wide, manned by 170 rowers stationed in banks of three on either side. They were also fragile, overcrowded, and difficult to maneuver. All Xerxes had to do was to wait until exhaustion, shortage of supplies, and the already overheated tempers of the various Greek factions eliminated the agreement Themistocles had forced upon them. He could then have wiped out the Greek fleet with ease. Instead he decided to attack.

It was September 22, 480 B.C.E. Xerxes sat on a silver-footed throne at the base of Mount Aegaleos overlooking the strait, waiting for the inevitable victory. Beside him sat his secretary to record the behavior of his subjects, just in case any of them decided to defect.

At dawn the Greeks moved out of the strait and engaged the combined Persian fleet. "The first thing we heard," declares the messenger in Aeschylus's *Persians* who brings the news of the defeat to Susa and who may well be echoing Aeschylus's own recollections of that day,

> *was a roar, a windhowl, Greeks*
> *Singing together, shouting for joy,*
> *And Echo at once hurled back*
> *that war cry*
> *loud and clear from island rocks.*
> *Fear churned in every Persian.*
> *We'd been led off the mark:*
> *The Greeks*
> *Weren't running, no*
> *But sang that eerie triumph-chant*
> *As men racing toward a fight*
> *And sure of winning.*[53]

At first the Persians had some initial success. But then, as the Athenian fleet began to push them back, the galleys in the rear, which had crowded forward in the hope of doing "some service before the eyes of the king" found themselves being rammed astern by their own vessels. The sheer size of the Persian fleet and the lack of any clearly conceived battle plan now proved fatal. "The Greek fleet worked together as a whole," remarked Herodotus with satisfaction, "while the Persians had lost formation and were no longer fighting on any plan." "At first the wave of Persia's fleet," declares the messenger,

> *Rolled firm, but next as our ships*
> *Jammed into the narrows and*
> *No one could help any other*
> *And our bronze teeth bit into our own stakes,*
> *Whole oar banks shattered.*
> *Then the Greek ships, seizing their chance,*
> *Swept in circling and struck and overturned*
> *Our hulls,*
> *And saltwater vanished before our eyes—*
> *Shipwrecks filled it, and drifting corpses.*

By then the Persians were in full flight and hopeless confusion. The triremes that attempted to fight were rammed by the Athenians, and an Aeginetan squadron that had positioned itself just outside the narrows waited for any that tried to escape.

"We might have been tuna or netted fish," wails the messenger,

> *for they kept on, spearing and gutting us*
> *With splintered oars and bits of wreckage,*
> *While moaning and screams drowned out*
> *The sea noise till*
> *Night's black face closed it all in.*[54]

TWO HUNDRED OF the Persian ships, a third of the entire armada, went down. Ariabignes, Xerxes' own brother, along with hundreds of Persian sailors and their allies, drowned because, unlike the Greeks, they could not swim. The long robes they wore further hindered their movement, so that they became entangled with the wreckage of their ships, their bodies then "submerged and tossed about lifeless." What remained of the Persian fleet limped back to Phalerum, where it was protected by the Persian army, to be greeted by the wrath of Xerxes.[55]

It was the greatest naval battle in ancient history. It was a victory gained by Athenian skill and largely by Athenian ships guided by an Athenian commander. It came, therefore, to be understood as a great Athenian battle. It was Athens above all, and the uniquely Athenian mode of government—democracy—that in the end saved all the Greeks from becoming the slaves of the great king, and Athens would emerge from Salamis as the preeminent state in the Greek world.

Two things helped to reinforce this image of Salamis as not merely a great Greek triumph but also a great democratic victory. In the first place it was, in large part, the work of a man who was probably closely associated with the "radical" democratic party in Athens. Themistocles would later be suspected of treason and condemned to death in his absence, and—the final irony—end up by seeking refuge with Xerxes. (Themistocles was one of the few Greeks, so far as we know, to have learned Persian.) But in the popular imagination, and in particular in the telling of the great Athenian historian Thucydides, he would remain the hero who had saved Greece from enslavement.[56]

Salamis had also been a naval battle. It had been won by sailors, not soldiers. Unlike the heavy infantry, the famed hoplites, who were all men wealthy enough to be able to provide their own equipment, the sailors—the "oarspeople"—were generally recruited from amongst the poor (which was why that disillusioned aristocrat Plato regarded the victory as altogether inferior to the one at Marathon, which had been won by aristocratic hoplites).[57] Salamis had been a battle in which those who had most to gain from democracy had succeeded in defending the whole of Europe from an Eastern tyrant. It would also transform Athens into the supreme naval power in the Mediterranean and lay the foundation for what has come to be called the "Athenian Empire."

It was destined to change forever not only the nature of the Greek world but with it the entire future both of Europe and of

Europe's relations with Asia. "The interest of the World's history hung trembling in the balance," wrote the German philosopher Georg Friedrich Hegel in 1830.

> Oriental despotism—a world united under one lord and sovereign—on the one side, and separate states— insignificant in extent and resources—but animated by free individuality, on the other side, stood front to front in array of battle. Never in History has the superiority of spiritual power over material bulk . . . been so gloriously manifest.[58]

After the battle the Greeks prepared themselves for what they assumed would be another Persian attack by land. None came. Xerxes' only act was to execute his Phoenician captains for alleged cowardice. Incensed, the remaining Phoenicians returned home, followed shortly afterward by the Egyptians, which effectively deprived Xerxes of his navy. The great king, fearing that the Greeks would now sail to the Hellespont to destroy his bridge of boats and thus cut off his retreat, handed the command of his armies over to Mardonius, while he himself returned to Susa.[59]

Xerxes' land forces under Mardonius remained behind, but, demoralized and divided, they were defeated, first at Plataea in 479 B.C.E. and then, in spring of the same year, at Mycale. Two of the six Persian forces under Mardonius's command were annihilated. One was recalled to fight disaffected populations in western Asia. The remainder struggled back to Persia as best they could. As Aelius Aristides would tell the Roman emperor Antoninus Pius centuries later, Xerxes had "evoked wonder less for his own greatness, than for the greatness of his defeat."[60] For with that defeat the Persian Wars had come, seemingly, to an end. To celebrate the final battle, the Greek allies set up a monument at Delphi made of three intertwined bronze serpents inscribed with the names of the thirty-one city-states that had resisted the Persians. Centuries later, the Roman emperor Constantine had it moved to his new eastern capital at Constantinople. The base is still there today in the Hippodrome in the center of Istanbul, which, with nice irony, was for centuries the capital of the new masters of Asia: the Ottoman Turks.

The battles of Thermopylae, Marathon, and Salamis had, as Lysias would later observe, "secured a permanence of freedom for Europe."[61] Had Xerxes succeeded, had the Persians overrun all of mainland Greece, had they then transformed the Greek city-states into satrapies of the Persian Empire, had Greek democracy been snuffed out, there would have been no Greek theater, no Greek science, no Plato, no Aristotle, no Sophocles, no Aeschylus. The incredible burst of creative energy that took place during the fifth and fourth centuries B.C.E. and that laid the foundation for all of later Western civilization would never have happened. What would have occurred in a Persian-governed Greece, it is obviously impossible to say. But one thing is certain: in the years between 490 and 479 B.C.E., the entire future of the Western world hung precariously in the balance.

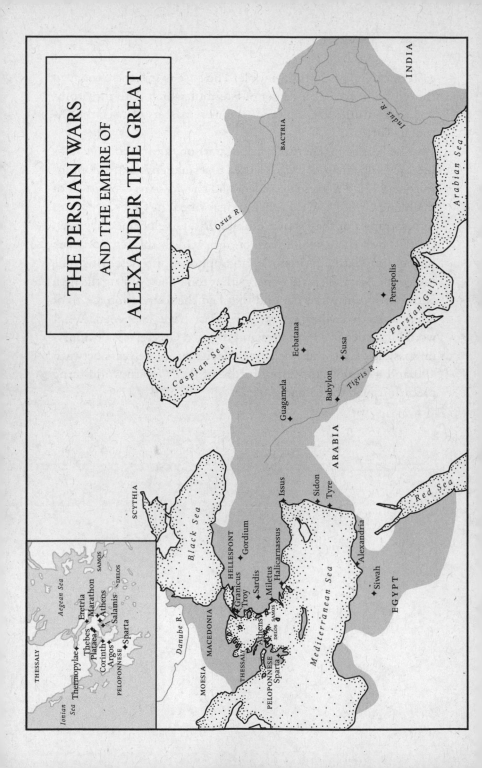

THE PERSIAN WARS
AND THE EMPIRE OF
ALEXANDER THE GREAT

# 2

## *IN THE SHADOW OF ALEXANDER*

**I**

The end of the Persian Wars changed the relationship between Europe and Asia in fundamental ways. Herodotus's story is the story of a struggle between two civilizations, two ways of understanding political authority, two modes of living, and ultimately two concepts of humanity. But his Persians are not "others" in anything like the sense that much-abused term is used today. For all their differences, his Greeks and Persians share a common humanity, which distinguishes them from the more outlandish races Herodotus elsewhere describes, creatures such as the Lotophagi, the fabulous "Lotus-Eaters" who lived exclusively on the fruit of the lotus, or the Atarantes, who have no proper names and curse the sun.[1] These were, and would remain, irreducibly "other." But not the Persians. For all that he abhorred Xerxes' tyranny, Herodotus had great admiration for Persian courage. He could even admire Persian grandeur, although it contrasted sharply with the much-vaunted simplicity of the Greeks, and he was particularly respectful of what he described as the Persians' insistence on always telling the truth—in contrast to the ever-wily Greeks.

The Persian Wars, however, brought about a hardening of the

distinction between the Greeks and all the others, the "barbarians" who populated the inhabited world. The word "barbarian"—*barbaros*—which was to have such a prominent place in the long history of the European perception of the non-European world, originally meant only those who spoke no Greek and whose many languages sounded, to Greek ears, merely like people stuttering: *bar bar.* Later generations, however, would take a much bleaker view of the possible meaning of the term. Since in the Greek view articulate speech was evidence of the capacity to reason (*logos*) the fact that the rest of humanity seemed to be made up of incoherent babblers could be understood to mean that only the Greeks were rational beings and thus only the Greeks were truly human. For the philosophers and physicians, this division reflected a division within the human personality. Ever since Plato, and then Aristotle, the mind—the psyche—had been, and would remain until the seventeenth century, divided into two: a rational part, located in the brain, and an irrational part, composed of all the passions and sensations, which were located somewhere in the body, sometimes in the liver, sometimes in the intestines or the genitals, most often in the heart. These two halves were at constant war with each other. In the truly civilized being—in the Greek, that is—the rational mind generally, if not always consistently, triumphed over the irrational. In barbarians, however, the reverse was frequently the case.

This is, broadly, how Aristotle seems to have understood the distinction. For Aristotle these barbarians might also—although he was never quite certain—be "slaves by nature," creatures with no moral qualities, no capacity for independent judgment, but brains enough to interpret their master's orders when required and brawn enough to carry them out. For this reason, he said (quoting the Greek tragedian Euripides), "It is fit that the Greeks should rule over the barbarians."[2] If Plutarch is to be believed, he also told his pupil Alexander the Great, when he began his campaigns in Asia, to treat only Greeks as human beings and to look upon all the other peoples he conquered as either animals or plants. This advice, Plutarch also says, Alexander wisely ignored, for had he accepted his mentor's council, he would have "filled his kingdoms with exiles and clandestine rebellions."[3] How widely shared this view of the "barbarian" was by most Greeks we cannot know. But certainly Plato, writing shortly before Aristotle, in his dialogue *The Statesman,* makes the main speaker, "Stranger from Elea," (a town in southern Italy) complain:

> In this country they separate the Hellenic races from all the rest as one, and to all the other races which are countless in number and have no relation in blood to one another, they give the single name "barbarian"; then because of this single name they think it a single species.[4]

For many later generations of Europeans it became—and for many has remained until this day—a matter of fact that the Greeks were extreme ethnocentrists. The Greeks' tendency to lump together the rest of humanity as "barbarians," argued the great eighteenth-century German philosopher Immanuel Kant, had been "a perfect source that contributed to the decline of their states."[5] Yet Kant was essentially wrong. Certainly the Greeks were ethnocentric. So are all peoples. But of all peoples the Greeks were among the least guilty. They possessed, after all, a word—*anthropos*—with which to describe not merely Greeks but all human beings. Other peoples may have had similar terms, but they are not easy to identify. As the great French anthropologist Claude Lévi-Strauss once observed, "A great number of primitive tribes simply refer to themselves by the term for 'men' in their language, showing that in their eyes an essential characteristic of man disappears outside the limits of the group."[6] Not so the Greeks. Yet for all their curiosity about the outside world, for all that the Greeks were in their own words "extreme travelers," *poluplanês,* the distinction between "them" and "us," between Greek and "barbarian"—which meant, in effect, between Greeks and Asians—hardened considerably after the end of the Persian Wars.

The Persian Wars changed the images of Europe over Asia in other ways, too. For many, even for Aeschylus, it would seem that the Greeks' astonishing victory at Salamis had been an act of divine retribution. It was not superiority of numbers, the messenger in *The Persians* tells Atossa, cunning, or courage that had given the Greeks their triumph:

> *It was some Power—*
> *Something not human—*
> *Whose weight tipped the scales of luck*
> *And cut our forces down.*[7]

"Zeus, the pruning sword of arrogance," the ghost of Darius the Great will later remind the audience, "is set over you a grim accountant."[8]

In Herodotus's version, however, the gods, although they lurk constantly offstage, play no direct role in human affairs. Salamis, like Marathon before it, is described as a victory won by men through their individual valor and, more enduringly, because of the virtues of the democratic political culture to which they belong.[9] For Thucydides, who conveyed all of this to the next generation, the gods played no part in the action at all. Thereafter, humanity would be held responsible for its own actions and be master of its own destiny. The gods—and later God—might be there to help, to guide, but never again to take an active, intrusive role—except on the most fantastic occasions—in human affairs. As we shall see later—much later—this will take the form, familiar to us but still unknown to the ancients, of a distinction between religion and society, between the authority of the ruler and that of the priest. It would, as we shall also see, ultimately prove to be the most powerful force in the shaping of the identity of the "West" and of securing its success.

## II

The defeat of Xerxes did not, however, put an end to the Persian menace. In order to avoid the near-disastrous lack of cohesion amongst the Greeks that had marked their response to previous Persian offensives, the Greeks formed a league in 478/7 B.C.E. "whose guiding principle," as Thucydides described it, "was to ravage the lands of the King in revenge for the wrongs suffered."[10] Known today as the Delian League because its various members agreed to meet in order to decide policy on the sacred island of Delos, where the treasury for the league was also stored, it was formed by the Ionian cities on the west coast of Asia Minor, the Hellespont, and the Propontis. Although on an equal standing with all the other members, Athens was, in fact, by far the most powerful member. Initially the League was very successful. Persian garrisons were driven out of Thrace and the Chersonesus, and Greek control was reestablished all along the west and south coast of Asia Minor.

Greek control, however, meant increasingly Athenian control. In 462 B.C.E., Thrasos attempted to abandon the League but was defeated two years later and forced to rejoin. When Naxos attempted to secede in about 467 B.C.E., it, too, was forced back in. Numerous other cities were "welcomed" into the League against their will. The final coup came in 454 B.C.E., when the treasury was moved from Delos to Athens—for safekeeping, it was claimed. But no one could doubt what the move implied. The Delian League had become—in all but name—the Athenian Empire. Like most empires, it brought benefits to some and misery to others. For Thucydides, as for many Athenian panegyrists, it had been justified by the need to continue the defense of Greece against the perpetual threat of the barbarian invaders.[11]

IN 466 B.C.E. the Athenian commander Cimon defeated a combined Persian force both on sea and on land at the mouth of the

River Eurymedon. It was a great triumph and since, like Salamis, the battle had been fought under Athenian leadership, it enhanced still further the power of Athens. In about 450 B.C.E., another Athenian commander, Cimon's brother-in-law Callias, negotiated a lasting peace with Xerxes' son and successor, Artaxerxes. The new great king agreed to limit all Persian activities to the east of Phaselis and outside the Euxine. The Greeks, for their part, agreed to withdraw from Cyprus and eastern Egypt. Although the Persians seem to have repeatedly violated the terms of the treaty in Asia Minor, the Persian threat to mainland Greece and the Aegean was now, for a while at least, seemingly at an end, and Athens was free to consolidate its hold over its "allies," who were now more frequently described as "the cities the Athenians rule."

Athens had become an empire by a simple process that would be repeated many times in world history. On the eve of the Peloponnesian Wars between the Delian League and a league of states led by Sparta, the Athenian ambassador told the Spartans that Athens had "been forced to advance our dominion to what it is out of the nature of the thing itself, as chiefly for fear, next for honor and lastly for profit."[12] Fear, honor, and profit—in that order—would remain the enduring motives behind the new forces of domination the Athenians introduced into Europe.

The Peloponnesian Wars lasted on and off from 431 B.C.E. until 404 B.C.E. In the end, Sparta and its allies emerged as the victor, but the conflict had so weakened both sides that it resulted in the collapse of the entire world of the Greek city-states. No sooner had hostilities ceased than war erupted once gain between the Greeks and the Persians. It continued until 387/6 B.C.E., when the Great King Artaxerxes II succeeded in imposing the so-called King's Peace, which re-established Persian control over the Greek city-states of Asia Minor. By the end of the fourth century B.C.E., however, Achaemenid power had been seriously weakened by rebellions in Egypt and Cyprus, and no large-scale invasion of the European mainland was attempted. In 361 B.C.E., the Persian regional governors—called "satraps," or "holders of power"—of the coastal cities of Asia rose in revolt against the great king, at the same time as the Egyptians declared a war on Artaxerxes that would last until 342 B.C.E.[13]

The Achaemenids managed to survive, and Artaxerxes III (358–338 B.C.E.) succeeded in crushing the revolts in Asia Minor and even reconquered Egypt. But his success was largely illusory. It was clear that the great Persian Empire had reached the possible limits of its power and was now slowly beginning to disintegrate piece by piece. The Greek city-states were, however, in no position to take advantage of the situation. The Peloponnesian Wars had seriously diminished all of those involved in them. By 346 B.C.E. Isocrates was able to describe Thebes, Argos, Sparta, and Athens as all equally "reduced to a common level of disaster."[14] Another rhetorical exaggeration, but although Athens was still powerful and its navy supreme in the Mediterranean, Thebes and Argos were already anachronisms and even the once-mighty Sparta had lost so many of its resources that its adult male citizen population now numbered fewer than a thousand.

Divisiveness had not prevented the Greeks from bringing Xerxes to his knees. But the great king had, after all, been right in claiming that one day they would finally exhaust themselves in interminable internecine conflict. "This," said the historian and minor state functionary Herodian, writing from the safety of the Roman world in the early third century C.E., "is an ancient condition among the Greeks, who exhausted Greece by quarrelling with each other and desiring to destroy those who seemed to be pre-eminent."[15]

BUT AS EVEN Athens and the cities of the Peloponnesus and Attica waned, a new and very different kind of Greek power was rising to the north. Macedon, a mountainous region that stretched from the foothills of Mount Olympus in the southwest along the shore of the Thermaic Gulf toward Thrace in the northeast, link-ing the Balkans to the Greek peninsula, had for a long time lain on the fringes of the Hellenic world. Although its peoples spoke a form of Greek and shared the same gods and many of the same cultural practices as other Greeks, they were widely despised as pastoral barbarians, overly fond of hunting, fighting, and drinking. Macedon was also unlike most of the other Greek states in being a monarchy, not a city-state. It was, however, populous and potentially rich, and had benefited greatly from Persian occupation between 512 and 476 B.C.E. Only dynastic feuding and foreign intervention had prevented the kingdom from playing any very significant role in Greek affairs. This changed in 359 B.C.E. with the succession of a new charismatic and ruthless monarch, Philip II. During the twenty-three years of his reign, Philip transformed Macedon into the most powerful of the Greek states. It was he who created the seemingly invincible Macedonian army, which in August 338 B.C.E., at Chaeronea in Boeotia, northwest of Thebes, won a crushing victory over an alliance of southern Greek cities led by Athens and Thebes. A magnificent stone lion was set up to commemorate the victory, and it is still there today. The battle of Chaeronea made Philip the effective master of the Greek world and Macedon an unchallenged superpower, which appeared to the Old Testament prophet Daniel in a dream as a beast "dreadful and terrible and strong exceedingly; and it had great iron teeth: it devoured and brake in pieces, and stamped the residue with the feet of it."[16]

Now, the ever-insistent Isocrates urged Philip, was the time to recall the ancient enmity that had always existed between Greeks and Persians. "So ingrained in our nature is our hostility to them," he declared,

> that even in the matter of our stories we linger most fondly over those which tell of the Trojan and Persian wars . . . even the poetry of Homer has won a greater renown because he has nobly glorified the men who fought against the barbarians, and on this account our an-

cestors determined to give his art a place of honour in our musical contests and in the education of our youth in order that we, hearing his verses over and over again, may learn by heart the enmity which stands from of old between us and them.[17]

Finally, as Isocrates hoped he would, Philip declared war on the Persian Empire, with the intention, so he claimed, of avenging the sacrilege Xerxes had perpetrated in the temple of the goddess Athena on the Acropolis and of ridding, once again, the Greek cities of Asia Minor of their alien rulers.

This, however, was to be no mere punitive expedition, no mere war of liberation. This was to be a war of attrition, in the name not of Macedon but of all Greece, a war that would put an end to Persian power forever. In the spring of 336 B.C.E. an expeditionary force made up of combined Macedonian and allied forces and numbering some 10,000 men crossed the Hellespont under Philip's command. It did not, however, get very far. By the autumn, and before any significant engagement with the Persian army had taken place, Philip was dead, cut down by an assassin, and the great army had retreated back to the Greek mainland.[18]

The man who now inherited Philip's throne was his son Alexander III, known throughout history as Alexander the Great. Two years after his father's death, he massed another army, which numbered 43,000 foot soldiers, armed with fearsome pikes, six meters long, and 5,500 horses, the largest army ever to leave Greek soil. Once again a Greek commander was preparing to follow Xerxes in reverse from west to east. To make the point, Alexander first went south to Elaeus, at the southernmost tip of the Chersonese. Here was the crossing point for the Troas, the mountainous northwest corner of Asia Minor that flanked the Hellespont, the place from which a united Greek force under Agamemnon, the first Panhellenic invasion of Asia, had launched its attack upon Troy.

Here, too, was the tomb of Protesilaus, supposedly the first of the Greeks to have disembarked at Troy and the first victim of the Trojan War. The Troas was also sacred to Athena, whose temple had been desecrated by Xerxes' men, just as a Persian contingent had plundered the shrine to Protesilaus in 480 B.C.E. Alexander now performed elaborate sacrifices at Protesilaus's tomb and, when his army was halfway across the Hellespont, sacrificed to Poseidon, the god of the seas, and to the Nereides, and poured a ritual libation into the waters of the Hellespont. The meaning of these acts could have been lost on no one: Alexander was bound on an expedition to revenge the atrocities committed against the entire Greek people and their gods. And he, the self-styled descendant of Hercules, Achilles, and Ajax, who is said to have always kept with him a copy of Homer's poem annotated by Aristotle, would repeat the deeds of the heroes of the *Iliad*, his ancestors, only now against a far deadlier enemy than even they had faced. His acts of obeisance before the deities who presided over the passage of his army may in some respects have been conventional. But the contrast between them and Xerxes' still unforgotten attempt to scourge and fetter the Hellespont could not have been more stark.

Once across the strait, Alexander, in imitation of Protesilaus, was the first ashore. Dressed in full armor, he stood before his troops, cast a ceremonial spear into the beach, and declared that the gods had given him Asia as "spear-won" territory. He then added, significantly, a prayer that "those lands should not be unwilling to accept him as king."[19] Alexander wished not merely to conquer but to rule, and he knew what the history of the long struggle between Persian and Greek had taught him: that ruling required the consent, however reluctantly acquired, of the ruled. To this end, the gods he invoked were not only those of Greece. They were also, as he understood them, those of Asia. Not the newfangled Ahura Mazda, to be certain, but the Trojan Athena, to whose sanctuary at Ilium he now made a visit. According to Plutarch, he ran a race with his companions naked, "as the custom is," around the column that supposedly marked the grave of

Achilles and offered elaborate honors to the dead heroes of the
Trojan War. He then exchanged his ceremonial armor for the
relics held in the shrine. Thereafter Alexander would always go
into battle carrying before him what he believed to be the arms of
his Homeric ancestors.

So far each step of his progress had been framed as a war to
right past wrongs, just as the Trojan campaign itself had been.
But Alexander had loftier ambitions than Agamemnon and his
followers. His armies had come not merely to conquer; they
had also come to reconcile and to unite; they had come not to
prolong but to put an end to Herodotus's perpetual enmity between Europe and Asia. In memory of the captive Trojan princess
Andromache, Alexander therefore bestowed lavish gifts upon
the community at Ilium and made an apotropaic sacrifice to
Priam, last of the kings of Troy, hoping thereby to expiate his
murder by Achilles' son—and Alexander's own supposed ancestor—
Neoptolemus.

The scene was now set for the Greek conquest of Asia, the final
act in a drama that had begun, centuries before, with the abduction by an Asian prince of Helen, queen of Sparta.


THE PERSIAN AND Macedonian armies met for the first time on
the banks of the River Granicus (now the Kocabas) in May 334
B.C.E. Alexander, in his now familiar role as the new Achilles,
dressed in the armor he had taken from the temple of Athena at
Ilium and with a curiously un-Homeric helmet with two great
white wings on his head, rushed headlong into the battle. He attacked two Persian nobles, Rhoesaces and Spithridates, at once
and would have been cut down by the unseen sword of a third had
it not been for the intervention of the veteran Cleitus, known as
"Cleitus the Black," the brother of Alexander's nurse. In the ensuing struggle the massively superior Persian forces were effectively
annihilated, largely due to Alexander's rash but inspired tactics.
Darius lost his son-in-law Mithridates, his son Arbules, his wife's

uncle Pharnaces, and numerous others of the Persian nobility. The Persian high command had been seriously depleted. The way into Asia was now wide open. The Greek Callisthenes, Aristotle's nephew, who accompanied the expedition as its official historian, linked the place of the battle to the Greek goddess of revenge, Nemesis, and claimed that it had taken place in the same month as the fall of Troy.[20]

Alexander buried his dead, ordered a commemorative statue in bronze from the great sculptor Lysippus, and sent three hundred Persian panoplies to the Parthenon to honor Athena. These, he said in his dedication, were to be but the first fruits of his war of revenge. The surviving Greek mercenaries who had served under Darius, Hellenes who had fought against Hellenes, were dispatched back to Macedon to end their lives as agricultural laborers, a far more degrading—and possibly far worse—fate than dying on the field of battle.

A policy of reconciliation intended to make the Persian nobility "not unwilling" to accept Alexander as their true ruler then began to emerge. The Persian officers who had fallen at Granicus were given proper burial alongside the Greeks. The army was forbidden to loot and pillage the defeated Asian cities, as would normally have been the custom. The peasants who came down from the hills to offer themselves as slaves to their new lord were instead sent back as free subjects, "each to his own property."[21]

Now, one by one, the cities of the Achaemenid Empire— Sardis, Ephesus, Miletus, Phaselis [Lycia], Aspendus [Pamphylia], Celaenae, and Gordium—surrendered to the seemingly invincible young king. At Gordium, Alexander paused long enough to perform one of those symbolic acts whose memory survived long after his conquests had been forgotten. In the temple of Zeus Basileus, he was shown the legendary wagon of Gordius, the mythical founder of the Phrygian dynasty. The yoke of the wagon was fastened to its pole by thongs of cornel bark in an elaborate knot tied in such a way that no ends were visible and it was therefore seemingly impossible to untie. An ancient prophecy had foretold that he who was able to untie the "Gordian knot" would become lord of Asia. Accompanied by his courtiers and a crowd of Phrygian and Macedonian onlookers, Alexander ascended the Acropolis. For a while he struggled to find a way to untie the knot. He must have been well aware that a failure, in front of so many eager witnesses, would have been disastrous. Finally, angry and frustrated, he is said to have declared, "What difference does it make *how* I loose it?" He then drew his sword and slashed through the knot. That night there was a violent storm, which Alexander and his soothsayers took to be sign of Zeus's approval, and immediately thereafter Alexander made his first public claim to be the rightful ruler of all Asia. For later historians this act became a sign of the divine endorsement for the entire campaign, and to this day, "cutting the Gordian knot" has remained a metaphor for taking bold and violent action to resolve an exceedingly difficult dilemma.[22]

In the early winter of 333 B.C.E., Alexander defeated a vastly superior Persian army at Issus in Cilicia, which gave him control of what is now the Near East as far as the Euphrates. It also left him in possession of Darius's wife, Stateira, her mother, Sisygambis, and a number of the princesses of the royal house. Alexander's capture of the Persian women became a subject of legend and, much later, a favorite topic of Renaissance mythological paintings. In all of these the royal women, still proud but prostrate, are shown offering obeisance to their new master. Instead of dispatching them to a life of slavery or worse, the young king stands down from his throne and offers Stateira his hand. Alexander, he told them, had not come to destroy them but "had made legitimate war for the sovereignty of Asia." It was staged as a scene of the magnanimous conqueror, the true hero who knows how to be generous in victory. And so it was. But Alexander's magnanimity was part of a far grander plan. Instead of ransoming the women, as would have been customary, he kept them with him. He must have learned from his old tutor Aristotle a great deal about Persian customs and, as his later actions make clear, he certainly knew that succession in the Achaemenid royal house could be transmitted through the female line.[23] In addition, by virtually adopting Darius's family as his own, Alexander was not only depriving the great king of his kingdoms, he was also depriving him of his status, which in time Alexander himself would assume. Alexander went so far as to address Sisygambis as "Mother," and legend has it that so devoted did she become to her new "son" that when he died ten years later she starved herself to death out of grief.

Alexander confirmed every member of Darius's family—even his young son—in their former titles and insignia and provided royal dowries for Darius's daughters, and when Stateira died in 331 B.C.E., she was given a royal burial. More was to follow. Later that same year Alexander moved his army out of Cilicia and northern Syria and entered Phoenicia. However, he left the royal princesses behind in Susa with instructions that they be given an education in Greek. They were destined to play an important symbolic role in his bid to unite Europe and Asia.

Alexander also took from Darius a golden casket in which he now placed the scrolls that made up his copy of the *Iliad*, which became known as the "casket copy," and henceforth slept with it— and a dagger—under his bed.

The Macedonian army moved south along the Syrian coast. There, at what is today 'Amrit, Alexander received an ambassador from Darius. To the great king's somewhat meager overtures of peace, Alexander replied that not only was he avenging Persian wrongs, but that Darius was a usurper to the Persian throne and it was he, Alexander, who was now the true ruler of all Asia—as the Persian nobles whom he had won over were all too happy to acknowledge. Darius could have peace, but only if he were prepared to accept Alexander as his suzerain. Steadily Alexander was beginning to transform a war of conquest and revenge into the foundation of a new imperial state, which would fulfill the ancient Achaemenid ambition to unite the two halves of the known world. Only now it would be the unity not of masters and slaves but, as Alexander told each new city that surrendered to him, one of equal free men under one ruler, although that ruler would, henceforth, be Greek, not Persian.

Not surprisingly, Darius rejected these claims. Alexander then moved into Egypt, which had always been an uncertain and rebellious subject of the great king. There he was welcomed as a liberator and crowned with pharaoh's crown. On September 30 or October 1, 331 B.C.E., on the plain of Gaugamela, near what is now Iribil in northern Iraq, the Greeks defeated the second grand army that Darius had mustered against them. It was to be the decisive encounter. Alexander had now taken Darius's place in all the lands west of the Zagros Mountains. He immediately had himself proclaimed great king and sent word to all the Asiatic Greek states that "all the tyrannies were now abolished and henceforth they might live under their own laws."[24] In the winter of 331–330 B.C.E., Babylon surrendered without a struggle, and its peoples welcomed him with strewn flowers and incense, in a re-enactment of Cyrus's entry into the same city in 539 B.C.E. Like Cyrus, Alexander claimed to have come to "impose" peace, and like Cyrus he accepted the formal titulature of the old Babylonian kingship and intended to make the city the new capital of his Asian domains.

Alexander then turned his attention to the heartlands of the empire and the cities of Persis itself. In January 330 B.C.E., he marched into the great Persian capital at Persepolis.

Throughout his campaign Alexander had been relatively constrained in his handling of defeated populations, or at least of their possessions. His men had stood by as, one after another, the treasuries of the great Persian cities had passed into Alexander's coffers. But now they were growing restless. In order to ward off what could have turned into a mutiny, Persepolis was handed over to the victorious army. Alexander encouraged his troops to kill all the adult males they encountered, "thinking this would be to his advantage." For an entire day, the houses of the nobility were looted, the women enslaved, and everything that could not be carried off was destroyed. Alexander himself, meanwhile, was inspecting the great king's vaults, which turned out to contain no less than 120,000 talents dating back to the time of Cyrus the Great. Alexander kept a small portion of this massive wealth with him and sent the rest back to Susa and Ecbatana. To do this he had to commandeer every pack animal in the army and no fewer than three thousand camels. It is pointless to attempt to calculate how much this would be worth today, but it has been estimated that it represented the equivalent of the income of the Athenian Empire at its height in the fifth century for nearly three hundred years.[25]

The great king's magnificent palaces and temples, the great Chamber of Audience, the Hundred-Columned Hall of Xerxes, the entire complex of buildings that constituted the empire's spiritual center and which spread out like a vast stage set along the terraces before the Kuhn-i-Rahmet Mountains—all of this, breathtaking even in ruins, had been spared. But not for long. Alexander may have been waiting for the Persian New Year festival, in the hope that the cowered Persian nobles would duly recognize him as the new great king and Ahura Mazda's representative on Earth. If so he waited in vain. By May it was too late. The New Year had come and gone, and the Persian priesthood remained as unwelcoming as ever. Persepolis had become, in Alexander's mind, "the most hated city in Asia" and a possible rallying point for a revolt. In late May, he offered ceremonial sacrifices in honor of his recent triumphs and held a banquet in the palace. The story goes that he became blind drunk—something that, legend also claims, he did often. Thais, an Athenian courtesan and the mistress of one of his generals, Ptolemy, delivered a rousing harangue and called on Alexander to remember his vow to avenge Xerxes' burning of the Parthenon. What better way than to burn Xerxes' great palace? Alexander himself led Thais in a drunken dance up the great staircase while the courtesans, the *hetairai*, played pipes and flutes for the benefit of the guests. At the entrance to the Great Hall, Alexander is said to have hesitated briefly before committing what would be the greatest single act of vandalism in ancient history. But only for a moment; he then pitched the first torch into Xerxes' Great Hall. The vast cedar rafters of which the roof was made and the cedar paneling on the walls, the ashes of which were still there in the 1950s, were soon ablaze. The army, attracted by the conflagration, looted and vandalized everything that escaped the flames. They seized every coin Alexander had not already carried off, all the gold work and jewelry. They raided the armory for swords and daggers (but left thousands of bronze and iron arrowheads behind for later archaeologists). They destroyed everything they could not easily carry, decapitated statues, and defaced reliefs.[26]

The palace would remain a gutted and abandoned ruin, visited only by looters and, later, archaeologists, until it was restored, in part, by the last Persian great king in 1971. Mohammed Reza Shah Pahlavi, the son of an army officer who had seized the throne in a coup in 1925—and named his newly minted dynasty after an earlier form of the Persian language—undertook the restoration to celebrate what he hoped to persuade his increasingly disaffected subjects would be two millennia of Achaemenid rule. For a week, a collection of international dignitaries dined on roasted peacock stuffed with foie gras and consumed 25,000 bottles of champagne provided by Maxim in Paris. (The affair is said to have cost more than $200 million, a massive fortune in 1971.)

At the height of the festivities, and surrounded by soldiers dressed up as Achaemenid warriors, the shah proclaimed himself the heir of Darius and Xerxes:

> To you Cyrus, Great King, King of Kings, from myself Shahanshah of Iran, and from my people hail! . . . We are here at this moment when Iran renews its pledge to History to bear witness to the immense gratitude of an entire people to you, immortal Hero of History, founder of the world's oldest empire, great liberator of all time, worthy son of mankind.[27]

Five years later he replaced the lunar Islamic calendar with a version of the ancient Zoroastrian "imperial" solar calendar, which began with the supposed founding of the Achaemenid dynasty twenty-five centuries earlier. The Muslim year 1396 (1976) now became 2535. As it turned out, none of this had the desired effect. In 1979, the shah was deposed by a revolution masterminded from Paris by a disgruntled cleric named Khomeini. The Iranian monarchy was thus replaced by an Islamic republic—ironically, the first time that a Muslim theocracy openly adopted the name, if not quite the practices, of a Western pagan, political constitution. Iran's new master had no time for "Persian" nostalgia, and what remained of Persepolis narrowly escaped being bulldozed into the ground by Khomeini's right-hand man, Ayatollah Sadegh Khalkhali.


HERODOTUS, AS ALEXANDER would have been aware, had made Xerxes declare as he had set out to conquer Europe that "All we possess will pass to the Greeks, or all they possess will pass to the Persians."[28] In the final act of the drama, all that the Persians had possessed had indeed now passed to the Greeks. The great king himself was first abandoned by his confederates, then imprisoned in a gilded cage with gold chains, and finally assassinated, shortly before he fell into Alexander's hands near the modern town of Shahr-i-Qumis.[29] The war of revenge was now formally at an end. Alexander, as was to be expected, treated his defeated rival with due ceremony, publicly executed his assassins, and had his body transported back to Persepolis for royal burial.

The Macedonians then marched eastward to consolidate their hold over what remained of the former Persian Empire. During one of his campaigns Alexander captured the famous Roxana, daughter of the Iranian-Bactrian nobleman Oxyartes, known to contemporaries as "little star" and reputedly the most beautiful woman in Asia. Shortly afterward, Alexander married her. In verse and paint, the marriage has been celebrated down the centuries as a love match—the most famous between a European and an Asian before Antony and Cleopatra more than three centuries later. But it was also the first of what would be many political marriages between the Persians and their Western conquerors and, in the view of the great nineteenth-century German historian Johann Gustav Droysen, it was evidence of a policy not merely to unite Greek and Persian, but to bring about a "fusion" of the two races, to blend Europe and Asia into one, as we would say today, "multicultural" society.[30]

Marriage, even to the beautiful Roxana, did not, however, delay Alexander for long. Moving in a great swath through what is now eastern Iran and western Afghanistan, he crossed the Hindu Kush and invaded Bactria (modern Afghanistan). In the spring of 326, he defeated the Indian ruler Porus at the battle of Hydaspes despite being faced for the first time by elephant squadrons of legendary strength. There he founded two new cities, Nicaea and Bucephala, in honor of his famous horse Bucephalus, which had collapsed with exhaustion after the battle. He then moved on.[31] But by now the monsoon had arrived, and by the time he reached the Beas River, which separated him from the lands of the Ganges, it had been raining without cease for seventy days. Finally his army refused to go any farther. Like Achilles, his favorite Homeric hero, Alexander retreated to his tent. For three days he nursed his anger and waited for a change of heart. None came. His army refused to move. In a last attempt to save face, he made the regular sacrifice for a river crossing. The omens proved—conveniently—to be most inauspicious. Now that he could interpret a retreat as a concession not to his men but to the will of the gods, he agreed to turn back. Alexander returned to Persepolis and then moved on to Babylon. After he had gone, the legendary Indian hero Chandragupta drove out those of Alexander's army who had remained and recovered what Alexander had seized of the Punjab. The peoples of India would now remain largely untroubled for more than a millennium, until a Turkic people who in time would become known by the Persian word "Mughals"—meaning Mongols—

arrived from the same direction in 1526 C.E. As the historian Megasthenes, whom Alexander appointed satrap of Arachosia and Gedrosia, warned his readers, never believe anything you hear about the Indians, because they are people who have never been conquered—and in Greek eyes, an unconquered people was an unknown people.[32]

In 324 B.C.E., Alexander returned to Susa, where the Persian princesses whom he had left behind seven years before were waiting for him. He now staged a grandiose marriage ceremony for them, the celebrated "mass marriages at Susa." He himself, together with ninety-one members of his entourage, took wives from the Persian nobility and married them according to Persian rites. The ceremonies lasted five days and nights, with musicians, dancers, and actors brought in from all over the Greek world. Like his father, Alexander was relentlessly polygynous. Although he was already married to Roxana, he took two further brides, the eldest daughter of Darius and the youngest of Artaxerxes III. This linked him by marriage to both of his Achaemenid predecessors. Alexander had now, indeed, become, "the last of the Achaemenids." Like them, he assumed a claim to universal sovereignty that no Greek had ever claimed. Soon he would gather a new army to conquer India, and once that had been achieved, he would turn his attention westward as far as the shores of the Atlantic itself.

Nothing, however, came of these ambitions. Toward the end of May 323 B.C.E., he attended a banquet and, if the traditional accounts are to be believed, literally drank himself to death. The climax came in an exchange of toasts in which he is said to have downed twelve pints of undiluted wine in one steady draught. He doubled up with violent spasms and collapsed into a coma from which his doctors were unable to revive him. He was thirty-two years and ten months old. He left no obvious successor. In his dying moments he is supposed to have bequeathed his kingdom "to the strongest." It was a guarantee of disaster, and disaster duly followed. A series of internal wars, divided along ethnic and tribal lines, began as soon as he was dead. Athens rose in revolt against what it had long perceived as Macedonian tyranny, and Aristotle, fearing that he would end up as Socrates had and claiming that he wished to spare the city from committing another crime against philosophy, fled into exile.

All that Alexander had conquered was divided up among his former generals. Ptolemy seized Egypt, stole Alexander's body from Babylon, and carried it off first to Memphis and then to Alexandria, where centuries later another world conqueror, the Roman emperor Augustus, would crown it with gold. The Greek city-states and the north ended up in the hands of a succession of Macedonian warlords. Western Asia, by far the largest portion, was taken by one of Philip's veterans, the one-eyed Antigonus, and then by the former leader of the shield bearers, Seleucus, whose successors created an empire that, at its height, reached all the way from Thrace to the borders of India—almost the entire territory once occupied by the Achaemenids.[33]

## III

This is Alexander's story. For centuries, he has been looked upon as the archetypal empire builder. For centuries there has been a more or less continuous dispute over his legacy. Was he really the golden boy, the conquering genius, or, as the Roman philosopher Seneca saw him, only a brutal thug, "swollen beyond the limits of human arrogance," unedifying, intemperate, and wild, who beheaded his father's most trusted general, the seventy-year-old Parmenio, executed Parmenio's son Philotas on a trumped-up charge of treason, and stabbed Cleitus, who had saved his life at the Granicus, in a drunken fury?[34] Certainly he blazed a trail of death and desolation across Asia, and between 331 and 326 B.C.E. pioneered a form of what we today called "ethnic cleansing" in Afghanistan and Bactria.

Yet for all his casual brutality—and sometimes because of it—he rapidly became a model to be followed and an example to be surpassed. He was emulated by Julius Caesar, by Pompey and Marc Antony, by the Emperor Trajan, by Napoleon, and doubtless by countless other would-be imperialists. But was this empire truly an empire as all these men had understood it or merely what one modern historian has described as "a sort of continental Lebanon," a chaos of competing tribal systems and pitiless warlords that, under its Achaemenid rulers, had been a more or less harmonious world?[35]

Alexander became known as "The Great," not merely for his astonishing military successes. For later generations what made him far more than a particularly brilliant military commander was the belief that he was the first Greek, the first European, to pursue truly universal ambitions, to wish not merely to conquer the world but to make the world as one. Alexander, enthused the garrulous first-century C.E. Latin novelist Apuleius, was "the sole conqueror in the memory of mankind to have founded a universal empire."[36] It was, and remains, a widely held opinion. Whereas Darius I and then Xerxes had marched into Europe with the intention of enslaving her peoples, Alexander's achievement had been not merely to defeat the mighty Persian Empire but to unite Asia and Europe, Hellene and barbarian. In so doing, he introduced into Greece, and subsequently into the whole of Europe, an ambition for universalism that would determine the future of the continent until it finally fell apart in the mid–twentieth century— or, some would argue, was carried across the Atlantic to the United States. In 1926 the English jurist and historian W. W. Tarn wrote of Alexander that "He lifted the civilized world out of one groove and set it in another. He started a new epoch; nothing could again be as it had been. . . . Particularism was replaced by the idea of the 'inhabited world,' the common possession of civilized men."[37] Tarn was writing shortly after the creation of the League of Nations, and on a rising tide of hope that the enmities that had led to the First World War would soon be replaced by a perpetual universal peace. The League of Nations—like its successor, the United Nations—preached the Brotherhood of Man. For Tarn, Alexander had helped to make that idea a possibility.

In part the claim that Alexander had ambitions as, if not exactly the "unifier of mankind," then the unifier of Europe and Asia, is born out by his actions: his persistent allusion to himself as the true legitimate ruler of Persia, the integration of foreigners into his army and administration, the mass marriage at Susa, the banquet he held at Opis after a mutiny of his exhausted troops and at which he prayed for "harmony and fellowship of rule between Macedonians and Persians." His attempt to hellenize large areas of the old Persian world also long outlived him. The mighty Seleucid Empire was a Hellenistic monarchy dominated by a Greek-speaking Macedonian aristocracy, and it flourished from Alexander's death until it was overrun by the Romans between 190 and 64 B.C.E.

Alexander's vision of a unified Europe and Asia was evidently one in which Europe would have the undisputed upper hand. Yet, like most Greeks of his time, Alexander's contempt for Asia was at best ambiguous. Like his tutor, Aristotle, he evidently not only admired Persian monarchy but imported into Greece many of its features. During his own lifetime, he built up an elaborate court, complete with bodyguard, harem, and eunuchs, something that had been wholly absent from the traditional Greek *polis*. He installed Persian commanders into the highest military and administrative positions. He imported Persian troops into the formerly exclusive Macedonian regiments and created new regiments composed exclusively of Asians. He bestowed honorific Asian titles such as "kinsman" on both his Asian and Greek subjects.[38] He invented a royal diadem part Macedonian, part Persian in design, dressed his courtesans in purple, adopted a modified version of Achaemenid dress, and even attempted, without much lasting success, to introduce the most loathed of all Persian habits, that of abasement before the ruler, or *proskynesis,* among his Greek followers. When later his mantle was assumed by the emperors of Rome, they too would imitate not only his mode of dress but also the panoply and ceremony of imperial rule that he had introduced into Europe from Asia.

He also tried to establish some kind of bridge between Greek and Persian religious beliefs. Alexander could never reconcile Greek paganism with the worship of Ahura Mazda, and for all his adoption of Persian dress, Persian titles, and Persian wives, he was not about to become a Zoroastrian. (Indeed, legend has it that it was Alexander who destroyed the original copy of the writings of Zoroaster, the Avesta, which had been inscribed in gold letters on twelve thousand ox hides and stored in the royal library at Istakhr.) But if he could not adopt the religion of the Shâhânshâh, he did the next best thing. Like most rulers in the ancient world, Alexander believed himself to be descended from semidivine beings, from Andromache and Achilles on his mother's side and from Hercules on his father's. The final tribute a man can pay to himself, however, is divinity itself. Aristotle had taught him that a true king was a god among men, and therefore a god he would become. On a visit to the shrine of Ammon in the Libyan oasis of Siwah, Alexander told the chronicler Callisthenes that he had been greeted by the god as his son and requested that after his death his body be taken to Siwah for burial. Ammon may have been Libyan by origin, but the Egyptians recognized in him the ram god Amun, and the Greeks settled in nearby Cyrene knew him as Zeus. On public occasions Alexander assumed the guise of Ammon with purple robes and ram's horns, and the great Athenian painter Apelles depicted him with the thunderbolt of Zeus in a famous painting for the Artemisium at Ephesus.[39] In choosing Ammon/Zeus as his divine parent, Alexander not only chose the deity who had fathered Perseus and Hercules but also one who belonged to both cultures, Persian and Greek, Asian and European.

Alexander's bid for divine ancestry, and by implication divine sanction for his rule, introduced an entirely new dimension into European perceptions of rulership. The Greeks had always chosen their rulers from among the best men (*aristoi*), who might and often did, as individuals, claim descent from a god. But they remained, nevertheless, resolutely human in their dealings with one another. Divine kingship had, however, been a feature of

Achaemenid rule since at least the days of Cyrus and had always been an integral part of what the Greeks looked upon as Persian despotism. "They keep their souls in a state of abject and cringing fear," wrote Isocrates, "parading themselves at the door of the royal palace, prostrating themselves . . . falling on their knees before a mortal man, addressing him as divinity and thinking more lightly of the gods than of men."[40]

But even if Alexander professed no desire to keep his subjects in abject and cringing fear, he knew that no ruler who believed himself to be the destined lord of the universe could be anything other than divine, or at least the favored agent of the divinity. With Alexander, Hellenistic kingship became theocratic, a tradition that would be taken up by the emperors of Rome. Julius Caesar was the first to attempt to have himself proclaimed as a god but succeeded only in 42 B.C.E., after his assassination. Thereafter all subsequent pagan Roman emperors had themselves similarly deified, and with divinity, the imperium they exercised became not—as it had been for the Roman Senate—a merely mundane right to rule but a quasi-mystic power reserved for them alone. The emperor Aurelian, in the third century C.E., brought back to Rome the Persian worship of the unconquered sun, and Diocletian came to regard himself as Jovius, the earthly representation of Jupiter. The unity of mankind, which had been a Roman ambition since the days of the early republic, was, as Tarn put it, "ultimately satisfied in the official worship of the Roman Emperor which derived from the worship of Alexander after his death."[41]

In the myths that sprang up all over Europe and Asia after his death, he became the defender of civilizations, the hero who built a wall to shut out the twin giants Gog and Magog, rabid inhuman creatures who eat scorpions, kittens, miscarried fetuses, and human flesh, and identified variously with the Scythians and later, after the fall of Rome, with the Goths. Only at the very end of time, with the coming of the Antichrist, would these creatures succeed in breaking through Alexander's wall to devour a stricken

world. Today he is prayed to by fishermen in Greece and worshipped as a saint in the Coptic Church of Egypt.[42]

This Alexander, the defender and unifier of mankind, was not
only a Greek, Roman, or European myth. Even after the old
Achaemenid domain had been overrun by (another) alien religion,
Islam, Alexander, known in Arabic as "Iskandar," continued to
provide a source of legitimation for new powers with similar theocratic leanings and similar world-encompassing ambitions. He
appears in the Qur'an as Dh^l-qarnayn, "the twin horned one"
(Q. 18:82), presumably because of his famous helmet, who built a
giant copper wall at the very edge of the world to protect the
whole of "civilization"—now understood as the world of Islam—
from Gog and Magog. In the Persian, Indian, and, later, Ottoman
accounts of his life, he became a prophet, a seer, and a seeker after
eternal life. For generations of Muslim Persian monarchs, Iskandar remained a model of the world ruler, remarkable for his asceticism and his wisdom, the precursor of the universal kingdom of
Islam that would one day come to encompass the globe. "My
name is Shah Isma'il," wrote the founder of the Safavid dynasty,
which ruled Iran between 1502 and 1736, in one of his self-
aggrandizing verses,

> *I am the living Khizr and Jesus, son of Mary,*
> *I am the Alexander of my contemporaries.*

Above all we have the Alexander who brought to Asia the values
of Greek civilization, the respect for individual liberty and self-
government, the Alexander who finally gave to the Persians the
*isonomia* that Otanes had offered them and they had refused. This
Alexander, claimed Plutarch, had hoped that he would be remembered not as a conqueror at all but as "one sent by the gods to be
the conciliator and arbitrator of the Universe." It was he who,
"using force of arms against those whom he failed to bring
together by reason . . . united peoples of the most varied origin
and ordered . . . all men to look on the *oikoumene* as their father-

land. . . . He taught them that the proof of Hellenism lay in virtue and of barbarism in wickedness."[43]

This Alexander taught the Sogdians to support their aged parents instead of killing them, the Scythians to bury their dead, and the Arachosians how to till the soil; and prevented the Persians from marrying their mothers. If the role of the philosopher is to civilize the untutored and intractable human character, Alexander, who "has been shown to have changed the savage natures of countless tribes should be regarded as a very great philosopher."

Plutarch made Alexander into the practical embodiment of the Stoic philosopher Zeno's vision for a world made up of "one community and one polity." It was Alexander who "bade them [humankind] all consider as their fatherland the whole inhabited earth."[44] This fatherland was to be a world monarchy, but one ruled in accordance with the Greek political virtues of equality before the law, freedom, and individuality. True, Philip and then Alexander were responsible for finally erasing the old democratic culture of the Greek city-states. But in part at least, Herodotus's *isonomia* lived on. The rulers of Hellas might now be monarchs, but they still paid respect to the law, still looked upon their subjects as individuals who could not be enslaved. They might harbor desires to rule over the entire "inhabited world," but they still knew how to respect differences and recognize individual worth. It was for this reason that in 1748 the "father of modern sociology," Charles-Louis de Secondat, baron de Montesquieu, dedicated an entire chapter to Alexander's deeds in his great masterpiece, *The Spirit of the Laws*. Alexander, he said, had resisted those, notably Aristotle, who had urged him to treat the Greeks as masters and the Persians as slaves. Because he "thought only of uniting the two nations, of wiping out the distinction between conquerors and vanquished," he had "assumed the mores of the Persians in order not to distress the Persians in making them assume the mores of the Greeks." His had been a conquest of reconciliation, in which "the old traditions and everything that recorded the glory or the vanity of these peoples" had been studiously preserved, so that Alexander himself, or so it seemed, "conquered only to be the monarch of each nation and the first citizen of each town."[45]

It is, in its own way, as deeply improbable an image as any of the others, but it has endured in many different versions and endures until this day. When the Iranian jurist Shirin Ebadi, in her acceptance of the Nobel Peace Prize in 2003, declared that she, as an Iranian, was a descendant of Alexander the Great, who had, in his own way, been one of the earliest upholders of human rights, she was echoing a historical fable that ran, if in an often uneven course, all the way back to the death of Darius.

That we owe this image of Alexander as the civilizer, the unifier of East and West—which would become the stuff of legends on both sides of the Hellespont—to Plutarch, a Greek of the first century C.E., who lived under and celebrated Roman rule, is significant. For it was, of course, Rome that would refine and spread the culture Alexander had helped to create far further than even Alexander himself had imagined. "The Greece that taught Rome," wrote Tarn, "was the Hellenistic world which Alexander made; the old Greece counted for little until modern scholars re-created Periclean Athens. So far as the modern world derives its civilization from Greece, it largely owes it to Alexander that it had the opportunity."[46] It is, therefore, to Rome that we must now turn.

# 3

## A WORLD OF CITIZENS

### I

One spring day in 143 or 144 C.E., a young rhetorician named Aelius Aristides, from the small Greek city of Mysia in Asia Minor, stood in the Athenaeum, a grandiose building in the center of Rome, and delivered a long oration on the "grandeur and magnificence" of the Eternal City.[1] In the ancient world, rhetoricians, professional public orators, provided highly popular large-scale public entertainment, at least for the educated classes. Aristides was a particularly brilliant performer, so brilliant indeed that the emperor Marcus Aurelius himself was said to have been prepared to wait hours in order to hear him speak. On this occasion he spoke—in Greek—on the creation and glories of Rome and its empire to a massed audience that may have included the emperor Antoninus Pius himself.[2] In many respects Aristides' "Roman Oration," like so many others of its kind, was a rehash of familiar Greek sources: Plato, Isocrates, Polybius, and Plutarch. To modern readers it might seem, at first glance, little more than the expected sycophancy of a provincial boy with big ambitions. But his account of what Rome and the Roman cosmos stood for captured a vision of the world that was widely shared at the time and continued to be shared long after the empire itself had vanished. Aristides was also a Greek patriot and a member of the ruling elites that governed the Greek cities on behalf of their Roman masters. At the core of his celebration of Roman greatness was a very Greek question. Was it just, he was asking his listeners, who were the men who now ruled over him and his fellow Greeks, that Rome should now dominate the fortunes of the greatest civilization in the Western world, greater indeed, in all its achievements, than Rome itself? Did the Romans really deserve the obedience, even the reverence, of the heirs of Aristotle and Plato, of Pericles and Alexander? Aristides' answer was an unequivocal yes. Yes, because the Romans had brought peace and stability to an impossibly unsettled world. Yes, because in the end the Romans, and Roman legions, now protected Greece not only from the kind of internecine conflict that had brought it to ruin after the Peloponnesian War but also from its ancient hereditary enemies to the East.[3]

Aristides was only one of many Greeks who were conscious that they owed their security to the Roman legions and their careers to Roman patronage. The historian Polybius, who became a close associate of the great Roman general Scipio Aemilianus, the historian and senator Cassius Dio, the wit and belletrist Lucian, each in his own way anticipated the sentiments of the third-century poet Claudian, from Alexandria in Egypt, that Rome, far from being a conqueror, had to them been a mother, rather than an empress, who had "called 'citizens' all those whom she subdued and bound with her far-reaching and pious embrace."[4]

When Aristides arrived in the middle of the second century C.E., Rome was at its zenith. By 117 C.E., the emperor Trajan, having defeated the Dacians in modern Romania and annexed Arabia Petraea, Mesopotamia, and Armenia, had pushed the empire to the farthest limits it would ever attain. It now reached all the way from the Atlas Mountains in the south to Scotland in the north, and from the Indus valley in the east to the Atlantic in the west, a territory of about 5 million square miles (the continental United

States is a little more than 3.5 million) with a population that has been estimated at about 55 million. It was, as Aristides said, using an image that would be repeated many times in the future, an empire on which the sun never set.[5] Peace and the rule of law, which in the past had reigned more in the imagination than in fact, now seemed truly to have descended upon the face of the Roman earth, and for most Romans that earth was simply *the* earth, the *oikoumene,* to use the Greek word that Aristides himself used, the word Herodotus was the first to use: "the inhabited world."

"Now a clear and universal freedom from all fear has been granted both to the world and those who live in it," enthused Aristides.[6] That freedom had been achieved under what would become widely regarded in Europe as the very best form of government: the so-called mixed constitution—part democracy, part aristocracy, part monarchy, allowing all walks of life to exercise some influence over the government that ruled them. Rome, said Aristides, had "established a constitution not at all like any of those among the rest of mankind." For the rest of mankind had always been ruled by monarchs, aristocrats, or democrats, as "choice or chance prevailed." Only the Romans had managed to spread across the globe "a form of government as if it were a mixture of all the constitutions without the bad aspects of any." Herodotus's constitutional debate had, in Rome, found its perfect resolution. Little wonder, then, that Aristides could have addressed Antoninus and his forebears as "alone rulers . . . according to nature" and make it sound like something more than mere toadying.[7]

For the Age of the Antonines, of the "Five Good Emperors" as they have come to be known—Nerva (reigned 96–98 C.E.), Trajan (98–117 C.E.), Hadrian (117–138 C.E.), Antoninus Pius (138–161 C.E.), and Marcus Aurelius (161–180 C.E.)—was, on every count, a Golden Age. Centuries later the great English historian Edward Gibbon, looking back from well beyond the disasters that would befall this Eden after the death of Marcus Aurelius, declared that "If a man were called upon to fix the period in the history of the world during which the condition of the human race was the most happy and prosperous he would, without hesitation, name that which elapsed from the death of Domitian [96 C.E.] to the accession of Commodus [180 C.E.]." It was, he added, a time when "the Roman Empire comprehended the fairest part of the earth and the most civilized portion of mankind."[8]

## II

Of course, it had not always been like that.

Like Europe itself, Rome owed its origins to a myth. There are many different and overlapping versions, but in one, carefully crafted to flatter the emperor Augustus, Rome is yet another creature of the fusion of Europe and Asia and yet another mythic triumph of Europe *over* Asia. After the sack of Troy by the Greeks, Aeneas, a Trojan prince and son of the goddess Aphrodite, flees from the burning city carrying his aged father, Anchises, on his back and leading his son, Ascanius, by the hand. He then begins a series of westward journeys that will finally lead him to the shores of Latium, in what is now central Italy. There, after a long struggle with the native inhabitants—the Latins—he will found the city and the state of Rome. Rome will be the true creator of "Europe" and the "West." But Aeneas himself is a Trojan and therefore of Asian (if also divine) parentage, which means that, like Europe, Rome, too, will have to share its mythopoeic identity with Asia. The author of this story was the great Roman poet Virgil, and his poem— the *Aeneid*—the greatest of all the Latin epics, was intended to be the Roman counterpart to Homer's *Iliad*.

By the time Virgil composed the *Aeneid* in the first century B.C.E., however, the Romans, although they were still proud of their supposed descent from the mythic lords of the eastern Mediterranean—Augustus even claimed to be the direct heir of Aeneas himself—had also developed a distrust of the peoples of Asia and a strong desire to assert their own racial, linguistic, and cultural distinctiveness. Too close an association with an Asian people could now be interpreted as a threat to the integrity of the

Roman *patria*. In the twelfth and final book of the poem, Virgil therefore makes the gods decide that the war between Aeneas's invading Trojans and the native Latins must finally be brought to an end. The goddess Juno, who has supported the Latins, agrees to allow the two peoples to intermarry and thereby create a new race. But she insists that this new race will look like the Latins, dress like the Latins, and speak like the Latins, and their customs—their *mores*—will be Latin. All they will preserve of their oriental ancestors will now be their gods, for those gods were also the gods of the Greeks and the common patrimony of all mankind.[9] And just as the Latins had absorbed the Trojans, so the new people who would be created by this merger, the Romans, would go on to absorb, in their turn, all the peoples of the inhabitants of "the world," from Britain to Syria. Soon, like the Trojans, these peoples, too, would become Latins in their customs, their culture, their law, their religion, and frequently also their language.

The *Aeneid*, however, is a fable. The historical Rome had begun sometime during the seventh century B.C.E. as a small city-state of farmers and tradesmen occupying a territory of a few square miles on the lower Tiber, not unlike the Greek states, which provided it with its civic model. From there it began, slowly at first but inexorably, to encroach upon its neighbors. In this it was also following a Greek example, and from the late fourth century B.C.E., the histories of the two peoples of antiquity follow very similar paths. In 338 B.C.E., at roughly the same time that Philip and then Alexander were taking Greek-speaking armies into Asia, the so-called Latin League, which had held a fragile balance among the various races of Italy, collapsed. A half century of wars against Samnites, Etruscans, Celts, and Greeks followed, in which Rome seized and finally extinguished most of the other cultures of the peninsula. By the time Roman armies crossed into Sicily, then a Greek settlement, in 264 B.C.E. Rome was already a power to be reckoned with on the edges of the Greek world. There is even a story that Rome sent an embassy to Alexander while he was campaigning in Persia and that Alexander, on seeing how the Roman

ambassadors were dressed and their love of labor and obvious devotion to freedom, made inquiries about the political constitution of Rome and predicted great things for the new state.
Thereafter Greek and Roman culture would be fused into what
has come to be called the "Greco-Roman world," which, in turn,
would provide the cultural and political foundations of what we
today think of as the "West."

By 168 B.C.E., the Romans had transformed Greece into a
Roman province. Yet it was also from Greece that they derived not
only their political examples but also most of their culture, their
arts and sciences, their literary styles, their gods, even the fashions
in which Roman women wore their hair. "Conquered Greece,"
wrote the poet Horace, "conquers her wild victor and introduces
her arts into rustic Latium."[10] The Romans looked upon Greek
culture in such awe that, when in 17 B.C.E. Augustus held the great
celebrations for the New Age that marked the beginning of the
principate—and therefore of what later generations think of as the
beginning of the Roman Empire—he did so with hymns in both
Latin and Greek.[11]

That Aristides' "Roman Oration," the most high-flown, most
compelling of all the many high-flown and compelling tributes to
the empire that survive from its Golden Age, was delivered by a
Greek in Greek is itself a tribute to the debt Roman civilization
owed its Hellenic forebears.[12] Aristides speaks of the Greeks as the
"foster parents" of Rome, and, in a sense, that is what they were.[13]
The Greek-speaking regions of the empire were the only ones in
which Latin did not replace the local language. Instead, Greek tutors traveled all over the Roman world instructing Roman patricians in Greek. On hearing a visiting "barbarian" address him in
both Greek and Latin, the emperor Claudius is said to have
replied, "Since you come armed with both our languages . . ."[14]
Even the menus and the nicknames of the girls and boys on sale in
the brothels in Pompey were bilingual.

As with all forms of cultural indebtedness, this one was not
wholly unambiguous. The mere fact of imitation carries with it a

certain level of unease. Americans import European antiques and employ European chefs and have their houses—those who can afford it—decorated in what is generally a parody of lush European styles. But this does not make them any less suspicious of European cunning, European "sophistication," or, in the worlds Henry James inhabited at the end of the nineteenth century, European decadence and European senescence.

The Romans' infatuation with Greece was similarly double-edged, with the difference that for the Romans the Greeks occupied a middle ground between Europe—or Italy at least—and Asia. The Greek cities in Asia Minor, which ever since the days of Herodotus had had mixed populations, part true Greek, part Iranian, were particularly suspect. The image of these people as wily "Orientals" had been firmly fixed in the Roman imagination by the massacre of the Italian populations of some of the Greek cities of Asia on the orders of the hellenized Persian Mithradates VI of Pontus in 88 B.C.E. Some cities, aware perhaps that their future lay within the Roman world, tried to exculpate themselves. The people of Trelles even hired a native Iranian to do the killing for them. But in others, Ephesus and Pergamum, which were important religious sites, Italians were torn from sanctuaries and shrines and butchered in their thousands.

These events seared into the Roman mind the image of Greeks as deceptive and untrustworthy. Even Cicero, one of the most hellenized of Roman intellectuals, harbored a profound contempt for what he took to be the Greek character:

> With them deception is second nature. This I can say of the whole race of Greeks. I grant them literature, I grant them knowledge of many arts, I do not deny them the charm of their speech, the keenness of their intellects, the richness of their diction. . . . But truth and honour in giving testimony—that nation has never cherished: the meaning, the importance, the value of this whole matter they do not know. Whence comes the saying "testify for me and I'll

testify for you"? It isn't thought to be Gallic is it, or Spanish? It is so entirely Greek, that even those who do not
understand Greek, know the Greek word for this expression.[15]

Too clever by half, in other words. And lying and deceitful with it.
The Romans called it *graeca fides*—"Greek faith." "I fear the
Greeks," Virgil had famously made his Trojan high priest Lao-
coön declare as he surveyed the Trojan horse, "and the gifts they
bring." The Greeks were wily like Odysseus, lightweight, and
overindulgent, and they spent far too much of their time whoring,
drinking, and feasting. The Romans had a word for such behaviour: *pergraecari*, "to play the Greek." Greek men were also overin-
terested in small boys, a custom that, in Cicero's view, they had
been responsible for introducing, like an infectious disease, into
Rome.[16]

The general feeling in Rome was that the Greeks had absorbed
too much from their truly "Oriental" neighbors. Alexander the
Great may have successfully hellenized much of Asia, but in the
process, the Greeks had themselves been orientalized. Had Alexander ever moved westward as he had planned, and had he met
the Romans in battle, mused the Roman historian Livy, the Romans would surely have won, because by then, Alexander would
have been irredeemably corrupted by his stay in Persia.[17]

The Romans feared that Greece could be another kind of
Trojan horse, its belly now stuffed not with warriors but with all
manner of sinister and alluring corruptions. Greece may have provided Rome with its cultural and religious foundations, but already by Cicero's day most Romans believed that it was now Rome
that was the bearer of the true values of Europe. These, as they saw
them, were summed up in a single evocative but highly contentious word: "virtue," *virtus*. Today we think of virtue in terms of
what are essentially Christian moral qualities: probity, honesty,
loyalty, modesty, generosity, and so on. The Romans valued these
things too. But our understanding of the word is very largely the

creation of the fifth-century Christian philosopher Boethius, and, in keeping with early Christian views on morals, it places much greater emphasis on turning the other cheek than the Roman meaning of the term did. The word itself derives from the Latin *vir*, meaning "male," from which we, of course, get the word "virile." Indeed, *virtus* is simply "manliness," and this in Roman eyes meant the attributes of the true warrior. The true warrior was characterized by physical courage, fortitude, constancy, and dignity or gravity (*maiestas* and *gravitas*). He should display clemency—on which the Stoic philosopher and playwright Seneca wrote an entire treatise—gentleness, when gentleness was required, and magnanimity.[18] Above all—and here he shared much with his Christian successors—he had to possess faith or trust (*fides*), which was the will to maintain just relationships, to honor contracts, and at the same time to pay due allegiance to the emperor and the gods.[19] In Roman eyes, the Romans were, by and large, the only true bearers of virtue. The rest of the world was filled, as it was for the Greeks, with "barbarians," or as Cicero calls them "provincials."[20]

The Romans' perceptions of these peoples, and Cicero's in particular, were inevitably much influenced by previous Greek writing on the subject. But it was also marked by experience. As the Roman Empire extended eastward toward the frontiers with India; westward through Germany into Britain; north into the steppes and lands of the Scythians and Sarmatae, the Alans, and later the Huns; and south along the coast of Africa, so the Romans' understanding of just who the "barbarians" were changed. By the first century C.E., at least, a rough dichotomy had emerged, not merely between Romans and barbarians, but also between categories of barbarians, between those of northern and western Europe—represented by the Germans, the Goths, and the Gauls—and those from the East—the Namibians; the Egyptians; the Syrians; the Persians; the "soft Arabs," as the poet Claudian called them; and, above all, the Parthians.

For centuries the Romans had lived in close proximity with both these groups of peoples. Some they had been able to overpower; some, in particular in western Europe, they had drafted into the legions; many had, in turn, been absorbed and had become Roman citizens. The characteristics of the western barbarians, although they were in many respects outlandish and frenetic, were not always so very different from recognizable Roman virtues. The Gauls and the Germans were fierce but also courageous, cruel but also honest. They honored their gods and their families, and, like the Romans of old, they were always prepared to sacrifice themselves for the good of the *patria*.

The "Orientals," on the other hand, had almost nothing to recommend them except, perhaps, a certain slick artistic talent. They were, much as they had been for Herodotus, simultaneously soft and cruel, luxurious, lascivious men who married their sisters or their mothers, and who did not bury their dead properly and exposed the sick. They were also, despite their casual savagery, remarkably lush and effeminate. "I shrink from words evoking future ruin," wrote the first-century satiric poet Petronius, who had been one of Nero's entourage and thus knew whereof he spoke,

> *From tender boys adopting Persian ways*
> *. . . Hence all sought their joy*
> *In harlots, in effeminates' mincing steps,*
> *In flowing hair, in novel garb oft changed,*
> *In all that captivates men's minds.*[21]

All of this could be captured in a single Latin word: *vanitas*. This was much more than simple "vanity." It was a love of appearances and the insincerity, exemplified by cosmetics and overluxurious clothing, that masked the less than luxurious, unimpressive person within. It described emptiness, sterility, imprudence, vain eloquence, and a kind of nonbeing. Above all, it implied instability. The East was the land of *vanitas*—and of tyrannies, because barbarians, although they might be vigorous enemies, were all too easily swayed by ceremony and customs; being empty they could see only the trappings of power, not the virtues that sustained it. The stereotypes of unfreedom, of slavish obeisance, that Herodotus had applied to the Persians, the Romans now applied to all the peoples of Asia. Asians, scoffed the first-century C.E. Roman epic poet Lucan, could never know the bitterness of losing liberty since they had never enjoyed it. "Let Syria be slave," he added, "and Asia and the East which is accustomed to kings."[22] The image of the slavish Oriental was, in Rome, enforced still further by the impression, false but widespread, that the slaves who were sold in the empire all came from somewhere in the East.[23]

The expansion of the empire into Asia posed an ever-present threat to the values on which Rome supposedly rested. Like all empire builders, the Romans feared the corruption their empire had brought them, feared the lure of the other and the temptation of "going native," as the British would later call it. Like all empire builders, the Romans seemed to believe that while their values, in particular the virility and simplicity on which the republic had been founded, must be superior to all others, they were, nonetheless, constantly susceptible to infiltration and degradation. Asians, Greeks, Anatolians, and Syrians were not, furthermore, dimly threatening presences on remote borders. Immigrants from all over the East, "experts in flattery, dishonest, lecherous and promiscuous," protested the poet Juvenal, had found their way into the heart of the Roman world, so that now it seemed as if the Syrian river Orontes had "long since flowed into the Tiber bringing with it its habits, its flutes and loud harps."[24] It was one reason why the historian Sallust, writing in the mid–first century B.C.E., had denounced all expansion beyond Europe as a threat to Roman integrity.[25]

Asia, with its passion for wealth and luxury, its softness combined with seemingly senseless cruelty and sexual ambiguity, was also the source of all the religions, both mystic and sensual, epitomized by the cults that derived from Judaic, Syrian, and Egyptian "superstitions": the cult of Isis, the secret societies that gathered around the Sibylline Oracles, Mithraism, and finally—and ultimately most devastating of all—Christianity, which spread through the legions from Asia all the way to the frontiers with Scotland.

Little wonder that the cruelest and most dissolute of the emperors, Caligula and then Nero, should both have been entranced by Asia, its gods, its art, its multiple religions. Nero was said to have despised all the Roman deities and venerated only the Syrian goddess Astargatis (or Ishtar). When the Parthian king Tiridates visited him and addressed him as the incarnation of Mithras, Nero responded by adopting the "religion of the Magi," into which he was supposedly initiated by Tiridates himself.

Quite early in the history of its rise to world dominance, Rome had also had a traumatic and nearly fatal encounter with a great Asian power: Carthage. Carthage was a Phoenician settlement on the coast of Africa that occupied much of what is now north and central Tunisia, southern and western Sardinia, and parts of southern Spain and sought to control most of the sea routes to the west. Ever since they had first appeared in the *Iliad* and the *Odyssey,* the Phoenicians had acquired some of the features of the "Oriental." Odysseus, who knew something of deceit, called them liars and cheats, cunning, and "well-versed in guile."[26] In Cicero's view, it was they who had been largely responsible for introducing the Greeks to the luxury and greed that had eventually sapped their former manliness.[27]

For centuries the Carthaginians had been in intermittent conflict with the Etruscans and the Greeks. When the Romans reached Sicily in 264 B.C.E., they, too, inevitably found themselves caught up in a struggle upon whose outcome rested the future mastery of the entire Mediterranean. Everything about Carthage appalled them. The men wore their hair and beards artificially curled; their dress was distinctly effeminate; both sexes had an apparently insatiable passion for perfumes. They tattooed themselves and wore makeup and elaborate jewelry: necklaces, bracelets, amulets, earrings, and nose rings. They indulged in exaggerated modes of address that demanded that the social inferior should

prostrate himself before his superior, a custom the Greeks had abhorred in the Persians. Their language, with its strange guttural
sounds, was portrayed by the Roman comic playwright Plautus as
the chattering of beasts. Phoenician religion was similarly grotesque. The awesome pantheon of deities—Baal Hammon, his
lunar consort Tanit, Eshmound, and Melqart—demanded human
sacrifices and the regular slaughter of children as offerings and
imposed absurd food prohibitions upon their supplicants. Their
temples were ugly sprawling places, and when the Carthaginians
were not bowing down before their monstrous divinities, they
were worshipping sacred stones and equally obscene, and absurd,
household deities. Phoenician cities were towering terraced
citadels, overornate and chaotic in their profusion, quite unlike
the carefully measured proportions of Greco-Roman settlements.
Even the Carthaginian systems of government, which were not, in
fact, so far removed from the Roman and had been extravagantly
praised by Aristotle, seemed devious, overcomplicated, and wholly
dedicated to the pursuit of trade and profit—necessary but unvirtuous activities in Roman eyes. *Punica fides,* "Punic faith," became
a catchphrase, like "Greek faith," for lying, mendacity, and treaty
breaking.[28] By contrast, the Romans were always, of course, direct,
honest, upright, and straight-dealing. Here, then, at a short distance across the strait from the Italian peninsula itself, was a culture that to the Romans seemed to embody, as had the Persians to
the Greeks of Herodotus's day, all the vices of the East: corruption, immorality, pride combined with servility, and duplicity
combined with ferocity.

    In 264 B.C.E., war erupted between the two powers. Known as
the First Punic War, it lasted until 241 B.C.E., by which time Rome
had consolidated its hold over Sicily and seized Sardinia and Corsica. In response the Carthaginians set about establishing a new
empire in Spain and building an army capable of defeating the
Romans. In 216 B.C.E., the Carthaginian general Hannibal, who
had been made to swear an oath of undying enmity to Rome by

his father, Hamilcar Barca, at the age of nine, marched across the Alps with a huge force drawn from all of Rome's traditional barbarian foes. It was also accompanied by a squadron of thirty-eight elephants that were probably more terrifying to look at than they were in action. Thus began the Second Punic War, which would last until 201, seventeen years, during which time, in Livy's words, the Romans "breathed not a word of peace."[29]

In 218 B.C.E., Hannibal defeated an army led by Publius Cornelius Scipio, the first of a legendary family of Roman generals, on the banks of the Ticino. The following year the Romans were defeated again on the shores of Lake Trasimenus, and on August 2, 216 B.C.E., almost all the available Roman legions in Italy were annihilated at a famous battle at Cannae (today Monte di Canne).[30] Since his arrival in Italy, Hannibal had killed or captured some 100,000 legionaries together with hundreds from the senatorial and knightly class, including two consuls. In a space of twenty-four months, a third of Rome's frontline troops had been killed, wounded, or enslaved. It could have been the end of the republic.[31] One night Publius Cornelius Scipio went to sleep in his tent and dreamed that he saw coming out of Asia

> bronze-breasted forces and kings allied with one another and peoples of all kinds to go against Europe, and the din of horses and the sound of lances and bloody massacres and terrible pillaging, and the debris of towers, and the demolition of walls and the unspeakable devastation of territory.[32]

Once again, Asia stood poised, as it had at Marathon, to enslave Europe. Hannibal's army pitched camp within three miles of the great walls of Rome itself. And Rome waited for the final onslaught. But by then the Carthaginian forces were overextended and exhausted, and when, in a shrewd flanking movement, a Roman army attacked Carthage itself, Hannibal turned for home.

For a while, however, it had indeed seemed as if the "bronze-breasted forces" of Asia were going to sack the Eternal City, and had they done so, there would have been little to prevent them from sweeping on through Europe.

In 202 B.C.E., Rome went on the offensive. At the battle of Zama, near present-day Saqiyat Sidi Qusuf in Tunisia, the Carthaginian armies were thoroughly defeated by Scipio Africanus the Elder. Carthage managed to survive for the next fifty years. But in 149 B.C.E., after six days of street fighting, it was plundered and razed to the ground by another Roman army led by yet another Scipio, Scipio Aemilianus. The carnage was appalling. When he was done, Scipio ritually cursed the site and sowed the ground with salt, vowing that no house nor crop should rise there again. In one final act of annihilation, the main library of Carthage was donated, symbolically, to the Numidian kings, and with it Carthaginian culture, too, was finally extinguished.[33]

With the destruction of their archenemy the Romans had prevented—or so it seemed to them—what might well have been the "orientalization" of the entire Mediterranean. Carthage had also been the only power with sufficient strength to check Roman expansion. With its disappearance there was little to prevent Roman armies and, in their wake, Roman cities, Roman law, Roman administration, and the Latin language, from gradually securing the entire Mediterranean, which by the first century B.C.E. had become "our sea," *mare nostrum*. Rome then moved on into Asia itself. The vast Seleucid Empire, which had once stretched all the way from Anatolia via Syria and Babylonia to Iran and thence to central Asia, fell into Roman hands. And although the Seleucids were Macedonians, not Iranians, and the quasi-legitimate heirs of Alexander the Great, the Romans looked upon them as quite as decadent, overbearing, and effete as any true "Oriental." Macedon itself fared no better. The Romans had not forgotten that the Macedonian king Philip V had allied himself with Carthage in

214–205 B.C.E., and at the battle of Pydna in 168 B.C.E. all that remained of the crumbling monarchy of Philip and Alexander passed under Roman rule.

For the later Roman world, however, the people who came to embody all the vices of Asia were the Parthians. Originally the Parni, a seminomadic tribe of the Dahae confederacy north of Hyrcania, the Parthians were famous for their mailed cavalry and their horse archers; they bred Niseaen horses, which were prized as far east as China; and their leader, Arsaces, in the third century B.C.E., had driven the Seleucids out of what is now Syria and Iraq.

Like the Carthaginians, the Parthians, were, in Roman eyes, the very image of Asiatic barbarism. Despite their considerable military successes against Roman legions, despite the fact that authors such as the Greek geographer Strabo saw the Parthian Empire as the only rival to Rome, the dominant image was one of constant revolts against the central power and furious and bloody intrigues within the royal house, with its matricides, patricides, and fratricides (not dissimilar, as few Romans could have failed to notice, from the history of the Julio-Claudian dynasty). "Their national character," concluded the historian and zoologist Pompeius Trogus, "is impetuous, truculent, devious and insolent. . . . They obey their leaders from fear rather than respect. They are unrestrained in their sexual habits . . . and no reliance can be placed on their word or their promises."[34] Before their eyes the Romans had a gallery of cruel, irrational, luxurious, superstitious despots—Artabanus, Vardanes, Gotarzes, Vologases—in which the *luxuria* and the *vanitas* of the Achaemenids merged with the savagery of the Arsacides to create "another world," an *orbis alius,* a world inverted, the very antithesis of Roman virtue.[35]

What marks all these descriptions is the insistence once again on the lack of freedom among the Asians. Like their Achaemenid predecessors, with whom they closely identified, the Parthians were seen to be a people driven only by fear, not by choice. "You hold your subjects in check by fear," Seneca told the Parthian king, "they never allow you to relax your bow; they are your bitterest enemies, open to bribes and eager for a new master."[36] Here again was a horde, a mob, not a nation. A horde, as Xerxes' armies had shown, might be capable of great feats, and there were many in Rome who were prepared to acknowledge the Parthians' skills in battle and their courage, not to mention their ferocity. But in the end, most were convinced, as Herodotus had been, that free men make the best warriors and that, therefore, the Parthian Empire would eventually go the way of all Eastern despotisms. As indeed it did, destroyed first by the Romans under Septimius Severus in 198 C.E. and then by the Sassanids. Consumed by their own vices, driven by a slavish inability to fashion themselves into anything resembling a people, they were the natural "provincials" who, in Cicero's eyes, belonged, for their own good as much as anyone else's, under the wise tutelage of Rome.

The Roman perception of the ultimate frailty of all Eastern tyrannies derived, as had the Greek before it, not merely from a supposedly "natural" love of freedom but, more concretely, from the confidence invested in a specific form of government. As Aristides pointed out, all that made Rome the legitimate ruler of the Greek world was "a form of government as if it were a mixture of all the constitutions without the bad aspects of any"—in a word, Roman republicanism.

## III

During the entire period of its steady absorption of large areas of Asia, Rome had been a republic, a *respublica* (literally the "public thing," perhaps best translated by the old English term "commonwealth"). It was not, certainly, as Athens had been, a democracy, but it had always been a state whose constitution relied upon maintaining a balance between the common people—the plebs—and the patrician classes who dominated the Senate. The empire itself was—and would remain in name, at least, until the end—the "Empire of the Roman People," ruled by the "Senate and the Roman People"—*Senatus Populus Que Romanus*—SPQR, the symbols that all the legions carried before them into battle, adorned every public building, and can still be found engraved on manhole covers in the city to this day. The people were always immensely, if aimlessly, powerful. Controlling Rome often meant controlling the mob, and the mob could always be mobilized by a popular general against his political enemies. When, in 184 B.C.E., Scipio Africanus the Elder, the popular victor of the battle of Zama, was accused by Naevius, the tribune of the Senate, of accepting bribes from the Seleucid emperor Antiochus, in what is now Syria, he made no effort to refute the charges (of which, in any case, he was probably guilty). Instead he turned to the huge crowd that had gathered to hear him judged and declared, "This is the anniversary of the great battle on African soil in which I defeated Hannibal, the Carthaginian, the most determined enemy of *your empire*." Pointing at Naevius, he added, "Let us not be ungrateful to the gods. Let us have done with that wretch and offer thanks to Jove." He then walked to the Capitol. The crowd followed him, and Naevius found himself alone and defeated.[37] To control Rome, as all these military commanders knew, one had to have the love of the people.

For the Romans, plebs and patricians alike, the empire and the republic were one. Today we all too often assume that empires have always been, in one guise or another, monarchies. But that has never consistently been the case. Democratic Athens, as we have seen, created what was in fact an empire. So, too, did the republic of Venice in the fifteenth and sixteenth centuries; so, too, did the United States in the nineteenth (and, many would argue, continues to do so today). So, too, did the USSR—which was, after all, the Union of Soviet Socialist *Republics*. So, in some respects, has modern China, which also describes itself as a republic; and so, too, did ancient Rome.

In time, however, as has often happened in republics, the commanders of the Roman army grew overpowerful and increasingly reluctant to follow the wishes of the Senate. In 48 B.C.E., a struggle between two consuls, Julius Caesar and Pompey, which resulted in the defeat of Pompey at the battle of Pharsalus and his subsequent death in Egypt, left Caesar in effective control of the entire empire. Caesar was a brilliant military commander and master of the Latin language, a spellbinding orator, a famous dandy (the inventor of a fancy frilled toga), a womanizer, and an epileptic. He was also overpoweringly ambitious. He now claimed the position of dictator (an office generally assumed only at moments of crisis, for a limited duration) and consul for life, and, after much bullying, succeeded in having himself declared a god. With these moves, the republic had become, in all but name, a monarchy. Caesar also passionately wished to be made king, an ancient title associated particularly in the minds of the people with the days of disorder and repression before the creation of the republic. The more sycophantic members of his entourage apparently spread rumors that the Sibylline Books had prophesied that Rome would never rid itself of its oldest and fiercest Asian enemy, the Parthians, until it became a monarchy.[38] The story goes that when at last the representative of the people reluctantly offered him the crown, Caesar, claiming to be unworthy, refused it, fully expecting that it would be offered to him again. But no second offer came. Furious, he retired, but, having been taken at his word, could do nothing. The outcome of this maneuvering, claimed Plutarch, was to make him "openly and mortally hated" by the common people and to provide a "useful pretext for those others who had long hated him but up to now disguised their feelings."

King or not, Caesar's behavior toward the Senate, in the eyes of many, had made him a menace to the freedom of the republic, and on March 15, 44 B.C.E.—the Ides of March—he was stabbed to death on the steps of the Senate by a group of embittered Republicans and ex-Pompeyans led by Brutus and Cassius, two of his former companions. Caesar's death, however, did not restore the republic, as his assassins had hoped. Instead, it plunged Rome into a succession of civil wars that nearly destroyed the empire and with it the Roman world altogether—something no later Roman would ever forget.

The civil wars became a defining moment in the history of

Rome and in the entire history of the West. They not only pitted Roman against Roman, they threatened to divide the empire for all time into western and eastern halves. They were also represented, by the victors, as the last titanic phase of a struggle between East and West, which would establish forever the dominance of Roman, European virtues over those of the languid, corrupt, and lascivious Orient.

After the assassination of Julius Caesar, two men emerged as his potential successors: Caesar's nephew and heir designate, Octavian, and one of the most successful and most powerful of his generals, Marc Antony. In November 43 B.C.E., the Senate, in the hope of avoiding war between the two, appointed Antony, Octavian, and Aemilius Lepidus, the governor of Spain and parts of Gaul, "triumvirs for the restoration of the state" for a period of five years. The administration of the empire was divided among them, and Antony, with the agreement of the other two, undertook the reorganization of the eastern half.

In 41 B.C.E., Antony summoned Cleopatra VII, queen of Egypt, to meet him at Tarsus. It was to be a fateful move. Cleopatra was a famous temptress, who had already learned that her charm could be employed to good effect in defense of her kingdom. Six years earlier, Julius Caesar had besieged and captured Alexandria but had then been seduced by Cleopatra, who bore him a son and persuaded him to allow her to remain as queen of a quasi-independent kingdom under Roman control. On Antony's arrival she decided to pursue the same strategy. Now that she was, as Plutarch put it, "at the age when a woman's beauty is at its most superb, and her mind at its most mature" (she was twenty-eight), she could be confident of success. After ignoring a number of Antony's demands that she come to him, she finally made her entry, on her own terms and in her own time. She came sailing up the River Cydnus in a barge with a stern of gold, "its purple sails billowing in the wind, while her rowers caressed the waters with oars of silver, which dipped in time to the music of the flute." Cleopatra herself lay in the stern beneath a canopy of cloth of gold, dressed as Venus "as we see her in paintings," remarked Plutarch, while to either side of her stood boys costumed as Cupids, who cooled her with their fans. Instead of a crew the galley was lined with the most beautiful of her waiting women attired as Nereides and Graces, some at the rudder, others at the tackle and sails, and all the while an indescribably rich perfume, exhaled from innumerable censers, wafted from the vessel to the riverbank.

When finally this cortege reached Antony in the marketplace, "enthroned on his tribunal," the word spread throughout the city that "Venus had come to revel with Bacchus for the happiness of Asia."[39]

The impact of this Oriental luxuriousness had the desired effect, and Antony fell in love, prompting the seventeenth-century French savant Blaise Pascal to make his famous comment on the role of chance in human history that "the face of the earth would have been different if Cleopatra's nose had been a little shorter." We do not know whether it was her nose or, as many have suggested, her conversation—she was by all accounts a brilliant woman—that held Antony in her thrall. But whatever it was, Cleopatra employed it on Antony to devastating effect. "Plato," observed Plutarch dryly, "speaks of four kinds of flattery, but Cleopatra knew a thousand." She used them all, and she never left his side for a moment.

Antony spent the following winter with her in the Egyptian capital, Alexandria, and a year later she gave birth to twins, Alexander Helios ("the Sun") and Cleopatra Selene ("the Moon"), who by all accounts inherited her mother's looks, charm, and feisty nature. By then it had become clear that Antony hoped to secure the eastern empire either as his personal domain or as a base for seizing Rome from Octavian. In 39 B.C.E., he visited Athens, where he was greeted enthusiastically and declared to be the living presence of the god Dionysius. He and Cleopatra now offered themselves as the embodiment of two Egyptian deities, Osiris and Isis, bound together in a sacred marriage for the prosperity of all Asia. In 36 B.C.E., Cleopatra gave birth to yet another son and named him Ptolemy Philadelphus ("Sibling—Loving").

In 34 B.C.E., Antony annexed Armenia, returned to Egypt with its king, Artavasdes, in chains, and led him in triumphal procession through the streets of Alexandria. This was a nearly unprecedented action, since traditionally triumphs were held only in Rome itself and dedicated to the city's tutelary deity, Jupiter Capitolinus. To celebrate a triumph in Alexandria was to suggest that this was now the capital of the empire. To make matters worse, Antony then staged a magnificent ceremony at which Cleopatra, seated on a silver throne (Antony's was made of gold), appeared in the guise of the Egyptian goddess Isis, "mistress of the house of life." In what has come to be called the "Donations of Alexandria," Antony proclaimed Cleopatra "Queen of Kings and King of Kings" and sovereign of Egypt, Cyprus, Libya, and Syria, together with Caesarion, her son by Julius Caesar, thus, in effect, challenging Octavian's position as Caesar's heir. The rest of the kingdom to the east and the west of the Euphrates was divided between Alexander Helios, who was dressed for the occasion in Median costume crowned by a Parthian tiara, and Ptolemy Philadephus, who was dressed as a Macedonian. Cleopatra Selene was made queen of Cyrene.

Symbolically, at least, the Roman Empire had now been divided into east and west. That, at least, was how it was perceived when news of the ceremony reached Rome. "People regarded this as an arrogant and theatrical gesture," wrote Plutarch, "which seemed to indicate a hatred for his own country."[40] In the vision created by later historians, most of them apologists for Octavian and his successors, Antony had, for the love of a woman, transformed himself into an oriental satrap.

Seneca, whose contempt for all things Eastern extended even to Alexander the Great, claimed to believe that although Antony had certainly once been "a great man and a fine intellect," he had been responsible for allowing "foreign customs to reach into the empire, and vices of which the Romans knew nothing." Lascivious, uncontrolled, drunken, he had become the willing instrument of an Eastern woman. "Henceforth then, let no one consider him to be a Roman citizen," sneered the Greek senator Cassius Dio, more than half a century later, "but rather an Egyptian: let us not call him Antony but rather Sarapis [Osiris], nor think of him as ever having been consul or imperator, but only gymnasiarch."

Back in Rome, Octavian began a propaganda campaign against Antony that would continue long after he was dead. In 32 B.C.E., he succeeded in intimidating most of Antony's Roman supporters into fleeing the city. He then seized and published Antony's will, which, because Antony had entrusted it to the Vestal Virgins, constituted a semisacrilegious act. In it Antony was supposed to have bequeathed the empire to his children by Cleopatra and to have asked to be buried in Alexandria—a clause that further added to the rumors already circulating that his intention was to transfer the capital of the empire from Italy to Egypt.

Octavian now secured from the Senate a formal annulment of Antony's remaining power and a declaration of war against Cleopatra, which in effect made Antony a traitor, unless he immediately abandoned her, which Octavian knew he would never do. Octavian then set out in pursuit of his rival. After prolonged conflict in western Greece, in which Antony's initial advantages were slowly eroded, largely through the skills of Octavian's admiral, Agrippa, the forces commanded by the two men met on September 2, 31 B.C.E., at Actium, a sandy promontory at the entrance to the Ambracian Gulf, in northwest Greece. Each side had assembled nearly 40,000 legionaries. The forces of Octavian were composed not only of Italians but also of Germans, Gauls, and Dacians. Facing them was Antony's army, gathered from Egypt, Libya, Ethiopia, and Arabia. Octavian is supposed to have addressed his troops with a typically denigratory piece of rhetoric. "Alexandrians and Egyptians," he assured them, "who worship reptiles and beasts as gods, who embalm their own bodies to give them the semblance of immortality, who are most reckless in effrontery but most feeble in courage, and who, worst of all, are slaves to a woman and not to a man," would be no match when it came to battle for true Romans.[41]

In the eighth book of the *Aeneid,* Aeneas is given a shield by his mother, Venus, in whose surface he is allowed to see forward into the future to the final triumph of Augustan Rome. In this version, Antony's army and Antony's fleet are made up of

> *Barbarian aids, and troops of eastern kings;*
> *Th'Arabians near, and the Bactrian from afar,*
> *Of tongues discordant and a mingled war;*
> *And rich in gaudy robes, amidst the strife,*
> *His ill fate follows him—th'Egyptian wife!*[42]

It is almost possible to hear the hiss of disgust in that last line—*sequiturque nefas—Aegyptia coniunx*—the hiss, too, of the asp, invoked a few lines later, that would be Cleopatra's final undoing.

Antony's troops far outnumbered those of Octavian. He also had twice as many ships as Octavian, heavier, larger, and better armed. And in line astern of them were sixty more provided by Cleopatra, who lay in another gilded barge and waited for victory.

At first the battle at sea seemed to be going Antony's way. Then, realizing where his opponent's true weakness lay, Agrippa forced Antony to extend his battle line, broke through, and began an assault upon Cleopatra's ships. The strategy succeeded. Instead of encircling Agrippa's squadron, Cleopatra, unused to real combat, took fright and fled, sailing straight before a following wind through Antony's line, thus throwing what was left of his formation into complete disorder. At that moment, says Plutarch, Antony "proved the truth of a saying which was once uttered in jest, that a lover's soul dwells in the body of another." Instead of regrouping his ships, he immediately abandoned them, commandeered a five-banked galley, and "hurried after the woman who had already ruined him and would soon complete his destruction."[43]

By nightfall, Antony's now-leaderless armada had either defected to Agrippa or been destroyed. Antony's troops, encamped on the shore, waited a week for their commander to return. When he failed to do so, they went over to Octavian. The battle, and with it Antony's bid to transform Rome into an Oriental despotate, was at an end.

Or that, at least, is the account that the victors have left us.

CLEOPATRA AND HER lover managed to escape eastward to Alexandria, with sixty galleys and Antony's treasure. It took Octavian more than a year to catch up with them, but when he did, it became clear that resistance was hopeless. Both took their own lives as the victorious Roman army entered the city. She did so in one of the best-known suicides in history, by clasping an asp "to her snow-white breast." Octavian was apparently so anxious to carry her back in triumph to Rome that, according to the historian Suetonius, he summoned Pysllian snake charmers to suck the poison from her wound—but to no avail. He had to make do instead with Cleopatra Selene, who walked in chains behind his chariot when he returned to Rome, master now of what his propagandists would repeat in prose and verse time and time again was the entire world.

Before leaving Alexandria, Octavian had Alexander's mummy removed from its shrine. After contemplating it for a long time, he crowned the shriveled head with a golden diadem and strew flowers on the body. When asked by the keepers of the mausoleum if he would now like to view the mummies of the Ptolemies, he replied, "I came to see a king, not a row of corpses."

Octavian had made himself master of the entire Roman world, and, for a while at least, the menace of the East had been laid to rest. Egypt would now go the way of Macedon itself and become a Roman province, and all memory of its Hellenistic and Pharaonic past would be erased.[44] Centuries later, as we shall see, Napoleon Bonaparte would arrive on the same shores, declare himself a new Alexander, and vow to reverse all the damage Octavian had done.


MUCH OF MARC Antony's "Orientalism" is the fabrication of subsequent Augustan propaganda, enthusiastically endorsed by the Roman epic poet Lucan and then by Plutarch. It is certainly true that he had made his base in Egypt and, with Cleopatra's help, and doubtless also at her behest, had extended the power of her kingdom, although he had also refused to seize the territories of Herod of Judea as Cleopatra had apparently asked. But Antony's "East" was a long way from being the East of the Persians; it was not even the East of the Egypt of the pharaohs. Ever since Ptolemy I (367–282 B.C.E.), known as "Savior," a Macedonian and one of Alexander's generals, had seized control of Egypt, it had been a hellenized monarchy. True, the Ptolemies were treated as both Egyptian pharaohs and Hellenistic monarchs. They supported native cults, cooperated with the all-powerful priests of Memphis in lower Egypt, and, ever since Ptolemy V had initiated the custom, had been crowned in the ancient Egyptian style. They even introduced a Greek version of the Egyptian deity Osiris under the name of Sarapis, to provide themselves with a divine Greco-Egyptian patron. But such fusions were commonplace in the ancient world, and they served political ends that the Romans themselves had no difficulty in understanding. The Ptolemies themselves remained Greeks, who lived by Greek customs and observed Greek laws. Cleopatra was reported to have known Egyptian (as well as, if Plutarch is to be believed, Mede, Ethiopian, Hebrew, Arabic, Parthian, and "Troglodyte"), but she was the first Ptolemy to do so. Antony may have made his base in Alexandria, he may have had three children by Egypt's queen, but there is very little evidence that he ever married her, and it seems clear that his ultimate ambition was not to become an Oriental despot but the emperor, Caesar, of all Rome. In fact, leaving aside his possibly apocryphal associations with Osiris, Antony's worst offense against Roman proprieties seems to have been wearing flimsy Greek sandals instead of decent solid Roman shoes.

But the historical record was easily brushed aside. In later accounts of the titanic clash between Octavian and Antony, Octavian's victory in the civil wars and the creation of the principate is represented as the purging of the Roman world of the stain of Orientalism, of the *vanitas* displayed in Antony's dalliances at the Egyptian court and his adoption of Egyptian deities. The victory at Actium had resulted in the full annexation of the Kingdom of Egypt, and with it much of Asia had now come under the thrall of Rome. Like Salamis before it, Actium had also been a naval battle, one in which the future of a free and virtuous West had been preserved from extinction at the hands of a tyrannical and corrupt East.

## IV

Behind this attempt to demonize Antony lay a deeper unease. Unwilling, after his uncle's humiliating failure, to adopt the title of king, Octavian had nevertheless conferred upon himself the name emperor—*imperator*. In 27 B.C.E. he took another title, Augustus ("revered one"), by which he is now known, and established a new imperial state under one ruler, now referred to as the principate.

Augustus always insisted that rather than creating a new state on the ruins of the republic, all he had done was to restore what came to be called "the people's cause"; and in the official propaganda of what now became in effect a monarchical empire, it was always the people who had granted to the emperor their *imperium et potestas,* their "authority and power."[45] In the accounts of his life, the "Deeds of the Divine Augustus," which he had inscribed on bronze pillars and set up in every major city in the empire, Augustus declared that he had "set free the republic which was oppressed by the domination of a faction." Even as late as the fourth century C.E. the last great Latin historian of the Roman Empire, Ammianus Marcellinus, could still describe the Roman emperors as servants of the *respublica,* which had "entrusted the management of her inheritance to the Caesars, as to her children." There was indeed always some ambiguity in the titles that the Roman emperors assumed. *Augustus*—which was adopted by all subsequent emperors—alone implies the semisacral powers assumed by the Hellenistic monarchs. *Princeps*, "first amongst men," suggests, at least, that there were others like him; and the word *imperator*, meaning "he who exercises imperium," merely describes the sphere of executive authority possessed by all Roman magistrates.

Despite this, despite the continuing political power of the Senate and the constant allusion, until the very end, to the *Senatus Populus Que Romanus*, Augustus's new Principate rapidly became a facade behind which a new kind of oligarchic regime took shape. Although equity and civil liberties survived—for a while at least—the older forms of power sharing that had been the source of the republic's success were dismantled, so much so that by the early third century C.E. the jurists Gaius and Ulpian could claim, without fear of contradiction, that the imperium of what was now being called the "prince emperor" had absorbed that of the Roman people.

For all that it marked the steady erosion of all the liberties that had been so highly prized under the republic, Augustus's new order did indeed seem to offer the prospect of unparalleled power: a vision of armies that were seemingly invincible and of limitless economic growth opened before the Roman people. To go with the new wealth and security came a Golden Age of Latin literature—the age of the epic poet Virgil; of the poet Ovid (although Augustus banished him to Tomis on the Black Sea, probably because of his part in some scandal involving the imperial house), whose writings have had more of an impact on later European literature than those of perhaps any other classical author; of the poets Horace, Tibullus, and Propertius; and of Livy, perhaps the greatest of the Roman historians. In their different ways, all of these writers celebrated the achievements, past, present, and to be, of the new order, when Rome had truly become, in Livy's words, *caput orbis terrarum*, "the head of the world," when Roman armies and Roman law had secured a Roman peace—the Pax Romana—throughout the world.[46]

But the promises of Augustus's reign were relatively short-lived. After his death, the empire fell into the hands of a succession of corrupt, incompetent rulers, all loosely related to one another, who formed a dynasty known as the Julio-Claudian imperial house. For although the Roman emperor wielded absolute and personal power, the empire itself was, until the end, a delegation of the people. This meant that the problem that besets all monarchies—how to transfer power from one generation to the next—could never be resolved. The Roman emperors were never, as the later rulers of Europe would be, "kings," hereditary monarchs whose persons, as well as offices, were thought to be hedged with some kind of divinity. The decisions taken by one emperor could never be binding on his successors. The continuity previously ensured by the presence of a senatorial class under the republic faltered and then vanished altogether, and what had been, under Augustus, an oligarchy, became under his successors a tyranny and finally, in the opinion of many Romans, something close to an Oriental monarchy.

It was under the Julio-Claudians that imperial Rome acquired the image on which so much popular fiction and filmmaking has been based: drunken orgies, cruel and fantastic sexual practices, gladiatorial combats, the interminable slaughter of innocent victims who had in some way offended the regime, all stage-managed by effete and lascivious emperors happy to leave the government of their vast domains to corrupt and sycophantic subordinates. Much of this, like all popular images, is sheer, and usually salacious, fantasy. Much was fueled, when not simply invented, by its most notable victims, the Christians. But by no means all. Some, it has been suggested, might have been caused by lead poisoning. Eleven aqueducts around Rome delivered 250,000 gallons of water a day to the city, which were then distributed to individual homes through a system of lead pipes. Physical anthropologists have discovered ten times the normal level of lead in bones excavated from this period. The effect of the contaminated water on those who drank it can only be guessed.

Whether suffering from lead poisoning, hereditary mental instability, or some other less tangible affliction, few of the emperors from Augustus to Nerva had very much to recommend them. First there was Tiberius, who, after having earned himself a reputation as a drunkard and lecher in Rome, retired to the island of Capri, where he built himself a magnificent villa. Its ruins are still there on the cliff top from which he used to watch the bodies of those who had offended him, "after prolonged and exquisite tortures," tumble into the sea far below. Here he also built what the historian Suetonius calls a "private sporting house," where bevies of young girls and boys, collected from all over the empire, would indulge in "unnatural practices" to excite his waning passions. Meanwhile, the state was left in the incapable hands of the commander of the imperial bodyguard, Sejanus. For years Spain and Syria had no governors. The Parthians overran Armenia, the Dacians and Sarmatians ravaged Moesia, and the Germans invaded Gaul.

Tiberius was followed by Gaius, known as Caligula, "little boot," a fidgety, neurotic sadist with a penchant for incest—which he committed publicly with each of this three sisters at lavish banquets, while his wife and accomplice, Milonia Caesonia, watched him from a balcony. Even Tiberius could see just what kind of creature his successor was likely to be. "I am nursing a viper for the Roman people," he once declared. His foreboding was fulfilled in full. Caligula delighted in watching tortures and executions and therefore organized large numbers of them. He expressed his contempt for the Senate by having his horse made a consul and is said to have shouted in a fit of rage that he wished only that the Roman people, in whose name he ruled, had a single throat so that he might cut it with one stroke. In the end he was, as were so many other intemperate emperors, assassinated by the palace guard, along with his wife and daughter.

Caligula was succeeded by his uncle Claudius, a weak and sickly man who probably suffered from cerebral palsy and who, because of his shaking hands and slurred speech, had been the butt of innumerable practical jokes during his nephew's reign. But Claudius, for all the humiliations of his youth, was a learned and shrewd man and something of a historian and grammarian. He was fluent in Greek and succeeded in having three new letters added, if only temporarily, to the Latin alphabet.[47] Although seemingly as cruel and bloodthirsty as his predecessors once he gained power, he turned out to be the only member of the dynasty to engage seriously, if often erratically, in the business of government, and it was he who was responsible for extending Roman citizenship to the provinces of the empire. Claudius made the mistake, however, of marrying two murderous wives, first the infamous Messalina, and then Agrippina, who finished him off by mixing toadstools in with his favourite dish of mushrooms.

This made way for the most destructive of them all: Nero, who murdered both his mother and his aunt and had his wife, Octavia, executed on a trumped-up charge of adultery. Nero fancied himself a poet, musician, and athlete, and he competed in theatrical competitions and chariot races (all of which he won). These activities, although they amused the people, scandalized the Senate. He was also accused, probably unjustly, of having sat in his villa at Antium during the great fire of 64 C.E. playing his lyre and, in the historian Tacitus's words, "singing of the destruction of Troy, comparing present misfortunes with the calamities of antiquity" as Rome burned. Afterward, some 125 acres of the gutted center of the city were leveled to make room for a magnificent, luxurious palace filled with gold and jewels—"quite vulgarized by extravagance," sneered Tacitus—and aptly named the "Golden House," Domus Aurea.[48] Later Trajan, in disgust, buried the entire structure beneath a massive mound, where it is still today.

Not content with building architectural follies where his citizens had once lived, Nero also, according at least to Suetonius's gossipy account, "practiced every kind of obscenity" and defiled "almost every part of his body." He attempted to transform an unfortunate boy named Sporus into a girl by castrating him, after which he married him, "dowry, bridal veil and all," took him to the imperial palace and set him up as his wife. How much happier Rome would have been, caustically remarked one senator, if Nero's own father, Domitius, had taken the same kind of bride.

Finally the Senate could take it no longer. In a rare gesture of solidarity, it declared him to be a public enemy. Nero took refuge in the villa of his freedman Phaon. But knowing that he would be assassinated the moment he stepped outside, he chose suicide instead. If Suetonius is to be believed, his dying words were "What an artist dies with me!"

With the death of Nero, greeted with "widespread general rejoicing" in the streets of Rome, the Julio-Claudian dynasty was finally extinguished. Nero's immediate successors, from Galba (68–69 C.E.) to Domitian (81–96 C.E.), rose and fell through persistent conflict with the Senate and internal divisions within the Roman legions. In one famous year, 69 C.E., there were no fewer than four emperors. This dismal tale came to an end in 98 C.E. with the election of the first of the so-called Antonine emperors, Trajan, who not only extended the empire significantly for the first time since the fall of the republic but also restored a degree of peace and stability throughout most of its vast extent.

With the accession of Antoninus Pius in 138 C.E., it seemed as if, after centuries of struggles and intermittent civil war, Rome had come to rest. With its frontiers secure at what were believed to be the furthest limits of, if not quite literally the "inhabitable world," then the world in which any possibility of civilization might exist, it had become at last what, since the days of the early republic, it had always claimed to be: the embodiment of a divine harmony, which had brought to humanity, in Aristides' own analogy, the "universal order . . . as a brilliant light over the private and public affairs of man," which Zeus, father of the Gods, had once conferred upon the natural world. In this new dawn, "a clear and universal freedom from all fear has been granted both to the world and all who live in it."[49]

For Aristides, Rome was not only the greatest of all civilizations, it was also the final, the most enduring, in a succession of empires. First, at least in his reckoning, there had been the empire of the Achaemenids, then of Alexander. Great and extensive though these had been, they had been built on sand. Alexander might have finally "abolished the rule of the Persians, but he himself all but never ruled."[50] Finally there was Rome, which had absorbed all that had preceded it and would last, or so Aristides believed, forever. Like many of his contemporaries, Aristides understood time as a progression that had led up to the present moment, which would now be projected unchangingly into the future; a world that in all respects represented the perfection that humanity could attain could not, in any meaningful sense, progress any further. Technological changes there might be, although in fact there had been remarkably few since the fifth century B.C.E.; but it was impossible to imagine that a different political order, different customs, or another religion might one day come to replace those that now existed. Much less was it possible to imagine that any might be preferable to those that now existed. Aristides, and with him much of the educated Roman world, was perhaps the first to embrace such an injudicious and in the end implausible view, but he would certainly not be the last.

If Rome was the last of the world empires, and if Rome had brought history to a close, then Rome must also have embraced the entire globe. The idea that the Roman Empire and "the world" were in some sense identical had begun to take shape at least since the days of the republic. By 75 B.C.E., coins were being struck with images of a scepter, a globe, a wreath, and a rudder, symbols of Rome's power over all the lands and oceans of the world.[51] Two decades later, Cicero was speaking of "our own people whose empire now holds the whole world,"[52] and by the time Augustus came to power, "the world"—the *orbis terrarum*—and the empire had come to be identified as one world state bounded, in Virgil's words, only by the Ocean—*Oceanus*—which was conceived of as a god, and a massive river encircling the three continents of Asia, Europe, and Africa.

If Rome truly was, in the words of the Pliny the Elder, author of the greatest natural history of the ancient world, the "nurse and mother of all lands," chosen by the gods "to give humanity to mankind, to become the single native land of all the peoples of the world"[53] what had made it so? Force of arms, combined with a capacity for mobilization and organization unprecedented in the ancient world, was part of the answer. The technological goods that the Roman Empire had to offer were all too obvious: Roman architecture, Roman baths, the ability to bring fresh water from distant hills or to heat the marble-lined rooms in villas in the wilds of Northumberland. All of those things, and others besides, spoke a universal language that proved irresistible. For centuries they persuaded non-Roman patricians from Africa to Scotland to identify themselves with the empire.[54] But there was always something more: a way of living, what Cicero identified as "our wise grasp of a single truth."[55]

Such sweeping claims were propaganda, certainly, but propaganda that remained highly effective as long as the Roman state could be seen to be living up to its own, always grandiloquent, image of itself. Rome was, as Aristides conceived it, a world that promised not only the possibility for advancement but ultimately security and protection from the kind of "faction uproar and disorder" that had prevailed before the arrival of the legions and that prevailed still in the wild reaches beyond Rome's frontiers. Rome was power. Rome was splendor. But Rome, "mother of cities," was also love. Even the name of the city itself, Roma, could be reversed so as to spell *amor,* "love." *Roma summus amor* was scribbled on the walls of the church of Santa Maria Maggiore in the third century C.E.: "Rome the supreme love."[56]

# V

For the Romans had learned that if an empire was to outlast its founder and be proof against intruders, it had to inspire, if not always love, then certainly self-interested loyalty, in its conquered peoples. No one should underestimate the sheer brutality, nor the ruthlessness and efficiency, of the Roman military machine. The Roman occupation of Europe was the most bloody colonizing venture ever undertaken by a European power. Nothing else, not Hernán Cortés's slaughters in Mexico nor Pizarro's in Peru in the sixteenth century nor Cecil Rhodes's in Matabeleland in the nineteenth, come close. In the last twenty years of Julius Caesar's conquest of Gaul, more ferocious perhaps than most but hardly unparalleled, a million Gauls lost their lives, a million more were carried off into captivity, and with them an entire generation was obliterated. But if military power had been all there was, Rome would have fallen as fast as it had risen. "An empire," the historian Livy had declared, "remains powerful so long as its subjects rejoice in it."[57] To survive, an empire had to make friends, not slaves. It had to persuade its conquered peoples that the way of life they were likely to live under the conqueror would ultimately be far better than the life they had lived before. The Romans had learned, too, that if peace were to be maintained throughout what they thought of as the world, if the murderous enmities that had divided Greek and Persian, Greek and barbarian, were finally to be laid to rest, one rule, one culture had to prevail. This had, of course, also been Alexander's ambition. But Alexander had not lived long enough, and his successors had not been skilled enough, to transform this insight into a reality.

Aristides had no doubt that it had been Roman virtue, and above all the Roman conception of government, that had been responsible for Roman greatness. The Persians, he told his audience, much as Herodotus had done, "did not know how to rule, and their subjects did not cooperate, since it is impossible to be a good subject if the rulers are bad rulers." The Persian great kings had looked upon those who served them as slaves and despised them, "while those who were free they punished as enemies. Consequently they passed their lives in giving and receiving hatred."[58] By contrast, the Romans alone "rule over men who are free . . . [and] conduct public business in the whole civilized world exactly as if it were one city-state."[59] Rome had thus gathered to itself all that was best and most useful on Earth, all the manufactured goods, arts, and architecture, all the crops, textiles, precious ornaments, "so that if one would look at these things, he must needs behold them either by visiting the entire civilized world or by coming to this city."[60] Rome had achieved all this because, as Aristides stresses time and again, the diverse peoples and nations of which the empire was composed were all free, all governed according to their own rights, respected, and defended, all loyal, all proud to count themselves not only Phrygians or Egyptians or Gauls but also, and perhaps above all else, Romans. "For the eternal duration of this empire," claimed Aristides, "the whole civilized world prays all together."

Rome was always more than an empire. It was, for those who were drawn into it, what the Romans called a *civitas,* the word from which, much later, the far more ambiguous modern term "civilization" would be derived. It was a society that, although it had always looked to the city of Rome, had no fixed place and indeed would one day gather all humanity into what Cicero called a single community "of gods and men." In this way it depended, in practice, on a process of reciprocation and assimilation. For the Romans had learned that their rule, and with it their very identity, would survive, as Aristides had it, "until stones float upon the seas, and trees cease to put forth shoots in spring," only if their subject peoples, the "barbarians," both East and West, could be persuaded to absorb what in the second century C.E., the Christian theologian Tertullian called *romanitas—*"Romanness."

And absorb it they did. From northern Britain to North Africa, from Spain to what is now Syria and Iraq, local elites adapted themselves to Roman ways of life. For its most skillful and most fortunate subjects, the empire became a vast resource, vastly more enriching than the narrow limits offered by the original communities from which they came. Living in Roman villas, they took up Roman dress, Roman customs, and the Latin language, and in time, they came to think of themselves as Romans. Peoples from all over the empire could be found in almost every Roman province. Rome itself had already by the close of the republic become a vast cosmopolitan metropolis, much as London or Paris or New York would be in their day.

Even the emperors themselves, after the second and third centuries C.E., were sometimes neither Roman nor even Italian, at least by birth. Trajan had been born in Spain; Septimius Severus, who became emperor in 198 C.E., was a recently romanized man of Punic origin from Leptis Magna in what is now Libya who, on all accounts, spoke Latin with a strong regional accent. The great reforming emperor of the third century, Diocletian, was the son of a freedman from Dalmatia, and his successor, Galerius, had begun life herding cattle in the Carpathians. And all of these, like so many lesser functionaries of the Roman state, while proud Romans who conversed in Latin (or Greek), were also keenly aware of their non-Roman origins. Septimius Severus even had the tomb of Hannibal, the fiercest and most successful of all Rome's historic enemies, restored in honor of his own supposed Carthaginian ancestors.[61]

By the time Aristides arrived in Rome, Roman culture, expressed in Greek and Latin, reached from the Tigris to the Atlantic and from the Elephantine on the upper Nile to Hadrian's Wall in northern Britain. The officers of the legions in Britain were housed in an Italian villa facing the Grampians, in Scotland; and a Roman town, complete with amphitheater, library, and statues of the classical philosophers, looked out over the Hodna Mountains at Timgad, in southern Algeria.[62] We do not know how much of the older pre-Roman world survived romanization. But it is remarkable that we do *not* know. Almost no trace now remains of any pre-Roman literature, oral or written, or any pre-

Roman history of the peoples at the heart of the empire in the western Mediterranean and northwestern and central Europe. Below the level of the urban aristocracy, the older ways and earlier languages must have survived. Several Celtic languages, some of which, in some version, survive to this day, must have gone on being spoken throughout the nearly four centuries of the Roman occupation of Britain. But if so, they have left no written trace.[63] The same was true of the Roman settlements in North Africa. The only language to survive was, of course, Greek, and Greek, as we have seen, was the second language of the empire and the language of the administration of what would become, after Diocletian divided the empire in the late third century C.E., the empire in the East.

So powerful, so pervasive, was Roman culture that it also survived the centuries of steady erosion and dismemberment that in time became the prolonged death throes of the Roman Empire. More than a century after the Roman Empire in the West had crumbled away, the Byzantine jurist Modestinus could still, without any sense that the Prince of Cities itself was no more, declare, somewhat forlornly, that "Rome is the common homeland [*patria*] of us all."[64] For Modestinus, who had never seen it, Rome had become far more than a place; it had become a way of life, a civilization.

And no one wished to see it end. The fall of Rome has been attributed to many causes. But the kind of hatred that many subject peoples have felt for their distant imperial overlords was only rarely one of them. The "barbarians" who eventually destroyed the empire did so from within. They did not wish to bring an end to Roman rule so much as to appropriate a sizable share of it for themselves. When at last, in August 410 C.E., the Visigoths of Alaric seized and plundered Rome, they did so not because they wished to destroy the great city but because its emperor had refused to allow them to settle in his domains. Even in defeat the lure of the grandeur that was Rome was seemingly irresistible. At the end of the century, when the Roman Empire in the West had fallen into "barbarian" hands, Theodoric, king of the Ostrogoths (493–526 C.E.), reflected that "An able Goth wants to be like a Roman; only a poor Roman would want to be like a Goth."[65] As James Wilson observed in 1790, while musing upon the possible future of the United States as the new Rome in the West, "it might be said, not that the Romans extended themselves over the whole globe, but that the inhabitants of the globe poured themselves upon the Romans."[66] As with Rome, he prophesied—correctly, as it turned out—so would it be with the United States.

Not that all this should compel us to take all the rhetoricians' words at their face value or blind us to the brutality and the cruelty that lay only a short way beneath the surface. The power of Rome was always present to shore up the benefits of Roman civilization. And at times that power could appear, to modern—and not only modern—minds, monstrous in the extreme.

There were the notorious games, the gladiatorial combats, the ritual slaughter of prisoners, both men and women, by wild animals. These had been held throughout the empire, before the emperor Domitian in the second half of the first century C.E. restricted them to Rome, to celebrate a victory or a marriage, or to mourn the death of a (male) relative. By the fall of the republic, they had already become a means of winning favor with the people for whoever paid for them. With the coming of the principate they became ever more extensive, ever more lavish, and ever more bloodthirsty. Power in Rome could all too easily degenerate from the lofty ideals praised by Aristides to a simple matter of offering gruesome entertainment—and bread—to a restless, violent mob.

Then, Rome, like Greece, had also been created by the labor of slaves. Hordes of the unfree tilled the fields, built the massive buildings whose ruins still astound even today. They manned the fleets, dug gold and silver out of the mines, provided the often highly skilled labor required to manufacture everything from shoes to swords. And, of course, they worked in every aspect of domestic life in all patrician, and a great many more humble, households. It has been calculated that there were some 2 million slaves in Italy alone at the end of the first century B.C.E., a ratio of about three slaves to every free person. Their activities also reached deep into the everyday administration of the state. Greek slaves worked as tutors and household administrators. Cicero's confidential secretary, who invented a form of shorthand that was named after him, was a slave named Tiro. Even doctors might be slaves, and they expected to be paid for their services, which were regulated by law.[67] These men and women may, in many respects, have been better treated and more highly regarded than the Africans shipped by the thousands across the Atlantic between the fifteenth and nineteenth centuries to labor on the tobacco and sugar plantations of the Americas. The conduct of their masters was controlled by law, and they could even make appeals before the courts. But they were still slaves; still the property of others; still, in legal terms, "things," not persons.

The goods the empire had to offer may have been shared by conqueror and conquered alike—or at least by a good many of them. But that did not always make life as predicable, as tranquil, or as safe as many might have wished. The garrulous first-century C.E. Roman encyclopedist Aulus Gellius tells a story that demonstrates all too vividly just how precarious life for the provincial self-made man under Roman rule could become. In 123 B.C.E., a consul, the highest-ranking of all the Roman officials, paid an official visit to the town of Teanum Sidicinium in Latium, south of Rome. His wife expressed a desire to wash herself in the men's baths—always more sumptuous than those for the women. The *quaestor,* or local administrator of the town, one Marcus Marius, was called for and ordered to instruct the men who were using the baths to be dressed and leave. He hurried off to do as he was told. But he did not move fast enough for the consul's wife who complained to her husband that she had been made to wait and that when she was finally admitted, the baths were not clean enough for her liking. The consul ordered a stake to be planted in the middle of the forum. The unfortunate *quaestor,* "the most illustrious man of his city," was then tied to it, his clothes were stripped off him, and he was beaten with rods in full view of all the citizens. Since corporal punishment, both private and public, was restricted in almost all cases to slaves, such treatment meant that the poor man's life would have been ruined, and all for the whim of a consul's wife.[68] Such high-handed behavior was not exceptional. At nearby Ferentinum one *quaestor* threw himself from the city wall to escape a similar humiliating punishment, and another abandoned his home and family and fled into exile.[69]

True, these events date from the earlier period, when the Latins were still formally allies, not full citizens, of the Roman *civitas*. But such autocratic behavior on the part of the high officials of the state was always to be feared throughout the whole period of Roman domination. Rome had absorbed the world, and Rome had imposed its laws and its institutions upon the world. The functionaries of Rome were the embodiment of its power, and nothing that they (or apparently their wives) desired could be denied to them.

But if the realities of life under Roman rule were rarely as rosy as someone like Aristides presented them, we must also remember that slavery was commonplace in the ancient world. Its origins were at least prehistoric, and for most if not all Romans, as for most Greeks, it was a part of the natural order. The games, by contrast, were a uniquely Roman pastime. But if they were unimaginably cruel, they were also indisputably popular. Roman imperial rule could certainly be brutal, but it was not uncharacteristically so. Roman officials were often capricious and autocratic, but hardly more so than many later servants of even democratic states. And if their modern counterparts are compelled to employ more gentle methods and can only surreptitiously exploit their office to please their wives, they can sometimes be no less destructive of individual lives. In the ancient world there simply was no society that was less cruel, less capricious than Rome, but there were many that were a very great deal more so. What Rome offered was inclusion, security, a way of life able to offer the likes of Aristides the opportunity to flourish in places far removed from the narrow provincial circle to which, under any other society in the ancient world, he would have been confined.

At the heart of the *civitas,* what ultimately bound it all together and gave it its name, was citizenship. Citizenship, in its modern form, was a Roman creation, and it has had more influence on shaping the civic values of the Western world than almost any other. In 212 C.E., the emperor Caracalla issued an edict granting citizenship to all free men living within the empire. "I give the citizenship [of the Romans] to all the world," it declared.[70] The Edict of Caracalla—or Constitutio Antoniniana, to give it its true name—was greeted as a spectacular act of generosity in its day and ever since has been looked upon as the supreme act of Roman universalism. As the crotchety contemporary theologian Tertullian warned Christians who might be tempted to see their faith as a reason for political dissent, "This empire of which you are servants is a lordship over citizens, not a tyranny."[71]

The Edict of Caracalla conferred citizenship upon the entire empire—or, what was much the same thing, the entire world. But what Caracalla had in fact done in 212 C.E. was to seal a policy that had been one of the great strengths of the Roman world since at least the early days of the republic. Rome had always been more open to foreigners than any other state in the ancient world. The historian Tacitus recounts how in 40 C.E., the emperor Claudius attempted to extend citizenship beyond the traditional limits of Italy by proposing that certain Gauls be admitted to the Senate. There was an outcry from the more conservative older members of the assembly. "Once," cried one senator, "our native-born citizens sufficed for peoples of our own kin . . . now a nation of foreigners, a troop of captives, so to say, is forced upon us."

To this Claudius himself responded:

What else proved fatal to Sparta and Athens, in spite of their power in arms, but their policy of holding the con quered aloof as alien-born? But the sagacity of our own founder Romulus was such that several times he fought and naturalized a people in the course of a single day."[72]

The argument won the day, and not only because it had been the emperor's argument but because Claudius knew that he was appealing to a tradition that was believed to be as ancient as the city itself.

With the accession of the emperor Hadrian in 117 C.E. the imperial policy of unification was intensified by emphasizing the all-inclusive benefits of Roman rule, by minimizing differences in culture and class, and by stressing the similarities of each individual's relationship to the emperor himself. Everywhere the Roman citizen looked he could see, or experience, the symbols of imperial rule: milestones, imperial portraits, military standards, holidays, and the ubiquitous Roman roads.[73] Rome itself was often described as the "moderator"—a word that is far milder than "ruler" or "governor," something closer to "regulator" or "supervisor." Roman imperialism came to be seen not as a form of oppression, as the seizure by one people of the lands, goods, and persons of others, but as a form of beneficent rule that involved not conquest but patronage. As Cicero said of the imperial republic he served, "We could more truly have been titled a protectorate [*patrocinium*] than an empire of the world."[74] Or, as the emperor Antoninus Pius put it later, "I am the guardian [*custos*] of the world."

It was precisely this, what Aristides called "your wonderful citizenship" and of which he said there was "nothing like it in the records of all mankind," that had united "the better part of the world's talent, courage and leadership." Even the oldest, most bitter division of all, the frontier between Europe and Asia, had finally been dissolved. He went on:

Neither sea nor intervening continent, are bars to citizenship, nor are Asia and Europe divided in their treatment here. In all your empire all paths are open to all. No one worthy of rule or trust remains an alien, but a civil community of the World has been established as a Free Republic under one, the best, ruler and teacher of order; and all come together as into a common civic centre; in order to receive each man his due.[75]

Like so much else in the Roman world, citizenship had Greek antecedents. The Greek citizen, however, was a *polites*—a member of a *polis,* a word that in time came to mean something like the modern "state" and from which, of course, our term "political" derives. But its original meaning was "citadel." Greek citizenship was therefore based on a specific location, the *polis*—the city—itself and had no meaning beyond its walls. When Diogenes the Cynic was asked of what place he was a citizen, he replied, "I am a citizen of the world"—*cosmopolites,* a "cosmopolitan." It was intended to be a snub, an insult to all forms of civility, not an expression of universalism. And if Lucian's satire on philosophy, in which the *cosmopolites* is depicted as a ludicrous figure, is anything to go by, as late as the first century C.E. the idea that one could be a citizen without a city was, for the Greeks, still unimaginable.

The Latin word *civis,* by contrast, derives from an Indo-European root connoting the idea of the family, and in particular of an outsider admitted into the family: in other words, a guest. It is perhaps best translated not as "citizen" but as "fellow citizen." The very vocabulary itself, therefore, carried with it the idea of a society composed of kin but always open to outsiders.[76] As the *civitas* was not a place but a body of the rights and duties of the citizen, it could be extended anywhere. Rome had not merely united Asia and Europe—it had transformed both into a single civilization. And because it designated neither place nor race nor people, it meant that a Gaul or a Spaniard or an Egyptian could declare, in that celebrated phrase, "I am a Roman citizen"—*Civis Romanus sum*—while at no time being under any obligation to shed his other local identity.

No matter where he was within the legal confines of the empire, a citizen threatened with punishment who had not been given a fair trial could "appeal to the people," in the person of the emperor. The Roman citizen, alone in the ancient world, enjoyed something like habeas corpus and protection against arbitrary justice by high imperial functionaries. No one demonstrates both the extent and the force of this more clearly than that persistent irritant to the authorities of the Roman East: Saint Paul.

On one occasion, after preaching his new religion, in Jerusalem, Paul had had to be rescued from an angry crowd by a praetorian cohort. He was flung into prison and after some time brought before the tribune. "May I say something to you?" Paul asked the tribune in Greek. "Do you know Greek?" replied the tribune in some surprise. "Are you not then the Egyptian who recently stirred up a revolt and led the four thousand men of the Assassins out into the wilderness?" "I am a Jew from Tarsus in Cilicia," retorted Paul, "a citizen of no mean city."[77] The tribune had him taken to the barracks and ordered him to be whipped until he revealed why the crowd had "shouted thus against him." Paul now played his trump card.

When they had tied him with the thongs, Paul said to the centurion who was standing by, "Is it lawful for you to scourge a man who is a Roman citizen and uncondemned?" When the centurion heard that, he went to the tribune and said to him, "What are you about to do? For this man is a Roman citizen." So the tribune came and said to him, "Tell me are you a Roman citizen?" And he said, "Yes." The tribune answered, "I bought this citizenship for a large sum." But Paul said, "I was born a citizen." So those who were about to examine him withdrew from him instantly; and the tribune also was afraid, for he realized that Paul was a Roman citizen and he had bound him.[78]

Paul was then taken to Caesarea to have his case investigated by Felix, the procurator of Judea. Felix seems to have treated Paul well and kept him under a form of house arrest for two years in order to avoid offending the Jewish community, but he took no further action against him. Felix's successor, Porcius Festus, however, was somewhat less benign, and when the Jews of Jerusalem, exasperated by this meddlesome theological deviant in their midst, again "laid many grievous complaints against Paul which they could not prove," Festus decided to put him on trial, an experience not unlike Christ's own and one that, had it proceeded, would probably have ended for Paul much as it had for his master. But Paul was not a carpenter from Nazareth.

> "I am standing before Caesar's tribunal," he told Festus sternly, "where I ought to be tried; to the Jews I have done no wrong, as you know very well. If then I am a wrongdoer, and have committed anything for which I deserve to die, I do not seek to escape death; but if there is nothing in their charges against me, no one can give me up to them. I appeal to Caesar." Then Festus, when he had conferred with his council, answered, "You have appealed to Caesar; to Caesar you shall go."[79]

The trial was stopped, and Paul was taken to Rome under military escort. There he would eventually be tried and condemned, and finally beheaded.[80] He could not escape his fate, but justice, of a sort, had been done.

## VI

Paul would become the bearer of a new kind of universalism, whose pretensions went far beyond the jurisdiction of the Caesars, but no one could have been more sensitive to the importance of the protection—and the dignity—offered by membership in the Roman *civitas*. Few, too, could have been more conscious of how much the image of the new borderless Christian order owed to the civic vision of what Cicero had called the "republic of all the world."

Citizenship, as Paul's case makes clear, was primarily a legal status, and it could never have existed without the creation of an extensive, complex, and all-embracing legal system. In an obvious historical sense it was Rome that created what today in the West is understood by that much-abused term "the rule of law." The Greeks had also, of course, been bound by the law. It was law, as Demaratus had warned Xerxes, and law alone, that the Greeks recognized as their master and that they feared "far more than Xerxes' subjects fear him." But Roman law was far more than a restraint or even the guarantee of equality that Herodotus makes of it; it was what held all the diverse peoples of the Roman world together.

In law, Roman citizenship was a universal claim to a unitary status. It entailed what we today would call rights. Equality before the law was, of course, the basis of the Greeks' claim to freedom and justice. It was what distinguished Greeks from Persians and, by extension, all other barbarians. But the concept of a right—perhaps the single most important term in the political and legal vocabularies of the West—is ultimately a Roman invention. The Romans, out of the prolonged experience of having to administer a far-flung empire that, unlike the leagues of the Greek world, was conceived as a single *patria,* evolved a highly complex system of law. As the empire spread, this became the law of the whole of Europe, and despite being greatly modified by the legal customs of the Germanic tribes, which overran the empire in the fifth century C.E., until this day it has remained the basis of most of our understanding of what the law is. It was the Romans' great intellectual achievement, as moral philosophy and the natural sciences had been the Greeks'.

The history of Roman law begins with the Twelve Tables, said to have been composed between 451 and 450 B.C.E. in an attempt to put an end to the manipulation of the law by the priests and patricians. The effect of these was to ensure that henceforth all customary law would be given a legislative basis and enacted by statute. They also ensured that in the Roman world, and subsequently throughout all of the West from Britain to the United States, the law would be secular and independent, despite many subsequent attempts to interfere with it, of divine command. Unlike the laws that governed most of the peoples of Asia from the Shari'a of the Islamic world to the imperial decrees of China, Roman law, although venerated, was not sacrosanct. It could be changed and modified to meet changing circumstances, for the law derived, as the later Roman jurists never tired of saying, from the fact. The law was based on both custom and practice. It was the voice of the people speaking as god—*"vox populi, vox dei,"* as the saying went—not the voice of God speaking to the people. It relied not on theory but on "the experience of things." And because it did that, it was irreducibly existential.[81]

As the Roman world and its needs expanded, so the body of customary law increased. Later Roman jurists attempted to gather all these enactments together in a series of codes, of which the most enduring was compiled under the emperor of the Eastern Empire, Justinian, in the sixth century C.E. Justinian's codification is divided into four great books, running to more than a million words: the Codex, the Digest, the Pandect, and the Institutes. Until then, Roman law, like all common and customary law, had been codified only incompletely. Justinian's massive project was an attempt to arrest time. Like most great legislators, he hoped that his definition of the law would prove to be so authoritative that there would be no further need for lawyers to interpret it, no place for what he called "the vain discord of posterity."[82] He turned out, of course, to be wrong. His codification was only the beginning of a vast proliferation of later commentaries and interpretations that, from the eleventh century on, became the basis of all legal education and administration throughout Europe. "The vain titles of the victories of Justinian are crumbled into dust," wrote an admiring Gibbon, "but the name of the legislator is inscribed on a fair and everlasting monument. . . . The public reason of the Romans has been silently or studiously transfused into the domestic institutions of Europe; and the laws of Justinian still command the respect and obedience of independent nations."[83]

And "public reason" is indeed precisely what the Roman law was: civil law, law derived out of custom by rational induction. Initially the civil law had applied only to the peoples of Rome and, as Rome spread, of the empire as a whole. But the Romans also created a legal category called the *ius gentium*, or "law of nations." In practical terms, this was the part of the Roman civil law that was open to Roman citizens and foreigners alike. In a wider sense, however, it was taken to be what the second-century C.E. jurist Gaius called "the law observed by all nations." This concept would have a prolonged and powerful impact on all subsequent European legal thinking. As the European powers reached outward into other areas of the globe, many of which the Romans had never ever imagined, it became the basis for what is now called "public international law," and it still governs all the actions, in theory if never consistently in practice, of the "international community."

The Romans were also the first to devise laws to govern the conduct and the legitimacy of warfare. For the Greeks and indeed for most ancient peoples, warfare was a simple fact of need and survival. The most that Alexander had done to justify his conquests was to throw a spear into enemy territory and declare it to be his. No legal grounds had been required. The Romans, however, developed the notion, to which all Western peoples have subsequently clung, that all warfare must in some sense be defensive.[84] War was, at least in theory, always a means of last resort, which meant that it could be waged only in pursuit of compensation for some alleged act of aggression against either the Romans themselves or their allies. "The best state," as Cicero observed, "never undertakes war except to keep faith or in defence of its safety."[85] War was thus only a means of punishing an aggressor and of seizing compensation for damages suffered. From this basic assumption a theory developed whose moral force survives to this day, of the "just war." A just war conferred upon the aggressor a right to wage war. This was known as the *ius ad bellum*. Warfare itself was governed by a set of agreements about how it should be waged and the benefits the victor was entitled to derive from it. This was known as the *ius in bello*.

Most Romans, Cicero among them, fully recognized that Rome—like the United States and the Soviet Union in the 1950s and '60s—had frequently acquired clients with the sole purpose of "defending" them against enemies, real or imaginary, whose territories they wished to acquire. But the often devious uses to which the laws of war have been, and continue to be, put does not diminish the importance of their existence. And that we owe to the Romans.

For Roman jurists, however, the law was more than a set of rules by which the *civitas* was to be governed. It was also the supreme expression of human rationality. As Cicero wrote:

> However we may define man, a single definition will apply to all. . . . For those creatures who have received the gift of reason from Nature have also received right reason, and therefore they have also received the gift of law. . . . And if they have received law they have received justice. Now all men have received reason; therefore all men have received justice.[86]

"All men." By this Cicero meant not only those who were actually Roman citizens but potentially the whole of mankind, the inhabitants of what he called—in a phrase that would be quoted time and again by his Christian heirs—the "republic of all the world"— *respublica totius orbis*. For Cicero this universal human republic was the living embodiment of a common universal law for all mankind.[87] This concept, like so much else in Western thought, was initially an idea of Plato refined by Aristotle. But the form in which Cicero adopted it—and in which it became, in a sense, the ideology of the Roman Empire, at least under the Antonines—was a creation of another ancient school of philosophers: the Stoics, who were named after the open, colonnaded buildings in which they met.

Stoicism has perhaps been the most consistently influential of the ancient schools of philosophy. It is certainly the one that has left the most lasting impact on the cultures of the West. Although founded by Zeno of Citium in the early fourth century B.C.E. as a rigorous philosophical system, it came in time, like most such systems, to mean many things to many different people. It is best known in the figure of the Stoic wise man, he who looks upon all sufferings (and most forms of pleasure that are not purely intellectual) as external to himself, who cultivates an inner peace, a sense of himself that is beyond the range of harm others might be able to inflict on him. This the Greeks called *ataraxia*—the doctrine of freedom from anxiety and care. When we today speak of someone being stoical, this is roughly what we mean. (We also mean much the same thing when we say that someone is being "philosophical," which is an indication of how central Stoic thinking has been to the Western understanding of what "philosophy" is.)

But Stoicism was more than simple resignation. It also embraced the notion that the natural world was a harmonious whole with a distinct and transcendental purpose. And at the core of this belief lay a claim that all human beings, no matter what their cultures or beliefs, share a common identity as humans. This begins as parental love but soon reaches out, in the truly wise person, to embrace first family and friends, then members of the same community or nation, and finally the whole of humanity. As Cicero put it: "From [parental love] there originates also a form of natural concern shared by humans among humans, that simply on account of the fact that he is human, one human should be thought to be not alien from another."[88] It is this sentiment—called by the Greek Stoics *oikeiosis*—that lies at the heart of the Stoic rejection of the idea that humans should define themselves only by the kin or society into which they happened to have been born but that, instead, we should identify ourselves with the race as a whole. This is what today is called somewhat loosely "cosmopolitanism."

What this means is best expressed by one of the most compelling of the Roman Stoics, the emperor Marcus Aurelius himself. In a series of notes "To Myself"—written in Greek—which have come down to us under the title *The Meditations of Marcus Aurelius,* he reflected that

> As the Emperor Antoninus Rome is my city and my country; but as a man, I am a citizen of the world. . . . Asia and Europe are mere corners of the globe, the Great Ocean, a mere drop of water, Mount Athos is a grain of sand in the universe. The present instant of time is only a point compared to eternity.[89]

Zeno himself had already proposed just such a vision of the true destiny of the race. He told his followers:

> We should all live not in cities and demes [tribal groups], each distinguished by separate rules of justice, but should regard all men as members of the same tribe and fellow citizens; and that there should be one [*koinos*] life and order as of a single flock feeding together on a common pasture.

On the face of it this sounds very ecumenical. All mankind should live together as one regardless of race, place of birth, or nationality. It became the basis of Christ's vision of the new church, and the earliest English translator of the Gospels echoed Zeno, consciously or unconsciously, when he made Saint John demand that in Christ "there shall be one fold and one shepherd" (10:16).

But there is another, somewhat less ecumenical, way of understanding such sentiments. Zeno's words have survived for us only because they were recorded by Plutarch. And Plutarch bothered to repeat them only because, as we have seen, he saw Alexander the Great as embodying Zeno's "dream or, as it were shadowy picture of a well-ordered and philosophical community."[90] If all humanity were to be one, humanity should not, as the modern cosmopolitan would insist, reject the idea of belonging to any nation. Instead, all humanity should belong not to many nations but to only one. For Zeno, possibly, and certainly for Plutarch, that nation had been Alexander's empire. For the Romans it could clearly be only Rome or, more precisely, the Roman *civitas*.

Aristides had understood this well.

> You threw wide all the gates of the civilized world and gave those who so wished the opportunity to see for themselves; you assigned common laws for all and you put an end to the previous conditions which were amusing to describe but which if one looked at them from the standpoint of reason were intolerable; you made it possible for the peoples of the world to marry anywhere, and you organized all the civilized world into one family.[91]

By bringing these gifts to the whole of the inhabited world, Rome had not merely brought peace, posterity, order, and justice: it had, at least in Aristides' understanding, transformed humanity itself. As the British political philosopher Ernest Barker declared in 1923—before the "low dishonest decade" of the 1930s had precipitated Europe, and the world, into the most ferocious division of humanity that humanity has ever experienced—"the thought on which the best of the Romans fed, was a thought of a World-State, the universal law of nature, the brotherhood and the equality of men."[92]

Aristides ended his oration with a resounding vision of the world, before and after the rise of Rome. Before Rome the world had been, as the cosmos itself had once been before Zeus imposed a celestial order upon it, "full of strife, confusion and disorder." Now, under Roman rule, humanity has left the "Iron Age" in which it had previously dwelt. Now "Cities gleam with radiance and charm and the whole earth has been beautified like a garden. Smoke rising from plains and fire signals for friend and foe have disappeared, as if a breath had blown them away, beyond land and sea." The war triremes that once roamed the seas have given way to merchant ships. The gods have received their dues and thus "lend a friendly hand to your empire in its achievements and confirm you in its possession."[93] Every road now leads to Rome. The harbors are choked with ships. Goods come from as far away as India and Arabia. "All meet here, trade, shipping, agriculture, metallurgy, all the arts and crafts there are or ever have been, all the things that are engendered or grown from the earth. And whatever one does not see here neither did nor does exist."[94]

The Antonines had embodied the Stoical imperial vision as no other emperors had or would again. Rome, when Aristides stood there that day in the Athenaeum, was in all respects a universal society, open to all who would accept the rule of the Caesars. It was largely indifferent to variations in customs and paid no heed to differences in religion as long as these did not interfere with the proper respect due to the emperor. But as with all golden ages, this one, too, was soon to be extinguished. With the accession in 180 C.E. of Commodus, who renamed all the months, and even for a while the city of Rome, after himself and who became obsessively concerned with performing as a gladiator (although at no immediate danger to his own person), the bad old days of the worst of the Julio-Claudians returned. Commodus was strangled in December 192 C.E. and his memory at once condemned. But in all important respects the empire of the Antonines had gone, never to return. Commodus was succeeded by Septimius Severus, who for eighteen years restored order to the empire and maintained the peace. When he died in York in February 211 C.E., he told his sons "not to disagree, give money to the soldiers and ignore the rest." Crass, instrumental, and ultimately unable to sustain a world that looked to far more than military power to hold it together, this was, in its own way, the tragic epitaph of Aristides'

"Oration." It was also ignored. The army effectively seized control of the state, and from 211 C.E. until the accession of Diocletian in 284 C.E. there were more than seventy contenders for the throne. "The rise and perpetual transitions from the cottage to the throne and from the throne to the grave might have amused an indifferent philosopher," wrote Gibbon, "were it possible for a philosopher to remain indifferent amidst the general calamities of human kind."[95]

In 260 C.E., the emperor Valerian and all his general staff were captured by the Parthian warlord Sapor I. In Gaul, Britain, and Spain, a breakaway "Gallic Empire" was created by one of the commanders of Valerian's son, Gallienus, aptly named Marcus Cassianius Postumus. Another, under the legendary queen Zenobia, with its capital in the oasis city of Palmyra in what is now Syria, controlled much of Asia Minor from 267 to 272 C.E. Palmyra was Armenian, but its peoples, who came from all over Asia, looked, to Roman eyes, all too much like Persians; and Zenobia, "Mother of the King of Kings" as she was called, in unmistakable Oriental fashion, resembled nothing so much as yet another would-be successor to the Achaemenids. Zenobia had never intended to separate Palmyra from Rome. Her ambition, after the succession of the emperor Aurelian in 270 C.E., had been to seek a division of the empire, with Aurelian as emperor in the West and her own son, Whaballat—known to the Romans as Septiminius Vaballathus—in the East. Early in 272 C.E., Aurelian marched east and defeated Zenobia at Antioch. She then withdrew south and proclaimed her son sole emperor and Augustus. By the summer, however, Palmyra itself had fallen to the victorious Roman army, and Zenobia, caught while seeking help from the dreaded Persians, had been captured.[96]

Aurelian was a successful and efficient ruler who succeeded in uniting the empire, but the effects of his reforms were short-lived. The Roman *civitas* would emerge from its nightmare of precipitous decline only in 312 C.E. under the iron hand of Constantine the Great. But by then it was shrunken and divided, and although it would survive for nearly another century in the West and more than a millennium in the East, it had become quite another world.

Yet for all that, the dream of universal citizenship that the Antonines had most clearly embodied would survive. It would survive the rise in the East of the Sassanids, who, like the Parthians, represented themselves as the heirs of the Achaemenids and who had hoped to finish off what Xerxes had begun. It would survive the forces of the Huns, Visigoths, Ostrogoths, Vandals, and Mongols. It would survive even the final dissolution of the empire in the West. It would survive, as it survives to this day, as one of the most powerfully enduring features of European, "Western" civilization. But what in the end would carry it out of the Roman Empire and into the fragmented disordered states that would inherit the cultural mantle of the "republic of all the world" was a new and, as with so much else in this story, originally Asian, religion: Christianity.

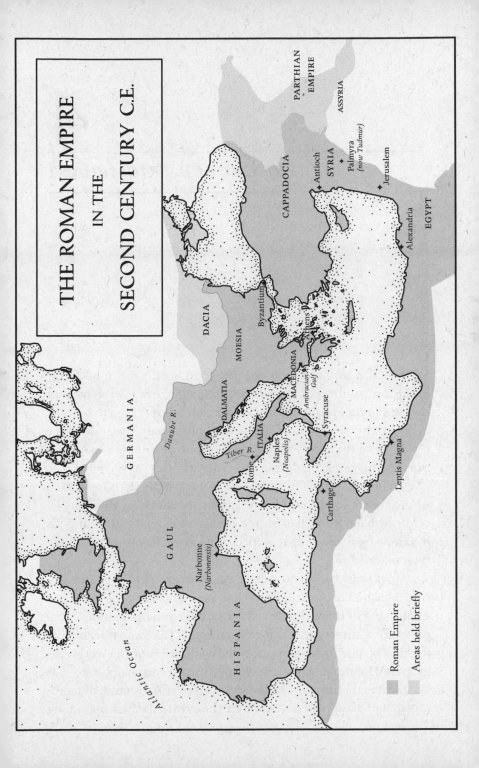

THE ROMAN EMPIRE

IN THE

SECOND CENTURY C.E.

PARTHIAN EMPIRE

ASSYRIA

CAPPADOCIA

Antioch
SYRIA
Palmyra
*(now Tudmur)*
Jerusalem

EGYPT

Alexandria

DACIA

MOESIA

Byzantium

Athens

DALMATIA

Danube R.

GERMANIA

MACEDONIA

Ambracian Gulf

Syracuse

ITALIA

Tiber R.

Rome
Naples
*(Neapolis)*

Leptis Magna

Carthage

Narbonne
*(Narbonensis)*

GAUL

Atlantic Ocean

HISPANIA

Roman Empire

Areas held briefly

# 4

## *THE CHURCH TRIUMPHANT*

**I**

In 413 C.E., a fifty-nine-year-old Christian bishop sat in his study in the Roman town of Hippo on the North African coast, in what is now Algeria. He was writing a book that was destined to become perhaps the single most influential Christian text, after the Gospels and the letters of St. Paul. The bishop's name was Augustine, and his book, his "great and arduous work" as he called it, is known in English as *The City of God*. It was a reply to those pagans who claimed, loudly and bitterly, that the disasters that had recently befallen Rome had been a punishment by the ancient gods for allowing a new vulgar, superstitious cult from Judea to take their place.

Three years before Augustine began writing his book, the unthinkable had happened: Rome had been sacked by the armies of Alaric, king of the Visigoths, who had done much the same thing to Athens fourteen years before. To the poet Claudian it seemed as if a new plague had fallen upon the civilization of the world, led by a new Hannibal.[1] For three days, Alaric's troops had behaved as Goths were traditionally believed to behave: they looted, pillaged, raped, and slaughtered. Although the great city had for a long time been exposed to barbarian incursions and had already been starved into cannibalism twice in the past two years, that had been as nothing compared to what took place during three intolerably hot August days.[2] Pelagius, a visiting monk from Britain who saw it all, described how as "the blaring of trumpets and the howling of the Goths" shattered the air, Rome, "the mistress of the world, had shivered, crushed with fear." All the natural order that had existed in the world, that had divided the lesser from the greater, that had maintained hierarchy and status, that had preserved families and ensured that time would flow on uninterrupted—all of this, of which Rome had stood for centuries as the ultimate guarantor—had now perished in three short days. Barbarism had penetrated to the very heart of the civilized world. "Where were the certain and distinct ranks of dignity?" wailed Pelagius. "Everyone was mingled together and shaken with fear; every household had its grief and an all-pervading terror gripped us all. Slave and noble were one. The same specter of death stalked before us all."[3]

In the months following the sack, thousands of refugees flooded southward. Many of these, for the most part angry and resentful patricians, turned up in North Africa, demanding an explanation for the disaster. Saint Augustine set out to offer them one. God, he argued, had not unleashed his fury against Rome because, as the Christians believed, so many of its inhabitants were still loyal to ancient pagan deities; nor had Rome been punished, as the pagans for their part believed, by the ancient gods for abandoning them. Certainly Rome had been punished for its sins, but then so, in time, would all mankind. Rome, said Augustine, had been "shaken but not transformed," as it had been "at other periods before the preaching of Christ's name." There was, he consoled his readers, thus "no need for despair of its recovery at this present time."[4] The city may have fallen into the hands of the barbarians, whom Augustine clearly looked upon as the least of all human beings. But it would survive. It was, it had to be, eternal, because it had now become synonymous with Christianity itself.

For Augustine, as for most subsequent Christians, the civilized world had moved inexorably from Greece to pagan Rome and thence to Christianity. At each step it became more universal; at each step it came closer to the values not only of the only true religion but of the only virtuous, sustainable, equitable, and just way of life. For Aristides, the Roman world of the Antonines had been the culmination of a steady growth of civilization. The march of empire westward from Achaemenid Persia to Hellenistic Greece to Rome itself had finally come to rest by the banks of the Tiber. For Augustine, this history had been but the prelude to the universal kingdom that Christ had ushered in and that would now flow on in unbroken time until the second—and last—coming. As God had sent his only begotten son to Earth during the reign of the first of the true emperors, Augustus, God must clearly have intended the Roman Empire to last for as long as his creation would last. The prophet Daniel, as the Christians rarely tired of pointing out, had already foreseen this. In Daniel's vision, four world empires, represented by four beasts—a lion with eagle's wings, a bear, a leopard, and a monster with "great iron teeth and ten horns"—would succeed one another until "the God of Heaven shall set up a kingdom which shall never be destroyed."5 That kingdom, the kingdom of Christ's followers on Earth, which Augustine called "The City"—the *civitas*—"of God," was now at hand.

Christianity had been a heresy of Judaism, although few at the time would have admitted as much. Like so much else that became a defining part of the Western world, it, too, had begun in the East. The Christ who was inescapably present in the words of the Gospels was, in most respects, a typical Eastern holy man, and some of the deeper sources of the most persistent of Christian beliefs—the slain god, the virgin birth, the incarnation—were more Asian still. Despite the Church's subsequent attempt to gloss over the ancestry of its founders, the Asianness of Christ returned persistently to haunt it. Much later, in the mid-seventeenth century, Samuel Purchas, the chronicler of the European voyages of exploration, with his characteristically quirky insight into the historical formation of Europe, could still say, "Jesus Christ, who is the way, the truth, and the life has long since given the Bill of Divorce to ingrateful Asia where he was born and of Africa the place of his flight and refuge, and has become almost wholly European."[6]

CHRIST'S OWN IMMEDIATE followers, however, and in particular that Roman *cives* Saint Paul, had wasted no time in attempting to transform a world-denying cult centered upon a wandering prophet into the state religion of a succession of world empires. They eagerly embraced both the political protection of the empire and its universal ambitions, and transformed it for their own ends. For the pagan Pliny it had been the numen of the gods that had been responsible for Rome's bid to "give humanity to man." For the Christians it was the will of their God.

As the fourth-century Christian panegyrist Aurelius Prudentius wrote:

> God taught nations everywhere to bow their heads beneath the same laws and all to become Roman. . . . A common law made them equals, bound them by a single name and brought them, though conquered, into bonds of brotherhood. We live in every conceivable region scarcely different than if a single city and fatherland enclosed fellow citizens with a single wall.[7]

By the time Augustine sat down to explain to pagans and Christians alike why God had allowed the Eternal City to fall into Alaric's unforgiving hands, Rome had been in most respects a Christian empire for more than a century. All that was captured by the Greek word *paideia*—classical learning, classical philosophy, the ethics of Plato and Aristotle—which was already being used to prop up the somewhat meager moral teachings to be found in the New Testament, could, in Christian eyes, be preserved only within and through the Roman world. As Augustine had told his friend

Paulinus of Nola in 408, the Heavenly City would be one where Christians would all be "not passers by, not resident aliens, but full citizens." The Roman language of citizenship extended now even to the Kingdom of God.[8] It was the true, the final "translation of empire" that Aristides could never have foreseen.

Augustine, however, was aware that he was looking back over at least two centuries of a deeply troubled history. For if the Christians had been willing to embrace Rome, Rome had been rather more reluctant to embrace them. The rootlessness of Christianity, the fact that it was precisely a universal sect, which showed minimal respect for ancestors and had no firm, local attachments, no *patria,* horrified most pagans. Christianity was also, as are all monotheist religions, intolerant. There can be only one god, only one way of venerating him, and only one way of understanding the place that humankind has in his creation. The pagan Romans, by contrast, were prepared to tolerate all manner of religions, as long as they did not directly offend the gods of Rome. There was also the question of the cult of the emperor. Veneration of the divine Augustus amounted to little more than an oath of allegiance to the state. But for Christians it was a blasphemy, to which they would not submit. Pagans were also revolted by the Christians' seeming delight in martyrdom. However much Christians themselves might choose to interpret this as evidence of "great *virtus* [of a] miraculous power in us," to most Romans, and pagans generally, it displayed merely a callous disregard for wives, children, parents, and friends. "Driveling idiocy," Lucian called it.[9] Yet despite persecution—although there were never as many victims as Christian apologists liked to make out—and despite the increasing moral and cultural gulf that divided Christianity from paganism, the number of converts to the new religion continued to increase.[10]

Why is still something of a mystery. As the religion of an empire, Christianity, with its insistence on turning the other cheek, on renunciation and forgiveness, on the final triumph of the weak and the poor over the powerful and the successful, would not have seemed to have had very much going for it. True, it promised instant rebirth to its adherents and some kind of proximity to God himself, something that was entirely absent from paganism. But it was not alone in that. Its chief rival was Mithraism, a Persian-Hellenistic mystery cult that by the middle of the second century C.E. had spread from the borders of the Persian Sassanid Empire throughout almost the entire Roman world from the Black Sea to Britain and from Egypt to the Rhine. Mithras was a god of contract and loyalty and thus particularly congenial to the Roman elite (although they never seem to have supported it with public funds). It was also closely associated with both Apollo and the Sol Invictus, "the Invincible Sun," both gods whom, as we shall see, Constantine the Great had worshipped and continued to tolerate even after his spectacular conversion to Christianity. It also had some clear associations with Zoroastrianism. Mithraism would seem to be far more obviously appealing than Christianity. It was sufficiently multiple in form to appeal to pagans, who had little time or taste for monotheism, and it fervently supported the divine nature of the emperor. It also had the attraction, which it shared with Christianity, of being a mystery cult: anyone who wished could join, but had to be initiated, and initiation set the initiate apart; it offered him, and also crucially, her, membership of an elite.

## II

There was another reason for, if not the triumph of Christianity, then at least the demise of paganism. The world in which the early process of Christianization took place was slowly, but irrevocably, coming to pieces. Gone now was the Golden Age that Aristides had believed to be the end of history. Already by 200 a serious trade recession had hit the Mediterranean. In the middle years of the century, the Roman legions had suffered terrible defeats at the hands of the Sassanids, the Goths, and other Germanic tribes, and civil war had brought the imperial government to the verge of disintegration.[11] "My kingdom is not of this world," Christ had said. As the greatest kingdom that was of this world, the seemingly indestructible, eternal Roman imperium, came to pieces, the prospect of another kind of kingdom offering the sanctity of bliss in a life ever after—something largely absent from paganism—clearly had immense appeal.

By the end of the century, Rome's mighty—and overextended—empire had become impossible to hold together as a whole. In a last attempt to keep what still remained intact, the emperor Diocletian divided the empire into western and eastern halves, and created two emperors, Maximian in the East and Constantius in the West. When Constantius died at York in 306, he was succeeded by his son Constantine. After a complex series of dynastic maneuvers, however, Maximian's son Maxentius seized hold of most of Italy, including the imperial capital itself.

In 312, in a bid to recover what he now looked upon as his hereditary rights, Constantine invaded Italy. Having defeated Maxentius's advance guard near Turin and then again at Verona, he marched south to Rome. Maxentius himself, at the head of a far larger force than Constantine's, now prepared to meet him face-to-face. On October 28, the two armies camped at Saxa Rubra near the Mulvian Bridge—now known as the Ponte Milvio—which carries the Via Flaminia across the Tiber to the north of Rome. Years later, Constantine told his biographer, Eusebius of Caesarea, upon oath, that while planning his campaign against Maxentius he had become concerned about his rival's supposed skill as a necromancer (Constantine was nothing if not superstitious). At this point Constantine had adopted the sun god Apollo as his tutelary deity and, for good measure, also favored the cult of the Invincible Sun. He prayed eagerly to these deities for assistance. One afternoon while drilling his army, he saw a cross of light in the sky superimposed upon the sun with the words "In this Sign you shall conquer" wreathed about it.[12] Constantine had prayed to the Sun, and the Sun had sent him a sign: the symbol of the still disreputable Christian religion. Some years earlier, he had received a not dissimilar message from the god Apollo, who had offered him a laurel wreath and promised that he would reign for thirty years. Whether this new apparition meant that the Christian god was a manifestation of Apollo or that the Sun was a symbol of the power wielded by Christ was immaterial at this stage. What was clear was that the god of the Christians had promised him victory in his name. On the eve of the battle Constantine also had a prophetic dream, as commanders in the ancient world often did. In this Christ himself appeared to clarify any misunderstandings. He had with him a sign and ordered Constantine to make a copy of it and carry it into battle on his standard. This became the famous Labarum. As Eusebius describes it, it consisted of a tall pole with a crossbar plated with gold and with a monogram composed of the first two letters of Christ's name in Greek, the *chi* and the *rho*. To reinforce the magic, Constantine also had it painted on his soldiers' shields and on his own helmet.

When the two armies met on the following day, Maxentius's forces, despite their superiority and the magical powers of their leader, were annihilated. The slaughter was terrible, and the panic-stricken mob, which included Maxentius himself, attempted to retreat back across the bridge to the comparative safety of the city. In the confusion, Maxentius was pushed over the edge and drowned in the murky waters of the Tiber. Constantine entered Rome in triumph, recuperated Maxentius's body from the mud, cut off the head, and carried it through the streets on a lance. The Senate obediently condemned the memory of the tyrant and duly elected Constantine senior Augustus. Thereafter he became known to all his Christian panegyrists as "Constantine the Great."

Constantine had made himself master of the Western Roman world. The Christian god had kept his promises, and, in gratitude, Constantine immediately set about rewarding his followers. In 313, together with Licinius, who was then the Augustus of the Eastern Empire, he issued an edict at Milan that restored to the Christians all the buildings that had been seized or damaged during the intermittent persecutions that had marked previous reigns. The "Edict of Milan," as it came to be called, also set out in forthright terms a policy of universal religious freedom. "No one whatsoever," it ran, "should be denied freedom to devote himself either to the cult of the Christians or to such religions as he deems best suited to himself." Despite the broadly tolerant sweep of this decree, it was already clear that Christianity was poised to take over the Roman state, as Saint Paul had always hoped it would. And when that occurred, the freedom to choose one's religion guaranteed by the second part of the edict would be silenced forever. In the years that followed, Constantine granted exceptional privileges to the clergy and showered benefactions on the Church. He built a massive basilica—a word that means "royal hall"—at Rome for the pope, on the site of the barracks of the Praetorian Guard, and laid the foundations of the church of St. Peter, which was to become the center of all (Western) Christendom. He built another at San Giovanni in Laterano. He created a huge golden-domed basilica at Antioch and built the Church of the Holy Sepulcher at Jerusalem. The Christians responded to this new regime with understandable enthusiasm. Provincial bishops attached themselves to the imperial court. The theologian Lactantius—one of the most virulent scourges of the ancient religion—became tutor to Constantine's son Crispus. Eusebius of Caesarea became not only the emperor's official biographer but also the earliest historian of the Church, with an enthusiasm for the Roman world unmatched since the days of Aristides.

Constantine's conversion was not as many later Christians have presented it, nor as many have expected it should have been. He may have professed himself a Christian, but he never displayed any interest in Christian ethics, a Christian way of life, Christian forms of worship, or indeed anything to be found in the Gospels. He banned crucifixion and the public branding of captives on the face, since this offended the Christian notion of the integrity of the human body. He may even have put a stop to pagan sacrifices, and he certainly forbade any sacrificial victims to sully the imperial cult, which survived, but now only as a secular celebration of a dynasty. But little else.

For Constantine, however, Christianity was, above all, a means to unite a divided world, and it is not insignificant that in a letter he issued to the peoples of the Eastern Empire in 324, he described it simply as "the Law."[13] He was baptized only on his deathbed, and although this was not uncommon at the time, he certainly never attempted to suppress paganism with the zeal that most Christians, avid for revenge for centuries of harassment, would have wished. He accepted pagan honors from the citizens of Athens. He paid the traveling expenses of a pagan priest who wished to visit the pagan monuments of Egypt and greeted one pagan philosopher as his colleague. Three years after his victory over Maxentius, a triumphal arch was erected in Rome to mark the event. The Arch of Constantine, today one of the best preserved of all the triumphal arches in the city, displays no trace of Christian imagery. When in March 321 Constantine decreed that on "the venerable day of the Sun" the law courts and all workshops should be closed and the urban population should rest, he still seems to have been in some uncertainty about the relationship between the Invincible Sun and the Christian God—and today in English and in German, God's day, the day of rest, the day dedicated to his worship, is still the "Sun's day."

Constantine decreed a strict toleration of the old beliefs. "Let all those who are in error," he reassured the pagans in 324, "enjoy peace. Everyone shall preserve what his soul may desire and no one shall torment anyone."[14] He kept his promises, and paganism remained a potent, if dwindling, presence at the center of Roman life in the West until 410. In the East, where it drew more direct inspiration from its Greek sources, it continued to be the religion of a staunch collection of intellectuals and administrators until the end of the sixth century, when the emperor Justinian, the great architect of the Roman law, began a program to convert all the remaining pagans, as well as the Jews. It is something of a tribute to this late Roman policy of toleration and persuasion that while persecution of Christians by pagans, of Christians by Christians, of Jews by Christians, of Muslims by Christians, and of Christians

by Muslims rarely produced any significant long-term consequences, by the end of the sixth century paganism was, in effect,
extinct throughout the whole extent of the former Roman Empire, both West and East.

But for all that, Constantine was no mere "god-fearing emperor" who merely tolerated Christianity and sometimes courted
the Christian clergy, as some of his predecessors had been. He was
a true "crowned Christian apologist."[15] To make the point plain
to all, shortly after the battle of the Mulvian Bridge, he had a
statue of himself erected in Rome holding in his right hand a
cross. On the base was an inscription that, according to Eusebius,
ran, "By this sign of salvation, the true mark of valor, I saved your
city and freed it from the yoke of tyranny, and moreover having
freed the Senate and the People of Rome, restored them to their
ancient glory." The first part, the association of Christ with military valor, sounds suspiciously pagan, but the second reiterates
what would for later generations become the only possible understanding of human history: that Christianity had restored Rome
and through Rome Christianity had come to embrace the world.[16]
His conversion, Constantine claimed, had saved the empire. It
had, as Lactantius and Eusebius argued so fiercely, made Christianity, in fact as well as in theory, the sole protector of the ancient
classical order in a new world. What Cicero had imagined as the
"republic of all the world" had now become—in a phrase coined
much later by Pope Gregory the Great in the sixth century—the
"Holy Republic."

In 324, Constantine forced Licinius to abdicate. In so doing,
he reunited the eastern and western halves of the Roman Empire.
This was still ruled from Rome and was still a Latin empire. But
there was no escaping the fact that Christianity was, in origin, an
Asian religion, and by the fourth century it had already dug deep
roots for itself in ancient Greek philosophical culture. (There is
much truth in the modern claim that Christianity is little more
than "hellenized Judaism.") Perhaps in recognition of this, and
because his position in Rome itself had always been precarious de-

spite his victory over Maxentius, Constantine began to construct a new city, a "New Rome," as it came to be called. He chose the site of the Greek town of Byzantium on the shores of the Bosporus, and thus on the ancient frontier between Europe and Asia, and called it—in Greek—Constantinople, "Constantine's city." He rebuilt the hippodrome of the original settlement in the likeness of the Circus Maximus in Rome, and, as Saint Jerome sarcastically remarked, denuded every other city of the empire of its statues in order to adorn its streets and squares. When all this was dedicated on May 11, 330, Constantine had a gold coin struck to mark the occasion. On it he was represented, appropriately, in the pose of Alexander the Great. A new Greek monarch had come to defend the civilized world from its foes and—once again—to unite Europe and Asia.

Constantine became a Christian because he saw in Christianity a means of uniting a deeply fractured state. His choice of creed, however, displayed very little understanding of how intolerant and factious Christians were capable of being. During his reign he attempted, without much lasting success, to resolve the differences between the main body of the Church and two splinter groups: the Donatists and the Arians. The first of these were the members of a puritanical church of martyrs in North Africa. The second were followers of Arius, an Alexandrian cleric, who argued that, as Jesus was the son of God, he must be distinct and apart from God as his father. God, furthermore, had existed throughout eternity, whereas Jesus had had a precise place and time of birth, and therefore there must have been "a time when he was not." For many within the Church, the Arian controversy was a source of great unease, not least of all because it pointed to some serious logical flaws in the conception of the Trinity.

As far as Constantine was concerned, however, the moral and theological squabbles that divided the new religion were literally without significance. What mattered was that the Church should be one and closely identified with the state, and that the state should be its undisputed master. In 325 he summoned all the

bishops of the Church to a council at Nicaea. It was the first council of its kind, but, unlike those that followed, it was presided over
not by a priest but by the emperor. It issued the first formal doc-
trinal statement, known as the Nicene Creed, which created a
Church-wide orthodoxy, offered some kind of answer to Arius's
seemingly irrefutable reading of scripture, and condemned Arianism as a heresy. By convening a council of Church leaders and con-
firming their decrees, Constantine had established himself as the
supreme authority over the Church. He claimed the right to adju-
dicate in its affairs and to summon councils of bishops if he so
chose. He claimed the right to exile clergy, seize churches, and ban
religious meetings if they struck him as potentially divisive. He
had made himself head of a Church that was to be universal,
"Catholic" as it came to be called, an empire like Rome itself, and,
like Rome, eternal.[17] From now on the concept of a Western civilization, which had been associated first with Alexander and then
with the Roman world empire, would be transformed from a
sense of shared, but also fluid, values into a credo.

## III

The vision of the Church as one and universal was clearly in keeping with both Christ's fragmented and uncertain teachings and
with Saint Paul's more robust and programmatic interpretations
of their theological and social implications. But Paul's universalism in particular—and it was Paul who was the true founder of the
Catholic Church—was not based on any political understanding
of how the future might unfold. The Christian "new man" was a
true cosmopolitan. He lived anywhere, under any kind of ruler. He
was united to his fellows not by common custom or a common
law but by his fellowship in Christ. Pauline Christianity was a
spiritualization, or reification, of the Stoic dream of a single order
in the world. Under the new order, difference itself would vanish
into the brotherhood of the God who was also the son of Man.
This "man reborn," Saint Paul informed the Colossians, the in-

habitants of a city on the banks of the river Lycus in western Asia Minor,

> which is being renewed unto knowledge after the image of him that created him, where there cannot be Greek or Jew, circumcision and uncircumcision, barbarian, Scythian, bondman, freeman: but Christ is all and in all. (3:11)

The universality that had once been embodied in the Roman notion of citizenship would now be carried by the Christian idea of the community of all believers. But by making the body that would now hold the *oikoumene* together a spiritual one, Saint Paul denied the Christian new man any political or even social identity. Entry into the Church implied no change in status for the neophyte. On this earth the bondman would retain his bonds, the freeman his freedom. Even slavery, the most extreme instance of social difference, remained unchanged by Christianity. Slavery, Augustine had argued, was clearly a punishment for some kind of sin. The Christian who found himself enslaved might not know quite what he had done, but the fact of his enslavement was sufficient proof that he had done something. Nothing could change that, and the slave consequently could not even claim a natural right to attempt to free himself.

The meeting of all these different ethnic, religious, and cultural groups—the Scythians, the Jews, the Greeks, the barbarians, and the Romans—would take place on an entirely different, and more elevated, plane. Paul had been a good Roman citizen, Christ an obedient Roman subject. Neither had chosen to defy the power of Rome; neither had seen any future for their creed outside Rome. Both had also drawn a clear distinction between the Church and the state, between the spiritual and the secular. When asked by the Pharisees, in the expectation that he would betray himself, whether Jews should pay taxes to the Roman state, Jesus asked to be shown a Roman coin. He was handed one with an effigy of the deified Tiberius. "Whose is this image and superscription?" he asked. "Caesar's," they replied. Then he retorted, "Render therefore unto Caesar the things which are Caesar's; and unto God the things that are God's" (Matthew 22:21). It was a clever answer. And it got Jesus off a potentially rather nasty hook. But it is also one of the most significant utterances in the Gospels. It was intended by Christ to indicate just how clearly his kingdom was "not of this world." He had no quarrel with Caesar because Caesar made no claims upon the state of his soul. Jesus himself, or whoever uttered these words, could have had no inkling of what the religion created in his name would become. Nor, of course, could he have predicted quite how his simple statement would shape its future.

For Augustine the matter was one of the spirit. The Christians had concluded the historical process the Greeks had begun. But in doing so they had also manifestly changed it. For the pagans, Greek and Roman, were creatures of only one kind: humans, with virtues but devoid of spiritual merits and, faultless though many of them may have been, unredeemed by Christ. They were citizens of only one world. Since the coming of Christ, however, there had existed two worlds or, as he expressed it, two cities. There was the earthly city, which had been founded by Cain, in which all humans are compelled to live out their days; and then there was the "City of God," which is the Church of Christ and its members here on Earth. The word *civitas* described them both. There was no other language in which to capture the preordained unity of God's anointed. But the two, the secular and the spiritual, were for Augustine, although inescapably interrelated, also irremediably separate, at least until the Second Coming dissolved all human history. The "Earthly City," of which the Roman Empire had been the supreme manifestation, although it was ultimately destined to founder, was a necessary presence in a world that, because of original sin, would otherwise have been one of unrelenting violence and disorder. For Augustine, the Roman Empire would not, as Aristides had so passionately declared, last for eternity; only the City of God would do that. But human, frail, and transitory though it was, it was still a far better polity than its predecessors had been, and it would survive as long as it remained true to the virtues on which it had been founded. The Romans, he said, had been given "the earthly glory of an empire which surpassed all others" as a reward for

> the virtues by which they pursued the hard road which brought them at last to such glory . . . They disregarded private wealth for the sake of the common wealth. They stood firm against avarice, gave advice to their country with an unshakeable mind and were not guilty of any crime against the laws, nor of any unlawful desire.[18]

The Heavenly City, by contrast, is eternal, "for no one is born there because no one dies. This true felicity which is not goodness, but the gift of God . . . In that City the sun does not shine 'on the good and on the evil'; the 'sun of righteousness' spreads its light only on the good." Augustine was concerned solely with the values these two cities embodied, not the administration of their respective spheres of jurisdiction. But by making such a stark and insistent division between the inner and outer lives of humankind, something that was largely unintelligible to the pagan world, *The City of God* created a breach between the temporal and the spiritual that would crystallize into a bitter and lasting struggle between the heirs presumptive of the two cities: the pope and the emperor; a battle that would endure until the secular rulers of Europe finally succeeded in wresting authority from the churches on their territories. Later, as we shall see, this division would have lasting consequences for the subsequent history of the struggle of the West with a new—and terrifying—power from out of Asia.

When Augustine sat down to write *The City of God,* he was witnessing the beginning of the end of the Roman Empire in the West. He was also aware of another kind of division, one within the City of God itself, although he could never have foreseen the consequences it would have. In 395, the Roman Empire, fitfully united under Constantine, was once again divided, and this time for good. In the West, the eastern part came to be called, after the site on which its capital now stood, the "Byzantine Empire." It was richer, more powerful, and more extensive and would last far longer than the frequently anarchic West, which would steadily decline into a series of quarrelsome and backward petty kingdoms.

By the time Alaric's Goths appeared on the frontiers of the empire in the West, Constantinople had, indeed, become the New Rome, the largest city west of China, with a population of half a million and a mile and a half of wharves to dock the grain ships needed to feed them. Even Antioch and Alexandria, the second cities of the Byzantine Empire, were by then as large as Rome itself. Unlike the crowded huddled cities of the West, Constantinople was crossed by the kinds of grand avenues and wide squares that Rome had once had before tenement buildings had filled them in. Its palaces were not, as were most of those in Rome, partially demolished ruins, but magnificent buildings with arcades and colonnades, inner gardens, and fountains. And the city's inhabitants were as sumptuous as its buildings. "The Greeks," wrote one later Arab visitor in obvious awe, "are very rich in gold and precious stones. They dress in silk garments, fringed with gold and embroidery; to see them with their accouterments, mounted on their horses, one might think that they were all the sons of kings."[19]

Inevitably the culture that developed in the East became increasingly distant from the society that at the same time was slowly emerging in the Latin West. Byzantium would always look upon itself, until the Ottoman Turks finally swept it away in 1453, as the empire of Rome and its inhabitants as Romans. But the language of the court and of religion—although not, significantly, of the law—became increasingly Greek and the entire society increasingly hellenized. The Byzantine Basileus was also far closer to a Hellenic or Persian priest-king than his Western counterpart. In consequence, an elaborate court grew up around him, which from the perspective of the rougher, simpler, "barbarian" West seemed at times to be indistinguishable from the splendors of an Asian potentate: luxurious, ornamental, and intricate.

Much of Byzantium also lay geographically within what, since antiquity, had been thought of as Asia. On its eastern borders lay the other superpower of the late antique world: the Sassanid Persian Empire. And as with all neighbors, however hostile to each other, there existed a good deal of traffic back and forth across frontiers that were fluid and uncertain. By the early seventh century, many of the Eastern Christians actually lived in what is now Iraq and thus politically under the authority of the Sassanid king of kings. To those in the distant Latin West the Greeks seemed to have more in common with their formal enemies than with their barbarian coreligionists beyond the Bosporus. It is not for nothing that the modern word "Byzantine" tends to mean not so much something or someone from Byzantium but something that is convoluted, distorted, overcomplex, and probably also mendacious.

THE BRAND OF Christianity that evolved in the East rapidly began to differ from the one that emerged in the West. The differences were many, but one of the most marked, and the one that in the centuries to come would stamp itself across the cultural and political landscapes of both East and West, was the relationship between Church and state, between the sacred and the secular.

For Constantine the Great it had been obvious that Christianity, like paganism, like all forms of religious observance, was to be an instrument of government. The Romans, like the Greeks before them, had long understood this. Whatever they might personally think about their gods, whatever they might believe or disbelieve, piety was as much a political duty as a religious one, which is why the heroic founder of Rome, Aeneas, is described by Virgil as "outstanding in arms and piety." The early Roman Christian state saw no reason to change any of this. The pagan emperors had literally been gods—although they had had to wait until after their deaths to be deified. Their imperium had been not, as it had been for the Roman Senate, a merely mundane right to rule but a quasi-mystic power reserved for them alone. Neither Constantine nor any of his successors could claim such divinity— something of which the bishops often had to remind them. But if they were never more than mortal, they were nonetheless always more than mere men. And if they could not deify themselves, they did the next best thing: they made themselves God's representatives on Earth.

The pagan emperors had always held the title of "Supreme Pontiff"—Pontifex Maximus—a title that even Constantine had assumed. With the final triumph of Christianity this, too, had fallen into disuse, but the Byzantine emperor continued to look upon himself, and to be recognized by the ecclesiastical authorities, as the ruler of the inhabited world—the *oikoumene,* a word that now came increasingly to be identified with the Christian community—and thus God's living viceroy on Earth.[20] The emperor was crowned by a patriarch, whom he himself had chosen. He could, unlike other mortals, enter the sanctuary. Unlike other mortals, who throughout Christendom until the Reformation partook of only Christ's body at Communion, the Basileus took Communion in both kinds—both the body and the blood of Christ—like an ordained priest (which he was not). On certain feasts he preached the sermon in the great basilica of Hagia Sophia, the Church of the Holy Wisdom. As God had his palaces on Earth, so, too, did the emperor. In Theodosius II's Law Code of 438, the whole imperial palace, down to the stables, was declared to be sacred. The emperor was also the sole source of law. In the hallowed phrase of the Emperor Justinian, he was "not bound by the law," which was always the expression of his "frank goodwill." And that law was not only civil, it was also ecclesiastical.

This did not mean that there was no tension between Church and state, between patriarch and emperor. As in the West, the

Church maintained its independence. And since, as in the West, the Church interpreted God's laws and intervened on his behalf, its authority could only, in the opinion of most churchmen, be superior to the authority of the state. "The domain of royal power," declared Saint John Chrysostom bluntly, "is one thing and the domain of priestly power another; and the latter prevails over the former."[21] In theory, at least, Byzantium was ruled as a diarchy— a government with dual authority—which was frequently compared to the two parts of the human person: the soul and the body. The emperor chose the patriarch, but the patriarch could, and did, demand a declaration of orthodoxy from the emperor before he would crown him. He could also excommunicate him. Leo VI was excommunicated in 906 and Michael Paleologus in 1262. But for all that, the always fine line that divided the things of Caesar from those of Christ was much finer in the East than it was in the West. The "Great Church" continued until the bitter end to exercise its independence and to assert its authority over the laity.

## IV

In the West, the history of the Church and of what remained of the Roman Empire followed a very different course. By the time Alaric's Goths entered Rome, the empire was already fragmented and evidently in an advanced state of decline. What remained of Rome after the sack revived briefly, but when in 476 the German Odoacer deposed the last of the emperors, Romulus—referred to contemptuously as "Augustulus," or "Little Augustus"—the Western Empire was finally extinguished. Gradually the former Roman *civitas* was transformed into a succession of fiefdoms, principalities, duchies, city-states, and bishoprics. All of these were nominally Christian, and although the brands of Christianity they practiced were often contentious, all, except for brief periods of rebellion, were bound, spiritually at least, to the papacy, the one surviving international power of any significance. The pope was a prince in his own right, the titular head of a secular state that occupied large areas of southern and central Italy. But he was also

the leader of a religious community, which had always claimed
that one day it would cover the entire globe. Because of that, all
that remained of the status of the ancient Roman *imperium* clung
precariously to his person.

For most of the fifth and sixth centuries, Europe was in a state
of permanent chaos. Then, between 771 and 778, Charles I, king
of the Lombards and the Franks, subsequently known as Charles
the Great, or "Charlemagne," made himself sole ruler of the once
divided Frankish peoples, conquered the Lombard kingdom in
northern Italy, and subdued and Christianized the tribes of what
is today Lower Saxony and Westphalia. In 800, the pope, Leo III, in
the name of the people and the city of Rome, conferred upon him
the title of emperor. By this gesture, the Roman Empire in the
West was reborn—and significantly so—through an act of spiritual
benediction. At the far end of the Mediterranean, in Constantinople, this unilateral act of creation appeared to be a deliberate attempt to destroy the unity of Christendom. Since the deposition
of Romulus Augustulus three hundred years earlier, there had
been only one emperor—and he had lived in Constantinople.
Charlemagne's coronation had broken with the apostolic line of
succession that reached back to Constantine the Great himself. In
an attempt to salvage a situation about which it could do nothing, the Byzantine court reluctantly recognized Charlemagne as
co-emperor, thus, in effect, reestablishing the division between
East and West first made by Diocletian. But with Charlemagne's
coronation the schism between the two halves of the Christian
world, which had been simmering for some time, now emerged
into the open.

Thereafter Charlemagne would assume in the West the image
of the ideal emperor who "had extended the empire as far as
Jerusalem," a claim that would later become a major inspiration
for the Crusades. "The pious Charles," wrote Jocundus of Maas-
tricht in the eleventh century, "who for fatherland and church did
not fear death, journeyed round the whole world and combated
the enemies of Christ," although in keeping with the Christian

ideal of the just war, he had conquered them "with the sword" only when he had failed to "subdue them with the words of Christ."[22]

In fact, not only did Charlemagne not journey around the whole world, nor conquer Jerusalem, nor do very much to combat the enemies of Christ, he was also a very long way from restoring the former Roman *imperium,* even within the traditional frontiers of Europe. His success was also short-lived. By 924, the empire he had created had been dissolved in Italy, while both France and Germany were in the process of becoming separate kingdoms.

The forces, later attributed to a characteristic "individualism," that would eventually lead to the creation of what became known as the "Europe of Nations" were already too far advanced for any single power, however strong, to halt them for long. By the mid-twelfth century each of the kings of Europe was claiming to be the "emperor in his own kingdom," and all that was left of the Carolingian Empire was gradually confined to what are today Germany, Austria, Hungary, the Netherlands, and the Czech Republic.

Charlemagne may not have created a new Roman Empire, but he helped re-create the universalism once associated with Rome. The new emperor was, as Constantine had been, the defender of the Church. He was the "second sword"—the pope wielded the first—of all Christendom. In 1157, in recognition of this role, then emperor Frederick I, added the word "sacrum" to his title and the empire thus became not merely Roman but also "Holy." This rump of the ancient Roman imperium was, as Voltaire sarcastically remarked in the eighteenth century, "neither Holy nor Roman nor an Empire," but it commanded considerable international prestige and would survive for more than seven hundred years, until it was finally brought to an end by Napoleon in August 1806.

In theory, the emperor in the West was the formal champion of Christendom, the armed defender of the papacy. In practice, however, Charlemagne's coronation led to a struggle between pope and emperor, between sacred and secular, which would simmer for centuries. One of the great strengths of Christianity, as we have seen, was its ability to distinguish clearly between obligations to Caesar and obligations to God, to allow for the existence of a political and social order independent of divine scrutiny. But maintaining this distinction in face of the ambitions of an increasingly powerful Church was not easy. Paul may have been a loyal subject of the Roman Empire; his successors, however, saw themselves as subject to no one but God and came increasingly to believe that the secular rulers of Christendom should be subject to them. Whereas in the Eastern Empire it had been the emperor who had sought to appropriate the powers of the Church, in the West it was the pope who attempted to assume the powers of the emperor.

Matters came to a head in March 1075, when Pope Gregory VII published twenty-seven propositions known collectively as the "Dictate of the Pope." This document stated that the pontiff was to exercise supreme legislative and juridical power throughout Christendom. Furthermore, it claimed the right to depose all princes, both temporal and spiritual. "The Pontiff is permitted to depose bishops," ran the twelfth of the propositions. "No one can condemn a decision of the Holy See," declared the twentieth. And, politically the furthest-reaching of all, the last stated: "The Pontiff can release the vassals of unjust men from their oath of loyalty."[23] In effect, although the pope may not have claimed to rule territories other than those of the Church, he was claiming the right to decide who did and how. It was what came to be called "papal plenitude of power," and very soon it had been extended by the lawyers of the Church to embrace all rulers and their subjects, whether Christian or not. The pope had become, as the Roman emperors since Antoninus Pius in the mid–first century C.E. had claimed to be, "Lord of All the World" or "Custodian of the Cosmos." He had also, of course, come dangerously close to confusing the things of Caesar with those of Christ.

Gregory's decree was a direct challenge to all the crowned heads of Europe, but the person most immediately threatened was the single most significant secular ruler in Christendom, at least in name: the emperor, Henry IV. A struggle, which has come to be known as the "Investiture Controversy," now began. The ostensible issue at stake was the right of the emperor to "invest"— that is, appoint—bishops in the dioceses of the empire. But the real reason for the conflict, as Gregory's dictate made abundantly obvious, was the issue of who wielded absolute sovereignty within the Christian world: pope or emperor.

Henry's response to Gregory's proclamation was instant and dramatic. In January 1076, at the Diet the Imperial Council— held in the German city of Worms, he ordered the bishops of the empire to excommunicate Gregory, denouncing him as "no longer pope but a false monk." "I, Henry, king by the Grace of God," he declared, injudiciously, "with all my bishops say to you, come down, come down and be damned through all the ages." The pope replied to this by excommunicating the emperor, which, according to the terms of the dictate, now released the emperor's subjects from their allegiance to him. The German barons, whose ambitions had for long been curtailed by the empire, instantly seized the opportunity to shed their allegiance to their emperor and launched what came to be called the "Great Saxon Revolt."

Henry now found himself outmaneuvered and, unable to quell the uprising, decided that penitence was his only hope of survival. He set out with his wife and child in the middle of winter, across the Mont Cenis pass through the Alps, to meet the pope at the Castle of Canossa in northern Italy. There, the story goes, he stood outside the castle walls in a hair shirt and robes of a penitent, barefoot on the ice for three days, waiting for admittance. On the third day Gregory granted him an audience and, in January 1077, lifted the ban of excommunication in exchange for Henry's promise to observe the terms of the dictate.

The reconciliation was dramatic, but it was also short-lived. Once his rebellious barons had been defeated in 1081, Henry descended on Rome with an army, determined to depose Gregory and replace him with a more compliant pontiff. Gregory called upon the Normans in southern Italy for help. They succeeded in defeating Henry but sacked the city themselves. Incensed, the Roman people took matters into their own hands and forced the pope to flee south, where he died of fever in Salerno in May 1085.

Despite this and subsequent defeats at the hands of impatient secular rulers, the papacy would continue on and off for the next six centuries to attempt to make good on its fitful claims to the lordship of all the world. Christ may have made a division between his kingdom and the kingdoms of the earth, but according to Matthew he had also declared that "All power is given unto me in heaven and in earth" (28:18), and that, as the great thirteenth-century theologian Saint Thomas Aquinas, among others, concluded, meant that it had been Christ himself who had been the true "Lord of All the World" and the emperor Augustus merely his regent.

Yet, despite such forthright claims, the papacy was never again allowed seriously to threaten the authority of the state. Kings and princes did their best to appear to live Christian lives, for, as Machiavelli later observed, "there is no quality that [a prince] should not seem to have." Religiousness went with piety and expectations of honesty and integrity, and that is what every prince's subjects hope from him. But the laws a prince actually observed were resolutely secular ones.[24] The state had its own reasons, and where the state was concerned these went largely unquestioned. And they have continued to be. This has frequently placed the Church in difficult moral positions, most strikingly in its passivity in the face of the Nazi and Fascist persecution of the Jews and its overt endorsement of General Franco's regime in Spain. In what must be one of the most extreme—and most questionable—declarations of the independence of the laws of the state from those of the Church, and even apparently from those of God, in 1956, Pope Pius XII decreed that no Christian could refuse to serve his country on grounds of conscience. "A Catholic," he declared, "may not appeal to his conscience as grounds for refusing to serve and fulfill duties fixed by law"—even, it would seem, of those laws that were not, in the hallowed phrase, "binding in conscience."[25]

For all that the Church could not intervene directly in secular affairs, it nevertheless lent the state a powerful ideological support. By appropriating the Stoic notion of universal humanity that had been central to the Roman conception of the universal *civitas,* Christianity conferred upon the rising European monarchies a vision of potentially unlimited power that would finally carry them around the entire globe. Already by the fifth century, Pope Leo the Great had declared that what the Romans had referred to as the "terrestrial globe" (*orbis terrarum*) had become the "Christian globe," and it was this that, a century later, Gregory the Great would translate into the "Holy Republic."

In this way, it was the Church that provided the ideological power behind the great overseas empires of Spain, France, and Portugal. The Portuguese solemnly declared that they had bartered for slaves on the west coast of Africa so as to liberate them from their spiritual servitude to heathen deities and unclean habits, in the certainty "of the salvation of those souls that had before been lost."[26] It was good business but God's work nevertheless. Spain had taken its armies all the way to the gates of China, in order to propagate the faith, to encircle the globe, so that the way might be made ready for the Second Coming. French missionaries had swarmed over the wastes of Canada in the wake of the fur traders in the hope of transforming Indians into Catholics and thus into French subjects. Of course, all these powers had done very well economically and politically from their pursuit of God's worth. But there was nothing inherently contradictory in that. As the English Puritan Edward Winslow nicely phrased it in 1624, the lands of the heathen were places where "religion and profit jump together."[27]

What made it possible for religion and profit, for religion and the secular state, to cohabit in this way was the Christian emphasis on the ultimate freedom of the individual. God had laid down laws for mankind—all mankind, not just Christians. Ten of these he had dictated directly to Moses. Many more, most of them rather vague ethical claims, had been delivered by his son, Jesus. All the rest, however, had been inscribed in what was called the "Book of Nature," and in order to read that book mankind had only to apply the power God had given it. Only the most blinkered and unthinking of Christian fundamentalists would claim that God has dictated directly to his creatures the means to their own happiness and well-being. Instead he has set out the world in such a way that if reason is allowed to operate freely, the means by which every individual comes to achieve what the Greeks called *eudaimonia*—happiness and human fulfillment—will become self-evident. Ultimately what mattered was reason and free will, the individual's right and ability to choose, not divine command. Christianity created a distinction between the secular and the sacred, between the things of Caesar and those of God, which existed in neither its Judaic nor its pagan sources. Paradoxically, it had been this that had allowed it to preserve, at its very core, the essentially secular, pagan notions both of the universality of all humankind and of the dignity and autonomy of the individual.

Christianity absorbed and redirected its pagan past. True, it has also fiercely repudiated it. But if the gods of antiquity were all merely foul, corrupted superstitions, the culture the worshippers of those gods had built was widely recognized, as it was by Augustine, as the greatest the world had ever seen. Walk through the streets of Rome today, and you will see any number of pagan buildings and pagan monuments transformed into Christian places of worship and Christian symbols merely by the addition of a cross or the image of a saint. Standing near the Coliseum in Rome is Trajan's column. It was erected in 113 C.E. to mark the emperor Trajan's triumph over the Dacians—the peoples of modern Romania. Winding in a spiral around the shaft of the column are carefully sculpted images of the defeated barbarians, who make their way up to a platform at the top where originally there stood a statue of the emperor. At some point after the conversion of Rome, Trajan was removed and replaced with an equally impe-

rial statue of Saint Paul. Today, it looks as if all those toiling barbarians who struggle upward around the column had, all along,
been moving toward not only incorporation into the Roman *civitas* but also, unknown both to themselves and their conqueror,
toward absorption into the Church of Christ. It is the most powerful image I know of the Church Triumphant.

For all its outward power and uniformity, however, the Christian Church in the West had been, and would remain, plagued by
internal dissidence until finally, in the early sixteenth century, it
split first into two and then into many, the condition in which it
remains to this day. Some of these divisions would be over questions of theology, some over questions of behavior. Some were
over concerns about how to preserve what many of the faithful
saw as the primitive Apostolic Church against the powerful overlords—both secular and religious—who had, in effect, hijacked
Christ's mission for their own, often questionable, ends.

One of the most serious and lasting of the conflicts that beset
the Christian world derived from a problem Christianity had inherited from its ancient Asian origins, one of the most troubling
of all the many problems that beset all monotheistic creeds. This
was the conviction that the spiritual world is divided into good
and evil, into angels and demons, yet the source of all of this was
one indivisible deity. It is called the problem of evil. It was and remains insurmountable. The doctrine of original sin, of the Fall, of
the redeeming grace of God, of free will—all these were invented as
means of shifting the responsibility somehow from the creator to
his creation. None of them, however, has proved to be lastingly
successful.

The only solution, stark and severe, was a form of dualism. On
one side of the divide was God—or a god—the source of all that
was good in the world. On the other, in incessant conflict, was the
devil—or a devil or another god—responsible for all that was evil.
The most influential and widespread of the dualist religions of
antiquity is associated with the Iranian prophet Zoroaster, who
flourished probably between 660 and 583 B.C.E., about the time of

the creation of the empire of Cyrus the Great. We do not know how much Zoroaster borrowed from earlier Iranian beliefs, though it is certain that he borrowed something. Gods, all gods, have a marked tendency to repeat themselves. In the Gathâs, the oldest part of the Zoroastrian scriptures, known as the Avesta, Zoroaster is described as "he who possesses the sacred formulas"—in other words, as with all prophets, he was the only true certified one. Original or not, Zoroaster's teaching attracted an immense following, and Zoroastrianism gradually became the semiofficial religion of the Achaemenid Empire with the accession of Darius I in 522 B.C.E.

Zoroaster divided the universe between the principles of Light—Ahura Mazda—and Darkness—Ahriman. The cosmic struggle between these two would continue until the end of time, and it was the duty of every man to assist wherever he could in the eternal battle against Ahriman by crushing his creatures—such as scorpions—wherever he encountered them and always observing the path of the good, which consisted, very largely, of scrupulous truth telling. (Which may be one reason why one of the virtues Herodotus attributes to the Persians is a strict adherence to the truth.) Zoroaster also held that the elements—air, earth, water, and fire—were sacred. Zoroastrians, therefore, exposed their dead on high wooden platforms—known as "towers of silence"—to be picked clean by the vultures, so as not to pollute them. Although it would survive successive invasions of Iran, and survives today in exile amongst the Parsees (or Parsis) of Maharashtra and Gujarat in India, Zoroastrianism struck most monotheists as unsettlingly polytheistic.

In 241 C.E. another Iranian prophet, named Mani, received a message from God that he believed, as Muhammad would later, completed all earlier revelations. He traveled as far as India and collected a large and varied number of followers. Like Zoroaster before him, Mani attempted to place evil outside good and God, and like Zoroaster he, too, divided the universe into two principles: the "Father of Greatness" and the "Father of Darkness." To this basically Zoroastrian vision, however, he added myths borrowed from the Old and New Testaments, that of the Mother of Life and her son the Primal One. The light of goodness is scattered across the material world, which is the domain of Ahriman. All matter is thus damned. A succession of messengers has been sent by God—Jesus, Buddha, and Zoroaster—to help extricate Adam from the material world, to which he has been bound since the creation. The last of these was, in Mani's opinion, Mani himself. He was the "Seal of the Prophets," a title that Muhammad later gave himself in the Qur'an.[28] Mani enjoyed considerable success under the initial protection of the Sassanid shah Shapur I. The Zoroastrian priesthood, however, saw him as a threat and conducted a fierce campaign against him, and in 276 the shah Vahram I threw him into prison, where he died. Mani's syncretism was wide and eclectic, which was one of its appeals in a region where Christianity, Zoroastrianism, and Buddhism, together with gnostic sects, struggled and, on occasion, cooperated with one another. As Augustine noted, although for orthodox Christians such people were "so far gone in their folly that they do not listen to what the Lord has said," they were, nevertheless, "with us in recognizing the authority of the Gospel."[29]

Like all forms of dualism, Manichaeism was essentially a static religion, a religion that demanded only that the believer perform a certain number of rituals and wait for the end of time. "I could make no progress in it," Augustine remarked. And it was this immobility, this turning away from all the complexities of the human condition, of forgiveness and redemption, that Christians stood for that had ultimately made it abhorrent to him.[30] It was a typically Persian and thus "Oriental" religion, literal, inflexible, ritualistic, and inhuman. Diocletian, who had published an edict against Manichaeism in 297 C.E., had dismissed it as, significantly, the creation of "the Persians, who are our enemies."

Manichaeism itself, although persecuted throughout the Sassanid Empire and Christendom, made converts among the Uyghur Turks of Upper Mongolia, where it became for a while the state religion, and penetrated all the way to China, where it survived until the fourteenth century. It lingered, too, in a number of powerful, if relatively short-lived, Christian sects: the Paulicians in seventh-century Armenia and the Bogomils in the Balkans between the tenth and the fifteenth centuries. The most important, however, were the Cathars or Albigensians in twelfth- and thirteenth-century France, who waged a war against the Church and the king who supported it, and built for themselves massive hilltop fortresses whose ruins dot the rocky landscape of the Languedoc to this day.

With the destruction of the Cathars, all trace of any formal Manichaeism vanished from western Europe. But in the East it took a new form, one that would prove to be a more formidable and more enduring threat to the West than the Achaemenids, the Parthians, or the Sassanids had ever been: Islam.

# 5

## *THE COMING OF ISLAM*

**I**

In 628, a man calling himself Dihya bin Khalifa al-Kalbi presented himself to the Byzantine emperor, Heraclius, at Jerusalem. He was dressed in the robes of an Arab, "a people of sheep and camels," of whom the Byzantines had some experience as mercenary soldiers and as traders in hides, leather, clarified butter, and woolens, but for whom they had very little respect. In Christian eyes they were the descendants of Ishmael, Abraham's son by his slave concubine Hagar, and thus perpetually marginalized from the rest of humanity. This Dihya bin Khalifa al-Kalbi had brought with him a letter containing a simple message from his master, a prophet named Muhammad, the self-appointed leader of an obscure community in the Arabian Peninsula. The message stated that if the emperor accepted the religion of this prophet, which was called, in Arabic, simply "Islam" or "obedience," he and his kingdom would be safe and God would give him "a double reward." It added that if he agreed to pay the poll tax demanded by the prophet he would also be able to avoid war with the Arabs. If not, he would be destroyed.[1]

There is no record of Heraclius's reaction to this piece of effrontery. Muhammad's biographers claim that he secretly recognized Muhammad as the prophet "mentioned by name in our gospels" but dared not act for fear of what his own people would do to him. This is unlikely to be true. The Persian emperor, Kushran II, who had been sent a similar letter, angrily tore his up. The negus of Ethiopia, the recipient of a third letter, immediately converted to Islam, although the sixty messengers he sent to Muhammad with the glad tidings all perished at sea.[2]

If any such letters were ever sent—and the entire story is probably a fiction—neither the Byzantine nor the Persian emperors—nor even the negus of Ethiopia—would have had very much idea who this Muhammad was. The Arabian peninsula had for centuries been a remote, barren place inhabited only by peoples already familiar to the Assyrians for their military strength. Others of their kind were scattered along the frontier between Persia and Byzantium and on the fringes of the Syrian Desert, where like most frontier peoples, like the desert Arabs of Jordan to this day, they drove caravans of camels, traded, and enlisted in the imperial armies of both sides. "They rove continually over wide and extensive tracts without a home, without fixed abodes or laws," wrote the Latin historian Ammianus Marcellinus at the end of the fourth century, with all the disgust of the cultivated city dweller for the nomad. "They wander so widely that a woman marries in one place, gives birth in another and rears her children far away."[3]

The Byzantine Empire had been protected from the worst depredations of these birds of prey by a border state called Ghassân, populated by Christian Arabs, who were paid an annual subsidy for their services. The frontiers of the other superpower of the late antique world, the Sassanid Empire, had been similarly protected by a vassal state called Hîra, also formed by Christian Arabs. It was, in part, from these groups that the Arabs of the peninsula had learned their military techniques, which they employed in ceaseless intertribal warfare, as they had also learned the use of textiles, the habit of drinking wine—which they consumed in large quantities, later prompting Muhammad's ban on all alcohol—and probably also writing. Information flowed back and forth across these borders, providing the Arabs with valuable military information about the disposition of the Byzantine and Sassanid empires. But very little information would have been available either in Constantinople or in the Persian capital of Ctesiphon about this upstart prophet and what would have seemed, to both Byzantines and Persians alike, his preposterous claims to have been vouchsafed a new religious vision that was destined to replace the fiercely guarded pieties of both Orthodox Christianity and Zoroastrianism.

We know rather more. But since everything we know about the life of Muhammad, as with the life of Jesus, was written down by his devout followers, sometimes long after the events they describe, most of it is, at best, suspect. According to the accepted biography (*Sîra*) of the Prophet, Muhammad was born in the "Year of the Elephant," sometime between 570 and 580, in Mecca, a prosperous mercantile town in the Hijaz, the northwestern part of the Arabian Peninsula. His father belonged to the Banu Hâshim, a respectable but undistinguished family of the all-powerful north Arab tribe of the Quraysh. Nothing much is said of his mother except that she was supposedly named Amina ("the believing woman" ) bint Wahb and belonged to the clan of the Banu Zuhra. As Muhammad emerged into the world, he emitted a light by which she was able to see the castles of Bostra in faraway Syria. From this she concluded that he would be an unusual child. Muhammad, however, lost both his parents early in life and was brought up by his uncle, Abû Tâlib. He was an orphan living in poverty—a fitting beginning for every prophet. As a young man, he earned a reputation for honesty and acted on occasion as an arbiter in local disputes, for which he acquired the sobriquet *al-amin,* "the sure one." When he was about twenty-five, he married a rich widow named Khadija bint Khuwaylid, who was considerably older than he. This stroke of good fortune raised him to a position of prosperity and some social standing.

In 610, one night sometime around his fortieth birthday,

toward the end of the month of Ramadan, while he was sleeping in a cave on Mount Hira, he was awakened by a celestial voice that informed him that he was the Messenger of God. Terrified, he fled to his wife, crying "Cover me. Cover me." The voice then came again, declaring itself to be the angel Gabriel and commanding Muhammad to "recite." "What shall I recite?" he replied. Instead of an answer the angel took him in his arms and forced the words of God out of his mouth: "Recite in the name of the Lord who has created you, and created man out of a curd of blood. Recite, because the Lord is very generous, and teaches man that which he knows not" (Q. 96:1–5). Muhammad emerged from this experience uncertain whether what he had endured had been a visitation from God or Satan. By one account, it was a Christian monk named Bahira who recognized the similarity between his experiences and those of Moses and finally persuaded Muhammad that the vision had been a divine one.

Thereafter, throughout the rest of his life, he continued to receive regular messages from God, via Gabriel. He also made a celebrated visit to Heaven in person. One night he was taken by Gabriel to Jerusalem mounted on a winged horse. Here he met Abraham, Moses, and Jesus—the three prophets whom he had succeeded and superseded—and led them in prayer. He then mounted a ladder "finer than any he had ever seen" into Heaven, where he was granted a picturesque vision of Hell; a brief encounter with Jesus, Joseph, Aaron, and Moses; and a glimpse of Paradise. This contained a "damsel with dark red lips"—but not apparently very much else.[4]

In 650, eighteen years after Muhammad's death, the content of these various revelations was collected and written down by one of Muhammad's disciples, Zaid ibn Thabit, to form the Qur'an. The word "Qur'an" means "reading" or "recitation," and it is believed to be composed, quite literally, of God's words. These were also in Arabic. "Surely We have revealed to you an Arabic Qur'an—that you may understand," says the twelfth *sura,* or verse.[5] The fact that God had spoken to Muhammad in Arabic created not only a new sacred text but also a new sacred language—to the degree that it is held to be blasphemous to tamper even with the now-archaic orthography in which it was first transcribed. But although the recitation was given to Muhammad in this way—at a given time and in a specific language—it is regarded by believers as uncreated, eternal, divine, and immutable.

The Qur'an is divided into verses of varying lengths and contains laws; prayers; threats; prohibitions; descriptions of Heaven, Hell, and the day of judgment; advice on how to behave toward one's wives; how much of oneself to wash before prayer (up to the elbows and from the feet to the ankles); how to act on pilgrimage; and how to deal with inheritance, homicide, and theft, and so on. Its overwhelming message, however, is that there is only one God, who has revealed himself through a series of prophets, and that Muhammad is the last and greatest of them. He is the "Seal of the Prophets" the *rasul Allah,* the "Messenger of God" whose words are, therefore, quite literally, God's words.

The Qur'an purported to offer, among other things, guidance for the future conduct of mankind—as did the Old and New Testaments. Since it is a sometimes baffling document, scrappy, and, in many places from a legal point of view, highly ambiguous, it rapidly became clear, after Muhammad's death, that its rulings had to be supplemented by some other source. Christians had faced a similar problem: the Gospels were no more up to dealing with the contingencies of a rapidly changing world than was the Qur'an. The Christians, in an attempt to resolve their problem, had turned for guidance on nonsacred matters to the political and ethical writings of the classical pagan world. Since for Muslims there were no matters that were not sacred, they could not easily do this, although a much later generation of scholars did draw heavily on classical sources. Everything that was authoritative had to come, in some way or another, directly from the last and only authentic intercessor with God: the *rasul Allah* himself. Within a few generations after Muhammad's death, a vast corpus of Hadith or "traditions" had come to fulfill this role. These are the sayings and actions attributed to the Prophet. Each takes the form of a chain of authorities: "I heard from . . . who heard from . . . who heard from . . . who heard the Prophet say or do . . ." Some are obviously false, some less so; but all are deeply unreliable. A method of criticism known as "wounding and authentication" (*al-jarh wa'l-ta'dil*) was developed in order to establish which were, and which were not, to be believed. In general this consists of examining the links—*isnad*—of the chain of authorities. To anyone even remotely acquainted with the process of oral transmission in nonliterate societies, this must seem at best a deeply disingenuous method of procedure. (But it is certainly no less so than the claim that the words recorded in the Gospels, taken down and substantially manipulated for ideological purposes a generation after his death, reflect those actually spoken by Jesus.) There exists no official codification of the Hadith, but six collections made in the ninth and early tenth centuries are now widely regarded as authoritative.

Most of the people of Mecca were polytheists. Like most pagans, they seemed to have been largely tolerant of the beliefs of others. There was, as one of them put it, no reason why a man should not choose a religion for himself as he pleased—and presumably also create one if he so wished. They tolerated Muhammad's early attempts to get Gabriel's message across to them, although in the opinion of many, it placed rather too much emphasis on the singularity of God, the wickedness of idolatry, and the imminence of divine judgment. It was only when Muhammad began to disparage the pagan deities that they started to take offense, mockingly offering either to make him king or to provide someone who would cure him of his delusions. Mockery, however, like poverty, is a necessary rite of passage for all prophets. "Apostles have been mocked before thee," God consoled Muhammad, somewhat opaquely, "but that which they mocked at, hemmed them in" (Q. 6:10).

The Meccans had economic as well as religious reasons to be wary of Muhammad and his new brand of monotheistic intransigence. Mecca was a city of pilgrimage. At its center stood—and still stands—the Ka'ba ("the cube"), a black shrine that before the advent of Islam housed an idol called Hubal. In the eastern corner a black stone, the *al-hajar al-aswad*—possibly of meteoric origin—was built into the wall and had become an object of veneration. The ruling oligarchs of Mecca made a great deal from the pilgrim route, and although Muhammad never raised any objection to the pilgrimage, they feared, not unreasonably, that his fiery brand of monotheism could not fail eventually to undermine their livelihood. They then began to persecute him. For a while Muhammad's own position was relatively secure, thanks to the protection of his pagan uncle, Abu Talib. But his followers, now calling themselves "Muslims" or "those who have submitted" to God, were more exposed, and at one point Muhammad sent a party of them to take refuge in Ethiopia.

In 622, by which time the number of his followers had grown sufficiently to constitute an independent community, Muhammad transferred himself to an oasis settlement some 280 miles north of Mecca, then called Yathrib. He was accompanied by some seventy of his companions, called the Muhâjirûn, whose descendants acquired a privileged status in the subsequent history of Islam. This migration or "withdrawal"—*hijra,* as it came to be known—is the first uncontested date in Muhammad's life and in the early history of Islam and now marks the beginning of the Muslim lunar calendar: the year one.[6] It also marked a revolution in the history of the Prophet's fledgling religion.

In Mecca, Muhammad had never been anything more than a not very prominent or significant private citizen. In Yathrib, now renamed Medina, or "City of the Prophet," he became the chief arbitrator of a community. In Mecca he had preached Islam; in Medina he set about making it the religion of the city. And it was in Medina that the future distinctive shape of the Islamic community, the Umma, took shape. Here Muhammad issued what has come to be called the "Constitution of Medina," although the name is late and misleading. It is not a constitution as much as a number of proclamations, and in many respects it is little more than a reaffirmation of the customs regulating property, marriage, and other relations within the traditional Arab community. What was new about the Umma, and what would have long-term consequences not only for Islamic society but eventually for the entire world, was its focus on loyalty. Arab tribes, like tribes everywhere, had been defined by blood, which meant that their size was, at best, confined. The Muslim Umma, however, was defined by faith, which meant it could be literally unlimited in its extent.

Muhammad had thus created a single politico-religious community, "religion and state" (*din wa dawla*), in a phrase used by the medieval jurists. He had also, as Christ had done, created the possibility of a universal community, and his followers, if not he, would come to look upon this as embracing the whole of humankind. Despite the fact that the policies, in particular of Muhammad's successors, the caliphs, were more concerned with the Arab world and its immediate neighbors than they were with the cosmos, the universalism of Muhammad's message was one of the sources of its power and appeal, and it was inescapable. It survives to this day, and inevitably it has, over the centuries, brought Islam into close and persistent conflict with Christianity. Christians have reluctantly come to accept that it is unlikely that the world will ever in fact be converted, as had once been predicted, thus bringing about the end of human time. Most Muslims today would also probably accept that the vision of the end of time contained in the Hadith, when Jesus will return in armor, slay the Antichrist at the gate of Lydda in Palestine, exterminate the pig, and break all the crosses of the Christians, may be a long time coming and can perhaps be quietly ignored.[7] But neither religion has ever formally abandoned its claim to be the sole true one, condemning all those who cling to other faiths to eternal damnation. In the centuries since the disappearance of the empires of the ancient world, it has been the universalistic legacy both religions inherited from Judaism that has been the single most determining feature of the conflict between them.

There was, however, one major distinction between Christianity and Islam in this respect. As we have seen, for Christ, even for Saint Paul, the religious community, although universal, could claim no social or political authority as such. The things of Christ remained resolutely those of Christ, but they were so—indeed they could *only* be so—as long as the things of Caesar remained resolutely those of Caesar. Even the laws of infidels, however tyrannical they might be, were binding upon Christians who lived under their jurisdiction.

Muhammad, however, chose a different path. In Medina he could ensure the survival of his new faith only by creating not merely an autonomous religious community but a political authority to go with it. He became, in effect, the sheikh of a new tribe. But whereas previous sheikhs, like the leaders of most warrior societies, had been men with very limited authority who derived their power exclusively from the community that had appointed them and that could replace them when it chose, Muhammad claimed to derive his power directly from God. "Whenever you disagree about something, the matter should be referred to God and Muhammad," ran one of the articles of the Constitution of Medina.[8] The Prophet was absolute and would accept no mediation, no defiance. "O mankind," God had instructed him to say, "I am the Messenger of God, to you all, of Him who belongs the kingdom of the heavens and of the earth" (Q. 7:157). The individual Muslim has an obligation to "command right and forbid wrong," and his or her chief happiness is to be gained through service, *'ibâda*, to God, which, in turn, requires absolute obedience. He or she is part of a community of believers—"Let there be one community of you, calling to good and commanding right and forbidding wrong" (3:104)[9]—within which there can be no separation between religion and politics. Muhammad had achieved exactly what the emperor Constantine had hoped for and ultimately failed to achieve at the Council of Nicaea: the complete identification of the secular realm with the sacred and the corresponding elevation of the ruler to the position of divinely instated and divinely inspired being. The founder of Islam, it has been said, "was his own Constantine."[10]

In Islam, therefore, there can be only one law. It is called the Shari'a, a word whose primary meaning is "the road to the watering place"—in the desert society in which Islam was born, the most sought after, the most blessed of places. The Shari'a is the direct expression of the will of God, derived from the Qur'an and the Hadith by the community of scholars known as the Ulema. It relates to all of God's commands concerning the activities of man. It is a doctrine of duties covering the whole of the religious, political, social, domestic, and private life of all Muslims. In practice it is the creation of the Ulema, but, despite its purely human origins, it, like everything that derives from Muhammad, also derives from God. The idea, therefore, that there might exist, as there does in the West, a secular law, a law that is the creation of human intelligence rather than divine command—and is subject not merely to change, at least as far as its prescriptions are concerned, but also to abrogation—is, for many Muslims, merely senseless.

As the Shari'a is God's law, it is held to be eternal and thus unchanging. This does not mean that it cannot be subject to rational evaluation. The *fiqh*—the science that determines how the Shari'a is to be applied in particular cases—is divided into two distinct areas: the ritual (*ibâdât*), which is made up of instructions as to how to pray, fast, go on pilgrimage, and so on; and the *mu'âmalât*, which applies to social relations. From about the end of the eighth century the latter became established as a distinct juridical science, not unlike Western jurisprudence, which relied heavily upon analogous reasoning (*qiyâs*) and on consensual judgment (*ijmâ*). Since ultimately what was at stake was the interpretation of some sometimes very opaque texts, several competing schools of law (called *madhhab*) grew up, as they did in the West. By the ninth century, however, four of these, each based upon the ruling one of the great specialists (*fuqahâ*), had succeeded in excluding all the others. They were Hanafism, named for Abû Hanîfa, who died in 767; Malikism (Mâlik, died in 795); Shafism (al-Shâfi'î, died in 820), and Hanbalism (Ibn Hanbal, died in 855).

In the West the law has, since antiquity, been looked upon as the creation of man for the needs of man. That law is civil, it is existential, and it is based on fact. And because facts and the very nature of existence change, so the law can, and indeed must, change. The Shari'a is also the creation of man; but it is based not, as Western law is, upon a codification of customary practices but on the supposed word of God. As such, its capacity for change is severely limited. Gods, particularly unique ones, are not accustomed to having second thoughts.

The Umma was a theocracy. In the world into which Muhammad had been born, this was very unusual. None of the peoples with whom Muhammad was familiar and neither of the great imperial states that flanked the Arabian Peninsula, Byzantium and Persia, although both their rulers claimed some measure of divine support, insisted on the indivisibility of their rulers' commands and those of God. Both Christ and Zoroaster—although the latter somewhat less unambiguously than the former—had explicitly recognized a distinction between Church and state. Of all the peoples of the ancient Near and Middle East, only the Sumerians are said to have begun their history with temple communities ruled by priest-kings. But if that were the case, it had been a very long time ago, and by the time of Muhammad's birth no one remembered them any longer. The unusualness of this aspect of Muhammad's message, and of the kind of society he had created in Medina, seems, however, to have gone largely unnoticed.

Muhammad soon proved himself to be as able a war leader as he was a political strategist. Once the Umma was securely established at Medina, he turned his attention to his birthplace. In March 624, a Muslim war party surprised a heavily laden Meccan caravan at Badr. The caravan itself escaped, but the army sent out from Mecca to defend it was destroyed by three hundred Muslims, assisted by an invisible army of angels. The success of the operation and the presence of the angels were taken as signs of divine favor. "And Allah did certainly assist you at Badr when you were weak," says the Qur'an. "Be careful of your duty to Allah, that you may give thanks" (3:123).

The victory at Badr boosted the reputation and authority of the Muslim community enormously and gave Muhammad undisputed control of Medina. He was now strong enough to deal with the last remaining independent groups, the Jews and the Christians. Initially Muhammad had hoped that these would easily be won over to Islam, and he also seems to have believed that the theological differences between them were either minimal or nonexistent. He was mistaken on both counts. The Christians had their own prophet, one, furthermore, who was nothing less than the son of God, and they were in no hurry to surrender his divinity for the claims of a mere man. The Jews, for their part, resented the idea that their God, the God of Israel, might have chosen the last and greatest of his prophets from among the Arabs. Muhammad now accused both groups of having falsified their scriptures to conceal the prophecies contained in them of the revelations made of the coming of the "Seal of the Prophets." Jesus—called Isa in the Qur'an—was for Muhammad a true prophet, who had indeed, as the Gospels claimed, performed miracles (2:253) and, according to one verse in the Qur'an, had even ascended into Heaven (3:55). But the Christians, not content with these marks of divine approval, had distorted his legacy by making him into a god and by insisting, bizarrely, that he had been crucified.

Similar accusations were made against the Jews. Muhammad recognized Abraham as "one of the true religion who had submitted to God" (Q. 3:60)—and thus the ultimate source of the origin of all three monotheistic faiths. Although the Jews had taken no action against the Muslims, their obduracy in the "City of the Prophet" was intolerable. Of the three Jewish tribes living in Medina, one, the Banu Qaynuqa, was expelled after an unsuccessful uprising against Muhammad's authority in 625. The second, the Banu Nadir, soon followed, after it was accused of plotting to assassinate Muhammad. Both went to join the Jewish settlement in Khaybar, some hundred miles to the north of Mecca. The third, the Banu Qurayzah, was not so fortunate. Having been accused of acting as spies for the Meccans after a Meccan army had laid siege to Medina in the spring of 627, the men were slaughtered, the women and children sold into slavery. Later, Muhammad concluded a treaty with the Jews of Khaybar as he did with the Christians of Najrân. But this was merely a temporary lull in what would become a perpetual hostility between Islam and both Jews and Christians. "And the Jews say: Uzair is the son of Allah; and the Christians say: The Messiah is the son of Allah; these are the words of their mouths; they imitate the saying of those who disbelieved before [i.e., they are polytheists]; may Allah destroy them; how they are turned away!" (Q. 9:30). Increasingly, as the traces of its previous indebtedness were erased, Islam became not merely a set of modifications of preexistent monotheistic traditions, fed from Jewish and Christian sources, but a wholly new religion, which would eventually unite a vast array of ethnically and linguistically diverse peoples, from Arabia itself as far as Indonesia and the west coast of Africa.

IN JANUARY 630, after a series of inconclusive negotiations between the Quraysh and Muhammad's representatives, a Muslim army attacked and finally occupied Mecca. The Quraysh surrendered almost without a fight. The other Meccans, apart from those accused of crimes against the Prophet and his followers, were spared their lives and properties, and the city itself was transformed from a pagan pilgrim site into one of the holy places of Islam. Muhammad ritually cleansed the Ka'ba of the 360 idols that stood around the building, all of which shattered at the touch of his rod. He also adopted the Ka'ba as the place of pilgrimage that every Muslim is obliged to visit at least once during his lifetime—walking seven times around the building and if possible touching or kissing the black stone.

Compared with the other two monotheisms on which it is partially dependent, Islam is a simple faith, which is certainly one reason for its success. To the outsider, Islam presents few intellectual obstacles—beyond the need to believe in a god in the first place, and furthermore in a god who has entrusted a crucial message to one man in one place and one time and thereafter lapsed into silence. But then, all monotheisms, and indeed, most religions with any kind of sacred scripture, pose that problem. There is an Islamic theology, but it has none of the inbuilt paradoxes and glaring inconsistencies of Christian theology. There are sects within Islam, forms of spiritualism, mysticism, and asceticism, as there are in Christianity. But there has never been a schism of the kind that threatened Christianity from its very beginnings and finally divided it into two irreconcilable camps in the sixteenth century. The division between Sunni and Shiite—which I shall come to shortly—has some of these features. But the divide is nothing so marked, nor for the most part, so ferocious, as the divisions between, say, Calvinists and Catholics.

There is nothing to suggest to the faithful, as the confessional conflicts of the sixteenth century suggested to so many Christians, that disagreement might derive, in fact, from the implausibility and incoherence of the entire creed. There is also in Islam none of the elaborate ritualization and hierarchy to be found in Christianity. There are rites and rituals. But there is no Church, in the sense of a religious organization. There are mosques, but no "Mosque." There is the Ulema (in Persian, *Mulllas*). But there are no licensed mediators between God and man, no caste, as in (Catholic) Christianity, endowed with special God-conferred powers. The imam present in every mosque is merely one who leads the faithful in prayer. The muftis (interpreters of the Shari'a) in the Ottoman Empire, and more recently the mujtahids and ayatollahs in Iran, have positions that in some ways resemble those of Church officials. But even the Iranian clergy—as it is so often called—has none of the religious authority vested in the Christian priesthood.

In the absence of any developed ritual or liturgy, conversion to Islam is remarkably simple and direct. All the neophyte has to do is to repeat once, before two Muslim witnesses, the profession of faith, the *shaha'dah*. This consists of a single and now famous sentence: "There is no God except God, and Muhammad is the Messenger of God." Having said this, the believer—the Muslim—is asked only to submit to the will of this one God—which means abiding by the Shari'a—and to observe the five pillars of Islam imposed upon all Muslims by God through revelation: the shaha'dah, prayer (*salât*), to be performed five times a day; the giving of alms (*zakât* or *sadaqa*); fasting (*sawm*) from dawn to dusk during the month of Ramadan; and, finally, the pilgrimage to Mecca (*hajj*)—which all those with the means and the abilities are required to perform at least once in a lifetime.

None of these should present any difficulties, or cause any offense, to non-Muslims. There are, however, a number of the communal obligations that do, of which by far the most important, and the most troubling, is jihad. The literal meaning of the word in Arabic is "effort," "striving," or "struggle" and it is generally followed by the words "in the path of God." Some, in particular classical Shiite theologians and more modern reformers attempting to find an accommodation with the West, have interpreted it to mean a spiritual or moral striving. Just as it could be argued that every true Christian has a duty (although it is not an article of faith) to persuade every non-Christian to convert, so every Muslim has a duty to persuade every non-Muslim to accept Islam. Muhammad himself, however, took a deeply Manichaean view of these, as of most other, matters. God's message, as it comes across in the Qur'an, is often difficult to understand. On this issue, however, it is crystal clear: you are either with him or against him. "Now is the way of truth manifestly distinguished from error," runs the second *sura*. "God is the friend of the believers, bringing them out of darkness into light; but the friends of unbelievers are idols who bring them out of light into darkness" (2:257–59). If the unbeliever does not willingly forsake his idols, he must be compelled to do so by force. "O Prophet! Urge the believers to war, if there are twenty patient ones of you they shall overcome two hundred, and if there are a hundred they shall overcome a thousand" (Q. 8:65). (The following verse, however, is rather less optimistic about the numbers.) The overwhelming majority of jurists have, therefore, interpreted jihad as a military obligation. "Learn to shoot," runs one of the Hadith, "for the space between the mark and the archer is one of the gardens of Paradise." "He who dies without having taken part in a campaign," runs another, "dies in a kind of unbelief."[11]

Traditionally the world is divided into two "houses": the House of Islam—*dar al-Islam*—and the House of War—*dar al-harb*—which includes all non-Muslims. Between the two there exists a permanent state of war that will endure until the whole world has come to accept the truth of Muhammad's revelation. That war is the jihad. The Ottomans, for instance, referred to the city of Belgrade, which was on the front line in their struggle against Christendom, as the "House of Jihad." And jihad is God's war. Opinions about just what this means can vary. But one thing is certain: no peace treaty between the two houses is legally possible. Although truces, never to exceed ten years, are acceptable, the war itself can never be terminated before the final—and to devout Muslims, inevitable—victory arrives.

Yet, although all non-Muslims are the enemies of Islam and must remain so until in the fullness of God's time they are converted, or at least brought under Muslim rule, not all non-Muslims are alike. One group stands out. This is the *ahl al-kitâb*, or "peoples of the Book"—Jews and Christians. They are distinguished from simple pagans and idolaters by having had a revelation from the same God who revealed the Qur'an to Muhammad and thus, in some sense, observing a recognized religion. Toward these peoples, Islam, in principle, if not always in practice, has made a number of far-reaching concessions. In Islamic law they are given the status of *ahl al-dhimmah*, or "protected peoples." They are allowed to govern themselves according to their own customs and by their own rulers. This became the basis of the *millet* system, which would play a crucial role in the government of the Ottoman Empire.

The "protected peoples" were, however, very much what today would be called "second-class citizens." They had to pay an annual poll tax (the *jizya*), in exchange for which they were entitled to practice their religion, although they were forbidden from indulging in outward observances, such as the ringing of bells or public prayers. They were also forbidden to build any new sacred buildings or—the most serious crimes of all—to attempt to seduce any Muslim from the true faith or to insult Islam, both of which acts were punishable by death. The law forbade them to ride on horseback, to build houses higher than those occupied by their Muslim neighbors, and to possess Muslim slaves. It also required them to wear a distinctive sash known as a *zunnar*. In some cases, *dhimm* status was widened to include Zoroastrians and, in Mughal India, Hindus. (Two schools of law, the Hanafis and the Malikis, went so far as to extend this status to all non-Muslims.) The application of these rules varied considerably from periods of great tolerance under the early Abbasid caliphs to periods of severe persecution under such rulers as al-Mutawakkil (847–861) and al-Hâkim (996–1021). This is the effective limit of the toleration of other creeds that many modern Western and fundamentally secular intellectuals, and even a number of Muslims, impute to Islam. It does not mean, as these people often seem to suppose, being prepared to accept other interpretations of God's intentions as being equally as valid as one's own. It means, in this case, being prepared to put up with the presence of those whom you *know* to be wrong.[12] This, however, was a great deal more than most Christian communities before the seventeenth century were prepared to accept. Yet, despite all the restrictions of life as a member of the *dhimmah*, for some Christian groups—in particular the members of the breakaway Greek churches, the Nestorians and the Monophysites, who had been subjected to harassment and persecution by the government in Constantinople—life under a Muslim prince, although restraining, might even be preferable to life under an Orthodox Christian one. And the Jews generally fared much better under Muslim rule than they did in any Christian state before the late nineteenth century.

## II

The integrity of Islam, the fusion of religion and politics, the notion that there cannot exist a law that is not somehow the expression of God's will, gave the intensely unruly tribes of the Arabian Peninsula a unity they had never possessed before. Muhammad had taken the monotheism of the Jews and the Christians and transformed it into an entirely Arab religion. In doing so he had created what the Arabs as a people had never enjoyed before: a single cultural identity. It would not last forever. But while it did, it made of the tribes of Arabia a formidable conquering force. The coming of Islam had also transformed the ancient struggle between Europe and Asia, which by the Greeks and their Roman heirs had been conceived as one between competing views of life, into one between two competing faiths.

After the occupation of Mecca, Muhammad controlled the oases and the markets of the wealthiest areas of the Arabian Peninsula. The leaders of the other tribes in the region needed his assistance, and by the time of his death on June 8, 632, many of them had pledged their allegiance to him and some had also converted to Islam. Muhammad's death, however, created a severe problem for the Umma. No provisions had been made for a successor. During Muhammad's life there had been no government in Medina, no chain of command, no political institutions, only Muhammad himself. How, in any case, could the "Seal of the Prophets" possibly have a successor? Many of the desert tribes took a similar view: their allegiance had been to a man, not an institution, and when he died, their allegiance ceased.

One thing, however, was clear; the future rulers of Islam, whoever they might be, had, like its founder, to derive their authority, even if at several removes, from the manifest will of God. This meant they had to be, in some degree, a member of the Prophet's own family. After some infighting among the obvious candidates, the choice fell upon Abû Bakr, one of the Companions who had made the Hijra with Muhammad and whose daughter 'A'isha had been one of the Prophet's many wives. Abû Bakr was designated *khalîfa,* or "successor"—hence caliph—and may even have styled himself *khalîfat rasûl Allâh,* "successor of the messenger of God."¹³

Abû Bakr's first task was to bring back to the fold, by force, the tribes that had defected on Muhammad's death. This resulted in the "wars of the *ridda* [apostasy]." When these came to an end in 633, the new caliph found himself in possession of a formidable army and nothing immediate to do with it. Arabia itself was now at peace; but peace had never been a part of the historical experience of the Arabs. The year he made his last visit to Mecca, Muhammad is supposed to have declared, "Know that every Muslim is a Muslim's brother and that the Muslims are brethren; fighting between them should be avoided." Then he added, "Muslims should fight all men until they say, 'There is no god but God.' "¹⁴ If the Arabs were to observe the first part of this injunction, they had now to begin in earnest to pursue the second. They therefore turned their attention away from the always inhospitable wastes of the desert for the far richer prizes they knew to lie to the north and the east.

In the mid-seventh century, the entire Middle East was divided up between the Byzantine and Sassanid empires—"the greatest powers of their time in the world" as the North African historian Ibn Khaldûn, one of the world's great historical thinkers, called them.¹⁵ In some respects these two superpowers closely resembled each other, as superpowers often tend to. Both bore traces of the legacy of Alexander. But whereas Byzantium was a hellenized, Christian state, the Sassanids, who in 226 had succeeded the Parthians, had done their best to restore the older Achaemenid legacies that Alexander and his Ptolemaic successors had destroyed. They had developed the somewhat loose set of beliefs associated with the prophet Zoroaster into a system of formal worship known as Mazdaism, the worship of the god Ahura Mazda. Like the Parthians before them and the Achaemenids before them, the Sassanids were locked in almost unceasing conflict with their western neighbors, first Greece, then Rome, and now Byzantium.

Between 602 and 628, Byzantium and Persia began a series of wars that would finally exhaust them both. In 615, a Persian army under Shah Khusrau II took Jerusalem and carried off the true cross to his capital at Ctesiphon. In 619, the Persians entered Egypt and captured Alexandria. By now the old Achaemenid Empire had in effect been restored. But not for long. In 628, the Byzantine emperor Heraclius marched an army as far as Dastragird, where Khusrau had a palace, and repossessed the cross. That same year, a final peace was agreed between the two powers. But the Sassanids were now divided and diminished in the eyes of their own nobility. Their expansionist ambitions had drained all loyalty from the court, which had constituted the sole source of cohesion within the empire. Weak and exposed, they were no match for the steadily encroaching Muslim forces.

By early 635, the Arabs pushed the Persians back across the Euphrates and annexed Hîra. In September a victorious and jubilant Arab army entered the Byzantine city of Damascus. A year later Heraclius abandoned Syria, and the Arabs rapidly took possession of the major cities of Antioch and Aleppo. Everywhere the victorious Arab armies imposed conditions that would become the norm for all future Muslim conquests. As long as the population surrendered, no action was taken against them. As long as they agreed to pay the poll tax, they were permitted to observe their religion without interference and would receive, as Khâlid ibn al-Walîd, the conqueror of Damascus, told its Christian inhabitants, "the pact of Allah and the protection of His Prophet, the caliphs and the Believers."[16] By the end of 636, a millennium of Greco-Roman occupation had been brought to an end by what one ninth-century Syrian Christian described as "the most despised and disregarded of the peoples of the earth."[17]

In 637, the new Sassanid shah, Yezdegird III, decided to take the offensive and, he hoped, rid his lands once and for all of what he looked upon as an unruly, uncivilized rabble. His ablest general, Rustam, advised him to remain securely on the eastern bank of the Euphrates and to draw the Arabs out of the desert, where they enjoyed an overwhelming advantage. Once across the river and into the plains beyond, crisscrossed with canals, on wholly unfamiliar terrain offering no hope of retreat, the Arabs would be easy prey.

Yezdegird, however, would have none of this. For him the Arabs were a trifling people and it was beneath his imperial dignity to wait for them to come to him. Thus in early 637, a Persian army numbering some 20,000 men crossed the Euphrates, entered the desert, and met a far smaller Arab force at al-Qadisiyya, just south of the modern city of Najaf. As Rustam had predicted, the Arabs used the desert to their advantage and at the end of three days the Persian army had been annihilated. The defeat meant not only the end of Sassanid rule but the beginning of the introduction of Islam into Persia. The battle of al-Qadisiyya itself became, and has remained, a highly symbolic moment in Arab history. For it was not only a battle in which Islam triumphed over paganism. It was also one that, for the Arabs, marked the final victory of their peoples, their tribes, over a once-formidable and often ruthless opponent. It was celebrated, too, as the moment when the Arabs freed the Persians from the servitude in which their kings had always held them. "We Arabs are equal," one of them told the Persians. "We do not enslave one another, except in the case of somebody at war with another."[18]

Centuries later, the name of the battle and what it represented could still arouse powerful memories. In 1980, Saddam Hussein's war against Iran was depicted by the Ba'athist propaganda machine as a struggle between virtuous (and still, improbably in Hussein's Iraq, "equal") Arabs and vicious Iranians, and described as "Saddam's Qadisiyya" or the "Second Qadisiyya." Shortly before the war broke out, an epic film of the battle, of Hollywood proportions, began filming on a lot outside Baghdad. Its purpose, the deputy chairman of the Revolutionary Command Council explained, was to present past history as a living thing, to encourage the youth of Iraq to sacrifice themselves for the nation and for the entire Arab world, just as their supposed forefathers had done against the selfsame enemy 1,400 years earlier. At the historical al-Qadisiyya, the Persian forces had been larger and better equipped than the Arabs, but the Arabs, with God, skill, courage, and determination on their side, had won. So, in the coming months, would Saddam's forces destroy the new "Persian" army. Eight blind men from the province of al-Qadisiyya itself turned up to enlist in the Iraqi army, following the stirring example of the one blind Arab who was said to have carried a banner in the original battle.

The Persians themselves, needless to say, did not always see things that way. Islam may have brought them to the truth and the Arabs may have freed them from servitude, but in the process a centuries-old civilization had been reduced to nothing. Some four centuries later the great Persian poet Firdausi, author of what became and has remained Iran's national epic, the *Shâh-Namâh*, or *Book of Kings*, wrote of the battle:

> *Damn this world, damn this time, damn this fate,*
> *That uncivilized Arabs have come to*
> *Make me Muslim*
> *Where are your valiant warriors and your priests,*
> *Where are your hunting parties and your feats?*
> *Where is that warlike mien and where are those*
> *Great armies that destroyed our country's foes*
> *Count Persia as a ruin, as the lair*
> *Of lions and leopards. Look now and despair.*

The battle of al-Qadisiyya destroyed all further resistance. The "uncivilized Arabs" then swept across the Euphrates and fell upon Ctesiphon. Its treasures were sent to Umar, who had succeeded

Abû Bakr on his death in 634 and who reputedly put the crown of Khusrau on display in the Ka'ba in Mecca. It now seemed that the Islamic conquest of the world was imminent. The Hadith, which were probably written around this date, are filled with certainty. "You will certainly conquer Constantinople," says one, "excellent will be the emir and the army who will take possession of it." In another the Prophet goes on to predict not only the fall of Constantinople but even that of Rome itself.[19]

The new master of the former Sassanid Empire, the caliph Umar, was a giant of a man with a long beard, who wore rough simple clothes in imitation of the Prophet himself and walked about the streets of Medina with a hide whip that he used on anyone he found transgressing the holy law. This made him respected but not popular, and in 644 he was assassinated because of a personal grievance, by a Persian slave. His successor was Uthmân, a member of the Meccan tribe of Umayyah and one of Muhammad's sons-in-law. Uthmân pushed the ever-expanding Arab domains westward as far as Egypt and Libya, eastward as far as Khorasan, and northward into the Caucasus as far as what is now Tbilisi. By now, however, fissures had begun to appear in the surface of what had only ever been a rough alliance held together by force of religious conviction and the desire for plunder. The families of the Muhârjirûn, the Meccan emigrants in Medina, watched with growing apprehension and newly acquired aristocratic disgust as more recent converts began to assume power in the conquered territories, while the peoples of the Arabian heartlands saw their power drained northward to the richer, more populous lands of Syria and Iraq.

Uthmân had been chosen by a small group of the members of the Quraysh, and in Medina this looked all too much like a bid for power on the part of the Meccans. He was also widely suspected of nepotism, favoritism, and having introduced innovations into both Muslim rites and the administration of property that ran counter to the dictates of the Qur'an and the Hadith. Meanwhile the soldiers of the victorious Muslim armies, who were now assembled from all over the Islamic world, had been compelled to watch as much of the wealth they had won on the battlefield was sent back to enrich the notables of remote Mecca and Medina.

Opposition to Uthmân began to grow. One of its leaders was 'Alî ibn Abî Tâlib, an early convert, Muhammad's cousin, husband of Muhammad's daughter Fâtima, and an obvious candidate for the caliphate who had been waiting, by most accounts, impatiently in the wings ever since the Prophet's death. Finally on June 17, 656, a group of rebel soldiers from Medina and Egypt, one of them a son of Abû Bakr, murdered Uthmân as he sat reading the Qur'an. Alî then established himself as caliph in Kufa in Iraq. By this time, however, the governor of Syria, Mu'âwiyah, who like Uthmân was an Umayyad, had gathered together another group of dissidents in Damascus. In May 660, he declared Alî unfit and himself caliph. In January of the following year, while preparing to march against Damascus, Alî was stabbed to death in the mosque in Kufa. After persuading Alî's son, Hasan, to abandon his claim to the throne, Mu'âwiyah was confirmed as sole caliph. Because he, unlike Uthmân, claimed to possess a dynastic right to the caliphate, he is generally regarded as the first of the Umayyad caliphs.

Most Muslims now accepted Mu'âwiyah as Muhammad's heir. But not all. The conflict between Alî and Mu'âwiyah would result in the creation of a division within Islam that has lasted ever since. Those, the vast majority who followed Mu'âwiyah, became *ahl al-sunna*—or Sunnis—that is, those who followed the *sunna,* or way of the Prophet. Those who remained loyal to Alî established what in time would become a religious-political group called the *shi'at 'Alî,* or the party of Alî. These are the Shiites. The Shiites' disagreement with the Sunnis had begun as one over the exercise of power, not religious doctrine. But in 681, at Karbala in Iraq, the Umayyad Caliph Yazîd I killed Alî's last remaining son, Hussein, along with all the members of his family. The slaughter became a defining moment in Shiite history. Hussein was represented as a martyr, whose sacrifice, not unlike that of Jesus, had paved the way for mankind to Paradise, and what had begun as a political party became, in effect, a religious sect.

Shiism appealed largely, although by no means exclusively, to non-Arab Muslims. Muhammad had created what was intended to be, in theory at least, a universal religion. Once a man became a Muslim, he attained equality of status with all other Muslims. But in practice, and in particular administrative and fiscal practice, Islam was, until the collapse of the Umayyad Caliphate in 750, heavily reliant upon Arab tribal alliances. Non-Arab Muslims— Persians, Armenians, Egyptians, and Berbers—were all classified as *mawâli*, a word originally used to describe a liberated slave. They were also sometimes unflatteringly alluded to as the "booty which God has granted to us along with these [i.e., their] lands" and were taxed as if they were not Muslims at all but the followers of one of the "protected religions." The *mawâli* saw in Shiism a means of resistance to the ruling elites and the sect found its most dedicated adherents above all in Iran, in Berber North Africa, and in Egypt, where a rival Shiite Caliphate was established in Cairo in 969.

Because, on conversion, the *mawâli* had inevitably brought with them some of their previous religious traditions, in time Shiism developed some highly unorthodox doctrines of its own, which incorporated elements of surviving pre-Islamic beliefs. For the Shiites there can be only one rightful ruler of Islam, the imam, a direct descendant of Alî through the Prophet's daughter Fâtima. He alone possesses the ability to provide the definitive interpretation of the Qur'an and the Hadith, for which he relies not upon human wisdom but upon miraculous guidance from God, called *ta'yid*. This introduced into Islam a notion that had been wholly absent from all previous teaching, that the revealed scriptures had not only a literal meaning (*zâhir*) but also a hidden one (*bâtin*). Because of his unique and God-conferred power, the imam was held to be both without sin and infallible—something that made him very like a pontiff. There was, inevitably, much debate over the centuries as to just who this person might be. But when the line of Alî finally vanished from sight in the ninth century, there arose the conception of an "occluded" or hidden imam. On this account, the twelfth and last plausible candidate, Muhammad al-Mahdî, who disappeared as a five-year-old in 874, had not really died but had been concealed by God and would one day emerge when the world had become irredeemably corrupt, to lead the righteous to victory. When he does, he will be the *mahdî*, the messiah, the "rightly guided one," and it is he who will lead Islam in its final victory over the *dar-al-harb*. (Although the concept of the *mahdî* is common to both branches of Islam, it has more precise meaning for Shiites than it does for Sunnis.)

With the succession of Mu'âwiyah, the caliphate passed from the Arabian Peninsula to Syria and Iraq, never to return, and with that move it underwent a radical transformation. In time, the original Arab armies were replaced with regular, and multiethnic, troops. The caliphs evolved into Near Eastern monarchs, borrowing court practices and much else besides from their Byzantine and Persian predecessors. Tribal loyalties were slowly replaced by political associations, and an empire, still somewhat ramshackle, began to emerge.

Like the Gothic tribes who had settled in the Western Roman Empire, and the Mongols in China, the Arabs lived somewhat uneasily amid the high cultures of the ancient worlds they had invaded. Unlike the Goths and the Mongols, however, who had carried very little with them and thus adapted rapidly to their new cultural surroundings, becoming romanized, in the case of the Goths, and sinicized, in the case of the Mongols, the Arabs came with a religion and a conception of how man should live his life if he were to appease God, which they were determined to impose, by persuasion or by force, upon the entire world.

## III

Under the caliph al-Walîd, who came to power in 705, the Arabs, having weathered the internal conflicts of the past half century, began to advance once again. Umayyad armies moved beyond the Oxus and occupied Bokhara and Samarqand, while another Arab force occupied the Indian province of Sind. Large numbers of the Berber tribes of North Africa converted to Islam, and most of northwestern Africa came under Muslim rule. But the most significant move for the future history of the relationship between Islam and Christendom was the invasion of Spain.

In 711, during a dynastic crisis among the Visigoth rulers of Spain, the Arab general Târiq ibn Ziyâd, the governor of Tangier, crossed the narrow strait that separates North Africa from Europe and landed an army on what subsequently became known as Jabal Târiq—"Târiq's mountain"—or, as it is called today, Gibraltar. From there he marched inland, routed a Visigoth army, probably near the River Guadalete, and killed Rodrigo, destined to be the last king of the Visigoths. Resistance was slight and ineffectual. The bands of serfs and runaway slaves who populated the countryside went over to the invaders. The Jews, who had suffered severe persecution in 616 and knew that they would fare much better under a Muslim ruler, handed over the city of Toledo to Târiq. Târiq and most of the men in his army were Berbers, and it was they who became known in Europe as Mauri, or "Moors." This term, which originally applied to all the inhabitants of North and West Africa, no matter what their religion or skin color, gradually came to be used to describe only the Muslim populations, making a confusing elision between "Muslim," "Turk," "Moor," and even "Negro" possible.

The invasion of Spain left the Christian Mediterranean world encircled by Muslim states. "Henceforth," wrote the great Belgian historian Henri Pirenne in 1935, "two different and hostile civilizations existed on the shores of *Mare nostrum*. And although in our own day the European has subjected the Asiatic, he has not assimilated him. The sea which had hitherto been the centre of Christianity became its frontier."[20]

Legend has it that the Muslim invasion of Spain had been engineered by a certain "Count Julian," a Christian from what is now the dismal Spanish garrison town of Ceuta on the North African coast. Julian seems to have been a highly successful commander who had succeeded in preserving Ceuta from Muslim occupation until 700 when, for murky reasons, he switched sides. The story goes that he sent his daughter for her education to the court school at Toledo, then the capital of Spain. There the Visigothic king Witiza, Rodrigo's predecessor—or perhaps Rodrigo himself; opinions differ—caught sight of her while she was washing in a river, and when she, virtuous lady that she was, resisted his advances, he, in proper Gothic fashion, raped her. In revenge, Julian, who knew something about the weakness of the Visigothic defenses, went to the governor of Muslim North Africa, Mûsa ibn Nusayr, and showed him how he might overrun the kingdom. This story became a favorite of the Spanish balladeers of the following centuries.

> If you sleep Don Rodrigo
> Awaken please
> And look upon your evil end
> Your last wicked hours
> You will see your people dead
> And your battles broken
> You will see your cities and your towns
> Destroyed in a single day.
> Another lord now holds
> Your castles and your fortresses
> If you ask me who did this?
> I will gladly tell you
> It was count don Julian
> For the love of his daughter
> Because you dishonored her
> And she has nothing more.
> He cursed you then
> And your life is gone.[21]

The story is almost certainly a fantasy. But it helped to explain to many among an otherwise perplexed Christian population, which now found itself under Muslim rule, why their God had deserted them. Witiza and/or Rodrigo had sinned and had been rewarded in kind. Hard, perhaps, on the people to be made to suffer in this way for a royal whim. But then the Christian god never was a democrat. The Muslim chroniclers, too, pointed to Witiza or Rodrigo's lust, together with their cruelty, impiety, and greed, as the principal cause for the collapse of the Visigoths. For Christians in Spain, the Muslim conquest remained for centuries a warning of what God was capable of inflicting upon an obdurate, sinful population. Spain, wrote Bartolomé de las Casas, the famous "Defender of the Indians," in 1552, in outraged protest against the depredations of the Spanish settlers in the Americas, had already been "destroyed once by Moors . . . and now we have heard many say, 'Pray God that He does not destroy Spain [again] for the many evils which we have heard are committed in the Indies.' "[22]

Certainly God seemed to favor the "Moors." By 720 most of the Iberian Peninsula was in Muslim hands. What remained of the old Visigothic realm retreated to the mountainous northwest and northeast of the peninsula and to the southern marches of the Frankish Empire of Charlemagne, and there split up into a number of different kingdoms. A Christian counteroffensive began a mere seven years after Târiq's initial invasion, led by a former member of Rodrigo's bodyguard named Pelayo. It does not seem to have amounted to very much, and the Arab accounts claim that rather than being defeated, the army sent to crush Pelayo had simply gone home, leaving the Christians to their mountains. "What are thirty barbarians perched on a rock?" asked one. "They must inevitably die."[23]

This again is the story. Someone called Pelayo certainly existed, but what he actually achieved is shrouded in elaborate fictions, most of them dating from the tenth century. Slowly, however, over the next seven centuries the descendants of those "thirty barbarians" would drive the "Moors" ever southward. In Spanish history this migration has, almost since it began, been referred to as "the Reconquest"—*la reconquista*. Most historical labels are willfully misleading. This one is no exception. It presumes that the Christians were the rightful owners of all Spain. But in this world of unceasing conquest and reconquest, of migration and deportation, ownership was established not by who got there first but by who could stay there the longest. The Visigoths had usurped the Romans, who in their turn had usurped the ancient Iberians, who had usurped who knows what early wandering band, and so on until the first humanoids came out of Africa. By the time the Nasrid Kingdom of Granada was finally brought to an end by the victorious armies of the Catholic monarchs Ferdinand and Isabella on January 6, 1492, the various Arab dynasties, who had swept through what are now Spain and Portugal since 720, had been in possession of the peninsula for far longer than their Visigothic antecedents. For more than seven hundred years, Islamic Iberia, al-Andalus, as it was called, was as much a part of the *dar al-Islam* as Syria or Persia.

The kingdom founded in 720 in the mountainous regions near Oviedo became the Christian kingdom of Asturias. In the following century its king, Alfonso III, transferred the capital to the Roman city of León, and the kingdom became the Kingdom of León. In the early ninth century it also acquired a very important patron saint. One morning on the rain-drenched coast of what is now Galicia, a stone coffin was washed ashore near the Roman town of Iria Flavia (now called Padrón). When it was opened, it was found to contain the body of the Apostle Saint James the Greater, miraculously transported from Jerusalem where, centuries before, King Herod Agrippa I had himself cut off its head.

Saint James became the patron saint not so much of Spain (as he is today)—at that time there was no such place—as the patron saint of the *reconquista*. At the battle of Clavijo in 844, the apostle appeared in the sky mounted on a white charger and led the (victorious) Christians into battle. Thereafter he acquired the sobriquet "Moor-Slayer" (*Matamoros*). He is invariably depicted on horseback, sword in hand, while a cringing turbaned Moor dies beneath his horse's hooves. *¡Santiago y cierra España!*—which might be roughly translated as "Saint James and have at them, Spain!"— became the battle cry of the increasingly successful armies of the Christian kingdoms, as the Reconquest moved inexorably south. And in each recaptured church a ferocious grinning, turbaned, and hirsute Moor's head, made of wax and horsehair, was hung from the rafters. They were still there in the 1960s, until taken down on the orders of General Francisco Franco, who was then trying to establish trading relationships with the Muslim kingdom of Morocco.

The great church of Santiago de Compostela, built to house the relic, became—and still is—one of the most important pilgrim sites in all Christendom. In 996, al-Mansûr, the vizier of the caliph Hishâm II, aware of the mounting importance of Santiago to the Christian opposition to Islam, sacked the city of Compostela, although the sacred relics escaped unscathed. If the attack had been intended to undermine the emotive power of the "Moor Slayer," it dramatically backfired. Far from causing fear and disenchantment, the sacking roused indignation and anger throughout the Christian world, transforming Saint James from a relatively local Iberian saint into the symbol of the worldwide struggle against Islam. An entire network of roads was now established all the way across Europe, from Germany, Italy, and France, which converged upon the passes across the Pyrenees. From there, the famous Camino de Santiago wound, perilously close to the frontier with Islam, through Navarre, Castile, León, and Asturias to Santiago itself. These roads were lined with minor shrines, hostels, and fraternities whose task it was to house and care for—and not infrequently rob from—the pilgrims. In the twelfth century, control of the route was taken over by the monks of the Benedictine abbey at Cluny in Burgundy, under the auspices of the zealous Abbot Odilone—known to his enemies as "King Odilone"—enormously enriching themselves and their abbey in the process. Because of its wealth, Cluny became the center of a renaissance in Christian learning that, paradoxically perhaps, would lead to a concerted attempt to understand (and refute) Islam on its own terms and to the first full translation of the Qur'an into Latin.

Spain was not only a permanent battlefield; it was also the final frontier, the place where, more than any other, Islam and Christendom, Europe and Asia met. In the early Middle Ages the frontiers between religions were often as porous as the frontiers between nations. In the centuries during which most of Spain was under Muslim rule, there were inevitably a very large number of conversions to Islam. Men and women also changed their religious allegiances for love, and marriages between Muslim and Christian were not infrequent. In the end, however, perhaps because of the proximity to Christian Europe, perhaps because fewer Muslims migrated into Spain than they did into the conquered regions of the East, the bulk of the population of Spain remained stubbornly Christian.[24]

Spanish Christians and Spanish Muslims were therefore compelled to find a mode of cohabitation that was unmatched anywhere else in the Islamic world. This has come to be known by the name *convivencia* or "living together." In the early twentieth century, much was made of this by nationalist Spanish historians, who sought to recast their homeland as an idealized multicultural society, long before the term itself had been coined or before anyone had seen any need for it. *Convivencia* is not, however, wholly the creation of romantic nostalgia. The cultural frontiers between Muslim and Christian were, if anything, even more fluid than the religious and political ones. There existed a type of poetry, known as Muwashashah, which was written in Spanish but used Arabic and sometimes Hebrew script, and Spanish itself became, and remains, rich in Arabic loanwords. Many Christian Spaniards were bilingual and deeply immersed in Arabic culture. As a Cordoban Christian named Álvaro lamented in the ninth century:

Many of my co-religionists read the poetry and tales of the Arabs, study the writings of Muhammadan theologians and philosophers, not in order to refute them, but to learn how to express themselves in Arabic with greater correctness and elegance. . . . Among thousands of us there is hardly one who can write a passable Latin sentence to a friend, but innumerable are those who can express themselves in Arabic and compose poetry in that language with greater art than the Arabs themselves.[25]

Christians put on Arab dress and took up Arab eating habits. They wore Arab cosmetics and adopted falconry—originally an Arab sport—Arab horsemanship, and Arab interior design. They even—on occasion—took to regular bathing. The Czech nobleman Leo of Rozmital, who visited King Henry IV of Castile—known as "Henry the Impotent"—in 1466 was shocked to discover that even the king "eats and drinks and is clothed and worships in the heathen [i.e., Muslim] manner," and that when Rozmital was received at court, "he and the Queen sat side by side on the ground."[26]

For nearly three centuries, life for both Christian and Jew throughout most of al-Andalus was, on a day-to-day basis, relatively easy. Many "Mozarabs"—Christians living under Muslim rule—rose to positions of eminence. Some even made their way to the other end of the Mediterranean and found employment in the great cities of the caliphate. In 723, one Willibald from the Anglo-Saxon kingdom of Wessex, about as far from the *dar al-Islam* as one could get, and the leader of a band of English pilgrims to the Holy Land, was arrested in Syria on charges of spying. The Christians were thrown into jail, where they were visited by a "man from Spain . . . who had a brother who was the chamberlain [presumably the vizier] of the king of the Saracens."[27] On the vizier's intervention, the Christians were freed and allowed to continue on their pilgrimage. We do not know who these brothers were, but as Willibald refers to them simply as "Spaniards," they must have been, at least originally, Christians. Had they been either Jews or Muslims, he would have said so. In the naming habits of medieval chroniclers, only Christians had nations to which they could belong.

Willibald's Spaniards were by no means unique. But they were unusual. The less able, less fortunate of the Mozarabs lived out their lives under a regime that they saw as alien, sacrilegious, and—or so they hoped—temporary. They suffered, not always in silence, and waited for inevitable salvation. In 953, a German monk named John from the Rhineland abbey of Görz traveled on a diplomatic mission from Otto I of Germany to 'Abd al-Rahmân III at Córdoba, then the capital of al-Andalus. Here he met a Spanish bishop, also named John, who explained to him the conditions in which his flock was compelled to live:

> We have been driven to this by our sins to be subjected to the rule of the pagans. We are forbidden by the Apostle's words to resist the civil power. Only one cause of solace is left to us, that in the depths of such a great calamity they do not forbid us to practice our faith. . . . For the time being then, we keep the following counsel: that provided no harm is done to our religion, we obey them in all else, and do their commands in all that affects our faith.

John of Görz seems to have been outraged by this piece of collaborationism. "It would be fitting for someone other than you, a bishop, to utter such sentiments," he reprimanded his coreligionist. "Since you are a propagandist for the faith," he continued, "your superior rank should have made you a defender of it." Warming to his subject, John indignantly accused the Mozarab community of practicing circumcision and of "rejecting certain foodstuffs to keep on good terms with the Muslims." Yet, as John of Córdoba's lament implied, the injunction to "render unto Caesar" made collaboration with the civil power a moral duty. "Otherwise," he protested lamely in the face of the overpowering

German, "there would be no way in which we could live among them."[28]

In the 850s, a number of Mozarabs, seeking martyrdom, publicly insulted Islam and were duly executed. They became known as the "Martyrs of Córdoba" and were celebrated throughout Christendom as witnesses—which is what the Greek word "martyr" means—to the true faith in a land given over to the savage tyrannies of the "Saracens." Yet no less a person than the senior bishop of al-Andalus, Reccafred of Seville, denounced their acts as false on the grounds that they had been sought deliberately. He was reviled for doing so. But, like Bishop John, he had a responsibility to his congregation, and he, too, knew that such melodramatic displays of piety would, in the end, lead only to ever-higher degrees of religious persecution. For John at least, *convivencia* was not so much a choice as a condition of survival.

MEN LIKE JOHN and Reccafred waited patiently for God's forgiveness for whatever sins they had committed and the final, inevitable annihilation of Islam. In fact, although no one at the time could have known it, the Arab Muslim conquest of Christendom had come to an end shortly after the occupation of Spain. In the annals of Christendom the turning point was marked by two famous battles—one in the East, the other in the West. Neither was, perhaps, as decisive or as significant as latter generations came to believe. But like all great victories, they offered a sense of hope to an embattled generation and, in time, a dizzying vision of what might have been had the outcome been otherwise.

In August 717, the caliph Suleyman took a massive Arab army numbering some 80,000 men and accompanied by a fleet of 1,800 ships and laid siege to Constantinople for an entire year. The fleet was prevented from entering the Golden Horn by a chain its defenders had stretched across the entrance to the city. The army did no better, suffering heavy losses due to "Greek fire" a potent mixture of naphtha, sulfur petroleum, and quicklime, which merged into a flaming inextinguishable jellied mass that was sprayed on the enemy much like napalm in later wars. In the spring of 718, the Byzantine emperor Leo the Isaurian persuaded Khan Tervel of the Bulgars to attack the Arabs from the rear. On August 15, 718, Suleyman's successor, Umar II, abandoned the campaign and took what was left of his army back to Syria. It was the most punishing defeat that the forces of Islam had suffered, and it put a stop to any further encroachment on Byzantine territory until the Ottoman Turks began moving westward from Anatolia in the thirteenth century.

Fourteen years later, on October 25, 732, a Frankish army commanded by Charles Martel, "Charles the Hammer"—the grandfather of Charlemagne—inflicted a crushing defeat upon the Arabs at a point on the road between Tours and Poitiers a few kilometers from the confluence of the Vienne and the Creuse, in central France. The battle of Poitiers was far less significant than it was later claimed to be by the victors and their, in this case Frankish and papal, propagandists. The Christians did not fight the massive army subsequent Western accounts of the battle have often described. The Arabs, under 'Abd ar-Rahmân, the governor of al-Andalus, although they had succeeded in defeating an army under the Duke of Aquitaine and in taking possession of the city of Bordeaux before encountering Charles's forces, were little more than a large raiding party whose prime objective was not to conquer but to plunder the famously wealthy sanctuary at Saint-Martin de Tours. In describing the battle, the medieval Arab historian 'Izz ad-Din ibn al-Athir (1160–1233) said that 'Abd ar-Rahmân "went out on a *ghaza*"—that is, a raid—"into the land of the Franks," and although he duly records the Muslim defeat he ascribes no epic importance to it. The northernmost frontier of Islam was the city of Narbonne, more than seven hundred kilometers to the south; but even that was at the very outer limits of the military capabilities of the Moors, and they knew it. There was even said to be a statue in the city bearing the inscription "Turn back, sons of Ishmael, this is as far as you go. If you question me,

I shall answer you, and if you do not go back, you will smite each other until the Day of Resurrection."²⁹

But no matter what actually happened on that day, in the subsequent Western historiography of the struggle against Islam, the battle of Poitiers was represented as another Marathon. It was, after all, a major encounter with a substantial Muslim army and one that the Christians, largely accustomed to defeat, had actually won. It was a moment from which the whole of Christendom could take some encouragement, which is perhaps why, in the mid-eighth century, an anonymous Toledan cleric described the victorious Franks as "Europeans." Although it is highly doubtful that Charles's troops would have identified themselves in that way, to later generations it seemed obvious that Poitiers represented a moment in the history of the West in which the whole of Europe had been saved from the forces of barbarism, which were forever poised to engulf her.

This, for instance, is how Edward Gibbon saw it:

A victorious line of march had prolonged above a thousand miles from the Rock of Gibraltar to the banks of the Loire; the repetition of an equal space would have carried the Saracens to the confines of Poland and the Highlands of Scotland; the Rhine is not more impassable than the Nile or the Euphrates, and the Arabian fleet might have sailed without a naval combat into the mouth of the Thames. Perhaps the interpretation of the Koran would now be taught in the schools of Oxford and her pupils might demonstrate to a circumcised people the sanctity and truth of the Revelation of Mahomet.

"From such calamities," Gibbon continues, "was Christendom delivered by the genius and fortune of one man."³⁰

Poitiers did not, however, entirely put an end to the Arab incursions into southern Europe. In 734, an Arab army occupied Avignon and sacked Arles. Three years later, another *ghaza* resulted in an attack on Burgundy that yielded an enormous quantity of slaves. In 827, the Arabs invaded Sicily, where they would remain until ousted by the Normans in 1091; and for a while they maintained bases at Bari and Taranto in southern Italy. In 846, they even sacked Saint Peter's in Rome, and in 881 the great Benedictine monastery of Montecassino.

But for all that, Gibbon's counterfactual imaginings were not entirely fanciful. After Poitiers, and despite repeated raids upon the ports of Italy and Spain, no Muslim army managed to establish a permanent base farther north than Narbonne, and even that was seized by its Christian population in 759 and handed over to the Frankish king Pippin.

## IV

In the years that followed the defeats first at Constantinople and then at Poitiers, massive changes had taken place in the caliphate itself. During the reign of the caliph Marwân II, mounting opposition to the Umayyads culminated in a revolution by the descendants of 'Abbâs ibn 'Abd al-Muttalib, one of the uncles of the Prophet. In 750, the armies of Marwân II were finally defeated at the Greater Zab River, and Abû al-'Abbâs, known as as-Saffah, "the Shedder of Blood," was proclaimed caliph by his brother Dawud from the pulpit steps of the mosque in Kofa.

The new dynasty became known as the Abbasids, and the revolution they brought about in Arab and Muslim society has been compared to the French and Russian revolutions, and was probably almost as consequential. The Abbasids came to power in part with the assistance of Persian forces and with a certain amount of support from the Shiites. The man who is thought of as the founder of the dynasty, Muhammad ibn 'Alî, the Prophet's uncle and the great-grandfather of al-'Abbâs, had begun sending emissaries into Iran as early as 718, during the reign of Umar II, in an attempt to build opposition to the Umayyad Caliphate among those who had most reason to hate it. The Abbasids also adopted the black standards associated with Persian messianic movements and became known as "the black-robed ones" as far away as China.

The Arab tribes that had dominated political life under the Umayyads receded in importance. Power was now vested in the person of the caliph and in the caliph's favorites, who were often men of humble origins. The distinction between Arabs and non-Arabs gradually ceased to matter. Islam replaced Arabism as the true mark of identity, and the Umma was launched upon its career as a true universal world community.

The Abbasids also transformed a network of familial and tribal alliances into a powerful absolutist monarchy. They modified the administration in conscious imitation of the Sassanids. They created *dîwâns,* or ministries, and placed them under the supreme control of a *wazîr,* or vizier, a position they seemed to have invented and that was to become all-important throughout the entire Muslim world.[31] The army came increasingly to rely upon specially trained slaves known as Mamluks, most of them Turks from central Asia—an innovation that was to have dramatic and long-lasting consequences for many subsequent Islamic dynasties, not least of all the Ottoman Turks.

Most significantly of all, perhaps, in 750, as-Saffah's successor, al-Mansûr, moved the capital of the empire from Syria to Iraq, and twelve years later built a new city, Baghdad, on a fertile plain at a point where the Tigris and the Euphrates are less than forty kilometers apart and linked by a series of canals. Here was a site at the very center of the Islamic Middle East. According to the great geographer al-Muqaddasi, the local people told al-Mansûr that by choosing their city, he would be in a place where "The caravans will come by way of the desert, and all kinds of goods will reach you from China on the sea and from the country of the Greeks [the Byzantine Empire] and from Mosul by the Tigris."[32] Baghdad was to be, as Constantinople had been before it, a new imperial capital, created to mark the foundation of a new state. Here, al-Mansûr built himself a royal citadel. Circular in shape and known as the "City of Peace," it became the home of a lavish and elaborate court, from which the Abbasids ruled over an empire that reached from southern Italy to the borders of China and India, until they were finally driven out by the Mongols in 1258.

The Abbasids not only transformed the Arab state, they also created the conditions for the great flowering of Muslim culture that lasted from the ninth to the eleventh centuries. For it was the Abbasids who presided over what is generally considered, in the West at least, to be the great era of Islamic cultural achievement, an era most closely identified with the reigns of al-Mansûr (712–775), Hârûn ar-Rashîd (786–809), and al-Ma'mûn (813–833).

For more than half a millennium, however, there remained one surviving outpost of Umayyad rule. In 755, 'Abd ar-Rahmân, an Umayyad prince fleeing from the destruction of his family, landed at Almuñecar on the coast of Spain with an army of Umayyad supporters. He rapidly defeated the governor of al-Andalus, who had recognized the Abbasids, and the following year entered Córdoba in triumph. There he established an independent emirate, which in 921 became a rival caliphate, which would last until 1031. Under Abd ar-Rahmân II, however, al-Andalus was itself reorganized on Abbasid lines, and both cultural and political connections were reestablished between Spain and the caliphate in the East.

UNDER THE EARLY Abbasids, the lands of the *dar al-Islam* were incomparably more sophisticated, more tolerant, more open, and richer in every conceivable aspect of life than the rude Christian kingdoms of the West. The Islamic world was a world of cities and of commerce, of the urban—and urbane—culture that cities inevitably foster. Christian Europe—with a few notable exceptions, most of them in Italy—was a world of villages and fortified hamlets, which the Christians called "towns," and an economy that was largely agrarian. The old Roman world, with its wealth and its ordered administration, its roads, its great villas, and its sheltering soldiery, had all vanished or crumbled into ruins. The hypocausts and the aqueducts had all, but for a few, been broken and abandoned. The great Roman buildings had become nothing more than quarries for materials with which to build and decorate the rough, unheated, unkempt castles from which an illiterate warrior nobility tyrannized the surrounding countryside. Even the empire of Charlemagne, which covered much of Italy, France, and what is today Lower Saxony and Westphalia, was nothing in size when compared with the realms governed by his Muslim contemporary Hârûn ar-Rashîd. In Carolingian Europe, there was very little to compare with the complex administrative systems of government that controlled the infinitely more extensive, vastly more dispersed, domains of the caliphate. The imperial capital at Aachen, or Aix-la-Chapelle, although sumptuous—not to say pompous, by contemporary Western standards—was a hamlet by comparison with Baghdad or Damascus.

In Spain, too, this was a period of great cultural revival. Under its Visigothic rulers the Iberian Peninsula had been a frankly chaotic, impoverished, and backward kingdom, far removed from the prosperous Roman province of Hispania, the birthplace of one emperor—Trajan—and of the ancestors of two others—Hadrian and Marcus Aurelius—and the home of some of Rome's greatest writers: Seneca, Columella, Quintilian, Martial. "Of all that she possessed once," wrote one early chronicler, "she retained only the name."[33] The Moors had transformed all this. They had rebuilt the great cities of Málaga, Córdoba, Granada, and Seville, given them running water, and adorned them with sumptuous palaces and gardens. They had introduced scientific irrigation and a number of new crops, including citrus fruits—the famous Seville oranges—cotton, and sugarcane (al-Andalus became the main source of sugar for much of Europe before the final extinction of Muslim Spain in 1492). They had created textile industries in Córdoba, Málaga, and Almería; pottery in Málaga and Valencia; and arms in Córdoba and, for centuries to come, in Toledo, where dismal simulacra of "Toledan steel," damascened and gilded, is still produced for the tourist market. Leather was made in Córdoba, carpets in Beza and Calcena, and paper—which the Arabs introduced into Europe from China in the eighth century—in Játiva and Valencia. By the end of the tenth century, when it was at its prime, the Muslim emirate of al-Andalus, with its capital at Córdoba, had become the most prosperous, most stable, wealthiest, and most cultured state in Europe.

Yet the world of Islam, this vast sprawling conglomerate united by religion and, at least among the clerical classes, by language, had almost no interest in and very little knowledge of the inhabitants of the western *dar al-harb*. That realm existed only on sufferance, waiting for the day when it would finally be incorporated into Islam. What united Arabs to Iranians and Berbers, the Fulani and Senegalese of West Africa, and countless other peoples around the globe, was not culture or language or some more amorphous understanding of a "civilization"; it was a reverence for the word of God as transmitted by his prophet. This carried with it certain implications. Muslims had very little direct linguistic access to what was frequently referred to as "the land of the Franks." Knowledge of languages other than Arabic, Persian, and later, Turkish—the official languages, that is, of the *dar al-Islam*—was considered to be unnecessary, possibly even impious. A translation of the Bible into Arabic by a Muslim would have been unimaginable. No Muslim scholar before the eighteenth century displayed the slightest interest in any European language. There was nothing to match the steady growth in the West of what would come to be called "Oriental studies" or the chairs of Arabic created at Oxford, Cambridge, Paris, and Leiden in the sixteenth and seventeenth centuries; nothing remotely to compare with the translation of the Bible into Arabic, Latin, and Castilian produced by John of Segovia in the fifteenth century. (Although it must be admitted that this remarkable work was seemingly of so little interest to the professors of the University of Salamanca to whom it was entrusted that they lost it, and it has never been found.) Such contacts that did take place between Muslims and "Franks" were largely confined to diplomacy and commerce, and these relied upon non-Muslim intermediaries who could speak Arabic, or on a pan-Mediterranean jargon composed of fragments of Portuguese and Italian, liberally scattered with Arabic and known as the "lingua francae"—or "speech of the Franks."

When an ambassador from the Frankish queen Bertha of Tuscany arrived in Baghdad in 906, carrying with him a letter written in Latin, which apparently proposed a marriage between Bertha and the caliph, there was no one at the court who could read it. They could not even identify the script in which it was written. It was, said one Arabic contemporary, composed in "a writing resembling the Greek writing, only straighter." Finally a Frank who worked in a clothing store was brought to the caliph, al-Mutawakkil. "He read the letter and translated it into Greek writing. Then Hunayn ibn Ishâq was summoned and he translated it from Greek into Arabic."[34] There is obviously something wrong with this account. How would *transcribing* a Latin text into the Greek alphabet assist a man like Hunayn ibn Ishâq, a Christian from Jundayshapur and the translator of Hippocrates and Galen, who knew Greek but no Latin, to translate it into Arabic? And how did a Frank possess a knowledge of both the Latin and Greek alphabets? One thing, however, that this encounter makes clear is that for the caliphs in Baghdad the Latin West was possibly even more remote than China.[35] (The caliph, it goes without saying, refused Bertha's offer of marriage, although how this was conveyed to her is not recorded.)

The Franks themselves were recognized to be a brave, often savage race, with a poor sense of personal hygiene. "The Fraguis," wrote one Muslim from South Asia in the seventeenth century, would be a great people, "but for their having three bad aspects: first they are Cafares—that is an infidel people—secondly they eat pork and thirdly they do not wash those parts from which replete nature expels the superfluous from the belly of the body."[36] As for the lands from which these "Franks" came, the attitude of the tenth-century geographer Ibn Hawqal was typical and would remain so for centuries to come. The land of the Franks, he said, was a good source of slaves. That was all that could be said about it.[37]

There was, however, one notable exception to this widespread indifference to the products and peoples of the West—although, like so many such exceptions, in the end it only confirms the rule. Muslim scholars may have had no interest in the culture of the West, still less in the philosophy of the Christians, their theology, or whatever science they might possess—such as it was, and it was not much. But like their Christian contemporaries, they were drawn by the immense intellectual power of the ancient Greek world. True, this was pagan, but it also predated Islam, which made it far less noxious, and very few Greek writings, or at least none of those to which the Arabs had access, were concerned directly with matters of religion. The writings of the ancients were mined, as they were in the West, for the privileged access they provided to the workings of nature. Aristotle's apparent argument for the existence of God, that all movement must have its origin in an "unmoved mover," suited the Muslim theological mind quite as well as it did the Christian.

During the reigns of al-Mansûr (712–775) and his successors, Hârûn ar-Rashîd (786–809) and al-Ma'mûn (813–833), translation from Greek, Syriac, and Coptic not only became respectable but was even patronized by the caliph himself (although the translators were, for the most part, Christians). Al-Ma'mûn went so far as to set up a school for translators at Baghdad with a regular staff and its own library. The caliphs sent scholars in search of manuscripts as far afield as Byzantium, and a number of works are today known only through the Arabic versions made during this period. Most of the works chosen were scientific or philosophical, above all the writings of Plato and Aristotle, together with some hermetic, gnostic, and neo-Platonic studies on medicine, astronomy, astrology, alchemy, chemistry, physics, and mathematics.

This activity gave rise to an entire school of hellenizing philosophers, jurists, and doctors, most of them Persians: the physician and alchemist ar-Râzî, known in the West as "Rhazes"; the surgeon Abû al-Qâsim az-Zahrâwi, known as "Albucasis"; the mathematician and astronomer Muhammad ibn Mûsâ al-Khwârizmî, after whom a crater on the far side of the moon is now named; the astronomer Thâbit ibn Qurrah; al-Kindî—the only Arab among them—and Abû Nasr al-Fârâbi, who tried to reconcile not only Islam and Greek philosophy but also Plato and Aristotle and transformed Plato's notion of the "philosopher-king" into the "imam-philosopher" who combines in himself the perfection of both religion and the "contemplative life."

For al-Fârâbi, the Greek view that "happiness" (or what the Greeks called *eudaimonia*) could only be achieved through life in a particular kind of community—in this case the *polis*, the Greek city-state, beyond whose borders, in Aristotle's phrase, lived only "beasts and heroes"—had an exact parallel in the Muslim claim that salvation could be found in the Umma of the Prophet. (The Christians found a similar parallel in the Christian Church.)[38]

Perhaps the greatest of these scholars, and certainly the most varied in his interests and accomplishments, was Muhammad ibn Ahmad al-Biruni, physician, astronomer, mathematician, physicist, chemist, geographer, and historian, who in 1018 made calculations, using instruments he had created himself, of the radius and circumference of the earth, which vary by as little as 15 and 200 kilometers from today's estimates. Sometime after 1022 he traveled to northern India with the armies of Mahmud, the Afghan ruler of Ghazna. There he learned Sanskrit and wrote a treatise, *The Book of India*, which, although hardly enamored of Hindu polytheism, offers a sympathetic view of Indian culture and, in particular, Indian philosophy.

The best known in the West, however, was Ibn Sînâ, called "Avicenna." An Iranian born in Bokhara in 980, Avicenna attempted to synthesize Aristotelianism, Platonism, and Neoplatonism in a work appropriately entitled the *Kitâb ash-Shifâ*', or *Book of the Cure for Ignorance*. His most lasting contribution, however, called *Qânûn fi at-tibb*, or *Canon of Medicine*, was a vast treatise that brought together all the medical knowledge of the ancient Greek world then available, from Aristotle, Hippocrates, and Galen, and enriched it from other sources, most significantly Persian and Hindu pharmacology. This work not only became a reference for all later Arab doctors but was widely used in the medical faculties of Christian Europe until replaced by a more experimental approach to the subject in the seventeenth century.

Spain, too, had its jurists, theologians, and philosophers whose works were known and respected in the West. There was Ibn Tufayl, called "Abû Bakr," the author of a philosophical novel that became one of Daniel Defoe's sources for *Robinson Crusoe*.[39] Ibn Bâjjah, known as "Avempace," developed a vision of the virtuous man as a solitary individual separated from the inevitably unvirtuous and imperfect world in which he is destined to live, a philosophy whose embodiment in the West was the monastic life. But the greatest of the Hispanic philosophers, and certainly the most influential of all the Muslim Hellenists, was Abû al-Walîd Muhammad ibn Rushd, or "Averroës," as he was called by his Latin readers. Born in Córdoba in 1126, he produced commentaries on Aristotle so highly regarded in the Christian world that he became known simply as "The Commentator," just as Aristotle was known simply as "The Philosopher," and this is how he appears, peering over the shoulder of Aristotle in Raphael's great fresco *The School of Athens* in the Sistine Chapel in Rome. Dante found a home for both him and Avicenna in Hell. But he gave them a relatively comfortable place in the second circle in congenial company with Aristotle himself, "the master of those that know," Socrates, Plato, Cicero, and Seneca, among others, all of whom were "virtuous pagans"—those, that is, who were there not because they were guilty of sin but because they had not been baptized. And thus because "they before the Gospel lived, they served not God aright."

> *For these defects*
> *And for no other evil, we are lost,*

*Only so far afflicted that we live*
*Desiring without hope.*[40]

Although what a pair of Muslims was doing in this company Dante does not say.

Averroës' commentaries, in particular on the *Physics, Metaphysics, De Caelo,* and the *De Anima,* works that constitute the basis of the Greek natural and human sciences, became a central component of the university curriculum in Europe until the end of the sixteenth century. From about 1230 until 1600, it was Averroës who, together with Aristotle himself, was responsible for introducing philosophical rationalism into the Christian West. His treatise *Fasl al-maqâl,* known in English as *On the Harmony of Religion and Philosophy,* sparked off a controversy involving no less a figure than Saint Thomas Aquinas, "Prince of Theologians" and "Doctor of the Universal Church." Philosophy—by which he meant syllogistic logic—was, Averroës argued, recommended, even required, by divine law, since man had an obligation to "know, by demonstration, God and all those things to which he has given being."[41] Philosophy was one of three ways to do this, of which dialectic and rhetoric were the two others. Each man must choose his own way, for as it says in Qur'an 16:125, "Call men to the road of the Lord by wisdom and exhortation and dispute with them in the best way possible." Only those whom Averroës describes contemptuously as a "small number of narrow-minded literalists" have denied this, and they can easily "be refuted by the most univocal questions from revealed scripture."[42]

The reaction of the more orthodox Muslims—and not merely "the narrow-minded literalists"—to the use of secular reason was varied. The impact of the writings of al-Kindî, al-Fârâbi, and Ibn Sînâ was profound and long-lasting. But the response of many of the theologians was suspicious when it was not hostile. (Something of the same response initially greeted Aquinas's turn to Aristotle. Theologians are habitually suspicious of innovation, in particular when it comes from secular or pagan sources.) The

most powerful—and most celebrated—onslaught on the philosophers came from Abû Hâmed al-Ghâzali (1058-1111)—known as
"Algazel" in the West—whose *Tahâfut al-falâsifah* (*Incoherence of the
Philosophers*) denounced the writings of the ancients as "incoherent" because they were contrary to the revealed wisdom of God,
from whom all truth must come. Those he called "the Philosophers," his main targets, had limited the knowledge of God to
universals, which constituted a denial of the qur'anic (and biblical) image of a God who cares for and knows about every creature
in his creation. Human intellect could have no direct access to
true knowledge without divine guidance from the prophets.
Ghâzali's view had the effect of reinforcing dogma in the face of
what seemed to many to be a purely human challenge to divine utterance. The sobriquet by which he was known, "Proof of Islam,"
was well earned. But it should be said that he had arrived at his
conclusions from a deep skepticism about the sources of all
human knowledge, which was itself philosophical and Greek in
origin if not in inspiration—the same skepticism that in the seventeenth century would lead many in Europe to abandon the idea of
a world controlled by a god altogether.

Great though this "Arab Renaissance" clearly was, at least as
seen from the West, it did not endure. Although it survived for the
better part of half a millennium and made a lasting impact on
subsequent Islamic thought—its works were still being cited (and
reviled) in the twentieth century—it had come to an end by the
close of the twelfth century. Averroës was not only the greatest of
the Arab Muslim scholars and perhaps the most influential of all
Muslim philosophers, he was also the last. When he died in 1198
in exile in Morocco, the victim of a war against "philosophy" and
its adherents throughout the Islamic world, the "Arab Renaissance" died with him.


FOR MANY LATER writers, both in and beyond the Islamic
world, in particular after the apparent decline of Muslim power in

the eighteenth and nineteenth centuries, this outburst of philosophical, legal, and scientific creativity also raised some difficult
questions. If Muslim society had been able to create and sustain a
culture that had surpassed anything in Europe at the time, why
had it failed to develop? How had the West been able to overtake
it? And perhaps rather more troubling, if it had scaled such intellectual heights once, might it not be able to do so again? In other
words, might it be a little rash to write off Muslim culture—as so
many later Western writers did—as incapable of modernization?

One of those who claimed to have an answer was the great
French theologian, historian, and polemicist Ernest Renan. In
March 1883, Renan delivered a lecture in the great hall of the Sorbonne in Paris, on the topic "Islam and Science." Renan wanted to
show that not only Islam but all the monotheistic religions were
incompatible with the progress of modern science. Yet the nominally Christian world had made immense progress, while no one
who had traveled in the "Orient" or in Africa could fail to be
struck by the backwardness, the decadence, the "intellectual nul-
lity" of all those who "derive their culture and their education
wholly from this religion"—that is, Islam. Renan's explanation for
the now-immense distances that separated the two heirs of Judaism was the now-familiar one: the difference between Islam and
Christianity lay not in the nature of their respective beliefs but in
the fact that Islam, unlike Christianity, had succeeded in making
itself master of all civil and political life. "Islam," Renan declared,
"is the indiscernible union of the spiritual and the temporal. It is
the kingdom of dogma, the heaviest burden that humanity has
ever had to bear." A sort of "circle of steel" bound the head of the
true believer, "making him absolutely closed to science, incapable
of learning anything, or of being open to any new idea." Only in
the Islamic nations and the Papal States had religion exercised
"such a domination over civil life." But whereas in the Papal States
this had brought oppression to only a very small number of people, Islam had oppressed "vast portions of the globe, and there has
maintained the idea which is most opposed to progress: a state

founded upon a supposed revelation, theology governing society." Christian theology had succeeded in crushing the human spirit in only one country—Spain—where a "terrible system of oppression has snuffed out the spirit of science." (Never fear, he added prophetically, "that noble country" would before long takes its revenge.) Islam, however, had crushed the "modern spirit" in every country it had conquered.

How, then, do we explain the so-called Arab Renaissance? Renan's reply was simple. If, he said, one were to follow, "century by century the civilization of the Orient," from 775 to the middle of the thirteenth century, what would one find? That the "momentary superiority" enjoyed by the Arab world had been an illusion that owed nothing to the Arabs and that had come about not because of Islam but very much despite it. Look closely at it, he argued, and the whole "Islamic Renaissance" can be made to disappear. The early Arabs, he declared, had been poets and warriors, simple men whose religion prevented them from undertaking any kind of rational inquiry. The Bedouin, "the most literary of men," were also the least given to reflection. The caliph Umar did not, as was widely believed, burn down the famous library at Alexandria; it had perished long before he arrived in Egypt in 642. But had he been able to do so, he surely would have. All that he stood for, all that he had helped spread across the known world, was "destructive of scholarly research and of the varied labors of the mind." But if these primitive tribesmen were simpleminded, they were also relatively tolerant, or at least too disorganized to be able to inflict very much misery. In the "second age," however, when Islam came under the sway of Tartars and Berbers, "races, heavy, brutal and spiritless," a reign of "absolute dogma" took over a system "that has been surpassed, in its persistent injustice and unjust persecution only by the Spanish Inquisition."[43]

Then, in 750, Sassanid Persia fell to the armies of the new Abbasid dynasty and the center of Islam was translated to the banks of the Tigris and the Euphrates. Here it became domesticated by what remained of the great court culture of the Sassanid emperors, in particular the last Zoroastrian shananshah, Khusrau II. When philosophy had been "chased out of Constantinople," Khusrau had given it a home in Persia. He had had books translated from Sanskrit, and his achievements had in large part been sustained by Christian, and in particular Nestorian, refugees. The city of Harran in Syria, the ancient site of the worship of the moon goddess Sin, known as the "heathen city" to the early Church Fathers, had managed, even under the Christian Roman emperors, to hang on to its pagan manners, if not exactly its paganism, and thus had "preserved all the scientific tradition of Greek antiquity."

The Abbasid caliphs who surrounded themselves with Persian advisers and Persian troops led to a partial revival of the glories of Khusrau II. "Their most intimate advisers, the tutors of their princes," claimed Renan, "their chief ministers all came from the Barmecide, an ancient and enlightened Persian family who remained loyal to the national cult of Zoroastrianism and converted to Islam only late and then without conviction." (None of this seems to be true. Khâlid al-Barmakî, one of the first Abbasid viziers, was an Islamized Central Asian whose family had originally been Buddhist, not Zoroastrian). Abbasid Baghdad became, in Renan's interpretation of it, a mixed society; although Arabic-speaking and confessionally Muslim, it was a culture sustained by Parsees and Christians. All the great caliphs of the time of Charlemagne—al-Mansûr, Hârûn ar-Rashîd, al-Ma'mûn—were, in Renan's view, "hardly Muslims at all. Outwardly they practiced the religion of which they were the leaders, the popes if you will, but their spirits were elsewhere." They were hardly even Arabs, but "a sort of resuscitated Sassanids." Sometimes, in order to placate their more puritanical subjects, these men were obliged to act as good Muslims should: ferocious, intolerant, and unthinking. A few impious friends and freethinkers would be sacrificed to the faithful, and then "the caliph would recall his wise men and the companions in pleasure and the free life would begin again." Only this, in Renan's view, could explain such a text as *The Hundred and*

*One Nights,* that "bizarre mixture of official rigor and secret laxity." Under the aegis of these men a culture flourished that by the twelfth century reached all the way from Baghdad to Córdoba. The works of Galen, Aristotle, Euclid, and Ptolemy were translated into Arabic. Men like al-Kindî began to speculate on the "eternal problems men raise without ever being able to solve." In other words, they began to philosophize, something no true Muslim, for whom there could never exist a problem for which there was not an answer in the Qur'an, could ever do. And two men, al-Fârâbi and Avicenna, "could be placed on a level with the greatest thinkers who ever lived." All this activity has been called Arabic because it was written in Arabic, and because it was written in Arabic it has been assumed that it was in some sense also Muslim. But in fact it was "Greco-Sassanid"; its creators were nominally Muslims, Christians, and Jews but, like all true intellectual endeavors, it was wholly untouched by any religious conviction.[44] It had been a momentary burst of light in an endless night. It would never come again. With Averroës' death "Arab philosophy lost its last representative, and the triumph of the Qur'an over freethinking was ensured for at least the next six hundred years."[45] "The greatest service we can offer the Muslims," Renan concluded to rapturous applause, "is to liberate them from Islam." The European scientific renaissance had been created in opposition to Christianity. Modernity, if it were ever to come to the East, would similarly have to come in opposition to Islam. Religions—all religions—should be treated as so many different "manifestations of the human spirit." But their followers should never be allowed to seize control of civil society. In his pugnacious onslaught upon the baleful effects not only of Islam but of all revealed religion on the progress of human reason, Renan was being, as always, contentious. But on the last point at least, he was also not far wrong.

Although the works of the great Muslim scholars had finally been snuffed out in the Islamic world, they nevertheless made a lasting contribution to Western Christian science. The work of the translators of Baghdad, declared the English translator Daniel of Morley, were the new "spoils of the Egyptians," just as the more literal treasure of the pharaohs, which the Jews took with them when they fled across the Red Sea, had been for Moses. "Let us then," he wrote, "in accordance with the commands of the Lord, and with his help . . . despoil these infidels that we might enrich our faith."[46]

And despoil them they did. In the eleventh century, the texts that had been translated out of Greek and Syriac into Arabic began to be translated from Arabic into Latin, together with the works of the more appealing, to Christian sensibilities, of the Arab Hellenists, in particular Avicenna, al-Kindî, and al-Fârâbi, and the medical writings of Averroës. The process of translation was, it must be said, a pretty hit-and-miss affair, in which texts were translated out of Greek into Syriac—a dialect of Aramaic—and from Syriac into Arabic, and then from Arabic into Spanish, and finally from Spanish into Latin. But clumsy and often hilariously inaccurate though such four-handed translations were, they had the effect of making a large number of otherwise unknown classical sources by Euclid, Aristotle, Galen, Ptolemy, and others available in the West. They also drove other scholars, in the following centuries, to attempt their own translations directly from the Greek. By the time of his death in 1286, the Flemish Dominican William of Moerbeke had made most of the essential writings of Aristotle available in reasonably reliable translations from the Greek originals. It was Moerbeke's versions of Aristotle's *Nichomachean Ethics, Politics,* and *Economics* (although this later turned out not to be by Aristotle) that Saint Thomas Aquinas would use to transform the theological and philosophical landscape of Europe. The scientific, literary, and philosophical activities of Western scholars during the twelfth and thirteenth centuries would also finally culminate in the attempt to edit and translate all that could be retrieved of the scientific, philosophical, and literary writings of the ancient world. And it was this that provided the foundations for the European Renaissance of the fifteenth and sixteenth centuries.

## V

While the Muslim response to the existence of Christendom was one of ignorance and indifference, the Western, Christian response to the rise of Islam was initially one of panic. Fear, curiosity, and loathing would mark most of Europe's dealings with the various Islamic peoples, first the Arabs, then the Mongols, the Ottoman Turks, the Safavids in Iran, and the Mughals in India, until they began to show signs of weakness and decay in the eighteenth century. From the perspective of the twenty-first century, it is perhaps hard to imagine a time when the societies of what we loosely call the West were not in the ascendancy. Yet for nearly a thousand years most Europeans, even in so remote a place as England, could never be entirely certain that if not they, then their children or their children's children, would not one day be compelled to live under a Muslim ruler. Gibbon's vision of minarets in Oxford was for him a speculative fantasy. For his not-so-distant ancestors, however, it had sometimes looked like an all-too-potential reality.

At first, however, the Christian West had very little understanding of who these barbarian interlopers were. They called them, invariably, not Arabs but Ishmaelites, that is, the descendants of Ishmael, one of the sons of Abraham, "a wild man," according to the Book of Genesis, whose "hand will be against every man, and every man's hand against him." More frequently they were the "Saracens," believed to be the descendants of Sarah, one of Abraham's wives. But whatever lineage they were given, they were always outcasts, the biblical scourge upon the civilized world. "What could be more dire," wrote Maximus the Confessor from Alexandria between 634 and 640, than to see "a barbarous nation of the desert overrunning another land as if it were their own, to see our civilization laid waste by wild and untamed beasts who have merely the shape of human form."[47]

Islam itself was equally baffling. Some of the stories that did the rounds of the taverns and the monasteries of early medieval Europe were delirious. Muslims believed that Muhammad was a god or, better still, one god in a pantheon that sometimes even included the Qur'an. His close associates were called Jupin, Apollon, and Tervagant, all corruptions of the names of classical deities. He was, or so the deacon Nicolas supposed, the founder of the Nicolaitans, an obscure sect condemned by Saint John in the Book of Revelation, and who, according to Saint Irenaeus, lived "lives of unrestrained indulgence."[48] He was even, on one account, an embittered cardinal who, having been passed over for pope, had established his own religion in the desert.[49]

These stories persisted well into the twelfth century and beyond. Slowly, however, as the frontiers of Islam advanced westward and direct contact between Christian and Muslim became more frequent, more information became available. From this it soon became evident that Muslims believed, as Christians did, in one god; most probably they also believed in the same god. Certainly he seemed to have some of the same characteristics as the god of the Old Testament. Allah was a warrior, a jealous god, vengeful and censorious. But then so too was Jehovah. He was, also, however, "the compassionate, the merciful," as was the god of the New Testament. There was no difference here of any real theological significance. Muslims also recognized the patriarchs, prophets, and kings of the Old Testament. They recognized Jesus, or Isa, as the last of the prophets before Muhammad and venerated his mother, Mary, to whom the entire nineteenth *sura* of the Qur'an is dedicated. For a few, at least, the similarities between the two faiths seemed to suggest that some kind of reconciliation between them might be possible. In 1076, Pope Gregory VII wrote to the Muslim ruler of Algeria, an-Nâsir, saying:

> There is a charity which we owe each other more than to other peoples, because we recognize and confess one sole God, although in different ways, and we praise and worship Him every day as creator and ruler of the world.[50]

The pope had reasons of his own for seeking an accord with an-Nasîr, reasons that had less to do with articles of faith than with a desire to protect the shrinking communities of Christians in North Africa who still accepted his sovereignty. Nevertheless, it was clear that as far as an understanding of the nature of the divinity was concerned, what mattered was the oneness of God and his role as both sole creator and sole source of all authority. On this both Christianity and Islam were in agreement. Where they clearly differed was over the central Christian doctrines. To Muslim eyes, the Trinity—God the Father, Son, and Holy Ghost as distinct yet indivisible—looked, not unreasonably, like polytheism, cloaked in obfuscation by Christian apologists. The Incarnation was taken to mean that the Christians had made a god of their prophet; and the Resurrection was an event that had never taken place. Jesus, who was not the Christ, had never been crucified. (The idea of a suffering god, or even a suffering son of God, is wholly alien to Islam, as indeed to most other religions. Gods do not suffer pain. They inflict it.) Jesus had not even died. He had instead been taken directly into Heaven (on this both sides were in agreement). There had therefore been no need for him to be resurrected.

The Christians had not dissimilar problems with what they understood to be the beliefs of the Muslims. But because there are no "mysteries" in Islam, and because they could more easily recognize their own divinity in that worshipped by Muslims than vice versa, they concentrated their polemics against the Prophet himself. At first, Islam seemed to be little more than yet another heresy. Muhammad was merely another false prophet, not unlike Mani (with whom he did indeed have much in common), who, in the opinion of Saint John Damascene, doctor of the Church and last of the Greek Fathers, "having been casually exposed to the Old and New Testament, and supposedly encountered an Arian monk, formed a heresy of his own."[51] Many of the subsequent accounts of Islam and its founder were of this kind, part misreading, part distortion, and part—in, for instance, Saint John's claim that a Muslim may have "four wives and one thousand concubines if he can, as many as he can maintain besides the four wives"—simple description made grotesque by comparison with Christian customs, which were assumed to be the God-ordained norms of all humankind.[52]

Perhaps the first attempt to come to grips with the substance rather than the myth of Islam was the earliest full translation into Latin of the Qur'an. In 1142, Peter the Venerable, abbot of the Benedictine monastery of Cluny, went on a tour of inspection of the various Cluniac foundations along the pilgrim route from Paris to Santiagó and traveled around for a while with the court of Alfonso VII, king of León and Castile. In the course of his journey he seems to have become aware for the first time of the presence of Islam and, perhaps more troubling, of the fascination many of the Spanish clergy seemed to have for Arabic culture. He decided, therefore, to start a war, not of arms, which hitherto had proved woefully unsuccessful, but of words. "I approached," he later recalled, "specialists in the Arabic language from which proceeds the deadly poison that has infected more than half the world and persuaded them . . . to translate from Arabic into Latin the origin, life, teachings, and laws of that dammed soul [Muhammad] that is called the Koran."[53] A year later, an Englishman named Robert of Ketton completed his translation of the Qur'an. True, given his patron's objectives, this was hardly disinterested. The entire work, Ketton declared, was evidence of the excellence and sanctity of Christianity, and to prove the point he equipped it with a commentary that repeated many of the more lurid fabrications of earlier writers. (It also lay, largely unread, in the great library at Cluny until rediscovered and published in the sixteenth century.)

The image that Peter's translators, commentators, and their successors created of Islam was of a heresy, indeed the greatest of all heresies. "Not if you were to enumerate all the heresies that sprang up by the Diabolical spirit through a thousand and one hundred years from the time of Christ," Peter told Saint Bernard of Clairvaux—who was partially responsible for the launching of the Second Crusade—"and place them all together on a scale could they equal this one."[54] Paramount evil though Islam clearly was, it was, nonetheless, in Christian eyes, theologically little more than a particularly virulent variant of Arianism, which had already been condemned—and, it had been hoped, annihilated—by the Council of Nicaea in 325.

Muhammad himself was clearly a false prophet. "The signs of a true prophet," wrote the eleventh-century Jewish convert Pedro de Alfonso, "are probity of life, the presentation of true miracles and the constant truth of all his sayings."[55] Muhammad could obviously be faulted on all three. No miracles had been performed by him during his lifetime, for the good reason that, unlike Christ, he had never claimed to be able to perform any. Yet for most Christians this was taken as proof from Muhammad's own mouth that he was an impostor. Muhammad was the self-appointed "Seal of the Prophets," successor to, and greater than, either Moses or Jesus. Yet both Moses and Jesus had been granted by God the power to perform miracles. Why, then, had not Muhammad? The miracles that overzealous Muslims had attributed to their prophet, despite his denial of the ability to perform any—a talking ox, a fig tree that prostrated itself and came when the Prophet called, the moon that divided into two and was then rejoined, a poisoned leg of lamb that warned Muhammad not to eat it—were, not surprisingly, all merely ridiculous to Christian eyes. As for the Qur'an, this was nothing more than fabrication, a parody of the Bible, filled with grotesque and absurd fables—"idle tales worthy of laughter," Saint John Damascene had called them—and it clearly contained nothing that was true.

It was Muhammad's life, however, that was always the prime target of the most virulent Christian attacks. Since they accepted Jesus as a true prophet, Muslims limited their derogatory comments on Christianity to the Christians' attempts to make a mere man into a divinity. Christians, however, were under no such restraints when it came to depicting Muhammad. Here, it seemed, was a low-born upstart, an impostor and a fraud, who had fabricated incredible and monstrous prophesies so as to further his own always ignoble and frequently disgusting ends, most of them political and sexual. It was Muhammad's sex life, in particular, and the Muslim failure to condemn the pleasures of sex—at least for men—that most fascinated and agitated Christian polemicists. Muhammad was a "shameless adulterer," "luxurious," "fetid," and "insatiable," who "burned with a libidinous ardor above that of all others," an unstable and uncontrolled womanizer who had dreamt up many of the laws in "his Koran" simply to sanction his own licentious behavior.[56] The most popular story about Muhammad's supposed sexual proclivities, which Christian polemicists told over and over again in ever more elaborate versions, was that of his marriage to Zahnab bint Jahsh after her divorce from Zayd ibn Haritha. Zayd ibn Haritha was an adopted son of Muhammad. He had a wife, Zahnab, said to be the most beautiful woman on Earth. As soon as Muhammad set eyes on her, he was seized with an uncontrollable passion. He therefore went to Zayd and told him, "God commanded me [to tell you] that you should divorce your wife." Zayd, being an obedient and believing man, duly did so. Some days later Muhammad returned. Now, he announced, "God has commanded me that I should take her." And take her he did. "And afterwards she used to go in glory in front of Muhammad's wives, saying, 'You were given to the Messenger of God as wives by your friends on Earth, but God married me to the Messenger of God from Heaven.' " Because of this little escapade, Muhammad came up with the law of *muhallil*, or repudiation, which, in Christian renderings of it, allowed any man to divorce his wife whenever he so chose, simply by renouncing her twice.[57]

Sometime after he had acquired Zahnab—so the story went— the Messenger of God was caught in the act by another of his wives, named Hafsah, with a woman called Mariah the Copt. It was not his infidelity—in any case a meaningless concept in a polygynous society—that troubled her, so much as the fact that Muhammad was doing it in her house, on their bed, and before her very eyes. In order to placate her, Muhammad swore not to "lie" with Mariah again. Afterward, however, "against this promise and oath, he lay with Mariah again, and said in his Koran, 'God appointed satisfaction of their oaths for the Muslims,' that is, if they wished to make some oath and then wished to go against it, they could do so without expiation, without satisfaction." Not content with using his spurious reputation as a prophet to seduce another's man's wife, Muhammad had then made marriage a mere matter of convenience, by allowing any husband to repudiate his wife at will, and had compounded this by claiming that lying and hypocrisy were God-given rights to him and his followers. Never ones for irony or understatement, his Christian detractors suggested that all this promiscuity notwithstanding, there was also more than a mere hint of the effeminate about him. "Muhammad put on purple and used scented oils," alleged San Pedro Pascal, "that he might smell sweet, and colored his lips and eyes, as the leaders of the Moors and many others, of both sexes are accustomed to do." Clearly, no such person could ever possibly have been entrusted with the word of God.

All of this was a parody, as Saint John, at least—who, as Ibn Mansûr, had lived much of his adult life as the loyal servant of the caliph at Damascus before retreating into the monastery of Saint Sabas in the desolate lands between Jerusalem and the Dead Sea— would certainly have known.[58] But such a vision of Islam would have a long, and often potent, life. The struggle between the virtuous paladins of Christendom and the monstrous hordes of the Saracens became a popular topic in both the sacred and the secular literature of medieval Europe. Here was Christendom— embattled, numerically disadvantaged, often weak, but always virtuous, always noble, always right—pitted in undying conflict against the mighty but corrupt and monstrous world of Islam. One of the most colorful examples is the great French eleventh-century epic poem *Le Chanson de Roland* (*The Song of Roland*), which reworks a relatively minor conflict in 778 in which a group of Basques cut off the rear guard of Charlemagne's army in the Ronscevalles pass in the Pyrenees—and was thus a battle between Christians—into a titanic struggle between Christianity and Islam. In this version, the "Admiral" of Babylon, who is older than

Virgil or Homer, gathers together an army from every corner of the East, from Hungary to Africa. The invading Muslims worship the now-familiar triad of idols, "Mahumet, Apollin and Tervagent," whose images adorn the banners the "Saracens" carry before them into battle. Needless to say, these deities prove to be powerless before the victorious Christian army. Roland himself dies heroically, although not before he has had time to cut off the right hand of the Saracen king "Marsile," for which spurious deed Dante allotted him a place in Paradise. But even without Roland, Charlemagne—he of the *barbe fleurie* and by this time two hundred years old—aided by the angel Gabriel, drives the Muslims back to Saragossa.

*The Song of Roland* is a blatant piece of Christian chauvinism, a Crusader text even if it may have been written before the preaching of the First Crusade in 1095. Yet even in this most extreme depiction of the struggle between good and evil, not everything is always quite as one might expect. Even here, there are good Muslims, good in the sense that they adhere to Western notions of valor and chivalry. "God, what a Knight," claims the poet of one, "if only he were a Christian!"[59]

Neither is *The Song of Roland* alone in this ambivalence. There were others who, while never losing sight of their Christian duty to bring the heretical Muslims to an understanding of the truth, worked hard to broker some kind of understanding between the two religions. One of the most remarkable of these was the Majorcan Ramón Lull (1232–1315), polymath, knight, poet, mystic novelist, tireless and intrepid traveler, and the author of more than two hundred writings of various kinds. He set up a college for training missionaries to Islam, and it was he who, in 1311 at the Council of Vienne, persuaded the papacy to establish schools at Paris, Oxford, Bologna, and Salamanca to teach not merely the Arabic language but also, in as nonpolemical a manner as possible, Islamic history, theology, and philosophy. In the end, however, Lull fell victim to his insistent attempts to put his beliefs into practice. Convinced that a reasonable Muslim audience would listen to his defense of Christianity, he was stoned to death in Tunisia in 1315.

In many respects Lull, with his insistence on the possibility of a polite exchange of views between the two faiths, and with his injunction "When we pray, let us remember the pagans"—by which he meant the Muslims—"who are of the same blood as we," was an exception born of a frontier society. But he was by no means unique. A century later, the great German humanist Nicholas of Cusa wrote a work called *Cribatio Alcorani*, or *Sieving of the Koran*. Originally commissioned by Pope Nicholas II in support of yet another crusade, Cusa, somewhat to the pope's alarm, argued that if the Qur'an were interpreted correctly—"sieved," that is—it would become clear that it was in most important respects compatible with the teachings of Christianity. Today Cusa is best known for his conception of "learned ignorance," the argument that all human knowledge can only ever be approximate or conjectural, and for his belief that there must exist other planets in the universe with other human populations on them. Perhaps because of this breadth of vision and his obvious willingness to believe not only in the plurality of worlds but also in their eventual compatibility, he was more inclined to see some worth in the other great deviant version of Judaism than were most Christians.

Exceptional though both Ramón Lull and Nicholas of Cusa may have been in the lengths to which they were prepared to go in their attempts to reconcile Islam and Christianity, their views were reasoned responses to prolonged exposure to Islam and the ever-increasing fear as to just where the Muslim conquests might finally lead. It took repeated defeat in battle to bear in upon the minds of the learned, at least, that Islam might be something rather more than a perverse and almost comical sect, and that while the kind of popular farrago represented by *Le Chanson de Roland* might be entertaining enough and useful for lifting the spirits of a demoralized population, survival, if nothing else, required a greater understanding of what Islam really was and what inspired its adherents to such seemingly irrepressible feats. Just as, centuries later, it would be defeat in battle that would finally compel contemptuous Muslims to look westward and make a more balanced assessment of the hitherto despised "Franks."

These were also the years in which, in Spain and then later in the southern Mediterranean and North Africa, the balance of power between Muslims and Christians began to shift. In 1031, the caliphate of Córdoba collapsed and al-Andalus fragmented into a series of petty *taifa*—or "party"—kingdoms, far less able to resist Christian encroachment. In 1085, Alfonso of Léon and Castile, then in league with the emir of Seville, whose daughter he kept as his concubine, captured the great city of Toledo. It was a highly significant victory. Toledo was not only the largest and most powerful of the *taifa* kingdoms, it had also been the capital of Visigothic Spain. Alfonso now styled himself "emperor of Toledo" and "emperor of Spain," and, at least according to Muslim sources, "emperor of the two religions." Steadily, if slowly and with innumerable reversals, the Christians pushed south. A year after the capture of Toledo, Norman knights seized al-Mhadiyya in North Africa, and in 1091 they drove the Arabs out of Sicily. In 1118, Saragossa fell to the armies of Alfonso I of Aragon. In 1147, Christian forces took Lisbon and Almería and, in the following year, Tortosa and Lérida. In 1212, a combined force of Spanish, French, and Templar knights, officially designated a Crusade by Pope Innocent III, met and destroyed a large Muslim army at the plain of Las Navas de Tolosa, some forty miles north of Jaén in Andalusia. It marked the end of Muslim power in most of the region, and with the capture of Córdoba in 1236 and Seville in 1248, most of the Iberian Peninsula was in Christian hands.

The Muslims, however, still held one large, prosperous, and culturally splendid stronghold in the West: the Nasrid kingdom of Granada, a triangle of southern Spain that reached from Gibraltar in the west to Cartagena in the east. Its capital, the "city of the pomegranate" with its hanging gardens and fountains, its sprawling palace compound known as the Alhambra, its mosques and gilded domes, was one of the most dazzling in Europe. On

January 2, 1492, this, too, would fall to a Christian army. At its head were Ferdinand and Isabella, the so-called Catholic Monarchs of the united kingdoms of Castile and Aragon, suitably dressed for the occasion in Moorish costume.

The conquest of Granada has come to be seen as a turning point in the long struggle between Christianity and Islam, the moment when Islam was finally pushed back across what for centuries had seemed to be some kind of natural frontier with Europe. It seemed like that to contemporaries, and the ecclesiastical propaganda machine of the Catholic monarchs ensured that it continued to do so until the nineteenth century. In fact, however, the last significant battle of the Reconquest had been fought in 1212 and Granada had been a tributary of Castile for more than a century. The Nasrids, factious and decadent, were no real threat to anyone. Campaigns along the frontier between the two kingdoms were regular events, only because it served both sides to have a battleground to which to send young men, who might otherwise create trouble at home. As Machiavelli shrewdly observed, such intermittent warfare not merely gave them something to do, it also greatly enhanced their sense of themselves. Granada also provided Queen Isabella, who had a famously sweet tooth, with an inexhaustible supply of sugar.

Between 1474 and 1479, Ferdinand and Isabella had fought a long and punishing civil war in order to secure Isabella's succession to the Castilian crown, and they now desperately needed land with which to reward and secure the continuing loyalty of their supporters. They also needed an image, an event that could be portrayed as momentous, to give their still fragile dynastic alliance some much-needed legitimacy. Isabella was a devout Christian, a model of frugality—except where sugar was concerned—who is said to have repaired a doublet of her husband's no fewer than seven times and to have imposed somber modes of dress upon the overindulgent Castilian court. She was also a woman of astute judgment, who fully understood the cohesive potential of religious conformity. By all accounts, Ferdinand, less obviously devout, was a sly, manipulative, and shrewd politician. Machiavelli, who had watched him closely at first hand, described him in *The Prince* as "always preaching peace and good faith, and he has not a shred of respect for either."[60] But he knew as well as Machiavelli the potential political profit to be had by imposing religious orthodoxy. Spain, then, was to become pure. The older ideologies of *convivencia,* however shabby in practice, were to be cast aside.

After the capture of the city, the last Muslim king, Abû 'Abd Allâh Muhammad XI, known to the Christians as Boabdil, and all those who had remained loyal to him were granted what proved to be temporary exile in the mountains of the Alpujarras and shortly after expelled from a now united and Christian Spain. (Jews who refused to convert were expelled immediately.) Ferdinand's propagandists then set about depicting him as a man of destiny, God having chosen him to give the final coup de grâce to the forces of Islam in the West and to restore the true religion to its rightful place.

One person who was there to witness the event and to soak up the propaganda was a hopeful, but temporarily unemployed, middle-aged Genoese navigator with graying red hair, named Christopher Columbus. He had, he recorded later, seen the "royal banner of your Highnesses, raised by force of arms on the towers of the Alhambra." This he took as a sign that his own great venture—to sail a Christian fleet westward to "the lands of India and of a prince called the Great Khan . . . to see the princes, peoples and lands and their disposition and all the rest and what should be undertaken for their conversion to our holy faith"— would, at Isabella's hands, and at long last, meet with the success it deserved.[61]

The fall of Granada became a popular topic for poems and plays well into the eighteenth century. It was hailed as a compensation for the loss of the Latin Kingdom of Jerusalem in 1187 and the subsequent loss of almost all of eastern Christendom. But although the capture of the city marked the formal end of Muslim rule in Spain, the Reconquest itself, as both Muslim and Christian contemporaries recognized, had been far more than a struggle to regain Spain. It had, in fact, been but one phase in a longer war between Islam and Christianity that would be fought out not only in Europe and the Mediterranean but also much farther east, in Constantinople and in Jerusalem itself. This was Christianity's own jihad: the Crusades.

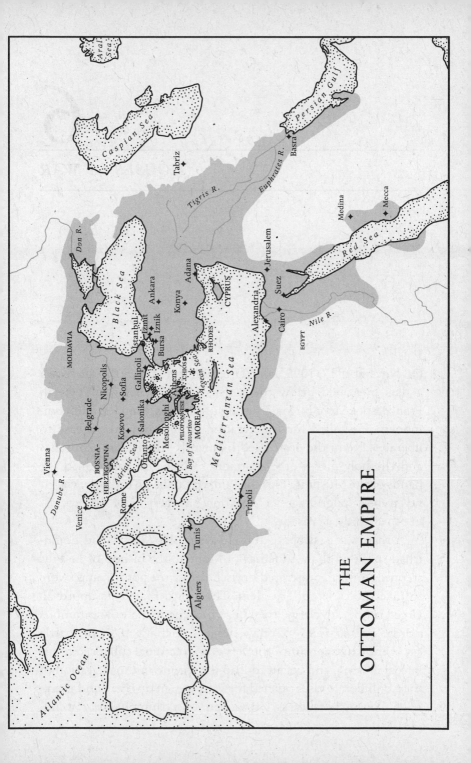

# THE
# OTTOMAN EMPIRE

*Aral Sea*

*Caspian Sea*

*Persian Gulf*

Basra

Tabriz

*Tigris R.*

*Euphrates R.*

Medina

Mecca

*Red Sea*

*Don R.*

Jerusalem

Adana

Suez

*Black Sea*

Ankara

Konya

CYPRUS

Alexandria

Cairo

*Nile R.*

MOLDAVIA

Istanbul
Izmit
Iznik
Bursa

RHODES

EGYPT

Nicopolis

Sofia

Gallipoli

CHIOS

*Aegean Sea*

Belgrade

Kosovo

Salonika

Athens

*Mediterranean Sea*

Vienna

*Danube R.*

BOSNIA-
HERZEGOVINA

Otranto

Mesolonghi
PELOPONNESE
MOREA
*Bay of Navarino*

Venice

*Adriatic Sea*

Rome

Tripoli

Tunis

Algiers

*Atlantic Ocean*

# 6

## *HOUSES OF WAR*

### I

On November 17, 1095, Pope Urban II, seated on a throne set on a dais in a field below the town of Notre-Dame-du-Porte in France, spoke to a huge gathering of bishops, knights, and commoners. He called for the creation of an army to go east to liberate—a term the pope used frequently—the Eastern churches and the holy places of Christianity, which had fallen into the hands of the Muslims. This, he said, was to be no ordinary war. It was to be a "pilgrimage," a "vocation," a "carrying of the cross"—in other words, a crusade.

Urban was passionate and eloquent.[1] No sooner had he finished speaking than Adhémar of Monteil, the bishop of Le Puy, stepped forward, prostrated himself before the pontiff, and swore to take the cross to Jerusalem. The crowd roared its approval. Urged on by the clergy, they began to chant the now infamous words *"Deus hoc vult"*—"God wishes it," which was to be the battle cry of the Crusaders until the very end. A cardinal collapsed to his knees, shaking, and began to lead the throng in the recitation of the Confiteor. Masses pushed forward to sign up there and then.

Following his success, Urban went on a triumphal tour, wending his way back and forth through southern France, preaching the Crusade. To those he could not reach in person, he wrote an endless stream of letters. "The barbarians in their frenzy," he told the knights of Flanders, urging them to join the Crusade, "have invaded and ravaged the churches of God in the eastern regions. Worse still, they have seized the Holy City of Christ embellished by his passion and resurrection, and—it is blasphemy to say it— they have sold her and her churches into abominable slavery."[2]

Popes had declared crusades many times before. But never before had a pope proclaimed a holy war, a war widely known as the *via Dei*—"the way of God"—a war on Christ's behalf, a war whose armies were to be the "armies of God," the "armies of the Lord," and whose warriors were to be the "knights of Christ," *milites Christi.*[3] Never before had a pope made it clear that participation in a war would be looked upon as an act of merit. *Recta oblatio,* Urban called it—a "right kind of sacrifice"—an act of devotion that would contribute toward the salvation of the participant's soul.[4] Every soldier would be a pilgrim, sworn to "slay for God's love," and he would carry a cross emblazoned on his chest, an allusion to Christ's call "If any man will come after me, let him deny himself, and take up his cross and follow me"—now given a meaning unlike anything Christ could possibly have intended.[5] As never before, the Church now turned her back on Christ's message to forgive one's enemies and "turn the other cheek."

For the first time in history, a European people was embarked upon an officially sanctified holy war. In some respects it was the exact Christian equivalent of the jihad. In others it was not. The Crusaders fought to recover lands that, in the view of the Church, were already an integral part of the Christian world. They were not, unlike the warriors of the jihad, bent upon the final conquest and conversion of the entire world. But as far as their sense of purpose and their conviction that they were doing God's work were concerned, these differences were insignificant. Christendom had never before had holy warriors. It had had only martyrs—those who had died, unresisting, so as to "bear witness" to the truth of their faith. Now the passive victim and the hero who died in battle had become one and, like their Muslim counterparts, would pass directly into Heaven. "All these entered Heaven in triumph," wrote one Crusader of his comrades who fell during the siege of Nicaea, "wearing the robe of martyrdom, saying with one voice 'Avenge, O Lord our blood which was shed for thee, for thou art blessed and worthy of praise for ever and ever.' "[6]

Urban's appeals, and the disorderly armies they unleashed upon the Middle East, also marked a radical change in the Western conception of warfare. Hitherto the Church had always looked upon war as a means, always in itself sinful but permitted in certain carefully defined circumstances because the natural world, the world of fallen mankind, was necessarily an imperfect and disordered place that could sometimes be set to right only through violence. A war was just if it was waged in pursuit of a just cause—invariably to right a previous wrong—and was sanctioned by a legitimate political authority. "War," Saint Augustine had written in what had become during the Middle Ages the most authoritative and the most stringent of definitions,

> should be waged only as a necessity, and waged only that God may by it deliver men from necessity and preserve them in peace. For peace is not sought in order to kindle war, but war is waged in order that peace may be secured. . . . So it should be necessity, not desire, that destroys the enemy in battle.[7]

Urban's sermons had changed all that. Henceforth Catholics would slaughter Protestants; Protestants would slaughter Catholics; and both would slaughter Jews and Muslims, not to regain territory, not to avenge injustice, not even from simple dynastic greed. They would do so because they believed, or claimed to believe, that God wished them to do so, because they were, as the Crusaders so often said, acting on God's behalf. Desire had replaced necessity. Later, after the wholly unexpected success of the First Crusade, the poetaster Graindor of Douai would make Christ on the Cross call down revenge upon the as yet unborn Muslims. Christ says to the first thief:

> *Friend, the people are not yet born*
> *Who will come to avenge me with their steel lances*
> *So they will come to kill the faithless pagans*
> *Who have always refused my commandments.*
> *Holy Christianity will be honored by them*
> *And my land conquered and my country freed.*[8]

As if in recognition of the horrors that lay ahead, Urban's tour was accompanied by a series of alarming portents. There were meteor showers, an eclipse of the moon, and a terrifying aurora around the sun. A severe drought that had brought about a succession of bad harvests and widespread famine suddenly came to an end. By the end of August, Urban was back in Rome. But as long as the Crusades were being preached up and down the length of Europe, these signs continued to appear. In the fall of 1097, a comet appeared in the sky; in February the following year, the sky glowed red. Next fall, there was light in the sky so bright that it seemed as if it were ablaze. In December, there was an eclipse of the sun, and in February 1099, another red aurora borealis filled the eastern sky.

Encouraged by such mysterious displays of God's approval, popular preachers spread the message even farther than Urban himself had. And they made it even more brutal and more hysterical. The Crusade was to be a war of terror, an act of revenge against the Muslims for the damage that, over the centuries, they had inflicted upon the Christian West. It was to be the final reckoning. Throughout Europe, from Britain to Italy, the "members of Christ's household" were called upon to rouse themselves and "seize firmly that city—Jerusalem—our commonwealth." "If an outsider were to strike any of your kin down, would you not avenge your blood relatives?" They were asked, "How much more ought you to avenge your God, your father, your brother, whom you see reproached, banished from his estates, crucified?"[9]

The blood feud had always been a central part of medieval European life. For centuries, however, the Church had struggled to suppress it. Now it had suddenly been given a central place in the Christian conception of the just war by no less a person than the pope himself.[10]

No matter how garbled the theology behind all this, it had an electrifying effect on a society that was in a state of conflict and crisis. Feudal Europe, and France in particular, was an unstable and unruly place. After the gradual demise of the Carolingian Empire, the local armed companies, bands of knights, and feudal barons who had once defended it, turned on their own subject populations, forcing them to produce more and more and forcing them in turn into destitution and driving them frequently to banditry. The French king controlled very little of what is now modern France. Counts and dukes, the descendants of the officials of the Carolingian court, became in effect sovereign rulers in their own territories. One of the objectives of the Crusade was to bring about what the Church called optimistically "The Peace of God," by substituting the sanctions of the Church for the limited and ineffectual authority of the king, and by siphoning off some of the kingdom's more unruly elements. The Crusade, it was hoped, would provide an outlet for the repressed energies and frustrated ambitions of scores of young men across Europe. "God," declared the Benedictine chronicler Guibert of Nogent, "has instituted in our time Holy Wars, so that the order of knights and the crowd running in their wake, who following the example of ancient pagans have been engaged in slaughtering one another, might find a new way of gaining salvation."[11] In this respect, if in no other, the Crusades could be said to have been a success.

## II

In the spring of 1096, a large army of itinerant knights, some major German nobles, and a horde of the dispossessed, with bands of women and children in its wake, gathered in southern France. This first wave was led by a wandering preacher named Peter the Hermit, who claimed to have been called to lead the Crusade by Christ in person and, to prove it, went about brandishing a letter of commission written, he said, by God himself. He was accompanied by a collection of maniacs, mountebanks, and desperadoes, among them one Abbot Baldwin, who had had a cross branded on his forehead and raised money from the gullible by claiming that it had been put there by an angel, and a sect whose members venerated a goose they believed to be filled with the Holy Spirit. Western fanaticism was now on the march.

Its first target was not, however, the Muslims but the Jews. The Crusaders had little if any understanding of theology or sacred history and none whatsoever of scripture, and they were probably unable to distinguish clearly between the various "enemies of Christ" about whom they had heard so many lurid tales from their equally ignorant priests. As one later writer put it, the Crusaders "held Jews, heretics and Muslims, all of whom they called the enemies of God, to be equally detestable." They went, in the words of one witness, to "wipe out or to convert" any non-Christian they found in their path.

Between May 25 and 29, the sizable and prosperous Jewish community of Mainz in the Rhineland was annihilated, "clearing the path," in the words of the German Crusaders, to Jerusalem. Satisfied with its work, the army now seems to have split up. Some of it moved north to Cologne. By this time the Jews had fled the city and taken refuge in the countryside as best they could. From June to July, the zealous pilgrims of Christ hunted them down, burning synagogues, destroying Torah scrolls, and desecrating cemeteries as they went. Another group marched southwest to Trier and Metz, where the slaughter continued. At Regensburg, a further contingent, probably under the leadership of Peter the Hermit himself, somewhat more lenient than the rest, forced the entire community to undergo baptism.

The Church hierarchy itself refused to condone such behavior even though local priests were certainly implicated. Some of the bishops, to their credit, went so far as to protect the Jews in their dioceses, offering them refuge in their fortified places and in Speyer, Mainz, and Cologne dispersing them into the countryside. But the pogroms that marked the launching of the first Crusade, and the anti-Semitic passion it aroused, were the worst in recorded history and would leave their mark on the Jewish communities of Europe for centuries to come. More than eight thousand people perished in what has come to be called—not without justification—the "first Holocaust," and dirges in honor of the German martyrs are recited in synagogues to this day.

Having butchered the Jews, the Crusaders regrouped and moved into the Balkans, where the army, now largely out of control, first attacked the town of Zemun and then sacked Belgrade. Although many perished in the subsequent reprisals, contingents made it as far as the Seljuq city of Nicaea (modern İznik), where they were wiped out by the Turks.

In August, however, a more orderly body of men under the command of the great nobles Godfrey of Bouillon, duke of Lower Lorraine; Hugh of Vermandois, the brother of the king of France; Raymond of Saint-Gilles, count of Toulouse; Robert, count of Flanders; and Robert, duke of Normandy began the long and hazardous journey across eastern Europe toward Constantinople. They arrived in groups between November 1086 and May of the following year, hungry, bespattered, saddle-sore, weary, and often despondent. They were then ferried by the Byzantine emperor Alexius hastily across the Bosporus. The "Franks," as the Muslims routinely called all Europeans, "a swarm of flies," "grasshoppers without wings," "howling savage dogs," had at last penetrated the sacred soil of the *dar al-Islam*.[12]

The world into which this ragtag army erupted so unexpectedly in the summer of 1097 was itself emerging slowly from a series of bitter internecine feuds. The Seljuq Turks, who ruled over what is now Iraq, Syria, and Palestine, despite having inflicted a famous victory over the Byzantine emperor Romanus Diogenes IV at Manzikert in 1071, were, in effect, divided into a number of semi-independent states, all nominally under the rule of the caliph in Baghdad but all deeply suspicious of one another. They had also, as good Sunnis, been locked in combat between 1063 and 1092 with the Shiite Fatimid Caliphate in Cairo. The Fatimids, who controlled most of the Holy Land, had survived this struggle intact but greatly weakened. For them at least, the disorderely, ill-equipped Franks seemed at first to be a far lesser threat than the Turks.

The Crusaders must have known something of this from their Greek informants. In any case, they moved swiftly and decisively, taking advantage of the Muslims' apparent failure to offer much effective resistance. On June 19, 1097, Nicaea, a major city on the old imperial road to the East, surrendered. A week later the Christians, on the verge of starvation and marching in heavy armor under a burning sun, set out for Antioch. They arrived before its walls in late October, and after a siege lasting seven months, and having somehow managed to defeat two relief armies, they entered the city on June 3, 1098. Antioch was no longer what it had been under Byzantine rule. But it was nonetheless the site of some of the great churches of the East, and it controlled the route from Asia Minor into Syria.

On June 28, another Muslim force arrived. The Crusaders managed to defeat this one, too, some claiming that a heavenly host of angels and saints was fighting on their side, accompanied by the ghosts of their dead comrades. Far-fetched though this was even to a devout ninth-century Christian, it seemed the only possible way to account for their success. By late January, the Crusaders were on the march once again. They moved swiftly along the coast, entering territories now under the control of the Fatimids. By nightfall on Tuesday, June 7, 1099, they had reached the walls of the holy city of Jerusalem—the "navel of the world."

Throughout all this time, as the armies moved steadily eastward, their ever-encouraging god marked their passage with a series of remarkable, if somewhat ambiguous, signs. In early October 1097, a comet with a tail shaped like a sword had been seen in the night sky. (This, at least, was no delusion; it is well documented in Chinese and Korean records.)[13] An earthquake followed on December 30; the sky glowed red, and a great light appeared in the shape of the cross, much as it had to Constantine centuries before. On the night of June 13, 1098, a meteor fell, significantly from the west, on the Muslim camp outside Antioch. On September 27, an aura appeared that was so bright it could be seen all across Europe and Asia. On June 5, 1099, as the battered Crusader army approached Jerusalem, there was an eclipse of the moon that was interpreted to foretell an end to the rule of the crescent, the symbol of Islam.

Jerusalem, however, proved to be far more resilient than either Nicaea or Antioch had been. For more than a month the Christians besieged what appeared to be a nearly impregnable city, making several futile assaults against its massive walls. Then, fearing the imminent arrival of an Egyptian army and having finally gathered together enough ladders and siege engines, they launched an all-out attack on the eastern section of the wall in the early morning of July 14. The legendary Tancred, a Norman from southern Italy and the hero of many later tales of the Crusades, was the first to break through into the city. The Muslim population fled, hoping to use the al-Aqsâ Mosque as a refuge. After a brief and futile resistance they surrendered to Tancred, who, as the true image of a chivalrous knight, promised to save their lives. He then raised his standard over the mosque, swearing that this would protect them from the fury of his coreligionists. By the afternoon, it was clear that all was lost. The Fatimid governor, Iftikhâr ad-Daulah, offered to hand over the city, with all the treasure it contained, to Raymond of Saint Gilles, count of Toulouse, in exchange for his life and the lives of his family and his personal bodyguard. Raymond agreed, and Iftikhâr was escorted out of the city to join the Egyptian garrison at Ascalon on the coast.

He was fortunate. The rest of the population—men, women, and children—were slaughtered. The Muslims sheltering in the al-Aqsâ Mosque soon discovered that Tancred's standard was no protection. They were dragged out and hacked to pieces. Tancred, it was said, was furious not so much at the sight of the killing but because his standard had not been respected. The Jews sought refuge in the main synagogue, where "the Franks burned it over their heads."[14] When the butchery was done, by the late evening, the Temple area of the city was piled high with corpses and blood flowed like a river through the streets.[15] "No one has ever seen or heard of such a slaughter of pagans," wrote one Christian witness, "for they were burned on pyres like pyramids and no one save God knows how many there were."[16]

When news of the victory reached the Latin West, it was greeted with astonishment and jubilation. "The Lord," wrote the pope, "has certainly revived his miracles of old." How else, he reasoned, could an army so badly led, so ill disciplined, so short of supplies, and so weighed down by noncombatants, have possibly prevailed against the seemingly invincible Muslim forces?[17]

With the seizure of Jerusalem the relationship between Western Christendom and the Muslim—and subsequently the Christian—East changed irrevocably. The leaders of the Crusade now carved out principalities and estates for themselves in the new territories—Baldwin of Boulogne in Edessa, Bohemond of Taranto at Antioch, Raymond of Toulouse in Tripoli. These made up the Crusader States of the Levant, or "Outremer"—"beyond the sea"—as it came to be known in Europe. Jerusalem itself became a kingdom that extended just north of Beirut and then along most of the coast south as far as Gaza and the Golan Heights. For nearly a century, from 1099 until 1187, a Latin Christian kingdom, with dependencies from Antioch to Acre, existed in the heart of the *dar al-Islam*.

In time the rulers of these states and the orders of armed monks, the Knights of Saint John, or Hospitalers, and the ill-famed Knights Templar, who fought with them, built a line of castles to protect their frontiers from their Muslim—and as often their Christian—neighbors. Many of these are still standing: Marquab, built by the Knights of Saint John; the Bagras Castle in the Amanus Mountains in Turkey; Tartus on the Syrian coast; Belvoir Castle, perched on the edge of the Jordan Valley; and the most imposing of all, Crak des Chevaliers in the hinterland of Tripoli. When T. E. Lawrence—"Lawrence of Arabia"—saw the Crak in 1909, he declared it to be "the best-preserved and most wholly admirable castle in the world . . . were Baibas [the Mamluk Sultan who besieged the Crak in 1271] to reappear he would think it as formidable as old."[18]

The Crusader States these huge fortresses were built to protect were inevitably multicultural and multireligious. The Europeans, although they were the most powerful and best armed of the various groups that lived in the region, were also heavily outnumbered by Muslims; Syrian Christians, who were at best uneasy with their new Latin rulers; and Jews, who had nothing whatever to gain and a great deal to lose by exchanging Muslim rule for Christian. Outremer was, and remained throughout its entire history, a frontier region, and its Roman Catholic population, like its counterparts in Spain, came in time to have more in common with their Muslim neighbors than with their coreligionists back in Europe. They adopted Muslim cuisine, frequently dressed in a curious variation of Arab-Turkic dress, and gradually came to accept the need for a form of religious tolerance on the whole unknown in the Christian West. A Muslim Syrian aristocrat visiting Jerusalem, Usamah Ibn Munqidh, recalled that he used to go to the former al-Asqâ Mosque (now occupied by the Templars and transformed into a Christian church) to pray. The Templars obligingly provided him with a small adjoining space, where he could lay out his prayer rug, and left him to his devotions. Usamah was, of course, facing west toward Mecca, and one day, "one of the Franks rushed on me, got hold of me and turned my face eastwards saying—'this is the way thou shouldst pray.' " The Templars ushered the man out and apologized to Usamah, explaining that "this is a stranger recently arrived from the land of the Franks." No more needed to be said.[19]

Outremer grew immensely wealthy, particularly at its height in the twelfth century, from the trade routes from India and the Far East to Europe, which passed through Damascus to the ports of Acre and Tyre. But for all its multiculturalism and all its riches, it remained an outpost of an alien civilization, a long splinter of the *dar al-harb* buried in the flank of Islam that, sooner or later, would have to be extracted by force. It managed to survive for nearly two centuries by creating a rough balance of power between the warring factions of the region, in which Christian Jerusalem faced Muslim Damascus, Christian Antioch faced Muslim Aleppo, and Christian Tripoli faced a group of lesser Muslim cities in the upper Orates Valley. Outremer was, in effect, absorbed into a system of Syrian politics that involved alliances with the Muslim warlords by whom it was surrounded, even, when necessary, against their fellow Christians. It was a delicate and inevitably precarious situation that could last only as long as the Muslim states in the region remained in a condition of near-permanent conflict with one another.

But it could not last forever. The first blow came in 1144, ironically on Christmas Eve. 'Imâd ad-Din Zangî, the ruler of Mosul and Aleppo, calling on all his allies to undertake a jihad against the Christian interlopers, captured and sacked the Crusader city of Edessa, killing all the Franks. "The troops set to pillaging," wrote one Arab contemporary, "slaying and capturing, ravishing and looting, and their hands were filled with such quantities of money, furnishings, animals, booty and captives as rejoiced their spirits and gladdened their hearts."[20] The news created a sensation throughout the Islamic world, and although Zangî was assassinated two years later, before he could consolidate his gains, it was clear to all, both Muslim and Christian, that the Muslim *Reconquista* of Palestine had begun.

The Christians' response was to launch another Crusade. Urged on by the fiery rhetoric of Saint Bernard of Clairvaux and led by no lesser persons than the Holy Roman Emperor, Conrad III, and Louis VII of France, its armies began to leave Europe in May 1147. It turned out to be a disaster. Conrad's troops, poorly disciplined and poorly provisioned, made it as far as Dorylaeum, where, on October 25, they were set upon and virtually annihilated. Those who survived, however, together with another and larger army led by Louis, arrived in Syria in the summer of the following year.

The army's commanders then decided to besiege Damascus. This proved to be a costly mistake. The siege lasted a mere five days, before the Crusaders were forced to flee by a relief army led by Zangî's son and successor, Nûr ad-Dîn. The survivors, including both the king of France and the emperor, limped back to Europe in humiliation. "The Sons of the Church and those who are counted Christians have been overthrown in the deserts," bemoaned Saint Bernard, "slain by the sword or consumed by hunger." The reason for this disaster, he could only conclude, was that God had been angry with his followers, for reasons they should not presume to question, for no human would "be so rash as to dare to pass judgment on something that they are not in the least able to understand."[21] Consoled perhaps by this thought, for the next forty years the Christian kingdoms of the West turned their backs on Outremer and the Holy Land and concentrated on the pursuit of their own internal rivalries.

Nûr ad-Dîn succeeded in consolidating his father's gains and in reducing to a coastal strip the principality of Antioch, the silver-encrusted skull of whose ruler, Prince Raymond, he sent as a gift to the caliph in Baghdad. But the man who would finally bring an end to the Crusade in the East was one of Nûr ad-Dîn's most gifted lieutenants, the Kurdish warrior Salâh ad-Din Yûsuf ibn Ayyûb, ("Righteousness of the Faith, Joseph, son of Job"), known in the West as Saladin.

Although a Sunni, Saladin had, by 1169, become sultan of Fatimid Egypt. Two years later, and without much opposition, Egypt passed under the nominal suzerainty of the Abbasid Caliphate in Baghdad, thus putting an end to Shiite political control to the west of the Fertile Crescent from that day to this.[22] In 1174, Saladin occupied Damascus, and a year later he became the formal ruler of Syria. He was now clearly bent upon jihad and the unification of Islam against its Christian enemies. "In the interest of Islam and its peoples," he wrote in reply to the demand from the emirs of Damascus that he think first of the good of his family, "we put first and foremost whatever will combine their forces and unite them in one purpose."[23] In 1183, Saladin laid siege to Kerak—known as the Pierre du Derent—the Crusader castle that dominated the caravan route that linked Aleppo and Damascus to the Red Sea. When Saladin's army arrived, the owner of Kerak, Reginald of Châtillon, was holding a wedding party for his son-in-law Humphrey of Toron. The party carried on, undeterred, as Saladin pounded away at the walls, and in one of those exchanges in which medieval chroniclers of both sides delighted, the bridegroom's mother, Lady Stephanie, sent dishes from the wedding feast to Saladin's tent. In response to this gesture of hospitality, the ever-chivalrous Saladin replied by asking where the young couple were passing their wedding night and ordered his troops to avoid that section of the castle walls. Before their honeymoon was over, however, Saladin's forces were driven back by a relief army from Jerusalem.

Saladin retreated to Damascus, but not for long. He was now in control of the Muslim world all the way from the Nile to the Euphrates, and, despite their occasional successes against him, it was clear that the days of the Crusaders in the East were now numbered. Early in 1187, and in violation of a truce between Saladin and Baldwin IV, the famous "Leper King" of Jerusalem, Reginald attacked a caravan from Cairo to Damascus and seized not only a considerable amount of booty but also a number of prisoners, including one of Saladin's own sisters. When Baldwin, in an attempt to maintain the peace, demanded that he make amends,

Reginald refused. It was the occasion Saladin had been waiting for. He gathered all his forces from Egypt, Damascus, Aleppo, Mesopotamia, Mosul, and Diyar-Bakr. The Crusader army that turned out to meet him was weary, overstretched, and desperately thirsty. The heat was overpowering, and they were so closely surrounded that, as one of them remarked, "not even a cat could have escaped."[24] As they pitched camp on July 3, Raymond, count of Tripoli, was heard to declare, "Alas, Lord God, the battle is over. We have been betrayed unto death. The Kingdom is finished." He was right. The following day, Saladin's army annihilated the Crusader forces near the village of Hattin. It was the most significant battle of the Crusades, and although it is now long forgotten in the West, except by historians, its memory lives on in the Arab imagination as the moment of triumph over the forces of Christendom, which most Muslims are firmly convinced no Christian has ever forgotten, nor will ever forgive.

Saladin executed Reginald of Châtillon for his treachery with his own hands, as he had promised, and the two hundred Templars and Hospitalers in the army were put to death. The other, secular, knights were ransomed. The common soldiery was sold into slavery in such numbers that they created a glut on the Syrian market for months to come. In the days that followed, the Christian strongholds of Acre, Toron, Sidon, Beirut, Nazareth, Caesarea, Nablus, Jaffa, and Ascalon, defended now by only skeleton forces, fell one after another to Saladin and his generals. "You have possessed the lands from east to west," sang the poet Ibn Sana' al-Mulk:

> You have embraced the horizon, plain and steppe
> God has said: Obey him;
> We have heard Our Lord and obeyed.[25]

In September Saladin gathered his entire army and camped before the walls of Jerusalem. The city, now under the command of Balain of Ibelin, son of the lord of Beirut and Cyprus, was ill provisioned and filled with refugees, most of them women and children. There were so few knights that Balain was forced to round up all the males over the age of sixteen he could find and put swords into their hands. The attack began on September 20. The city held out for six days. Balain then agreed to surrender on condition that the Christian inhabitants be allowed to pay for their lives. He would, he said, kill all the Muslims in the city and destroy all the Muslim holy places if Saladin did not agree. It was a powerful argument. After Mecca and Medina, the al-Aqsâ Mosque in Jerusalem is for the Muslims the third most holy site in Islam; and the Muslim population of the city was substantial. Saladin had no alternative but to agree. On October 2, he entered Jerusalem, unopposed and in striking contrast, as many, both Christian and Muslim, have not failed to point out down the centuries, to the Christians' entry eighty-eight years before. Now there was no bloodshed, no destruction, no looting. Those who could afford the ransom, including the patriarch Heraclius—who, ignoring the poverty-stricken masses now facing a life of servitude, fled to Tyre—paid their ten dinars and were allowed to leave. In a final act of magnanimity, Saladin liberated several thousand unconditionally. The rest, however, were marched off to the slave markets. The Christian Syrians were allowed to retain their churches and stay in Jerusalem if they chose to do so, which most of them did. The Jews, who had fled the city sometime before, were encouraged to return, and shortly thereafter Saladin concluded a treaty with the Byzantine emperor Isaac Angelus by which the Christian holy places in the city were to be returned to the care of the Greek Orthodox Church. All that now remained of the Crusader States in the Levant were three cities: Tyre, Tripoli, and Antioch.

Saladin has become, and remains, a legend in the Islamic world. He was, on all accounts, a chivalric hero in the Arab tradition, skilled in all the elegance and social refinements called *zarf* in Arabic, honorable, tolerant within the terms laid down by the Qur'an—which few Muslim rulers in practice were—generous, true to his word, and dedicated to the cause of Islam over and above his own personal ambitions or even those of his family.[26] He was an exceptionally good polo player, a sport that had something of the same standing as jousting had acquired in Europe. He was also, if his aide and biographer, Bahâ ad-Din ibn Shaddâd, is to believed, a man who had passed all the possible tests of his faith to which the very best doctors and jurists had subjected him, "so that he could converse with them on their own level."[27] And his ambitions both for himself and for Islam reached far beyond the desire to drive the hated Christians out of Palestine. In 1189, he told Bahâ ad-Din, "When God Almighty has enabled me to conquer the rest of the coast, I shall . . . cross the sea to their [the Christians'] islands to pursue them until there remains no one on the face of the earth who does not acknowledge Allah—or I die!"[28] Like other Muslim rulers of the time, however, he was prepared to make alliances with Christian rulers, when it suited him, and made war on rival Muslim warlords, few if any of whom could easily be described as "heretics." Yet he was, by most accounts, a man of deep convictions who towered morally and intellectually over the shabby rabble by which he was surrounded in both the Muslim and the Christian camps.[29]

In the West he also acquired the reputation of being the "worthy enemy" so beloved of medieval Christian chivalric romance. He was, Voltaire said of him much later, "at once a good man, a hero, and a philosopher"—although, he added, few of the "chroniclers with which Europe is overburdened" had done justice to his deeds.[30] "In a fanatic age" wrote Edward Gibbon, "himself a fanatic, the genuine virtues of Saladin commanded the esteem of the Christians."[31] Later still he was transformed into the archetypical Romantic hero, chivalrous, courteous, generous, and above all devoted to the defense of his homeland against a gang of marauders whose motives were as base as their conduct. "No greater name," said Sir Walter Scott, "is recorded in Eastern history."[32]

When in 1898 the German kaiser Wilhelm II paid an official visit to Istanbul and Syria in an attempt to increase German influence in the Ottoman Empire, he went to Damascus and, standing before Saladin's tomb, described the liberator of Jerusalem as "a knight without fear or blame, who often had to teach his opponents the right way to practice chivalry."[33] Wilhelm then declared himself to be the friend of "the three million Mohammedans" and laid a satin flag and bronze gilt laurel wreath on Saladin's tomb with the inscription "From one great emperor to another." In November 1918, this was brought to England as a war trophy by none other than T. E. Lawrence. It is now on display in the Imperial War Museum in London with a note in Lawrence's own handwriting explaining that he had removed it because, now that Jerusalem had been freed from the Ottomans, "Saladin no longer required it."[34]

Today Saladin is still a hero in the Muslim world. A vast equestrian statue, erected at municipal expense in 1992—two years after the Gulf War—stands in front of the citadel in Damascus. It represents its hero in much the same pose and much the same dress in which he appears in a number of nineteenth-century Western depictions of the Crusades (but then there is no independent tradition of statuary in the Muslim world). To either side of Saladin's horse stand two foot soldiers and a Sufi. Behind the horse slump two crusaders, Guy of Jerusalem and Reginald of Châtillon. As the sculptor, Abdallah al-Sayed, has explained, he wished Saladin to appear not as an individual warrior but as a leader embodying a wave of popular feeling against the "Franks." The Sufis represent (somewhat improbably) the simple religion of the people, the foot soldiers, the humble warriors, all united with their hero under the banner of Islam.[35]


THE LOSS OF Jerusalem was a terrible blow. But Christendom was not to be so easily defeated. In May 1189, the Emperor Frederick Barbarossa, now nearly seventy years old, left for Byzantium at the head of the largest crusading army ever gathered. On June

10, however, he drowned while attempting to swim across the River Saleph, and although many of his troops made it as far as Tyre, they were seriously demoralized. Saladin, who had received a sudden and unexpected reprieve, accepted the old emperor's death as the work of God. The reprieve did not last for long, however. In July of the following year, Philip Augustus of France and one of the best-known of all the Crusaders, Richard the Lionheart of England, gathered an army and sailed for the Holy Land.

From the Christian point of view, the Third Crusade, although ultimately ineffectual, proved to be something of an improvement on the Second. In July 1191, the Crusaders managed to retake Acre and then Jaffa and, in the following year, Ascalon. By then, however, Richard had come to realize that the conditions that had allowed Outremer to survive for so long were over. No Christian army would now be large enough or powerful enough to recapture Jerusalem, much less hold out against Saladin's forces for any length of time. In September, having secured from Saladin an agreement that the Christians would be allowed to remain in the coastal cities as far south as Jaffa, Richard retreated back to Acre and on October 9, 1192, he set sail for England.

In 1203, yet another attempt was made, initially at least, to reconquer Jerusalem. The armies of the Fourth Crusade left Venice in October 1202 and stopped first to recover Zara, a former Venetian outpost on the Adriatic, from the Hungarians. Here Alexius, the son of the deposed Byzantine emperor, in a bid to reclaim the throne, offered the Crusaders a contribution of 2,000 marks and the promise to station a force of five hundred Greek knights permanently in the Holy Land if they would help him regain the throne. They agreed, and on June 24, 1203, the fleet sailed into the Bosporus. By January of the following year, Alexius had been installed, had reneged on his promise of 2,000 marks, and had been strangled and deposed. The Crusaders, exasperated and angered by a hostile Greek city whose religion they looked upon with distrust, now decided to seize the decaying Roman Empire of the East for themselves. Constantinople fell, for the first time in its long history, on April 13. "The empire," wrote Edward Gibbon centuries later, "which still bore the name of Constantine and the title of Roman, was subverted by the arms of Latin pilgrims."[36]

The victorious army ransacked the city, killing all the inhabitants they could find, pillaging the churches, and destroying the icons, which they believed to be sacrilegious. The veil of the sanctuary in the great basilica of Hagia Sophia was torn down for the sake of its golden fringe, the magnificent gilded and bejeweled altar broken into pieces and parceled out among the soldiers. The doors and pulpits were stripped of their carvings. A prostitute was set on the throne of the patriarchs, who danced and sang garbled versions of Orthodox hymns. The booty was immense. Jewels, sculptures, paintings, and manuscripts flowed back to western Europe in the years following the sack, the most famous being the four bronze horses from the Hippodrome, which dated from the early third century B.C.E. and which were then placed on the facade of St. Mark's Basilica in Venice. Baldwin, count of Flanders, was duly crowned Basileus by the patriarch of Venice, and the Venetian Tommaso Morosini was installed as patriarch. The Latin Kingdom of Constantinople, called by its rulers the "Empire of Romania," would last until 1261. It was the greatest betrayal of one Christian community by another. "Thus," observed Voltaire dryly, "the only benefit that the Christians derived from their barbarous Crusade was to have slaughtered other Christians."[37] Even the pope, who had little love for his Greek opposite number, denounced it, and the schism between the Eastern and Western branches of Christendom it caused would never fully be healed.

From 1219 until 1270, three further Crusades were launched. None of them, however, got any closer to the Holy Land than Egypt or Tunis, where in 1270 King Louis IX of France, after having purchased most of the holy relics of the city of Constantinople from the last impoverished Latin "emperor of Romania," died of the plague, thereby earning himself sanctification. When, in 1291, Acre, the last Christian stronghold in the Holy Land, fell to Sultan al-Ashraf Khalîl, the Crusades in the East were finally at an end. Despite periodic and always unsuccessful attempts to launch new crusades in the intervening centuries, no Christian army would now return to the heartlands of Islam until the nineteenth century.

## III

The enormous historical significance given to the Crusades, in both the West and the Islamic world, is in many respects the creation of later generations. Throughout the sixteenth and seventeenth centuries, while frequent and unheeded calls to mount a new crusade were made by the papacy, it was still possible in Europe to look upon the Crusades as a heroic, if ultimately flawed, achievement against the common enemy of the entire Christian, European world. Torquato Tasso's immensely popular epic poem *Jerusalem Delivered* (*Gerusalemme liberata*) of 1574 reinvented the First Crusade as a tale of love, chivalry, magic, intrigue, and, implicit at least, sex. Noble Christian warriors battle with fierce but noble "Saracens." As with the *Aeneid*, by which it was inspired, the agents of the gods—or rather Christ and Satan—manipulate their respective champions. Armida, a beautiful witch, is sent by Satan via the wizard Hidraort, the ruler of Damascus, to sow discord in the Christian camp but is transformed into a devout Christian by her love for Eustace, "a bold and lovesome knight," youngest brother of Godfrey of Bouillon. In one of the most famous episodes, which became the subject of plays and operas until the nineteenth century, Clorinda, a Muslim princess, in order to defend her faith, dresses in armor like a man and finds herself fighting a melancholy Tancred, whom she secretly loves. When Tancred kills her, she dies in his arms and is finally received into the true faith. And so on.

With the decline of the Turkish threat and the increasing secularization of European society in the eighteenth century, however, the Crusades, and the whole idea of a "holy war," came to be seen as nothing more than the product of precisely the kind of misdirected religious fanaticism most enlightened Europeans deplored. The entire venture had been, said the Scottish philosopher David Hume, "the most signal and most durable monument of human folly that has yet appeared in any age or nation."[38] And he was not alone. "The principle of the crusades," wrote Gibbon, "was a savage fanaticism; and the most important effects were analogous to the cause." With the Crusades, he claimed—reacting with the horror of a man who had himself converted to Catholicism and had then converted back to Protestantism—the Church became ever more prone to superstition, "and the establishment of the Inquisition, the mendicant orders of monks and friars, the last abuse of indulgences, and the final progress of idolatry flowed from the baleful fountain of the holy war."[39]

In the nineteenth century, however, the Middle Ages, which had once seemed to be only the "Dark Ages," a long night separating classical antiquity from the Renaissance, were suddenly filled with romance, great feats of heroism, and unselfish love. Johann Gottfried von Herder, the precursor of German romanticism (and some would say German nationalism), looked upon the Crusades as one with "those monstrous institutions of priestly offices of honour, monasteries, monastic orders." They were part of some deeper, darker "monstrous Gothic structure, [so] over-freighted, oppressive, dark, tasteless," that the very earth "seemed to sink beneath it." Yet for all that, they were also a "miracle of the human spirit and certainly Providence's tool." This was this "spirit of northern knightly honor," and although the Enlightenment had dismissed it as merely being what stood "between the Romans and ourselves," Herder could see in it what he valued most in European civilization, the "fighting against shortcoming, struggling for improvement" that had made of the Gothic a "gigantic step in the course of human fate."[40]

For subsequent generations of both French and Germans (and even the occasional Englishman), the Crusades came to be seen as a moment when the nation—that great, glorious nineteenth-century fabrication—had first begun to emerge. In October 1807, François-René, vicomte de Chateaubriand, diplomat, minister of state, and one of the great French writers of the nineteenth century, was invested with the order of the Knights of the Holy Sepulcher, with what he imagined (falsely) to be the sword and spurs of Godfrey of Bouillon. "Touching that long and heavy iron sword, wielded by so noble a hand," he could not refrain from musing bombastically, "this ceremony was not, in the end, wholly in vain. I was a Frenchman. Godfrey of Bouillon was a Frenchman; touching his ancient arms communicated to me an in- creased love for glory and the honor of my country."[41]

The French had not, of course, been the only crusading na- tion; so had the Germans. In October 1898, the German kaiser Wilhelm II, two weeks after having paid homage to the ghost of Saladin, entered Jerusalem through a specially prepared breach in the city walls. Mounted on a black charger, wearing a ceremonial uniform vaguely reminiscent of that of a medieval knight, and sporting a plumed helmet, he declared that "from Jerusalem there came the light in the splendor of which the German nation has become great and glorious, and what the Germanic people have become, they became under the banner of the cross."[42]

The idea of a crusade has remained a potent image both in and outside the House of War. Napoleon's ill-fated *mission civil- isatrice* to Egypt in 1799 had been regarded by some as an attempt to make good the failure of the Crusaders to establish lasting Eu- ropean colonies in the Middle East.[43] In 1915, Pierre-Étienne Flandin, the leader of what was known as the "Syrian party" in the French Senate—whose objective was to make Syria into a depen- dency of France—issued a manifesto claiming that Syria and Palestine were in effect one country that had been a "France of the Near East" ever since the Crusades and that it was now France's "historic mission" to recover, if not quite the Latin Kingdom of Jerusalem, then at least some kind of sovereignty over the area. The same point was made again after the end of the First World War, when at the Paris Peace Conference, France claimed the right to a mandate in Syria, in part, at least, because of her role in the Crusades. In reply to this, Emir Faisal, fighting to keep what he had secured from the Ottomans out of the hands of the Western Allies, asked acidly, "Would you kindly tell me just which one of us won the crusades?"[44] No wonder that when the new Musée des Colonies at the Colonial Exposition was opened in Paris in 1931, the first room was dedicated to Syria and Cyprus at the time of the Crusades.

This sense that the Crusades were the beginning, not the end, of a historical process, as yet unfinished, was shared, in an acute form, in the Muslim world. Western perceptions of the past are generally short. Modernization has seen to that, for modernization demands a form of forgetting. Neither the European Union nor NATO would have been possible if the French still harbored grievances against the British for the Battle of Waterloo or the British against the Germans for the Blitz. History in the Islamic world, by contrast, has always moved at a different pace. The present is linked to the past by a continuous and still unfulfilled narrative, the story of the struggle against the "Infidel" for the ultimate Muslim conquest of the entire world.

Crusading and European imperialism thus became indissoluble in the Muslim vision of the West, in particular, and with good reason, after 1918. "Crusading was not confined to the clangor of arms but was, and before all else, an intellectual enmity," wrote Sayyid Qutb, the ideologue of the Egyptian Muslim Brotherhood and of much of modern "Islamism," in 1948. Qutb was convinced that "every imperialist state" had been "opposing and stifling this religion [Islam] for centuries." This, he claimed, could be attributed to various causes: "Anglo-Saxon cunning," the "Jewish financial influence in the United States," the "struggle between the Eastern and Western bloc." But all of these, though clearly significant, overlooked the "real element in the matter," which was simply "the Crusader spirit that all Westerners carry in their blood." It was this blood that had created the "European imperial interest," and this, he declared, can never forget that "the Islamic spirit is a bulwark resisting this spread of imperialism and that it must destroy this bulwark or at least shake it." This might seem an oddly contradictory statement in the light of his claim that the Arab conquests of the Byzantine and Persian Empires were the high points in a perpetual struggle between Islam and what he called the "polytheists," principally Christians and Jews. But as with so much of this rhetoric, it is apparently only Westerners who are guilty of "crusading" and "imperialism." The Muslim conquests were, by contrast, an act of liberation, a bringing of the true faith to the benighted infidel. (Europeans have, of course, frequently been guilty of the same double standards. "Imperialism," for its enemies, is always something practiced by others.)[45]

Qutb might also perhaps be forgiven for believing that only Westerners are "crusaders" and have never ceased to be, from the slipshod manner in which the terms "crusade" and "crusading" have come to be applied to any war believed by those waging it to be just. When, on September 16, 2001, five days after the destruction of the World Trade Center, George W. Bush unadvisedly declared, "We understand, and the American people understand, this crusade, this war on terrorism is going to take a while," he did not intend any specific allusion to the historical Crusades.[46] He meant simply that it was a good and noble cause. But for those in the Islamic world who heard him use that hated word, he seemed to be alluding to a perpetual war, one that had been in progress, with only brief interruptions, since the tenth century, a war not against anything as amorphous as "terrorism," but a war against Islam.[47]

The use of the term "crusade" was particularly infelicitous in view of the association that subsequently formed in the mind of the Bush administration between the "war on terrorism" and the largely unrelated war against the Iraqi dictator Saddam Hussein. For Saddam had his own take on the history of the Crusades in which, if the first and second George Bush bore some shaky resemblance to Godfrey of Bouillon or Reginald of Châtillon, he, Saddam, was quite unmistakably Saladin. When the Gulf War broke out in 1991, the Iraqi press immediately compared it to the battle of Hattin. "We smell the smell of Hattin," wrote one, "and of the battle of the innermost sanctum." And it was to Hattin that Saddam was alluding when he kept promising "the mother of all battles"—a common phrase in Arabic, although it sounded very odd to most Western ears—which would finally finish off the allied forces.[48] At the height of the war the Iraqi daily newspaper *al-Qadisiyya* printed a poem to remind its readers of what linked Hattin to the coming "mother of battles":

> *History is returning*
> *Yesterday the Crusaders' war*
> *Today the Zionists' offensive*
> *And tomorrow the coming victories*

In the kind of history the poet had in mind, the Crusades were followed by the Western seizure of the Ottoman territories after 1918; the creation of the state of Israel in 1948; the virtual creation of the autocratic, westernizing Pahlavi dynasty in Iran until it was swept from power by the Islamic revolution of the Ayatollah Khomeini in 1979; through the Gulf War of 1990, to the Afghan War of 1991 and, most recently, the war in Iraq. All were but phases in a single "Crusade."

Even the relatively secular Libyan leader Colonel Muammar Qaddafi—who seems to hanker more after the primitive life of the Bedouin than the rule of the caliphs—could depict himself in the 1980s as the leader of a new jihad against "the American Crusader Christians," who had launched "the offensive of the Cross against Islam." For Qaddafi, the battle lines were still where they had been nine hundred years before, between "Islam and the Christians, between East and West." Defending the Eastern, Muslim world stood the defiant Qaddafi, described in an odd blend of the rhetoric of Lenin and Muhammad as "he who has uncovered the conspirators, exposed the fascist reactionary rulers and made summons to the true holy jihad."[49] One cannot easily imagine a Western opposition group defining its objectives in such terms.

The Crusades had been one kind of struggle between the

Christian West and Muslim East. But they were, in the historical memories of both parties, but one phase in a far longer, deadlier tussle. The recapture of Acre in 1291 might have finished off any further European ambitions to reconquer the Holy Land. But within a decade of the final extinction of the Latin kingdoms in the East, the struggle between Christendom and Islam had already begun to take a new and, for the West, far more menacing turn.

# 7

## *THE PRESENT TERROR OF THE WORLD*

**I**

Ever since the end of the tenth century, the power of the caliphate had been on the wane. On February 10, 1258, the armies of the Mongol khan Hülegü sacked the city of Baghdad. It was said of the Mongol method of conquest that it left "no eye open to weep for the dead." Baghdad was no exception; the Mongols destroyed the system of dikes that controlled the waters of the Tigris, and the waters flooded the surrounding countryside and drowned the peasants by the thousands in their villages. When at last the city fell, the caliph, al-Mu'tasim, the last of the Abbasids, was forced to hand over all his treasure and ten days later was executed outside the city walls by being rolled in a carpet and trampled to death by horses. Hülegü, although a pagan, was apparently superstitious about spilling royal blood. The Mongols, as Mongols usually did, then put the city to the sword, burning the libraries and the academies, hated symbols of all that the Mongol horde was not; burning the mosques; and killing so many that the stench of their rotting corpses drove the khan and his court to take refuge in the countryside for fear of the plague.

The Mongol conquest of Syria and Iraq, although brutally

spectacular, was, however, as with all Mongol conquests, short-lived. Hülegü was soon forced to return east to confront another threat from another Mongol horde. In September 1260, another Mongol army was defeated by the Mamluks, a dynasty of former Turkish and Circassian slave soldiers, at Ain Jalut in Palestine. The battle, which put a definitive end to the Mongol advance, subsequently became one of the celebrated moments in the history of Islam and is routinely mentioned by modern Islamic militants along with the battle of Hattin and the fall of Jerusalem. With the defeat of the Mongols, what had been the Abbasid Empire now passed into the hands of the Mamluks. The entire area of what is now Syria, Iraq, Egypt, Lebanon, and Palestine was soon divided up among factious and divided Turkic tribes, none of which posed any real threat to the equally divided, equally factious, Christian kingdoms of the West.

At the beginning of the fourteenth century, however, a new Turkic people emerged from the hinterland of Anatolia that would finally destroy all that now remained of the Roman Empire in the East and for the next half millennium threaten the very existence of western Europe and Christendom itself: the Ottoman Turks.

The Ottomans were at first but one of the more successful of many Turcoman tribal groups from Central Asia then fighting for control of the lands bounded by the Black Sea, the Mediterranean, the Aegean, and the easternmost flank of the diminishing Byzantine Empire.[1] They derived their name from Osman, the legendary founder of the dynasty. In the early fifteenth century, Osman was provided with a legitimating mythic ancestry that ran back through the Turkic Oguz tribe to Noah himself, who, it was claimed, had given the entire East to his son Japheth and thus by descent to the Ottoman sultans. In reality, Osman was probably a peasant, although clearly one with considerable military skills and personal authority. By 1301, he had gathered enough followers to be able to defeat a Byzantine army on the southern shore of the Sea of Marmara, a few miles from Constantinople itself. This initial success gave the Ottomans considerable prestige, and by the time of Osman's death sometime around 1323, they had consolidated their authority over a substantial area of northwest Anatolia between the Byzantine Empire to the west and the empire of the Seljuq Turks, who at this stage were Osman's nominal overlords, to the east.

By the time they first appeared in the historical record, the Ottomans were already Muslims. Osman's son Orhan described himself as "Champion of the Faith," and by the late 1330s the Ottoman emirs began to assume the title "Sultan of the Gazes." A *gâzî* was one who went on a *gâzâ*, or "war for the faith," a formal equivalent of the Arabic jihad. In practice, however, most *gâzî*s were plunderers rather than true "holy warriors," and they included a substantial number of Christian renegades, both Greek and Arab, among their number. Like all marcher lords, the Ottomans were on relatively good terms with their neighbors when they were not fighting them, and the Christians in the areas under their control seem to have practiced their religion freely, and intermarriage was not uncommon. Because of their origins, the Ottomans, for all their reputation in the Christian West for ferocity and tyranny, would be the most tolerant and pragmatic, as well as the most extensive, of all the Muslim empires. They made frequent alliances with their Christian enemies in both the West and the East, and Orhan even married the Byzantine princess Theodora.

In 1326, Orhan captured the Byzantine city of Bursa, which became the capital of the rapidly expanding Ottoman domains and remained a burial place for members of the royal house even after the seizure of Adrianople in 1362 and of Constantinople in 1453. In 1331, the city of Nicaea (now İznik) fell after a prolonged siege that, according to the Moroccan traveler Ibn Battûtah, who visited the city soon after it had surrendered, had left it in "a moldering condition and uninhabited except for a few of the men in the Sultan's service."[2] By now it had become obvious to the Byzantine emperor Andronicus that unless he struck some kind of bargain with the Ottoman sultan, within a few years all that remained of his empire would be gone. In 1333, he therefore went to pay a humiliating visit to Orhan, who was then besieging Nicomedia (modern Kocaeli). In exchange for a tribute, the once mighty successor of Constantine the Great was allowed—for the time being—to hang on to what little he still possessed in Anatolia. (This, however, did not save Nicomedia, which was forced to capitulate four years later.) He did so, precariously, for another twenty-eight years. Then, however, in 1361, Sultan Murad I occupied the Thracian city of Adrianople, which now became Edirne, and the dissolution of the Byzantine world began again.

The Ottomans also began slowly to absorb the other Turcoman and Muslim emirates to the east. By the time Orhan died in 1362, the Ottoman state reached from southern Thrace as far as Ankara, the capital of modern Turkey. It was a formidable military machine. At its core was the Janissary Corps—the *yeniceri*, or "new force"—created by Orhan's successor, Murad I. The Janissaries were raised by means of a levy, called the *devsirme*, of young boys, often no more than babies, taken from Christian communities and then raised as strict Muslims, to become members of an elite warrior caste. They would be the main resource of the Ottoman armies until they came to be an impediment to progress in the nineteenth century and were finally abolished in 1826 by Sultan Mahmud II.

Meanwhile, the Byzantine emperors in their beleaguered enclave tried desperately to secure help from the Latin West. It was not an easy task. Since at least the coronation of Charlemagne in 800, the churches of East and West had been at loggerheads with each other. The Greeks accused the Latins of Judaic leanings because they fasted on Saturdays. They professed to be shocked by the fact that Latin priests shaved their beards and could see no reason why they had been forbidden to marry. The Greeks were also uneasy about the doctrine of Purgatory—a relatively late Latin invention—which seemed to suggest that mere humans could have very precise information about what God intended to do with errant souls after death. The Latins, for their part, found the Greek belief in the "uncreated Energies of God" too mystical to be comprehensible and Greek Orthodox ritual, like most aspects of Byzantine life, altogether too "Oriental."

The most heated debate, however, was over whether leavened or unleavened bread should be used for the Sacrament. The Latins insisted on unleavened bread on the not unreasonable grounds that this was what Christ himself must have used at the Last Supper. The Greeks, hostile to such historicizing, insisted that unleavened bread was an insult to the Holy Spirit. Only the best bread was worthy of becoming the flesh of the Christ. One popular Greek term of abuse for Latin Christians was "Azymites," "the unleavened ones," which implied not only that they ate an inferior food but also that they were untouched by the Holy Spirit.

Ever since 1054, when the pope had excommunicated the patriarch of Constantinople, Michael Cerularius, and Cerularius had replied by casting an anathema against the pope, the Greek Orthodox Church had been branded by the Latin West as schismatic and heterodox. Historically, of course, this was an absurdity. Christianity had been an Eastern, Greek religion long before it became a Latin, Western one. From the viewpoint of the Greek Orthodox Church, the "Great Church"—as its adherents called it—was *the* Church. It was Rome, despite the accepted primacy of the pope over all the other patriarchs, that was the deviant.

But these historical niceties mattered little. What was at stake in this struggle between pontiff and patriarch was authority. At regular intervals the papacy made insulting demands that the Great Church relinquish its independence and accept union with Rome to form a single, truly "Catholic" church, in which, inevitably, the pope would become the undisputed leader of all Christendom. The Byzantine emperors had ignored these demands as long as they could afford to. The basileus, unlike the emperor in the West, had retained the close association with the Church that had been created by Constantine. He was the anointed Viceroy of God on Earth, and to have accepted union with Rome would have meant, in effect, surrendering a large part of his political authority. If that were not sufficient, the still powerful memories of the Latin occupation of 1204–1261 were enough to make any Greek nervous about letting the unruly and marauding Latin knights anywhere near their territories. The patriarchs were even more fiercely opposed to union than the emperors were. Union with Rome would not only mean the eventual disappearance of their independence, it would also imply the ultimate extinction of the Great Church. If they had to be ruled by outsiders, on the whole they preferred them to be Muslims, who at least took no interest in what kind of bread was used in the Sacrament and could be relied upon to preserve the ecclesiastical hierarchy intact. As the last great minister of Byzantium, the megadux Lucas Notaras, is said to have declared, "better the Sultan's turban than the cardinal's hat."[3] (With nice irony, he was beheaded by Mehmed some months after the fall of Constantinople for—so it was said—refusing to allow his son to be given to the sultan "for his pleasure.")

In desperation, however, as the Ottomans moved steadily west, the Emperor John V Paleologus, the son of a Latin mother and therefore perhaps more sympathetic to Rome than his predecessors, turned to the pope for assistance against what he hoped to persuade the Latins was now a common enemy. In 1355, he wrote to Innocent VI, undertaking to convert all his subjects within six months if the pope would provide him with five galleys and a thousand foot soldiers. He even offered to send his second son, Manuel, to be educated at the papal court and to abdicate in his favor if he failed to bring about the promised union. But the pope had no ships or troops to send. All he could offer was a papal legate with a blessing. In 1364, John first sought help from Serbia, also a Christian Orthodox state; then, when that failed, he turned to King Louis of Hungary. That too came to nothing. In desperation, in 1369, he traveled to Rome in person and publicly submitted himself to the pope. But none of the clerics would follow him, and he returned home empty-handed.

Two years later, the king of Serbia raised an army and marched east, but he was taken by surprise by an Ottoman army at Chernomen (Cirmen) on the Maitsa River and his men were slaughtered in such numbers that the field became known as "Destruction of the Serbs." The battle put an end to the southern Serbian kingdom and turned the Serbs, and subjects of the three Bulgarian rulers who had joined them, into Ottoman vassals. The route to Macedonia and the Balkans now lay open, but it was eighteen years before the Ottomans were able to take advantage of the situation. On June 15, 1389, at Kosovo Polje, the "Field of Blackbirds," near the town of Priština, Sultan Murad I defeated a combined army of Serbs, Albanians, and Poles, and the whole of Macedonia was incorporated into the Ottoman state. The battle of Kosovo—in 1989 declared by the Serbian leader Slobodan Milošević, to be the "Cradle of Serbian Civilization"—has come to be seen as a turning point in the history of Serbia, the moment when an alien, Muslim, Eastern presence established itself in a Christian kingdom that, if not exactly Western, was certainly European. And it has not been forgotten. In subsequent Serbian history, Kosovo is represented as having placed a duty upon all Serbs to struggle against the hated Ottoman, Muslim yoke until the day when they are once again a free people. As the nationalist poet Vuk Karadžić wrote in 1814:

> *Whoever will not fight at Kosovo*
> *May nothing grow that his hand sows*
> *Neither the white wheat in the field*
> *Nor the grapes on his mountain.*

The crown prince, Lazar, who was executed after the battle, was claimed as a saint and even represented as Christ in nineteenth- and twentieth-century paintings, surrounded by twelve knight-disciples. During the civil wars that followed the collapse of Yugoslavia in the 1990s, the relics of the Christ-Prince Lazar were paraded around the province of Kosovo, and the terrible atrocities later committed by Christian Serbs against a minority of largely secularized Muslims—labeled significantly "Turkifiers" by Christian Serbs—were often hailed as revenge for Lazar's "martyrdom" and for the "Serbian Golgotha."[4]

The Serbs had been annihilated on the "Field of Blackbirds," but they managed to achieve one notable success. Murad himself was cut down by one Miloš Obilić, later compared, by Vuk Karadžić, to Achilles and held up as a future role model for all Serbs. When news of the sultan's death reached Europe, King Charles VI of France ordered a mass of thanks to be sung in Notre Dame. But if Charles thought that the sultan's death would stop the Ottoman advance toward Europe, he was mistaken. Murad was succeeded by his son Bayezid, and in the spring of 1394 the latter laid siege to the city of Constantinople itself. For a while it seemed as if the end of the Roman Empire in the East was now only a matter of time.

Time, however, was precisely what mattered. In the days before gunpowder a city as well defended as Constantinople, with two circles of massive and near-impregnable walls on its landward side, could be forced into submission only by cutting off its supplies. Although the great city could be isolated by land, it was very difficult, even for the increasingly powerful Ottoman navy, to prevent supplies getting through by sea. Bayezid was by no means the first Muslim ruler to discover this. No fewer than eleven previous attempts had been made to subdue the city since an early Arab siege in about 650 in which Ayyûb Ansari, one of the companions of the Prophet, had supposedly taken part. After eight years of encirclement by Bayezid's armies, the city had still not surrendered. How long the blockade would have continued, it is impossible to say, because Constantinople was saved not by its defenders but by an unexpected invasion from the east that very nearly put an end to Ottoman rule altogether.

\*   \*   \*

EARLY IN THE morning of July 28, 1402, the armies of the legendary Turkic-Mongol chieftain Timur-i-Lang, "Timur the Lame," known in Europe as Tamerlane or Tamburlaine, met a hastily assembled Ottoman army on the plain just outside Ankara. By evening, Timur, whose thirty-two elephants had rained Greek fire down upon the Ottomans, had destroyed Bayezid's forces, taken him and his son Musa captive, imprisoned his wives in his harem, and deprived him, in a single day, of most of the territory he had acquired during his lifetime. Bayezid died in somewhat mysterious circumstances in the town of Aksehir in March of the following year.

When news of the Mongol victory reached Europe, it was greeted with rejoicing. Timur, although a Muslim and as much an enemy of Christendom as Bayezid had been, now came to seem like a savior, and the story of his triumph over the Ottomans became a popular subject for years to come. Christopher Marlowe's play *Tamburlaine the Great,* which won its author instant popular appeal, opened in London in 1587, more than a century after the battle itself and three years after England had begun formal trade relations with the Ottomans. The subject was still topical enough in 1648 for the French playwright Jean Magnon to stage a fanciful version of the story that provided Bayezid in captivity with a wife and daughter. George Frideric Handel wrote an opera, *Tamerlano,* on the theme in 1724, and Antonio Vivaldi another, *Bajazet,* ten years later. Much of this was mere playacting with nothing more than a good story, with the kind of exotic setting sixteenth- and seventeenth-century audiences relished, as topical as the stories from ancient Rome and Greece. But not all. As long as the Ottoman sultanate remained high on the horizon of the European imagination, a threat to European safety and the very image of Oriental despotism, the tale of its near annihilation remained immensely gratifying.

The battle of Ankara allowed Byzantine Constantinople to survive for another half century. It turned out, however, to be not the end of the seemingly ineluctable rise of the Ottomans as much as an incentive for their consolidation. In 1403, Timur re-treated eastward, and died two years later on the road to China. By 1415, after a prolonged war of succession, Bayezid's son and suc-cessor, Mehmed I, had all but regained most of the former Ot-toman territories in Anatolia. Constantinople, which had enjoyed more than a decade of relative tranquillity in which the emperor had thrown over his vassal status, expelled the Ottoman mer-chants resident in the city, and torn down the mosque that had been built for them, was once again under threat.

But it was not until after Mehmed's death in 1421 that his son Murad II made another attempt to seize the city. In August 1422, his engineers constructed a massive stone embankment along the entire length of the landward walls, from which he pounded the city with artillery. The emperor, John VIII Paleologus, made the by now almost routine appeal for help from the West. As always, no help came. By early September, however, Murad was forced to abandon the siege. Short and unsuccessful though it had been, it had clearly demonstrated the degree to which the use of artillery had changed the rules of the game. The old strategy of encir-clement, which had proved so effective, was no longer necessary. Now it was clearly only a matter of amassing enough guns power-ful enough to breach the walls. Once this became clear, Constan-tinople was, in effect, doomed, unless the Latin West could be persuaded to mount a massive and sustained campaign for its de-fense. The Byzantine emperors knew this and redoubled their ef-forts to secure some kind of assurance from their coreligionists in Europe.

They had some initial success. In 1439, the long-delayed union between the Latin and Greek churches was finally achieved at the Council of Florence. After months of deliberation on ab-struse theological issues, while the Greek delegation was, it was said, kept short of food and comforts, the emperor John VIII fi-nally capitulated to most of the pope's demands—although the Greek Orthodox Church was permitted to go on using leavened bread. On his return to Constantinople, however, the emperor found it hard to persuade any of his subjects to accept the terms of the union. By the time he died, weary and disillusioned, in 1448, it was, in effect, a dead letter, and his successor, Constantine IX—who would, as it turned out, be the last of the Roman emperors in the East—made no attempt to press it upon those of his people who were still left to him until 1453, and by then it was far too late.

By the end of the fourteenth century, Byzantium lacked any strategic importance and certainly posed no real military threat to Ottoman ambitions. Constantine's great city, and what little remained of the crumbling Byzantine Empire, had never fully recovered from the Latin occupation from 1204 to 1261. When, in 1400, the emperor Manuel II Paleologus paid a desperate visit to Henry IV of England in the vain hope of raising troops to defend his vanishing domains, the lawyer Adam of Usk, who saw him and admired his learning and his spotless white robes, reflected sadly, "How grievous it was that this great Christian prince should be driven by the Saracens from the furthest East to these furthest Western islands to seek aid against them." "What dost thou now," he asked, "ancient glory of Rome?"[5]

By then the ancient glory of Rome, the once-mighty Byzantine Empire, had been reduced to little more than the city of Constantinople itself and its surrounding countryside. It was a melancholy, dying place, whose population by the end of the thirteenth century had shrunk to a mere hundred thousand and was shrinking still. When Ibn Battûtah visited it in the mid-fourteenth century, he counted thirteen separate hamlets within the walls, all remnants of what had once been prosperous districts of the great city itself. In many of these, he wrote, "you might have thought that you were in the open countryside with wild roses blooming in the hedgerows in spring and nightingales singing in the copses."[6]

But whatever its present condition, Constantinople was still the "Golden Apple"—*kizil elma*—the capital of the ancient Roman Empire, by Muslims and Christians alike reckoned to be the greatest power the world had ever known. For Murad's successor, Mehmed II, it was, for all its present dilapidation, the most treasured prize of all, whose possession would transform him into the master of the world. If he could not rule an empire that contained Constantinople, he told his ministers, then he would rather not rule an empire at all.[7]

Taking the city would not, however, be easy. Despite its depleted population, despite the decay of its internal defenses, the fourteen miles of walls that encircled the city were still believed to be almost impregnable, even in the face of artillery. Mehmed therefore prepared his onslaught with considerable care. He began in 1451 by building, with astonishing speed, a fortress some five kilometers north of the city. Called Bogazkesen, "Cutter of the Strait" (or, alternatively, "Cutter of the Throat"), and now renamed Rumeli Hisar, it was meant to provide a forward base for the siege and to secure the passage of the Ottoman forces across the Bosporus. The emperor Constantine responded to the sight of this fortress rising up under his very gaze by imprisoning all the Turks residing in Constantinople and then, realizing that this was at best a futile gesture, releasing them again. He then sent ambassadors to Mehmed in the hope of extracting an assurance that the new fortress was not, in fact, intended for the purpose it so clearly was. Mehmed had them thrown into jail and then decapitated. This amounted to a declaration of war.

Once again the emperor sent hurried pleas for assistance from the West. And once again he received only evasive and conditional answers. The English, French, and Burgundians explained that they needed all the troops they had, for fighting one another. The maritime republics of Genoa and Venice were reluctant to jeopardize their harmonious, if always precarious, trade relationships with the Muslim world by assisting an ally whom they distrusted and whose fate was, in any case, in their opinion, already sealed. Save for the Latin residents, mostly Genoese and Venetian, in and around Constantinople itself, the Greeks were on their own. Latin

Christendom, which had afflicted one kind of betrayal on them in 1204, was about to afflict another.

On April 5, 1453, Mehmed's army reached the outer walls of the city. His forces, according to the Venetian merchant Nicolò Barbaro, who saw them arrive, numbered some 160,000. But that was only an intelligent guess. Other accounts, all of them Christian, put the figure anywhere between 200,000 and 400,000. Most were Muslims, marshaled from all over the empire. But their ranks were swollen by renegades of all kinds in the expectation of rich pickings: Latins, a large contingent from George Branković, king of Serbia, even some Greeks.

Mehmed himself set up his tent behind the front line facing the Gate of Saint Romanus and waited.

Inside the city a state of terror now reigned, made worse by clear portents of the doom to come. An icon of the Mother of God carried in procession slipped from the hands of its bearer for no apparent reason and then proved to be so heavy that it took "great efforts and much shouting and prayers by all" to lift it up again. The following day, a dense fog engulfed the city, which, recorded the Greek chronicler Kristovoulos, "evidently indicated the departure of the Divine presence."[8] The portents were all too probable. The able-bodied male population of the city was some 30,000, but the Byzantine statesman George Sphrantes estimated that of these, fewer than 5,000 were able and willing to fight. (The Greeks, sneered later Latin historians, would prefer to discuss the sex of angels than make war, even for their own survival.)[9] There were also a handful of Latins, mostly Genoese from the colony at Galata on the western shore of the Bosporus, reinforced by some 5,000 additional troops under the command of the legendary corsair Giovanni Giustiniani.

Mehmed moved into place no fewer than fourteen batteries of artillery along almost the entire length of the outer line of walls, the Wall of Theodosius. Each of these guns had been made by a renegade Hungarian canon founder from Edirne named Orbain, and they were far larger than anything the Christians had ever seen. The largest could send a ball weighing twelve hundred-weight for more than a mile—a huge distance at the time. It required twenty pairs of oxen to haul the guns into place and two hundred men to steady the huge wheels of the gun carriages as they lurched over the uneven ground. On April 12, these behemoths, nicknamed "basilisks," began to bombard the walls as fast as the scores of men needed to load and reload them could manage. Day after day the massive stone balls of the guns carried away great chunks of masonry, sometimes entire towers; hour by hour, as the entire population turned out at night to rebuild what they could, the city's defenses crumbled steadily away.

The first assault took place on May 12. Some 50,000 men launched themselves at the Adrianople and Kaligaria gates of the city. But after a long day's fighting, which left corpses stacked so high that the roads became impassable, the Turks finally retreated. Six days later, four immense siege towers, veritable "wooden castles," which the Greeks dubbed "the city takers," their sides protected by three layers of leather curtains, were pushed into place against the walls. Once more, however, the attackers were beaten back, and during the night the Greeks crept out of the city and burned the towers to the ground.

Like Xerxes centuries before, Mehmed had hoped that the sheer size of his army, his navy, and his artillery batteries would grant him a relatively swift victory against a city that was, by all reports, sparsely defended. Now he wondered if he had, after all, been badly informed about the Christians' ability to resist. The Turkish troops, accustomed to easy victory, had been demoralized by their repeated failures and their massive losses, and rumors soon began to circulate though the Ottoman camp of the approach of a huge Christian army. On May 27, Mehmed called a Great Council to decide what course to take. The grand vizier, Halil Pasha, who was inclined to think that the city would be more use to the sultan as a Christian commercial entrepôt, urged his sovereign to abandon the siege. Mehmed, however, decided to launch one final, all-out assault. If that failed, he would withdraw.

The following day, he addressed his troops. He praised them for their zeal, their piety, and their bravery, but just to ensure that they did not falter in the final hours, he also offered them great wealth, "gold and silver and precious stones and costly pearls." He conjured up for them the prospect of a life surrounded by gardens and fine public buildings and magnificent houses, and in the company of "very beautiful women, young and good looking and virgins lovely for marriage, and even till now unseen by masculine eyes." And for those with other inclinations, there would be "boys, too, very many and very beautiful and of noble families." He promised them, "for your spoil and plunder," a "great and populous city, the capital of the ancient Romans, the head of the whole inhabited world."[10]

All of this was taken down by Kristovoulos, who, although he was not present during the siege, seems to have questioned many who were and to have known Mehmed himself well enough to secure an appointment as governor of the Aegean island of Imbos. It sounds too much tailored to Western views of the rapacious Turks—in particular the reference to virgins and boys—to be literal, but the gist of it rings true, in particular the allusion to Constantinople as the capital of the *oikoumene,* the "inhabited world," over which Mehmed, the *Amîr al-Mu'minîn,* "Commander of the Faithful," and his descendants would soon rule until the end of creation.

Mehmed now sent an envoy to Constantine. "Let us," he said, "leave this affair in the hands of God. You leave the city with all your superior and inferior dignitaries and all your goods and go where you please. In that way your people will suffer no harm from either me or you. Or do you prefer to resist and lose your life and all your goods, both your own and those of your court, and see your peoples made the slaves of the Turks and scattered throughout the world?"

It was the traditional offer every warrior of the jihad was required to make to his enemy before declaring war. Constantine refused it. He reminded Mehmed that his victory was by no means
assured and that, in any case, "it is not in my power or that of the
citizens who live in it to surrender the city to you, because we have
all freely decided to die and not to spare our lives" in its defense.[11]

About three hours before dawn on Tuesday—a day still regarded in the Greek world as ill omened—May 29, Mehmed gave
the order for the final assault. The Greeks managed to drive back
the first two waves of attackers. But the outer walls of the city were
now virtually in ruins, and, as Orbain's massive canons pounded
what remained of them, the Janissaries, the sultan's crack troops,
broke through the Kerkoporta, or "Gate of the Circus," and
poured into the city. The fighting was fierce, but "by the early part
of the forenoon," recalled the Ottoman historian Tursun Bey,
"the frenzy of the fiery tumult and the dust of strife had died
away."[12] The Byzantine emperor, aptly named for the imperial
city's own founder, the last "Roman" to rule in Constantinople,
fell in the fighting. No one knows where and how he died, but a
head the conquerors claimed to be his was later displayed on a
marble column in the Augustean Forum and then carried on a triumphal tour of Anatolia, Arabia, and Persia. (He had, however,
already sold the imperial title, basileus, to King Ferdinand of
Aragon, who then, in 1494, sold it to Charles VIII of France—
although neither ruler ever had the gall to use it.)

FOR THREE DAYS Mehmed's victorious army was allowed to pillage the city. "This crowd, made up of men from every race and nation, brought together by chance, like wild and ferocious beasts,"
lamented Kristovoulos, fell upon the defenseless population,
"stealing, robbing, plundering, killing, insulting, taking and enslaving men, women, and children, old and young, priests and
monks—in short every age and class."[13] The blood ran in the
streets "as if it had been raining," wrote Nicolò Barbaro, and "bod-

ies were tossed into the sea like melons into the canals in Venice."[14] Nothing so devastating had been seen in the Christian world since the sack of Rome by Alaric's victorious Goths in 410. When it was all over, the body count, including the civilian dead, according to Kristovoulos, was "well nigh four thousand."[15] (Although there were some who observed sarcastically that however bad Mehmed's Muslim soldiery might be, they were certainly no worse than the Crusader knights had been in 1204. Certainly, if the Turks carried away more than the Latins had, they also caused less lasting destruction.)

One of those who had participated in the defense of the city was Cardinal Isidore of Kiev, a Greek in the service of the papacy. He had been wounded in the fighting and then captured by the Turks, but he had managed to escape and then made his way to Candia on the Venetian-controlled island of Crete. There he wrote to his fellow Greek Cardinal Bessarion in horrified despair at what he had seen: "All the streets and avenues and even the alleyways ran with blood and gore and were choked with eviscerated and butchered bodies." Muhammad, he declared, was none other than the precursor of the Antichrist. Like Kristovoulos, and Pelagius during Alaric's sack of Rome, Isidore was particularly troubled by the invaders' complete disregard for social and sexual distinction. The looters had paid no respect to status, sex, or age, he complained. He had seen "noble and well-born women dragged from their houses with ropes around their necks." This brutality had then been compounded by another kind of sacrilege. "As soon as they were able," he recalled, "they entered the temple known as Hagia Sophia—and which is now a Turkish mosque—they tore down and shattered all the statues, the icons and images of Christ and of the saints."[16] It was the end. "The City of Constantinople," he cried, "is dead."

Late in the afternoon of the third day, when there was nothing more left to loot, Mehmed himself entered the "Golden Apple" mounted on a white horse. The "bronze-breasted forces" from Asia, which Publius Cornelius Scipio had seen in his dreams centuries before, had finally arrived. Escorted by the Janissary guards, he rode slowly through the by now almost deserted streets to the doors of Hagia Sophia. There he dismounted, picked up a handful of earth, and poured it over his turban as a gesture of humiliation before the god who had given him victory. Later he crossed the square and wandered through the half-ruined rooms of the Sacred Palace and is said to have murmured to himself the words from Firdausi's *Shah-Namah,* "The spider weaves the curtains of the palace of the Caesars; the owl calls the watches in Afrasiab's towers."[17]

Ever since the armies of the caliph Umar II had been forced to abandon the first sustained siege of Constantinople in 718, prophecies had spread throughout the Muslim world of the inevitable day when the great city, the last bastion of the ancient enemy, would pass into the *dar al-Islam.* Now, under a sultan who bore the name of the Prophet himself, these predictions had finally come to pass. Thereafter Sultan Mehmed II was known to both Muslims and Christians alike as "the Conqueror."

For the West the fall of Constantinople was a calamity. It marked, it seemed, the occlusion of that part of Asia from which so much of Western culture had derived, and it drew down a curtain over eastern Europe for more than four centuries. It was not only a great Christian city, the last bastion of Constantine's empire in the East, that had fallen. Gone, too, was the last living link with the ancient Greek world. "O famous Greece," wrote Pope Pius II in a celebrated and widely distributed lament for the city, "behold thy end! Who does not grieve for you? There remained up to this day in Constantinople remembrance of your ancient wisdom. . . . But now that the Turks have won and possess all that Greek power once held, I believe that Greek letters are finished." And all of this glittering past had been snuffed out by a horde of Muslim barbarians from the depths of Asia.[18] "A barbarous people," cried Lauro Quirini, a Venetian from Candia attached to the household of Cardinal Bessarion, "an uncultured people, living without laws or fixed customs, but scattered [i.e., nomadic], vague, arbitrary, perfidious, and fraudulent have trampled upon a Christian people in a manner that is infamous and ignominious." Worse even in his view than the killings was the wanton destruction of an entire culture. "The language and the literature of the Greeks," he claimed with a degree of exaggeration, "invented with such industry and such labor and perfected over such a long time, has perished, alas, perished!"[19]

THE GREAT CITY now began a rapid transformation into "Istanbul," a name that derives from a corruption of the Greek phrase *is tin polin*, meaning "to the city," although it continued to be known officially as "Konstantiniyye" in Turkish until 1930, when the name was changed by law.[20] Hagia Sophia, a church that had been built by Constantine the Great himself, was transformed into a mosque. Minarets were added and the paraphernalia of Christian worship were removed and replaced by a prayer niche and a pulpit. The banners that Mehmed's victorious army had carried into the city were hung on the walls, and prayer mats that had supposedly belonged to the Prophet himself were laid on the floors. The colossal statue of Constantine the Great with a golden ball in his hand, erected in 543, nine meters high and standing on top of a thirty-meter column, which, legend had it, would last as long as the Byzantine Empire itself, was now taken down. Parts of it ended up in the Topkapi Palace, which Mehmed built on the site of the ancient Byzantine acropolis. There, in the 1540s, the French humanist Pierre Gilles saw Constantine's leg, which, he recorded, was larger than he was, and his nose, which was "over nine inches long."[21]

Mehmed was now—save for the tiresome presence of the Timurid Persian Empire to the east—ruler of all Muslim Asia. He could also now claim to be not only the legitimate heir of the succession of emperors—Solomon, Constantine, and Justinian—who, in myth and reality, had built and rebuilt the city, but also to have fulfilled one part of the prophecy recorded in the Hadith that the day would come when a Muslim emir would take both Constantinople and Rome.

As the Cretan historian George Trapezountios told Mehmed in a more than usually sycophantic mood, "No-one doubts that you are the emperor of the Romans. Whoever is legally master of the capital of the empire is the emperor, and Constantinople is the capital of the Roman Empire." The Holy Roman emperors would never accept that. But with the fall of the "Golden Apple" the Ottomans became the only other state in the world to which the princes of Christendom were prepared to concede the title of "empire."

The Ottoman Empire was now the major power in the East, and the sultan, although it would be some time before he would formally assume the mantle of the caliphs, was the "commander of the faithful," the self-appointed leader of the Islamic world. Little wonder, perhaps, that Mehmed—who, if the Venetian historian Niccolò Sagundino is to be believed, had Herodotus and Livy read to him in Greek and Latin and had the *Iliad* and Arrian's *History of Alexander* copied for his library—should, like so many would-be world conquerors, have identified himself with Alexander the Great.[22] In 1462, on his way to attack Lesbos, then a Venetian colony, he, like Alexander, stopped off at the supposed site of Troy to pay homage to the heroes of the Trojan War. It was, however, with Hector that Mehmed identified himself, not Achilles. Kristovoulos, records him as saying,

It was the Greeks and the Macedonians and the Thessalians and Peloponnesians who ravaged this place in the past, and whose descendants have now through my efforts paid the right penalty, after a long period of years, for their injustice to us Asiatics at that time and so often in subsequent times.[23]

In one fanciful Latin account of the conquest of Constantinople, the "Great Turk" is said to have raped a virgin in the Church of

Hagia Sophia, crying out as he did so that he had thereby avenged the rape of Cassandra, daughter of the Trojan king Priam, by the Greeks.[24]

Mehmed's relationship with his new Greek subjects was not, however, quite as brutal and destructive as those who had fled to the West made it seem. He was, perhaps more than any subsequent sultan, determined to rule over a united, prosperous, and above all disciplined people. In January 1454, he summoned George Gennadios Scholarios from captivity in Edirne to become patriarch of Constantinople. He is said to have had regular disputes with Gennadios as to the respective merits of Islam and Christianity, on which Gennadios wrote a brief "objective" account for translation into Turkish. The philosopher George Amiroutzes also wrote a book for the sultan not only explaining how much in common there was between Christianity and Islam, but even suggesting that the two religions could be blended into one. The differences between the Bible and the Qur'an, he argued, had been exaggerated by bad translations and by the Jews, who had been responsible for exacerbating the subsequent false perceptions each faith had formed of the other. Mehmed was unimpressed by such extremes of ecumenism. But he did restore to the Orthodox Church the powers and privileges it had enjoyed under Byzantine rule, together with a large part of its property.[25]

From beyond the Dardanelles, however, it seemed as though Eastern Christendom had now vanished for good. In its place stood the most imposing power to threaten the liberties of the peoples of Europe since the days of Xerxes. All Christendom waited to see what would happen next. Would Mehmed remain where he was and consolidate his gains? Or were further conquests in the West to be expected? And if so, where would they stop? Everyone was aware that Constantinople had been the "Golden Apple" of Ottoman ambitions. But the sacred city of the West, the still-beating heart of Christianity, was, of course, Rome, and Muhammad himself had reportedly promised that one day Rome, too, would be incorporated into the *dar al-Islam*. Angelo

Giovanni Lomellino, leader (*podestà*) of the Genoese community at Galata, watching events from just across the Golden Horn, had no doubt as to what was to come. On the day that Constantinople fell, he wrote to his brother in Genoa, "The sultan now says that only two years will pass before he comes to Rome."[26] Isidore of Kiev was of the same opinion. In July, he wrote to Pope Nicholas V to warn him that Mehmed was threatening to eradicate "the very name of the Christians," and just in case Nicholas should think that this did not also apply to him, Isidore added that the sultan's ultimate objective was to "subdue by force of arms your city of Rome, the head of the empire of the Christians."

On September 30, 1453, Nicholas issued a bull to all the Christian princes of the West, enjoining them to shed their blood and the blood of their subjects in a new crusade against the anti-Christ now seated in Constantinople. Pleading insolvency or the pressure of domestic affairs, the princes of Christendom—Charles VII of France, Henry VI of England (now, in any case, out of his mind), King Alfonso V of Aragon, and the emperor Frederick III—all politely declined. The pope then turned to the richest ruler in Europe, Philip the Good, duke of Burgundy. In February 1454, Philip held a banquet at Liège at which a live pheasant studded with precious stones was released onto the royal table while a huge man dressed as a Turk pranced around the room threatening the guests with a toy elephant and a youth, Oliver de la Marche (who kept a diary of the whole affair), dressed as a damsel, mimed the sorrows of Our Lady Church. Deeply moved, the entire company rose as one and swore to go on a crusade. But it turned out to be nothing more than a pantomime. Not one of those who took the "Oath of the Pheasant," as it came to be called, ever left home.[27]

For Nicholas's successor but one, however, the Turkish menace became something of an obsession. Enea Silvio Piccolomini, who had become Pope Pius II in 1458, was a humanist, scholar, poet, and author of comedies in Latin. For him the fall of Constantinople had meant not only the destruction of a great Christian city but, more significantly perhaps, "the second death of Homer and Plato."[28] Pius was not only learned, he was also shrewd and well traveled. And he had a far broader vision of the possibilities and need for Christian unity than his predecessors. In 1459, he embarked on a tour of Italy, culminating in the Congress of Mantua in 1459 at which he proclaimed a new crusade to retake Constantinople. Nothing came, however, of this, except a resolution on which no one was prepared to act. Four years of squabbling and bickering passed. "We longed to declare war against the Turks," Pius wrote later, but "if we send envoys to ask aid of sovereigns, they are laughed at. If we impose tithes on the clergy they appeal for a future council. . . . People think our sole object is to amass gold. No one believes what we say."[29]

On September 23, 1463, Pius made a speech to the college of cardinals once again, demanding action before the Turks finally swept all of Europe before them. In October, he formally declared a new crusade, which he intended to lead in person in the vain hope that the sight of the frail, sick Vicar of Christ sailing off alone to confront the Infidel would shame the quarrelsome princes of Christendom into finally doing something. In June of the following year, he left Rome for Ancona, a seaport on the Adriatic coast from which the assembled Christian armies were due to sail. When he arrived, no one was there. On August 11, two Venetian ships showed up. But by now it was too late. Three days later the pope had died, still hoping for the fleet that would finally take him to the East.

Pius had also tried diplomacy and flattery. He had written the sultan a long letter—known as the "Letter to Mehmed" and distributed throughout Christendom—in which he proposed not only to recognize Mehmed's claim to be ruler of the Eastern Roman Empire but also to transfer to him the imperium of the West, just as six and a half centuries earlier his predecessor Leo III, by the coronation of Charlemagne, had transferred (or "translated," to use the technical term) it from the Greeks to the Franks. All the sultan had to do was to convert to Christianity. What, after all, asked the pope in somewhat unpapal terms, were "a few drops of baptismal water" in exchange for the right to rule over the entire Roman world?[30] It was an empty gesture, as he must have known.

As it turned out, Mehmed never did come to Rome. In fact, he spent most of the rest of his reign consolidating his hold over the Balkans and securing his eastern frontiers. When he died at the age of 49 on May 3, 1481, he ruled over an empire that reached to the Adriatic in the west, the Danube-Sava line in the north, and most of Anatolia in the east. The Black Sea, the outlet for the trade of a vast hinterland reaching as far as Poland, Lithuania, Muscovy, and Persia, had become, in effect, an Ottoman lake.[31] True, the Mediterranean still remained divided, but by taking Constantinople, Mehmed now controlled its eastern end and could claim to have fulfilled the old Islamic dream of becoming "Sovereign of the Two Lands and the Two Seas"—that is, "Rumelia" (the Byzantine Empire) and Asia, and the Mediterranean and the Black Seas.

In 1480, toward the end of his life, Mehmed commissioned the Venetian painter Gentile Bellini to paint a portrait of him—in itself something of an act of heresy for a Muslim ruler. This picture, which now hangs in the National Gallery in London, shows the Sultan in three-quarter profile, turbaned, bearded, and wearing a massive collar of what is probably wolf skin, the totemic animal of the Ottomans. At the base of the painting is an inscription that describes him as *imperator orbis*—"emperor of the world." He is framed by triumphal arches, to the left and right of which are three crowns symbolizing the number of kingdoms in his realms, while the four flowers made of precious stones on the carpet in the foreground may be a reference to the dream that Osman, the founder of the dynasty, was claimed to have had, in which the future Ottoman conquest of the world was foretold. They may also symbolize the subjugation of the world to Constantinople, the petals, or rubies, being the continents, the black and white gems the Black Sea and the Mediterranean.[32]

What is clear is that whoever was responsible for this iconography wanted the viewer to know that here was an emperor who was the heir to all the rulers of Asia and Europe. When his embalmed body was buried, an effigy of the dead sultan was carried on top of the coffin. No previous Muslim ruler had been interred like this. But in 337, the Emperor Constantine the Great had. Mehmed had made himself not only another Hector, another Alexander; he had also cast himself, even in death, as the new Constantine. This man, the "precursor of the Antichrist, the prince and lord of the Turks," now stood at the gates of the Christian, Western world.[33]

The failure of Mehmed to make good on his alleged promise to march on Rome did not, however, lessen the fear in the West that this remained the ultimate objective of his successors. Ever since 1480, when Ottoman forces had sacked and occupied the town of Otranto on the Puglian coast of Italy, the Turkish navy, backed by the pirates of the Barbary coast, had cruised the eastern and southern Mediterranean, creating an atmosphere of almost constant alarm. All along the coasts of southern Italy and Spain towers were constructed, many of which are still standing, to maintain a permanent watch for the marauders. To this day there is a Spanish saying, *"Hay moros en la costa"*—"There are Moors on the coast"—which means roughly "Watch your back." Calls for a new crusade, for a new Pausanias—the Spartan general who had led the combined Greek forces at the battle of Platea that had finally driven the Persians from Greek soil—or for a new Scipio Africanus to deal with this new Hannibal were frequent in humanistic and literary circles.

The fear was not limited to the Mediterranean or to the eastern borders of Christendom. As one anonymous English observer put it in 1597, "The terror of their name doth even now make the kings and princes of the west, with the weak and dismembered relics of their kingdoms and estates, to tremble and quake through the fear of their victorious forces."[34] He was exaggerating. Not all of the kingdoms and estates of the rulers of Christendom were quite so weak and dismembered as he believed. But there were certainly many who quaked with fear. And with good reason. Even as far away as Iceland, Christians prayed to be delivered from "the terror of the Turk." (Their fears were substantial enough. In 1627, Ottoman-backed corsairs from North Africa penetrated deep into the North Sea and carried off four hundred captives for sale in the slave markets of Algeria.) What the English historian Richard Knolles called "the glorious Empire of the Turks" had become, for all those who still lived beyond its borders, "the present terror of the world."[35]

ALMOST EVERY TURKISH victory was greeted with calls to mount a new crusade, with the objective this time not of recovering the Holy Land but of pushing the Turks out of Europe, out of Constantinople, even possibly out of all that had once been the Byzantine Empire. In 1517, Pope Leo X commissioned a report on the matter from his cardinals, who replied that since the Ottomans were clearly bent upon the destruction of Christendom, there could be no alternative to a crusade. Although the emperor Maximilian proposed a five-year truce among the princes of Christendom, so as to allow them to concentrate their efforts against the Turks, nothing ever came of this or any subsequent initiative. The Crusades—so called—fought after the end of the thirteenth century were directed against breakaway Christian sects the Church deemed to be heretical, rather than against Muslims.[36]

No pontiff could do anything much more than pontificate—and raise a certain amount of money. If there was to be a new crusade, it, like its predecessors, would have to be manned and financed by the secular rulers of Europe. And these preferred, whenever possible, diplomacy to conflict. While its leaders were busy wrangling among themselves, Christianity watched as the Ottomans slowly encroached upon Europe. By the end of 1461, all that remained of the Byzantine *oikoumene*—the Duchy of Athens, the Despotate of Morea, and the Empire of Trebizond—had passed into Turkish hands. Serbia capitulated in 1459 and Bosnia four years later. Albania was overrun in 1468. Across the Danube, the Transylvanian state of Wallachia, which had maintained a precarious independence under the infamous Prince Vlad Drakul, known as "the Impaler" because of his favorite method of disposing of his opponents and the source for Bram Stoker's vampire, Count Dracula, fell in 1462. The neighboring principality of Moldavia followed in 1504. And so it went on. In 1521, an Ottoman army seized the Hungarian city of Belgrade, having failed twice before, once in 1440 and again in 1456. And with Belgrade, lamented the historian and bishop of Nocera, Paolo Giovio, "the rampart not only of Hungary but of the whole of Christendom" had fallen, leaving the entire civilized world at the mercy of the "Turkish barbarian."[37]

In August 1526, Sultan Süleyman I defeated Louis II of Hungary and Bohemia in the marshes of Mohács, whose muddy waters closed over the unfortunate king's head before he could escape the pursuing Ottoman cavalry. At the time the battle of Mohács seemed like a magnificent victory for the Ottomans. But in the long run it was to be something of a pyrrhic one. For the death of Louis brought to the Hungarian throne Ferdinand II, the Hapsburg archduke of Vienna, brother of the Holy Roman emperor Charles V, and the ruler of Spain, Spanish America, much of Italy, the Netherlands, and a great swath of central Europe. The Ottomans were now faced by a far greater and more united Christian power than they had ever had to confront before. As the Italian poet Ludovico Ariosto put it, now there were "two suns" shining upon the globe and two rulers competing for universal supremacy: a Christian emperor in the West and a Muslim sultan in the East.

Süleyman I, called "the Magnificent" in Europe, fashioned an imperial image for himself and his dynasty that went beyond even that which Mehmed II had created. He saw himself as the heir of Alexander the Great, as the "last world emperor," who would de-

stroy his rival Charles V and then march west and conquer Rome.
From the moment he ascended the throne, he cast himself in the
role of the perfectly just ruler, the great codifier of the law. He was
accredited by the Ottoman jurist Kinalizade Ali Celebi—not, perhaps, an entirely impartial observer—with having made the "Virtuous City on Earth" a reality. He was the builder of bureaucratic
institutions along Western lines that would enable the Ottoman
Empire, for a while at least, to weather the kinds of internal ten-
sions between the sultan and his provincial governors that had
been the undoing of the Abbasid Caliphate. He stood as the upholder of orthodox Sunni Islam against the Holy Roman Empire
to the West and the heterodox Shiite empire of the Safavids to the
East. Like his competitors in the West, Süleyman was also eager to
see himself as the beneficiary of an apocalyptic tradition based
loosely on the Book of Daniel, which was read as having foretold
that sometime toward the end of the sixteenth century, the Great
Year would dawn in which one true religion (Catholic Christianity
for Charles V, Sunni Islam for Süleyman) would triumph over all
others and there would be only one empire on Earth, ruled by one
divinely appointed ruler—the *sahib-kiran,* "Emperor of the Last
Age."

In preparation for this transformation, Suleyman, although
he could make no claim to being descended from the Quraysh,
which all previous Sunni Caliphs had been, took the title of
caliph.[38] At the time, the gesture did not, in fact, mean very much,
and not much was made of it. "Caliph" was a purely political des-
ignation, and since Süleyman now ruled over most of the lands
(with the exception of Persia) previously occupied by the Abbasids, he had good political grounds for appropriating their titles. The grand vizier extraordinary, Ibrahim Pasha, would bow
down before the sultan and call him not only *zill Allah*—"shadow
of God"—which was a conventional Islamic title for the Commander of the Faithful, but also "universal ruler and refuge of the
world" and "universal ruler of the inhabited world"—which were
not. When, in 1560, he had the Great Mosque built in Istanbul,

Süleyman ordered an inscription to be placed over the portal of the main gate, which reads, "Conqueror of the Lands of the Orient and the Occident with the help of Almighty God and his victorious army, possessor of the Kingdoms of the World."[39]

In 1529, Süleyman marched west again, his sights now set on the emperor Ferdinand's capital at Vienna. This time, however, the sultan, like so many imperial commanders before and after him, overextended himself. The centralized nature of the Ottoman state demanded that the entire army, recruited from every province in the empire, muster outside Istanbul before it departed. This took months. Then, when it was finally on the march, it was dragged down by heavy rains and floods and took more than four months to reach Vienna. By the time it arrived, the troops were demoralized and exhausted and supplies were running out. After only three weeks, Süleyman called off the siege and retreated back to Istanbul. Although the weather had done more damage to the Ottoman forces than the Austrians, in subsequent Western accounts of the event this became another Marathon, another moment at which the tide of barbarism sweeping in from the East had been turned back by an outnumbered, heroic Western force.

But the joy following the failure of the siege was seriously offset in anxious Western minds by the sheer audacity of it and the devastation Süleyman's armies had left in their wake. If the sultan's forces could reach that deep into the heart of Christendom, across daunting terrain crossed by mighty rivers, the Danube among them, might they not, rather more easily, seize another Christian capital, such as Rome, which was easily accessible from the sea? Alarmed by this prospect, in 1534, Pope Paul III commissioned the architect Antonio da Sangallo to build a protective wall around the Eternal City with no fewer than eighteen bastions. Although lack of funds finally forced him to abandon the project, the fear that one day Süleyman would return to complete what Mehmed had begun remained.

For Süleyman, Vienna was merely a setback. In 1551, the port of Tripoli, held by the Knights Hospitalers for Charles V, fell to a joint attack by the Ottoman imperial fleet and the legendary corsair Turgud Reis. That same year, Piri Reis, the Ottoman admiral who had commissioned a map of the Americas (now in the Topkapi museum in Istanbul) so that his master might see what new realms still waited to be conquered, sacked the Portuguese settlement at Ormuz on the Persian Gulf. All too often, it seemed to the Europeans that one small Christian victory was followed by a far larger Christian defeat. "I tremble when I think of what the future must bring," wrote the emperor Ferdinand's ambassador to the Ottoman court, Baron Ogier Ghiselin de Busbecq, in 1560.

> When I compare the Turkish system with our own; one army must prevail and the other be destroyed, for certainly both cannot remain unscathed. On their side are the resources of a mighty empire, strength unimpaired, experience and practice in fighting, a veteran soldiery, habituation to victory, endurance of toil, unity, order, discipline, frugality and watchfulness. On our side is public poverty, private luxury, impaired strength, broken spirit, lack of endurance and training; our soldiers are insubordinate, the officers avaricious; there is a contempt for discipline; license, recklessness, drunkenness and debauchery are rife, and worst of all the enemy is accustomed to victory, we to defeat. Can we doubt what the result will be?[40]

True to Ghiselin de Busbecq's foreboding, the Ottoman advance continued largely unchecked. In 1565, an Ottoman fleet besieged the island of Malta. It was beaten off. But the success was short-lived. The following years, it was the turn of Chios and Naxos. In August 1571, another Ottoman force took Cyprus from the Venetians and massacred the Christian population of the city of Famagusta. Six years later, Samos suffered a similar fate. To despondent Christian observers in the eastern Mediterranean, the

Ottomans now seemed to be as invincible at sea as they appeared to be on land.

A month after Cyprus capitulated, however, Christendom secured one of its greatest victories over the Ottomans, near Nafpaktos, in what was then called the Gulf of Lepanto. In May 1571, Venice, Spain, and the papacy forged a somewhat shaky alliance in response to the attack on Cyprus and in the hope of preventing any further Ottoman incursions in the Mediterranean. A combined fleet was hastily assembled under the command of Don Juan of Austria, an illegitimate son of Charles V and half brother of Philip II of Spain. With 170 Venetian war galleys more than 160 feet long and as much as 30 feet in the beam and powered by as many as twenty to forty banks of oars, it was the largest single Christian fleet ever to venture into the Mediterranean. In the front line were also six massive bargelike oared ships known as galleasses, which the Ottomans had never encountered before, each of which carried nearly fifty guns and could deliver more than six times as much shot as any of the largest galleys of the time.[41]

In September, Don Juan sailed east from the port of Messina, in Sicily. His original intention had been to recapture Cyprus. But on the morning of Sunday, October 7, he surprised a massive Ottoman fleet in the Gulf of Patras at the mouth of the Gulf of Corinth. The battle lasted a little more than four hours.

The galleasses disabled, destroyed, or scattered as much as a third of the numerically superior Ottoman fleet before the battle even began. No sooner had the galleys engaged than *La Reale,* Don Juan's flagship, succeeded in ramming the Ottoman admiral Müezzinzade Ali Pasha's flagship, the *Sultana.* Ali Pasha himself was killed by a bullet to the brain. He was decapitated by the victorious Christians and his head exhibited on a pike on the quarterdeck of *La Reale.* The sacred banner of the Prophet was torn from the masthead of the *Sultana* and replaced with the papal pennant. When the rest of the Ottoman fleet realized that their admiral was dead and his ship in Christian hands, they scattered in panic. By early afternoon, it was all over. Some 40,000 men, both Christians and Muslims, had died in the carnage, making it one of the bloodiest encounters in the history of European warfare. More than two thirds of the mighty Ottoman fleet was sunk, in flames, or in the hands of Don Juan and his triumphant admirals.

The victory was hailed far and wide across Europe as a new Actium, a new Salamis. Even the future king James I of England, who as a Protestant monarch might have had some misgivings about a Catholic, and a papal, victory, tried his hand at a poetic celebration of the battle. A European Christian fleet had crushed an Eastern enemy and, once again, had saved Europe and all the values it represented from the yoke of a despotic power. The analogies were, of course, entirely empty. The forces of Don Juan did not represent either Greek democratic freedom or Roman civility. The Spain of Philip II was hardly less despotic than the Ottoman Empire and in many respects was a good deal more so. The men who had powered the galleys at Salamis had been free men fighting for their cities. Those at Lepanto, on both sides, were slaves.

From the Ottoman point of view, furthermore, Lepanto was far from being the victory the Christians claimed it to be. The imperial fleet was largely rebuilt within a year. Don Juan put to sea again in 1572, and although the two fleets skirmished off the Peloponnese, neither side could claim a victory. Venice accepted the loss of Cyprus and even agreed to pay the sultan an indemnity of 30,000 ducats. It soon became clear to even the most optimistic Christian observer that although Lepanto had admittedly been a setback for the Ottomans, it had been nothing more. The Turks still dominated the eastern Mediterranean and still controlled most of Hungary, and they were evidently both determined and able to make yet another major onslaught against the West. It was only a matter of time. "The fire advances little by little," warned the French Huguenot captain and military strategist François de la Noue in 1587. "It has already consumed the outskirts of Europe, that is, Hungary and the great litoral of the Adriatic."[42] Unless, he claimed, his own plans for a counteroffensive that would rid Europe of the Turk in four years were adopted immediately, the Turks would once again be at the gates of Vienna, and this time there would be no hope of turning them back.

THE ASSAULT THAT La Noue had predicted did not, however, come for nearly a century. With the death of Selim II in 1574, the Ottomans were more concerned with maintaining the peace within their own territories through years of unrest, palace intrigue, and a number of weak and incompetent sultans than they were with making any further advances against the West either at sea or on land. Then there was the Persian question. The struggle between the Sunni Ottoman Empire and the Shiite Safavid Persian one lasted off and on for most of the sixteenth and seventeenth centuries. The two behemoths shared a common frontier that stretched more than 1,500 miles from the Black Sea to the Persian Gulf. For a while, one of the greatest of the Safavid rulers, Shah Abbas, who was responsible for creating a great capital at Isfahan—which the English travelers who visited it in the late seventeenth century said rivaled London in size and opulence—actively sought support from the West. This promoted a number of Christian rulers to float vague schemes to use the Safavids in a pincer movement to put an end, once and for all, to what in the West was called the "Sublime Porte"—the gate to the Grand Vizier's palace, from which the Ottoman government took its name. Although most of this was politely resisted, the shah kept up his own onslaught on his western neighbors, and with the help of two English adventurers, the brothers Anthony and Robert Shirley (who had first found employment in Philip IV's Spain), he created a formidable and highly westernized military machine. In 1603, he took the Ottoman border cities of Tabriz and Yerevan, and the following year he drove the Ottomans from their remaining garrisons in the Caucasus and Azerbaijan and extended his power as far as the Armenian town of Kars. He was also able, in 1622, with the assistance of Robert Shirley's cannons and a number of English warships, to drive the Portuguese out of the famously wealthy island of Hormuz, in the Persian Gulf.

With the death of Shah Abbas in 1629, however, the empire fell into the hands of a series of weak and quarrelsome rulers and went into precipitous decline. Freed from the need to maintain a constant presence along their eastern borders, the Ottomans resumed their offensive in the Mediterranean. In 1645, the Ottoman fleet attacked Crete. Parts of Venetian Dalmatia were seized in 1646 and then lost again the following year. In 1665 a joint Maltese-Venetian fleet attacked the Ottomans off the Dardanelles, in what the Christians hoped would be another Lepanto. After a six-hour battle, however, the Ottomans withdrew, their forces still largely intact. Four years later, Crete, which had been Venetian for four and a half centuries, surrendered to the forces of Mehmed IV.

On August 26, 1682, Sultan Mehmed IV decided, somewhat reluctantly, to yield to the insistence of the grand vizier, Kara Mustafa Pasha, that the time had now come for a massive military campaign against the Hapsburgs. The sultan had signed a treaty with the emperor Leopold I in 1664 that was not due to expire until 1684; but treaties in the early modern world, in particular those between Christians and Muslims, were often flimsy affairs. The sultan also had the support of the Magyar rebel leader Thököly, whom he recognized as "king of central Hungary" and had placed under Ottoman protection. The French, who had long preferred the Turks to the Hapsburgs, had promised not to intervene. The other Christian power in the Ottomans' western flank, the Duchy of Muscovy, was eager to maintain the peace. The Hapsburgs, it would seem, were alone.

In October, the sultan's insignia, the *turgh*, or horsetails, was mounted outside the Grand Seraglio in Istanbul, publicly proclaiming his intention to leave the city. By early December, he had reached Adrianople. Here Mehmed camped for four months while his forces gathered from every corner of the empire. On March 30, the sultan and his ever-expanding army began to move west toward Belgrade. Some 100,000 people and the food needed to feed them (the Hapsburg envoy Albert Caprara, who accompanied the sultan, estimated that 32,000 pounds of meat and 60,000 loaves were consumed daily) were on the move. The going was tough. Torrential rains turned the roads to mud. Great flocks of sheep and herds of cattle, which frequently strayed or sank in the mud, followed the troops, together with the innumerable carts and wagons; and the inevitable train of hangers-on, wives, women, and concubines, which accompanied every army, trailed dejectedly behind.[43] The chronicler Silhadar Findikhh Mehmed Agha, who traveled with the expedition, complained bitterly of the terrible rains that hampered the army's movements from the moment it left Edirne on March 30. He was particularly incensed by the difficulties in conveying the sultan's favorite concubine, Rabia Gülnüs Emetullha, and her entourage, together with eighty coachloads of women from the harem, across an improvised bridge over the river near the town of Plovdiv.

On May 3, the army finally reached Belgrade and pitched camp at Zemun on the right bank of the Danube. By the end of the month, they moved out again. As they marched, they were joined by troops from Albania, Epirus, and Thessaly, even Egypt. "King" Thököly showed up with a sizable contingent, and some 80,000 Tartars came along for the pickings. On June 26, the army entered enemy territory and moved on the Hapsburg city of Györ. Caprara's opinion of this massive but disparate and ill-coordinated force was dismal. It was, he said, outstanding only for its "weakness, disorder, and almost ludicrous armament." (On this last point, at least, he may well have been right. One Turkish observer claimed that they had only sixty cannon and mortars.) The sultan had only about 20,000 decent fighting men; the rest were a rabble. Such a force, he concluded, could never hope to defeat "the men of Germany."[44]

The emperor Leopold, however, thought otherwise. By now he was in no doubt as to the sultan's ultimate intentions, and on July 7, the court abandoned Vienna and retreated to Passau with all the treasure it could carry, pursued by the Tartar cavalry. On July 14, the Ottoman army set up camp in front of the city. An Ottoman envoy appeared at the gates with the demand that the Christians "accept Islam and live in peace under the Sultan!" Ernest Rüdiger von Starhemberg, who had been left in command, cut him short, and a few hours later, the bombardment began. Within two days, the Turks had completely surrounded the city and, by one contemporary estimate, were within a mere 2,000 paces of the salient angles of the counterscarp. The grand vizier Kara Mustafa Pasha (Mehmed himself had stayed behind in Belgrade) set up a magnificent tent in the center of what was virtually another city outside the walls. There, in the company of an ostrich and a parakeet, he dispensed favors in complete confidence of an eventual victory and sauntered forth each day to inspect the Turkish trenches. The situation inside the city grew steadily more desperate as water ran low, garbage piled high in the streets, and little by little the familiar diseases of the besieged—cholera, typhus, dysentery, scurvy—took hold. Yet the defenders managed to hold out for two months. The Turks, as Caprara had rightly seen, had very little heavy artillery; what they had could kill people and damage buildings inside the city but made very little impact on the massive walls, bastions, ravelins, glacis, caponnières, palisades, counterscarps, and the other paraphernalia of sixteenth-century fortifications with which Vienna was ringed.

Meanwhile, a relief army of some 60,000 men under the command of King John III Sobieski of Poland and the emperor's brother-in-law, Charles of Lorraine, moved slowly toward the beleaguered city. Crossing the Danube at Tulln, they then marched through the Wienerwald in order to approach the city from the west. The Wienerwald was a no-man's-land, mountainous and covered in dense forest, and the Ottomans, assuming that no relief army of any size could possibly penetrate it, had left it largely undefended. It was to be a fatal mistake. The progress of the combined Christian army was slow, but by late on Saturday, September 11, it had all assembled along the ridges on the edge of the forest. The following morning, it swept down on the largely unprepared and poorly defended Turkish encampments below. By late afternoon, it was all over. "We came, we saw, and God conquered," wrote Sobieski to Pope Innocent XI, echoing Julius Caesar's famous remark after the conquest of the Parthians, "I came, I saw, I conquered." The Turks who had not been killed or taken captive fled as best they could, back toward Belgrade. Kara Mustafa's ostrich had died in the battle and his parakeet had flown away, but he succeeded in taking with him the Flag of the Prophet, which had been flying vainly from the tent pole, and most of his treasure. It did not do him much good. As so often happened to those who had failed to please the sultan, he was beheaded two months later. There is a skull in the city museum in Vienna that is said to be his, although Silhadar Findikhh Mehmed Agha claims that Mehmed, respectful if not compassionate at the end, had sent his vizier's body—plus the head—back to Istanbul for burial.

Vienna, wrote one despairing Ottoman historian, had been a defeat "so great that there has never been its like since the first appearance of the Ottoman state."[45] He was almost right (the battle of Ankara had been more devastating), and although neither he nor any of his contemporaries, Christian or Muslim, may have realized it fully, Mehmed's failure was to be the very first step in the steady but inexorable decline of what had for so long seemed, to Christian and Muslim alike, the unstoppable advance of the Ottoman Empire.

After Vienna, the relationship between Christendom and Islam began to change. For centuries, the Christians had attempted to keep the Muslims at bay and, if possible, to recapture areas, most obviously Palestine, that they considered to be sacred to their religion. Now, as Ottoman power visibly weakened, it become possible to imagine not merely the limitation of Muslim power but its eventual elimination.

The Hapsburgs were not slow to capitalize on their success. In March 1684, in an unusual show of solidarity, Austria, Venice, Poland-Lithuania, the Grand Duchy of Tuscany and Malta, and the papacy formed a Holy League against the Sublime Porte. Two years later, on September 2, 1686, they secured their first major success. The Hungarian city of Buda, which since 1526 had stood on the frontier between Christendom and Islam, fell to a besieging Hapsburg army. For the Ottomans the loss was of immense psychological significance. The failure to take Vienna had been a crushing humiliation for the mighty Ottoman armies, but Vienna had always been a European, Christian city that lay within the "House of War." Buda, by contrast, was considered to be a Muslim city and integral part of the *dar al-Islam*.

The real threat to the continuing survival of the Ottomans, however, came not from the Austrians. It came instead from a relatively new Christian imperial power: Russia. The conversion of the Russians to Christianity in 988 had been one of the triumphs of the Greek church. As the Byzantine Empire had slowly lost ground to the Turks, the Russians had been gaining territory from their former Mongol overlords, in a continuing struggle that, since it pitted Christians against Muslims, was also seen by the Christians as a "crusade." With the fall of Constantinople and the disappearance of the Roman Empire in the East, Moscow had become, in Russian eyes, the sole bearer of Orthodox Christianity and consequently the true heir to the Roman Empire, now ruled by a prince who styled himself "czar," the equivalent of "Caesar." "The Christian Empires have fallen," wrote the monk Philotheus in 1512 to Czar Basil III, "in their stead stands only the Empire of our ruler. . . . Two Romes have fallen, but the third stands and a fourth there will not be. . . . Thou art the only Christian sovereign in the world, the lord of all faithful Christians."[46] Now prophecies about a blond race of warriors emerging from the north to drive out the Muslims began to appear throughout the Eastern Christian world. In 1657, one Orthodox patriarch, rash enough to predict the end of Islam and the return of the Church Triumphant, was hanged for his perverse optimism.⁴⁷

Yet for all these claims the peoples of western Europe had never quite known what to make of the Russians. The vast size of Russia and the fact that so much of it had, for so long, been ruled by nomadic peoples who were clearly not European had placed it, in the minds of many Europeans, beyond the formal limits of "civilization." While it remained, in this way, stubbornly an Oriental despotism, Russia remained firmly within Asia—the "Turk of the North," as the Geman philosopher Leibniz called it. But once, beginning in the 1680s with Peter the Great, the creator of Saint Petersburg, who is described by Montesquieu as "having given the customs and manners of Europe to a European power,"⁴⁸ the czars began to "modernize"; once its aristocracy took to wearing silk brocade and conversing in French, what had hitherto been looked upon as the backward empire of the steppe gradually came to seem inescapably European. In 1760, Voltaire wrote *The History of Russia Under Peter the Great,* whose objective was precisely to demonstrate that Russia was now an inescapable part of the culture of Europe. By then, few would have disagreed with him; not, perhaps, a very refined or sophisticated part of Europe, but European nevertheless. When, in 1791, the British prime minister William Pitt, in a bid to limit the ever-expanding power of the czars, proposed sending British troops to help the Turks, the great Irish orator Edmund Burke angrily demanded to know "What have these worse than savages to do with the powers of Europe, but to spread war, devastation and pestilence among them?" Russia, he assured the House of Commons, belonged within Europe, and any attempt to assist the sultan, for no matter what political ends, would only threaten the integrity and security of what he called the "grand vicinage of Europe." And Russia would remain firmly a part of Europe, albeit a somewhat remote and exotic one, until the Bolshevik Revolution returned it—at least in the minds of many in the West—once again to Asia.

The modernization, or "Europeanization," initiated by Peter had not only transformed an Asiatic people into a central European one, it had also greatly enhanced the military capabilities of the czars. Now Russia began to move eastward to seize her portion of what was clearly a giant in distress.

On August 6, 1696, Peter the Great seized the Black Sea port of Azov. The Turks agreed for the first time in their history to discuss peace. In October, representatives of the two sides met at Carlowitz in the Voivodina. Finally, on January 26, 1699, with the help of British and Dutch mediators, a peace treaty was signed among the Ottomans, the Russians, and the various members of the Holy League.

The Treaty of Carlowitz was not a total surrender. But it deprived the Ottomans of territories in eastern Europe, almost all of Hungary and Transylvania, which they considered to be not European or Christian at all, but a fully integrated part of the Islamic world. More crushing still, it was the first time in history that the sultan, the "Commander of the Faithful" and heir presumptive to the caliphate, had ever been compelled to sign a treaty with his enemies. By so doing, the sultan had, in effect, agreed to abide by the—admittedly rough and ready—tenets of international law as it was understood in the West. It was an unprecedented move for a political and religious culture for which war against all unbelievers was a necessary duty and the permanent obligation of every ruler. Muslims could and did enter into treaties with non-Muslim rulers. These might, for the sake of convenience, last for a very long time. But no Muslim ruler could accept a permanent settlement with a non-Muslim state, if only because the obligation of the jihad prevented them from recognizing its very right to exist.

At Carlowitz, the sultan, the supreme leader of the Muslim world, had, implicitly at least, violated one of the precepts of the Shari'a. It would change the nature of the Ottoman state forever. As long as the Ottoman forces had been supreme, there had seemed to be little reason to question the established order. Now there was. Mehmed II might have adopted Byzantine and Latin modes of address and had himself depicted by Christian painters employing Western iconographical motifs. But no sultan before Mustafa II had had any compelling reason to suppose that the great empire over which he ruled would not go on until the day when the *dar al-Islam* would cover the entire world, which would be one under Ottoman rule. This vision of the future began to fade after the Treaty of Carlowitz and continued thereafter to get dimmer and dimmer.

More than any previous event, Carlowitz forced upon the Ottomans a new awareness of the potential might of the West and the recognition that, if the empire was to survive, it would have to adopt new ways of dealing with the "House of War," ways that would replace the simple force of the jihad by diplomacy. It marked, too, an unmistakable reversal of fortune. Now it would be the West—utterly transformed, culturally, religiously, politically, and militarily, from the squabbling assembly of states that had failed to stop first Mehmed and then Süleyman—that would take the offensive. And from 1699 until 1918, when a contingent of British troops entered Istanbul, it would be the West that would steadily but inexorably push back the frontiers of Islam until they were virtually no more.

DURING THE SAME time another transformation had been taking shape within Christendom itself. The great French historian Lucien Febvre once claimed, in the uncertain days following the liberation of France in 1945, that it had been another great French historian, Philippe de Commynes (c. 1447–1511), who had been the first to register the superiority of Europe over Asia—"that Asia which for so long had crushed barbarism by the weight of its superiority, the power of its culture, of its brilliance." Commynes, the historian of Louis XI's France, was certainly speaking with a new sense of cultural assurance, despite the looming presence of the Ottoman Empire. And Febvre was right to see in him a modern writer, a man "disengaged from Christianity and the Christian faith"—which is not to say that he was not a devout Christian.[49] But in his awareness of something called "Europe" he was not alone. Europe for Herodotus had ended where Greece had ended. For Rome "the West" had been identical with the *civitas* that had reached deep into Asia. Most Christians after the fifth century, although they used the terms "the West" and "Europe" interchangeably, thought of the realms they lived in as coterminous with "Christendom." A millennium later, a new conception of "the West" was beginning to emerge. It was one, as Commynes saw, that placed its identity not in religious loyalties but in a way of life, what much later would come to be called a "civilization." What made that possible was a radical transformation in the ways by which the nations of Europe derived their knowledge and understanding of the world.

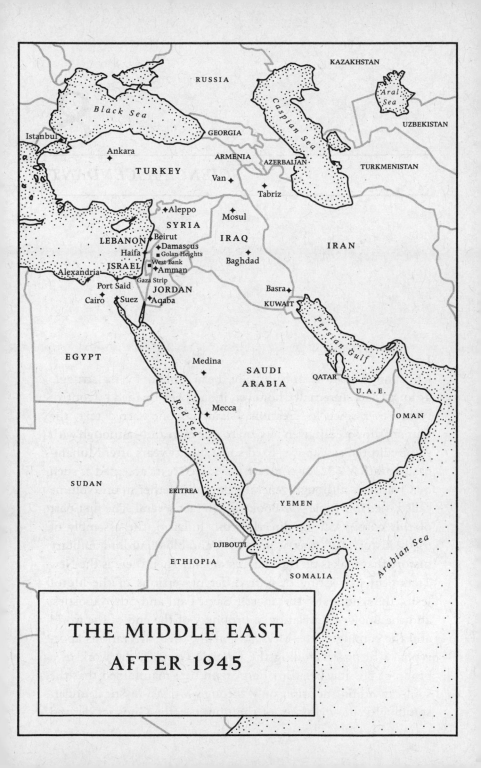

RUSSIA

Black Sea

Istanbul

Ankara

TURKEY

GEORGIA

ARMENIA

Van

Caspian Sea

KAZAKHSTAN

Aral Sea

UZBEKISTAN

TURKMENISTAN

AZERBAIJAN

Tabriz

Aleppo

Mosul

SYRIA

IRAQ

IRAN

LEBANON

Beirut

Damascus

Golan Heights

Baghdad

Haifa

West Bank

ISRAEL

Amman

Alexandria

Gaza Strip

Port Said

JORDAN

Basra

Cairo

Suez

Aqaba

KUWAIT

Persian Gulf

EGYPT

Medina

SAUDI
ARABIA

QATAR

U.A.E.

Red Sea

OMAN

Mecca

SUDAN

ERITREA

YEMEN

DJIBOUTI

ETHIOPIA

SOMALIA

Arabian Sea

# THE MIDDLE EAST
## AFTER 1945

# 8

## *SCIENCE ASCENDANT*

**I**

Christianity had, if only fitfully, been united by its struggle against Islam. Internally, however, it had been divided by controversies ever since its creation. Islam has one sacred text, the Qur'an, supposedly given by God to one man and—although with some editorial assistance—fixed some twenty years after Muhammad's death in 632. By contrast, the sacred texts accepted as such by Christians, although generally bound together in one volume called the Holy Bible—*the* Book—are in fact several. The first part of this corpus derives directly from Judaism, the assembly of myth, history, law, poetry, prophecy, and bloodcurdling military history that makes up the Old Testament. Then there is the New Testament, consisting of four different versions of the life of Jesus, the writings of the apostle Saint Paul and other apostles, and the Book of Revelation, a prophecy of the end of the world and the coming of the Messiah. Finally, there are various apocryphal writings, including the Gnostic Gospels, the work of a group of possibly pre-Christian origin that maintained that the soul can attain salvation only through a quasi-mystical understanding of the mysteries of the universe (the Gnostics derived their name from the Greek word *gnosis,* "knowledge"). All of these widely divergent bits and pieces are also written in a number of different languages: Hebrew, Greek, and Aramaic. Furthermore, although the most authoritative of these texts, the New Testament, recorded the words of God, it did not claim to have been dictated by him. It could not, therefore, escape the need for interpretation. And interpretation inevitably resulted in schism and confrontation.

In the Christian Church, for centuries the official orthodox version of the Bible was a translation into yet another language, Latin, made by Saint Jerome between 382 and 405 and known as the Vulgate. By the fifteenth century, a number of other texts had also acquired canonical or quasi-canonical status among Christians: the writings of the early Greek theologians, known as the "Church Fathers," and those of a select number of saints, called "Doctors of the Church" (there are thirty-three of them to date), plus some of the writings from classical antiquity, most notably, after the end of the thirteenth century, the works of Aristotle. By contrast, the Qur'an is, of course, written throughout in one language, Arabic, and although translations do exist, none has any official standing in the Muslim world. Furthermore, as we have seen, although interpretations of both the Qur'an and the Hadith constitute a legitimate part of Islamic theology, they are very restricted.

Little wonder, then, that in Christendom interminable disagreement, much of it acrimonious, should have arisen over how to understand this bewildering variety of utterances. Heresies flourished, particularly among the early Christians. Some of this raucous discord was purely theological in origin. Was Christ all man; part man, part god; or all god? What is the relationship among the three persons in the Holy Trinity—God the Father, God the Son, and the Holy Ghost? What could "son of God" possibly mean? How could God be his own son? Was the Trinity an absurdity, since God must have existed before he conceived Jesus and survived after his death? Did the Holy Ghost proceed from the Father and *from* the Son or from the Father *through* the Son? Should Communion be taken in both kinds—the body and the blood of Christ—or only in one by the laity and both by the clergy? And so on. Some were more obviously political. Did the papacy have power over all mankind or only over Christians? Could the pope command Christian princes in matters that were purely secular, or did his authority extend only over the spiritual domain and in matters of conscience? What right, if any, did secular rulers have to make ecclesiastical appointments or tax ecclesiastical properties? All of these had led to sometimes bloody conflicts in both Western and Eastern Christendom. The bitter dispute over Christ's humanity (or lack of it) had led to the creation in the East of two breakaway churches, the Nestorian (which held that Christ had a divided nature, part human, part divine) and the Monophysite (which claimed that Christ had a single nature). A host of splinter groups flourished throughout Europe. There were the Cathars and the Fraticelli and the Waldesians, the Taborite Brotherhood, the Hussites and the Lollards. All were branded by the Church as heretics and all were finally suppressed, sometimes, as in the case of the Cathars and the Hussites, after bloody civil wars.

The Church was also divided on social and political issues and divided bitterly over its role and its place in the world as an institution. Ever since its founding it had been riven by a paradox. Its founder, like all holy men—and Eastern holy men in particular—had made impossible demands on his followers. Abandon all—home, house, family, and personal possessions—if you wish to follow me, he had said. Yet the great institution that had been built in his name was founded precisely upon home, house, family, and personal possessions. The Gospels, despite Saint Paul's clever reading of them, provided very little grounds for the construction of an all-powerful state religion.

At its core Christianity was and remained divided against itself. It is a tribute to the early Church Fathers and to a succession of powerful popes that it survived as the sole source of authority within Europe for as long as it did.

In the early sixteenth century, however, nemesis finally struck. In 1518, an obscure German friar named Martin Luther, struggling to find a solution to his own personal crisis, sat meditating in the tower of the monastery of Wittenberg. He was reading a text, Saint Paul's first Epistle to the Romans, and what caught his eye—and would sear his soul—was a single phrase from verse 17: "The righteous shall live by faith." What exactly, he wondered, could it mean?

Large, dyspeptic, impulsive, and given to bouts of melancholy and anger, Luther, whose mother, Margaret, it was said, "had more than one occasion to believe that she had given birth to a flaming torch," was not a man to do things by halves.[1] As he pondered Paul's words, he was suddenly seized by the kind of revelation that comes to all self-declared bearers of the truth. The experience transformed his life, and eventually it would transform the lives of half of the populations of Europe. He had, he said, been "born again" and carried into Paradise. What these cryptic words meant, Luther now knew, was that God did not make demands upon his creatures but bestowed upon them the benefits won by the sacrifice of Christ through grace. Mankind did not need to labor to win God's favor, as the Church had always maintained; it could be vindicated by faith alone. To be justified in the eyes of God, one had only to believe and lead a true and godly life. One did not need to do penance, to make costly pilgrimages, to venerate the wasted carcasses of supposed saints; one did not need to make sacrifices. Above all, one did not need to buy the tawdry goods the Church peddled to its deluded flock in order to raise the cash it needed for its wars; for its vast buildings; for its paintings, sculptures, carved woodwork, rood screens, golden goblets, silver aspergilla, and bejeweled, inlaid cases in which the relics of the sanctified lay, which it relentlessly commissioned from all the finest and most expensive artists and craftsmen in Europe.

Luther had seen these things at first hand. In 1510, he had been to Rome on a pilgrimage, and he had been horrified. If he had not been some kind of fundamentalist when he left Germany, he certainly was by the time he returned. The papacy, in the person of Julius II, patron of Michelangelo, a man said to be better on horseback than at prayer, was enlarging Saint Peter's Basilica into the magnificent building it is today. To do this he needed money, and so, he had—or so it seemed to Luther—put Christianity up for sale, in the form of something called an indulgence. To put it simply—and simply was how Luther saw it—indulgences were documents issued by the Church that promised to eliminate the burden of human sin and thus reduce, by millions of years, the suffering the penitent invariably faced, however virtuous he or she might have been, in Purgatory. The more you paid, the more speedily you would gain admittance to Paradise.

Indulgences, however, terrible and cynical though they were, were only the last straw. What really offended Luther, what Saint Paul's words had illuminated for him, was the whole idea that anything we might do could influence how God behaved toward us. The whole Catholic doctrine of "works," the belief that we can curry favor with the Almighty by handing out cash to his representatives on Earth, even by performing outwardly good deeds while remaining inwardly, fundamentally unrepentant, was, for Luther, the worst kind of blasphemy. In a now-famous act of defiance, on October 31, 1517, he sent a text containing ninety-five theses against the sale of indulgences and other abuses to a number of friends and bishops he hoped might be sympathetic to his cause. By the end of the year, these had appeared in print in Leipzig, Nuremberg, and Basel, and Luther began to acquire both notoriety and a following. (Alas, the story that he nailed these to the door of the Castle Church in Wittenberg is a legend.) He began furiously to publish pamphlet after pamphlet against what was, in effect, one of the central doctrines of the church of which he was a member. That church, he insisted again and again in a language both powerful and direct, had been irredeemably corrupted by those who had transformed Christ's congregation on Earth—Saint Augustine's "City of God"—into a rich and powerful political institution. His main target, as the great

Dutch reformer—but staunch if unorthodox Catholic—Desiderius Erasmus put it bluntly, was the "Pope's crown and the Monks' bellies."[2]

Luther's revolt might have gone the way of all previous heresies had it not been for the precarious political situation in Germany in the early sixteenth century. In early modern Europe, religion was closely linked to the power of the state. Christ, Saint Paul, and after them the princes of the Holy Roman Catholic Church had created and maintained an unquestioning distinction between the things of God and those of Caesar. In the process, however, the princes of the Church had built up, at the center of every European state, what were, in effect, a set of separate self-governing authorities and, across the middle of Italy, a powerful independent state whose sovereign was like any other sovereign except that he was also God's anointed, the pope. State and Church might be separate powers, as Christ had decreed, but kings nevertheless believed themselves to be authorized by divine authority. Unlike Islam, in which the state serves the interests of religion, Christianity could, and often did, find itself in opposition to the state, and when that occurred, as the seventeenth-century English philosopher John Locke tartly put it, "the state religion" became "the state trouble."[3] The refusal of the Church to bow before his sovereign power was the main reason that in 1532, Henry VIII of England, in order to divorce one wife and marry another, rejected the authority of the papacy and made himself "Protector and Supreme Head of the Church of England," a position his heirs occupy to this day.

Unlike Islam, Christianity had been created in defiance of the established order; unlike Islam, it had been the religion of the poor; and unlike in Islam, the doctrines of its founder were fraught with all kinds of potentially revolutionary implications. Nowhere in either the Qur'an or the Hadith, for all their occasional egalitarian claims, is there anything to match Christ's troubling warning that it would be easier for "a camel to pass through the eye of a needle" than it would for the rich to enter the Kingdom of Heaven; nowhere is there anything like the world upside down invoked in the Sermon on the Mount, where the poor in spirit, the persecuted, the reviled, the downtrodden, and the martyred are to be blessed and the meek to inherit the earth. Because they contain no such message, neither the Qur'an nor the Hadith had any need to be shielded from the faithful lest they misunderstand them by claiming that they are saying precisely what they do say. The Gospels, however, did.

Almost from the moment it had allied itself with the state under Constantine and had begun to enjoy its ample favors, the Christian Church, both East and West, had done its best to hide from the eyes of the faithful just quite how subversive the unglossed, uninterrupted message of its founder could be. This is one reason why the Gospels remained in Saint Jerome's Latin, beyond the reach of the "vulgar," and Christ's message to his followers was transmitted to the faithful in terms that no simple Galilean fisherman could possibly have grasped. In Christendom, then, unlike in Islam, religious beliefs easily transformed themselves into political ideologies long before the term "ideology" itself was coined in the late eighteenth century.

In most of Europe, bishops exercised a strict control over their clergy and the state maintained a distant, but nevertheless firm, control over the bishops. In Germany, however, things were very different. What are now Germany, Austria, part of Hungary, and the Czech Republic were all that remained of Charlemagne's Frankish empire and, in theory at least, the successor state to the Roman Empire in the West. By the fifteenth century, in a wild mixture of contradictions, this had come to be called the "Holy Roman Empire of the German Nation." In reality, the entire area was divided up into a number of different political communities: towns, principalities, and prince-bishoprics, whose ecclesiastical overlords ruled their territories in the same manner as any secular prince and generally a great deal more harshly. Brooding over all of these political factions was the Holy Roman emperor. In keeping with ancient Germanic traditions of kingship, he was, technically, elected from among the crowned heads of Europe by a cabal of six "electors," the rulers of Brandenburg, Cologne, Mainz, the Palatinate, Saxony, and Trier. In fact, the empire had been in the hands of one family, the Austrian Hapsburgs, since 1438, and the election, although it still took place, had become little more than a formality. In 1519, the empire passed to Charles V, who, through the good fortune of dynastic inheritance, thus acquired not only what are now Austria, Hungary, the Czech Republic, the Netherlands, Belgium, and parts of modern France but also Spain, a good portion of Italy, and in due time Spain's conquests in America.

The imperial title, descendant of the Caesars and of Charlemagne, carried with it immense prestige, but very little else. The emperor was not even, except in name, absolute overlord of his realm but only what, in a nice term of compromise, was called the "first among equals." For centuries the emperors had been content to act in this way as a kind of judge among their warring subjects, calling regular assemblies—known as diets. Charles, however, was a great deal more powerful than his predecessors had been, and he determined to bring the German princes into line. At the Diet of Worms in 1521, he put an end to the idea that there could be a first among equals by declaring, "There shall be only one Emperor in this Empire." The princes, however, had no intention of taking this quietly.

Luther was present at Worms. In the fall of 1520, he had received a papal bull (decree), *Exsurge domine,* condemning forty-one statements from his writings and demanding recantation on pain of excommunication. On December 10, he replied by publicly burning the bull, along with a number of works of scholastic theology and the text of the canon law, before the Elstertor at Wittenberg. The whole university was invited to attend. When Luther's part in the conflagration was over, the students took to the streets, holding up a puppet of the pope and a mock papal bull, which they then burned along with all the books by Luther's opponents they could find. Excommunication swiftly followed.

Luther was then called to face his emperor, so as, he was informed, to be given one last opportunity to return to the fold. On April 2, he started out for Worms. His journey became something of a triumphal march, and he preached in Erfurt, Gotha, and Eisenach on the way, each time condemning not only the sale of indulgences but the entire Catholic doctrine of works. He reached Worms on April 16 and entered the city to rapturous applause. It was certainly not what Charles had intended. Twice he appeared before the emperor. Twice he was asked to recant. Twice he refused. Popes and councils of the Church, even if they were supposedly sovereign (although not yet infallible) in all matters of doctrine, could, he insisted, err. In any case, he declared in a sentence that was subsequently repeated up and down the length of Christendom, his conscience had been made "captive to the Holy Spirit itself." To deny what it had told him in his heart would have been to court eternal damnation. And what, in comparison to the flames of eternity, were the inquisitors' stake and faggots? As he left the chamber, he said, "I am finished." Charles then issued the Edict of Worms, which outlawed Luther, thus making it legal for anyone to kill him and all his followers, and proscribed his writings. He stood now condemned by both Church and the state. He might well have ended up in the imperial dungeons and then at the stake. Luther, however, had powerful backers, above all Frederick, the prince elector of Saxony, the ruler of the territories in which he resided, and Charles was not, at this stage, looking for an outright confrontation with the German princes. Luther was therefore issued with a limited safe conduct so that he might have time to reconsider his position. He set out for home. On the way, however, he was kidnapped on the orders of Frederick and held for his own protection in the Castle of Wartburg near Eisenach. There he spent his days translating the New Testament into German, the first such translation and a literary achievement that would have as lasting an impact on the German language as the King James Bible would have on English.

The Reformation—although it was never one movement but

many—had begun. It unleashed a conflagration upon Europe, far greater and far more radical than any Luther himself had ever intended. In Luther's vision of the true church, each person had to face his or her God directly. The priests of the Catholic Church, who had claimed quasi-supernatural powers and acted as intercessors between man and God, were replaced by "pastors," whose task was to guide and assist the faithful but who were themselves no different from ordinary mortals. This was what the more radical Calvinists would later name the "priesthood of all believers." This meant that priests were denied the power to transform bread into the body of Christ and, more crippling still, the authority to cast the disobedient out of the Church, if, for instance, they failed to pay their tithes. Worse still, one of Luther's claims was that the word of God, the Bible, should no longer be filtered through to his followers in a tongue known only to the learned. The Vulgate, Saint Jerome's Latin version, was gradually replaced by Luther's translation. Now, for the first time, the German people had direct access to the Word in a language they could understand.

The effect was devastating. In 1525, the German peasants rose in an increasingly well-orchestrated revolt against their overlords. The Peasants' War, as it came to be called, spread rapidly from Upper Swabia and the Black Forest into Alsace and the Tyrol, and onward to the Thuringian Forest and Saxony. Driven by centuries of economic and social oppression, fired by popular preachers who made Luther's and Christ's words even more radical than their authors had intended, they proved unstoppable. The prince abbot of Fulda, the bishops of Bamberg and Würzburg, and finally, on May 7, the archbishop of Mainz, ruler of the foremost principality in Germany, were all forced to accept their terms. One of these, significantly, was that where social and economic grievances could not be justified by scripture, they should be abolished. The uprising, ferocious though it was, did not, however, last long. Within months the war was over, crushed with astonishing ferocity. Luther, who had had no intention of disrupting the prevailing social order, was horrified by the uses to which his ideas had been put, and wrote a ferocious pamphlet entitled "Against the Robbing and Murdering Hordes of . . . Peasants," in which he condemned the uprising as the work of the Devil and called for the princes to punish the perpetrators without mercy. It makes ugly reading. But by now things were beyond Luther's control. Other more radical reformers—Zwingli in Switzerland, Bucer, Oecolampadius, and the Anabaptists, Mennonites, Hutterites, the Swiss Brethren—had arisen with even more extreme claims. The most important, and the only one whose doctrines would endure, and endure still, was the French humanist John Calvin, who created a new political community in the Swiss city of Geneva.

Luther's revolution also armed the German princes in their struggle against the emperor. By 1530, five princes and fourteen towns had declared themselves to be Protestants, not that their conversion should be taken to indicate any great change of heart. In 1531, they formed an alliance against Charles known as the Schmalkaldic League. Although the league was defeated the following year at Mühlberg, it had effectively launched a succession of wars that would ultimately divide Christendom in ways from which it would never recover. As the Reformation spread, it brought increasing internecine conflict in its wake. These struggles, which consumed almost every region of Europe from the mid-sixteenth century to the mid-seventeenth have come to be called the "Wars of Religion."

From 1559 until 1600, France was embroiled in a succession of conflicts between the Catholic Crown and its often undisciplined supporters, and the Calvinist nobility. In 1566, the nobility of the Netherlands, for centuries a self-governing part of the Hapsburg domain—Charles himself had been born in Ghent—converted to one or another brand of Protestantism and led a rebellion against their overbearing Spanish Hapsburg master, Philip II, which would last on and off for eighty years. The Eighty Years' War—or the Revolt of the Netherlands, as the Spanish called it—fought across the length and breadth of what was then called the Catholic Monarchy, which, at one time or another, reached from the Gulf of Mexico to the Philippines, has some claim to being the first world war. When it was all over, Holland, a modern Protestant republic, had been created in the north; it would go on to control many of the economic destinies of Europe and build itself a trading empire from Africa to the China Sea.

Then, in 1618 began the greatest conflagration of them all. On the morning of May 23, three figures were tossed out of the window of the Hradshin Castle in Prague in what was then Bohemia and what is now the Czech Republic. Throwing people out of windows—"defenestration"—was a traditional Czech way of expressing disapproval. The three men landed in a pile of garbage (probably placed there on purpose—defenestration was not intended to be lethal) and walked away filthy, evil-smelling, but alive. They were Martinitz and Slavata, the regents of Holy Roman Emperor Mathias II, and one of their clerks. The "Defenestration of Prague" became the signal for a revolt that led to the deposition of Mathias and the installation of the Protestant king Frederick V of the Palatinate. It was the opening salvo in a series of wars that would last uninterrupted until 1648. The Thirty Years' War, as it was subsequently called, was the final and the most bloody showdown between Catholics and Protestants. It raged across the whole of central and eastern Europe, drawing into its maw, at one time or another, all the major states of the continent, from Spain to Sweden. "All the wars of Europe are now blended into one," wrote the Swedish king Gustavus Adolphus. The huge armies that this war created, entire populations on the move, left behind them vast tracts of the continent in smoldering ruins. When it was finally over, a third of the population of central Europe was dead.

These turmoils left western Christendom permanently divided against itself. For the first time in history, the peoples of Europe were fighting not over dynastic claims, not for land or to defend the supposed rights of their rulers. They were fighting over beliefs. True, there was often more at stake than a disagreement over the nature of God's grace or even the authority of the

Church. As with most ideologies, Catholicism in its various forms and the several shades of Protestantism enforced older divisions and armed dissident groups across the continent with new arguments with which to buttress old claims.

But for all the cynicism and opportunism that any ideological conflict necessarily involves, what divided Europe, what finally, with the Treaty of Westphalia of 1648, drew a curtain across the continent between the Catholic south and the predominantly Protestant north that has remained until this day—what did all that was religion.

In 1644, the representatives of some two hundred Catholic and Protestant powers gathered in the northeastern German province of Westphalia to negotiate a settlement. Still not deigning to speak directly to each other, the Catholics set up camp in the town of Münster and the Protestants in Osnabrück, some thirty miles away. It proved to be a long, drawn-out business. For nearly four years the negotiations dragged on; petty squabbles over agenda and protocol, even over the shape of the debating table, held up negotiations for months on end, and all the while the war continued unchecked. ("In winter we negotiate, in summer we fight," one diplomat remarked.)[4] Then, on January 30, and again in October 24, 1648, a lasting agreement between the various representatives was finally reached. The Peace of Westphalia, as it has come to be known, was in effect the first modern treaty. The fighting straggled on in Germany for a further nine years, and from 1648 until 1656, Poland and Lithuania were invaded by waves of Swedes, Russians, and Ukrainian Cossacks, who left as much as a third of the population dead. The Poles still refer to it as "the Deluge" and look upon it as the worst calamity in their particularly calamitous history.

Yet for all that, the Peace of Westphalia was the first treaty between sovereign nations that aimed at creating a lasting peace and not merely, as all previous treaties had, a temporary cease-fire. It was also the first truly international gathering of European states and the first formally to recognize the existence of two new states:

the United Netherlands, which had, in effect, established its independence from Spain forty years earlier; and the Swiss Confederation, which now became a sovereign republic, independent of the Hapsburg Empire. Most important of all, however, Westphalia banished religion from the world political stage. No longer would European nations go to war because they disagreed over their understanding of God's intentions for mankind. (Perhaps the only exception, where confessional differences still have all too bloody consequences, is Ireland. But in Ireland religion became an anticolonial cause, a way of ridding the country of a hated ruler that was not only Protestant but also English.) A formula began to be used in the monarchies of Europe to describe the relationship between Church and state: *Cuius regio eius religio,* which translates roughly as "Whoever is king will decide what the religion in his kingdom should be." It was a neat, acceptable, and essentially secular solution to a problem that had littered the continent with corpses for nearly a century. If, as happened in England in 1688, a Catholic monarch succeeded to the throne of a Protestant country and refused to change his religion, he might be replaced by his Protestant next of kin. Little wonder, then, that the only power to reject the terms of the Treaty of Westphalia was the papacy. The treaty was, Innocent X said, mustering all the derogatory epithets he could think of, "null, void, invalid, iniquitous, unjust, damnable, reprobate, inane, empty of meaning and effect for all time."[5] But no one except his own bishops and a handful of the faithful were around to listen to him. Even their most Catholic majesties the kings of Spain and France quietly accepted that in the future, religion would play no role in international politics.

From 1648, the disordered, divided monarchies of Europe slowly began to transform themselves into the modern nation-states that most of them still are to this day. As John Stuart Mill wrote in 1859, the Reformation and the violence it had unleashed had created a situation in which no one party had emerged victorious and thus "Each church or sect was reduced to limit its hopes to retaining possession of the ground it already occupied; minorities, seeing that they had no chance of becoming majorities, were under the necessity of pleading to those whom they could not convert, for permission to differ." Toleration, as he knew, "is admitted with tacit reserve" by even the most tolerant of "religious persons." It was necessity—defeat on the battlefield—that ultimately compelled the Christian churches within Europe to relinquish their hold over the judgment of the individual.[6]

The finally irrevocable division of Christendom and the steady emergence of powerful Protestant states—England, Holland, Sweden, and various parts of Germany—resulted, however, in very little change in Christians' attitudes toward Islam. In the various works he wrote about Islam, most of them on the Turkish menace, neither Luther's language nor his opinions differed very much from those of his Catholic predecessors. The only significant change was that now there was not one enemy to the true religion but two, although often two united as one, with Islam as the body of the Antichrist and the Church of Rome as his head. "The Turk and the Pope do not differ at all in the form of religion," Luther declared, "they vary only in words and ceremonies."

## II

The most enduring consequence of the Wars of Religion, and the one that would have the most lasting significance for the relationship between East and West, was the emergence in Europe of an entirely new way of looking at the world. In the opinion of the English philosopher Thomas Hobbes, who had lived through and suffered from the English civil wars, the Reformation, and the conflicts it had unleashed, all had been the direct outcome of the squabbles between the theologians or, as they were commonly called, the "schoolmen."[7] In Hobbes's view, bad philosophy was the source of all ideological conflict; and the conflict between Protestant and Catholic had not only strewn the fields of Europe for decades with the dead and the dying, it had also destroyed all faith in the intellectual system that had hitherto sustained the authority of the Catholic Church. Now that there was no longer any certainty to be found in religion, the only security mankind could hope to discover would have to be of its own making. All across Europe people from all walks of life were coming to something of the same conclusion.

Religion and religious conflict were not, however, the only things that had jolted European society out of the complacence in which it had lived for so long. For more than a century a movement, now broadly described as "the Renaissance," had also been steadily gnawing away at the older certainties. The word "Renaissance" means, literally, "rebirth," and it was an attempt to recover, and to emulate, the massive achievements in the arts and sciences of the ancient world. Steadily Europe began to emerge from the period that the great fourteenth-century Italian poet Francesco Petrarca, known to English readers as "Petrarch," had called "the Dark Ages." What was called loosely "humanism," the study of the literature, philosophy, and science of Greece and Rome, although it offered no direct challenge to the established intellectual order, turned away from the more abstruse philosophical and theological problems that preoccupied the university professors toward history and literature. Philosophy, the humanists insisted, should above all be well written—in decent Latin and not the garbled technical, "academic" argot used in the universities—and it should be practical. True philosophy was there to teach one how to live in the world. The humanists, therefore, concentrated their studies on history, politics, ethics, and metaphysics.

At the same time, and quite unrelated to these developments, for much of the fifteenth century, Europeans, in particular in Italy, Spain, and Portugal, had been steadily probing the limits of geographical space. Ever since 1434, when a small fleet of Portuguese ships had successfully rounded Cape Bojador—a promontory jutting far into the Atlantic from the western Sahara and, at that time, believed to mark the limits of the navigable ocean—the Portuguese had sailed regularly down the coast of West Africa. Then, in 1492, as every schoolchild knows, while unsuccessfully attempting to demonstrate that it was possible to reach China

and India by sailing west, an obscure Genoese mariner named
Christopher Columbus stumbled across a continent no one in
Europe knew existed. Until his dying day Columbus himself
insisted that the island on which he had landed shortly after
dawn on October 12 lay on the westernmost rim of the East, of
"Cathay" or the "Lands of the Great Khan." Most of those in
Europe, however, who had any geographical and astronomical
knowledge, soon realized that he was wrong. This in itself was
alarming enough. But it was not all. Geographical knowledge
in fifteenth-century Europe relied heavily upon the speculations
of a first-century Greek astronomer and geographer named
Ptolemy. Columbus, despite himself, had proved Ptolemy to be
wrong. And if the ancients, on whose philosophy and science
so much of the learning of Europe relied, had been so totally
wrong in their descriptions of the globe, wrote Erasmus in 1518,
might they not have been wrong in other, equally dramatic, ways?
Gradually more and more information about this new continent
and, more important, about the often startling and puzzling behavior of its inhabitants began to filter back to Europe. The stories of the conquest of the Aztec and Inca Empires, which revealed
the existence of large, complex, sophisticated civilizations wholly
unlike anything anyone had ever imagined in Europe, became
bestsellers. Here were worlds in which there existed peoples who,
if credulous travelers were to be believed, ate and sacrificed one
another; who lived to be more than a hundred years old; who
mated with their siblings; who knew no gods; and who left the
bodies of their dead to rot in the open air. All of this flatly contradicted the traditional Christian view that there could exist only
one type of person and one kind of society, a society that, despite
enormous differences in customs, dress, languages, and even beliefs, nevertheless conformed unquestioningly to certain moral,
sexual, religious, and cultural rules. These rules made up the "laws
of nature," and the laws of nature were immutable. They had been
inscribed, as Saint Augustine had said, "into the hearts of men" at
the creation. You did not choose what sex to marry or whether or

not to believe in a god (although delusion and human weakness might lead you to choose the wrong god); you did not choose whether to welcome the stranger at your door or to eat him. It was nature's law that determined how you would act on these occasions.

By this logic, any people who did the kinds of things American Indians were rumored to do could not really be humans. But although there were some who were prepared to write them off as simply bestial, fit only for slavery, and doomed, like many other races, to eventual extinction, it was hard to see how their inhumanity could be reconciled with the conviction that God does not make serious mistakes with his creations. "The incapacity we attribute to the Indians," objected Bernardo de Mesa, later bishop of Cuba, in 1512, "contradicts the bounty of the creator, for it is certain that when a cause produces its effects so that it is unable to achieve its end, then there is some fault in the cause; and thus there must be some fault with God."[8] A few stray exceptions, the occasional madman or dwarf, might slip through. But to believe that an entire continent was full of semihumans was to deny the very goodness of the creation.

Faced, therefore, on the one hand, with incessant bloodletting on behalf of competing confessions and with irreconcilable diversity of belief and behavior on the other, the only possible conclusion to which any reflective person could come was that there could be no certainty in the world. God might indeed have created a pattern in the universe but it could not simply be equated with the customs and practices of the peoples of Europe. The general assumption that certain things were "natural," others "unnatural," was therefore an error. The natural law had no foundation in nature; it was only a matter of collective opinion. To call something unnatural was merely to condemn it as different, other, alien, and frightening. As the French philosopher and mathematician Blaise Pascal (1623–1662), put it, all that the word "natural" meant was that something was generally accepted "on this side of the Pyrenees"—if, that is, you were French. On the other side another kind of "nature" reigned. And if "nature" could no longer be relied upon to be consistent between France and Spain, who could know what might be natural in China or Ceylon? Or, as Hobbes sarcastically noted, all those who called "for right reason to decide any controversy, do mean their own."[9]

Europeans, who had once lived in a world of theological certainty, who had, at least at the level of belief, all shared a common culture, now found themselves adrift. And since theology had provided the basis not only of their understanding of their relationship with God but also of their moral and even their physical worlds, they were driven to reexamine not merely all the old certainties but, more significantly, all the older methods of inquiry. The English poet John Donne captured this poignantly and somewhat despairingly in 1611:

> And new Philosophy calls all in doubt,
> The Element of fire is quite put out;
> The Sun is lost, and th'earth, and no man's wit
> Can well direct him where to look for it . . .
> Tis' all in peeces, all coherence gone;
> All just supply, and all Relation:
> Prince, Subject, Father, Sonne, are things forgot,
> For every man alone thinkes he hath got
> Toe be a Phoenix, and that there can bee
> None of that kinde, of which he is, but hee.[10]

Every man, every individual, had now to raise himself, like the phoenix, from the ashes of the old world order, and he had to do so unaided. The task, however, as Donne knew, was an impossible one. We all need guidance if we are not to despair. But if the authority of the Church—and the confidence in the rightness of our cultural habits—could no longer be made to sustain the massive political, moral, and intellectual structures for which it had once claimed to legislate, what could?

The answer—very simply—was modern science.

In the seventeenth century, a group of thinkers—which included Hobbes himself, Francis Bacon and John Locke in England, René Descartes in France, Galileo Galilei in Italy, Gottfried Wilhelm Leibniz in Germany, and Hugo Grotius in Holland—set out, in their very different ways, to overthrow what was then contemptuously known as "Scholasticism." What was generally understood by this term was the theology based on the writings of Saint Thomas Aquinas, which, by the end of the sixteenth century had come to dominate the faculties of most of the great European universities. In the hands of a succession of Dominican and Jesuit writers, theology had been transformed into an all-embracing study of the natural world of God's creation, and above all of mankind's place within it. It was, in the traditional phrase, the "Mother of Sciences." The method used by the Scholastics had been essentially scriptural. Their science had consisted of the painstaking reading and rereading of a canon of supposedly authoritative texts, the Bible, the Church Fathers, and a body of classical writings, predominantly those of Aristotle. Traditionally, the criticism against Scholasticism had been the seemingly interminable fascination with utterly trivial concerns, the most celebrated being the—in fact apocryphal—debate about how many angels could dance on the head of a pin. But in the eyes of their critics this was not in fact its main failing. The theologians' real crime had been to confine knowledge, all possible knowledge, to what had been inscribed in ancient texts, to the point where, as Hobbes complained, philosophy had been reduced to what he scathingly called, mere "Aristotelity."

The first task of what Donne called "the new philosophy," therefore, had been to free the sciences from the clutches of theology and from the grip of ancient thought and ancient tradition. This "new philosophy," which had followed hard upon the death of Scholasticism—which was, in effect, also the death of theology—began with a simple question: "How can I know anything?" It was a question an ancient school of philosophers called the Skeptics had put most forcibly. Skepticism was often referred to as "the challenge of Carneades" because one of its best-known exponents was Carneades of Cyrene, a second- to first-century-B.C.E. orator who argued that if it were possible—as he frequently demonstrated it was—to argue just as forcibly for any case as against it, nothing of certainty could be known about the world. On one celebrated occasion, while on an embassy to Rome, he delivered a rousing speech in favor of justice. The following day, he returned and gave an equally convincing speech against justice. This led to his immediate expulsion from the city in order to protect the morals of Roman youth.

In its most extreme, Skepticism took the form of questioning whether one could be certain one existed at all. What if, asked Descartes, "I have convinced myself that there is absolutely nothing in the world, no sky, no earth, no minds, no bodies. Does it follow from that that I, too, do not exist?" No, he answered, because "if I convinced myself of something then I certainly exist." From this he concluded "that this proposition I am, I exist is necessarily true whenever it is put forward by me or conceived by my mind."[11] This, which was subsequently turned into the Latin phrase *Cogito, ergo sum*—"I think, therefore I am"—became one of the touchstones of the new philosophy. Not many Skeptics went so far as to doubt the existence of the world. But Descartes's point is much the same as John Donne's: the only things of which I can be certain emanate from myself. The implications for the traditional view of the world of even a moderate form of this kind of skeptical reasoning were quietly devastating. For centuries Skepticism had been largely ignored or avoided as an insidious form of nihilism. The Church, after all, had a simple answer to the question of doubt: you know because God, or God's interpreters, have told you so. Believe and obey. Now that that had lost its force, some degree of skepticism seemed inescapable. But if the world were to be reordered once again, some kind of lasting answer had to be found to the "challenge of Carneades."

The answer seemed to be to trust only knowledge that could be derived directly from the senses, not to what had been said by others before you, no matter how wise they might have been. True, the senses were frequently unreliable. But that could be corrected for. Ultimate knowledge of the kind supposedly available to God—what was called the "knowledge of first causes"—was, in any case, as the great mathematician Isaac Newton warned, inaccessible to mere humans. Human knowledge—the knowledge of "secondary causes"—however, was, and that, as John Locke nicely phrased it, was "sufficient for all our concernments." But human knowledge could be acquired only at first hand, through observation and experiment. Throw away all your books, urged Locke. Begin with the senses, with first principles. Much of this was simple rhetoric. Locke had a fine library and a deep knowledge of the ancients. But his point was that for all that books might help us to improve our lives, and help humans, as he phrased it, to "look abroad beyond the Smoke of their own Chimneys," true understanding of the world had to begin with the world itself, not with what generations of the dead had once thought about it.[12]

In the course of the seventeenth century, a revolution in science and philosophy took place. There have been innumerable conflicts over this, over its nature and its extent, over its influence and its significance. It has been pointed out that magic, alchemy, and astrology were still revered sciences alongside the new physics and the new astronomy. Isaac Newton, for instance, was not only the man who discovered the laws of gravity and the theoretical architect of so much of modern science, he also wrote extensively on theology, astrology, and the occult. But for all that some of the older ways of thinking lived on in an uneasy symbiosis with the new, by the end of the seventeenth century, what had once been a universal consensus had vanished in all but a few still resistant corners of Europe. What replaced it was a number of methods of inquiry based upon the rigorous examination of evidence. All that could be shown to be true had to be argued from the ground up, and this applied as much to the understanding of the workings of the human mind as it did to the movement of the planets. This science made no attempt to deny the existence of God. Indeed, many of its creators were practicing, if not always devout, Christians, and none was a self-confessed atheist—although Hobbes came very close. But it deprived the Church of any right to speak authoritatively of anything except the nature of God, of "theology" in the strictest sense of the word. The world could now be understood only in terms of those laws that could be discovered by induction, by observation and experiment. These were the laws that were written in what Galileo called "the great book of nature," not the scribblings of the followers of a Judaic divinity.

By the same logic, all that was human could be understood only in human terms. Morality, once believed to be the outcome of divine decree, became a matter of convention. "Men everywhere," wrote Locke, "give the Name of *Vertue* to those actions, which amongst them are judged praise worthy; and call that *Vice,* which they account blamable."[13] But why they believe certain things to be "praiseworthy" and others "blamable" was now recognized to be merely a matter of convention. In the opinion of Anthony Ashley Cooper, the third earl of Shaftesbury, a philosopher, Whig politician, and deist, who had been Locke's pupil as a child and retained no fond memories of him, " 'Twas Mr. Locke that struck at all the fundamentals, threw all order and virtue out of the world, and made the very ideas of these (which are the same as those of God) *unnatural* and without foundation in our minds."[14]

Shaftesbury's indignation is perhaps comprehensible. But he was wrong. Locke had indeed "struck at all the fundamentals." But his rejection of the absolute validity of moral laws did not make them any the less binding than the older theocentric codes had been. It also did not substantially change their content. For there could still be one thing we can know about human nature: that under normal circumstances, sane people wish, as far as is humanly possible, to avoid death for as long as possible. It was, said Hobbes, a law "as powerful as that by which a stone falls downwards." And if we accept this, which no reasonable person— be he a Finn or an American Indian, a Turk or a Hindu—could possibly deny, then it was clearly not "reprehensible, nor contrary to reason, if one makes every effort to defend his body and limbs from death and to preserve them. And what is not contrary to right reason, all agree is done justly and of Right."[15] The old "natural law" had thus been reduced to one simple proposition or, as the Dutch humanist Hugo Grotius—who had come to much the same conclusion as Hobbes—put it, to two: "It shall be permissible to defend one's own life and to shun that which threatens to prove injurious," and "It shall be permissible to acquire for oneself and to retain those things which are useful to life."[16] On this basis, you could demonstrate that it was wrong to kill just as easily as you could by ascribing it to the commandment "Thou shalt not kill." The point, now, was that every moral code had to be accepted for what it was: merely a solution to a problem, not an inscrutable, unquestionable divine decree, written by God's hand into human nature as onto a clean slate.

Neither Locke nor Grotius nor Hobbes nor any of the other creators of the "new philosophy" believed that by reducing the commands of nature to a simple impulse to self-preservation they were denying the possibility of passing judgment on human behavior. Quite the contrary. But Shaftesbury was right to see the lethal potential of making the idea of the good something other than an innate part of the human character. For in claims such as Locke's lie the origins of modern relativism, the idea that what is practiced, thought, believed, or valued by us—whoever *we* may happen to be—has a legitimacy that is merely relative to what is practiced, believed, or valued by others. No moral judgment can be made on the activities of anyone outside our particular world because we are in no position to understand it. Other is other and has simply to be respected as such. None of the philosophers of the seventeenth century, nor any of their Enlightenment heirs, was a relativist in the sense the term is understood today. The idea that the civil law might stand calmly by as girls were "circumcised" against their will (albeit with the active support of their mothers) to satisfy the superstitious manifestation of male anxieties would have seemed to them as repellent as condoning wife killing. The fallacy of modern relativism is to assume that because our Western, European culture cannot claim to be "natural" and therefore the model to which all others should aspire, it follows that all cultures, no matter what they practice, are equally valid.

For the argument put forward by the new science in the seventeenth century and developed in the Enlightenment of the eighteenth is not that if there are no laws in "nature" anything is valid. It is that nothing is valid unless it can be demonstrated to be so on the basis of some certain principle, to which all reasonable people could be supposed to agree. Female infibulation, or female circumcision, is an inhuman practice, as is, for instance, sati, the Hindu custom of immolating the wife along with the corpse of her dead husband, because it is a violation of the bodies and of the rights of the women involved. The purpose of skepticism was not to privilege ignorance and cruelty wherever it was found; it was to offer a basis for understanding what constituted ignorance and cruelty no matter where one might find it.

There was one further aspect of modern rationalism that would have a lasting impact on the development of the West. The division of Church and state had been a command of God. But for all that, Christians still clung to the belief that political power, although exercised by men, nevertheless derived from God. Kings were semidivine beings, because their authority had been conferred upon them by a divinity. In the seventeenth century a radically different theory of the origin and sources of political power emerged. This claimed that political authority could derive only from those over whom it was exercised, that is, the people themselves. Furthermore, it could be exercised only with their consent and in their interests. This is known as the contract theory of government. Its most powerful original exponents were English, Thomas Hobbes and then John Locke. But it was rapidly taken up in France, where it provided the ideological inspiration for what in 1789 became the greatest rebellion against the power of kings in the modern world: the French Revolution. The contract theory of government was not entirely new; no theory ever is. Medieval kings had also held power with the tacit consent of their subjects and ruled supposedly in their interests. But it was the central belief that government depended upon a contract, willingly entered into and binding upon both the ruler and the ruled, that made modern Western liberal democracy possible. Rulers ceased to be benevolent but despotic fathers and instead became servants. The people ceased to be subjects and instead became citizens.

The "scientific revolution" changed forever the nature of the Western world. It opened up an immense potential for scientific knowledge. But modern science was not, as so much of the old had been, merely theoretical knowledge. It was linked, directly and inescapably, to technology. In the old order a doctor had been someone who sat in a study in a university and pored over the works of the ancients: Galen, Hippocrates, Celsus, and Aristotle. These were not the men who set your broken leg or applied leeches to your flesh or opened your veins, in the vain hope of draining from you the "bad" blood that had caused whatever you were suffering from, be it a mild case of flu or pancreatic cancer. Those poor butchers, "barber surgeons" as they were known in England because they also curled your hair and shaved your beard, were jobbing laborers with almost no medical knowledge at all—and what they had was, as their diagnoses and prescriptions indicated, invariably wrong.

The "Scientific Revolution" changed all this. After the seventeenth century, medicine gradually became an integrated and respectable science. Astronomy, which had been largely indistinguishable from astrology, became with Copernicus and Galileo an exact science that by the middle of the eighteenth century had wholly transformed the image of the heavens and the place of the earth within it. Botanists and geologists began to map and chart the surfaces of the planet. Expeditions—themselves products of the new scientific inquisitiveness and of the new technologies of navigation—were sent out in ever-increasing numbers to every corner of the globe to bring back samples, so that a new breed of

scientists could study them at home. Botanical gardens, sumptuous, massive enclosures, a visible tribute to the care that power
lavished upon knowledge, were set up in the capitals of Europe:
the Jardin des Plantes in Paris, Kew Gardens in London, the Hor-
tus Botanicus in Leiden. All of this activity led to the creation of
new sciences, which greatly enhanced human control over the en-
vironment, navigation, geology, statistics, and modern economics. Indirectly but inevitably, it was also responsible for the
creation of immense wealth. More darkly, it led to the development of formidable military technologies from large-bore cannon, rifled barrels, and breech-loading weapons all the way to the
steam-driven battleship, which would, within a very short space of
time, make Europe and the United States masters of most of the
world.

It also paved the way for a revolution in the perception and
understanding of humanity.

## III

By the beginning of the eighteenth century, most of what we
today think of as the sciences, both natural and moral, had been
pried away from the clutches of the Church and the churchmen.
Europe, the West, remained, and predominantly remains, Christian in inspiration. But the conflict that had arisen in the sixteenth
century between reason and dogma had been, substantially if not
entirely, resolved in favor of reason. It was this that laid the
grounds for the European-wide intellectual movement that has
come to be called "the Enlightenment," a cosmopolitan assertion
of the virtues of informed and benign reason that, or so its advocates hoped, would transform the entire world. By the last two
decades of the eighteenth century, in almost every country in Europe, there were those who called themselves "enlightened" and
those who condemned them for being so. Even a Scottish cleric,
on one of the remotest islands in Europe, could protest to the urbane lowlander James Boswell that up there in "ultima Thule" he

and his companions were "more enlightened" than might be supposed.[17]

There have been many definitions given to this "Enlightenment" and what it meant for the history of Europe. Enlightenment could mean freedom from prejudice, "of which," said the great materialist the Baron d'Holbach, "the human race has been so long the victim";[18] freedom from constraint—the willingness, as the greatest of all the philosophers of the eighteenth century, Immanuel Kant, put it, for man to emerge from "his self-incurred immaturity."[19] Enlightenment implied greater social equality. It meant legal reform. It meant, too, a heightened awareness of the existence and needs of others. Above all, perhaps, it meant the right to subject everything to rational, disinterested scrutiny: to be able to criticize freely. As Kant declared in a famous passage in his most famous work, *Critique of Pure Reason:*

> Our Age is, in especial degree, the age of criticism, and to criticism everything must submit. Religion through its sanctity, and law-giving through its majesty, may seek to exempt themselves from it. But they then awaken just suspicion and cannot claim that sincere respect which reason accords only to that which has been able to sustain the test of free and open examination.[20]

No society, claimed Kant in some indignation, could "commit itself by oath to a certain unalterable set of doctrines, in order to secure for all times a constant guard-ship over each of its members." This, as Kant well knew, was precisely what most clerisies of most religions seek to do. But it was "a crime against human nature" for any group, however powerful, backed by no matter what authority, to "put the next age in a position where it would be impossible for it to extend and correct its knowledge, particularly on such important matters, or to make any progress whatsoever in enlightenment."[21] This was the basis of the Enlightenment values that spelled the end of religious dogma and the beginning of the secularism that has ensured the future progress, for all its many setbacks and deficiencies, of the West.

Perhaps the simplest, and the most poignant, description of just what "enlightenment" meant for those who saw themselves as its intellectual champions was provided by the Marquis de Condorcet, the former permanent secretary of the Academy of Sciences and father of the modern science of statistics. In 1793, in a tiny room in the Rue Servandoni in Paris, in hiding from the very revolution that had set out to transform and enlighten the world, Condorcet wrote a short and highly optimistic account of the progress of the human race. The modern age, he said, the age of reason and philosophy, was one in which a truth had taken hold of the peoples of Europe that was "independent of the dogmas of religion, of fundamentals and of sects," and with it a conviction "that it was in the moral constitution of man that one had to seek for the foundations of his obligations, the origin of his ideas about justice and virtue."[22] There was still, however, much to do. With the agents of the Jacobin Terror hunting him through the streets of Paris, Condorcet cannot have been unaware that enlightenment and reason were not always their own best masters. Yet, as he wrote furiously by the shaded light of a candle for fear of discovery, his faith in the onward progress of humanity seems never to have wavered. There would now be no future breaks in the forward progress of mankind, no ruptures of the kind that had dogged the march of civilization since the collapse of the Roman Empire. The day was not far off, he assured his readers, when "all the nations should one day reach the state of civilization at which the most enlightened of peoples, the most free from prejudice, such as the French and the Anglo-Americans, have already arrived."

Condorcet was one of the first to recognize that the culture of rational deliberation, of freedom and science, that had arisen in ancient Greece and flowed though Rome into the rest of Europe had now been transplanted across the Atlantic to North America

(South America remained mired in the superstition and tyranny of the world of the ancient regime), and he was one of the earliest to describe this not simply as "European" but as "Western." All that now remained, he believed, was for the Africans and Asians to welcome the enlightened Westerners as friends and liberators from their own "sacred despots" and the "stupid conquerors" who for centuries had kept them in the same darkness in which the peoples of Europe had lived, under the rule of priests and monarchs.[23]

This was the source of the great Enlightenment illusion, at once both ennobling and perilous. It assumed that all persons are individuals. Religion or custom could be used to blind them as to what their real interests might be or where their real happiness might lie, but ultimately, no matter how tenacious the hold deplorable local customs might be over the primitive peoples of the world, it would be loosened, inexorably, by science and education. Once the victims of the kings and priests of old had come to see with their own eyes the benefits only an enlightened—which inevitably also meant a Western—culture could bring, they could not fail to embrace it. The Europeans themselves had followed the same historical path. They too had lived for years in ignorance and poverty, ground down by ambitious autocrats and bigots who had befuddled their reason with the absurdities of revealed religion. They, however, had succeeded in shaking off these monstrous deformities and were now in a position to offer a waiting world the benefit of their enlightenment.

Science and human understanding, then, went hand in hand. Both served to liberate the human mind, and the liberation of the mind offered power, progress, and, ultimately, a better life. In Samuel Johnson's curious little Oriental tale *Rasselas* of 1759, Rasselas, an "Abyssinian Prince," has been confined to a spacious palace in a valley in "the kingdom of Amhara," which for all that it is called the Happy Valley is, in reality, a prison. He has everything he needs in life, except access to knowledge. Here he is visited by a poet named Imlac "from the kingdom of Goiama," who

tries to explain to Rasselas the nature of the world that lies beyond the valley. He tells Rasselas of all he has seen in India, Arabia,
Persia, Syria, and Palestine. But what, unsurprisingly, is of particular interest both to him and to Rasselas is the Europeans. They
are, declares Imlac, irresistible. They are now "in possession of all
power and knowledge." They have established settlements everywhere they go, and their fleets are "in command of the remotest
parts of the world." "When I compared them," declares Imlac,
"with the natives of our own kingdom and those that surround
us, they appeared almost another order of beings." But how, asks
the puzzled Rasselas, can this be? Surely, if the Europeans can
travel to the far corners of the earth, "for trade or conquest," cannot the Asiatics and Africans repay the compliments? Cannot
they, too, "plant colonies in their ports, and give laws to their natural princes? The same wind that carries them back would bring
us thither." To this Imlac has a simple reply. The Europeans are
more powerful simply because they are wiser; and "knowledge will
always predominate over ignorance, as man governs the other animals." And this knowledge, although Imlac attributes it ultimately to the "unsearchable will of the Supreme Being" (Johnson
was nothing if not a good Anglican Tory), had been acquired
through "the progress of the human mind, the gradual improvement of reason, the successive advances of science."[24]

Today it is easy to be scornful of such confidence in the beneficent nature of reason and science. The dismal, tragic history of
the nineteenth and twentieth centuries in Europe and the West
generally has shown us that such a vision of enlightenment was
all too short-lived. We now know that the kings and priests of old
Europe were replaced not by Condorcet's enlightened scientists
but by other kinds of dogmatists in the service of other kinds of
religion and by other breeds of tyrants. The Africans and the
Asians, had they indeed been so unwise as to have welcomed their
European brothers to their shores, would have only found themselves pressed into forced labor even more speedily than they
eventually were.

There can be no denying the perfidious uses to which both Western rationalism and Western science have been put. But for all the scorn that has been poured on the Enlightenment, for all that the faults of Western culture, from colonialism to National Socialism, have been laid, somewhat haphazardly and generally by self-indulgent Western intellectuals, at its door, it remains more compelling, more compassionate, more humane than anything that has attempted to replace it.[25] Even so free a spirit as the German philosopher Friedrich Nietzsche could urge his fellow Germans against what he saw as the obscurantism ushered in by the Romantics:

> This Enlightenment, we must now carry further forward: let us not worry about the "great revolution" and the "great reaction" against it which has taken place—they are no more than the sporting of waves in comparison with the truly great flood which carries *us* along.[26]

And carry them along it did, and with them all the modern democratic states of the West, whose values, derided, coerced, threatened by religious fundamentalisms and extremes of cultural relativism, have yet survived, battered but still recognizable, to this day.

# 9

## *ENLIGHTENED ORIENTALISM*

### I

By the beginning of the seventeenth century, the "East," identified with the lands of the former empires of Alexander and Byzantium from the Bosporus to the Himalayas, was firmly in the hands of three great Muslim empires, the Ottoman Turks, the Safavids of Iran, and the Mughals of India. Yet despite that, all three maintained regular, if sometimes uneven, trading and diplomatic relationships with most of the states of Christian Europe, and European merchants, diplomats and simple adventurers traveled in increasing numbers to Asia. Some wrote books about their experiences, and three in particular became immensely popular: Paul Rycaut's *The Present State of the Ottoman Empire* (1665), Jean-Baptiste Tavernier's *Six Voyages to Turkey, Persia, and India* (1676–1677), and Jean Chardin's *Travels of Sir John Chardin in Persia and the East Indies* (1686). They were a huge improvement on the scattered fantasies and lurid tales that had hitherto fed the European idea of the "Orient," and if their often disingenuous accounts failed to make Asia any less exotic and menacing than it had been for previous generations, they made it a great deal more familiar.

The direction of travel was, however, almost entirely one way.

From the late fifteenth century, increasing numbers of Europeans went to the various regions of Asia. Very few Asians, however, ever made it to Europe. For the most part, the East, from Turkey to China, displayed very little curiosity about the West until the mid–eighteenth century. There were exceptions. The Persian legate Muhammad Riza Beg, who was received by Louis XIV in 1715, aroused a good deal of short-lived curiosity at the French court. An Indian of Persian origin named Mirza Abu Talib Khan Isfahani paid a visit to England in 1799 and his account of his journeys, *Talib's Travels in the Lands of the Franks,* enjoyed considerable popular success when it was translated into English in 1810. There was the famous Ottoman ambassador Mehmed Said Efendi, who, in 1719, had gone to discover what he could of the secrets of Western success and wrote a much printed—in both Turkish and French—account of his time in France. And there was the Persian Mirza I 'Tisam ad-Din, who visited Britain in 1766 and wrote an account of his experiences, *The Wonder Book of England.*[1] From yet farther east, there was Hu, an unfortunate Chinese convert to Christianity who traveled to Europe in 1722 and spent three years there, most of them in the hospital for the insane at Charenton outside Paris, before being returned home.[2]

BY FAR THE best-known "Oriental" visitors to Europe, however, never existed at all. In 1721, a book was published in Holland claiming to be the record, in the form of letters, of a journey made by two Persians to France. Their near-impossible names were Usbek and Rica. The book was entitled *The Persian Letters.* Its author chose to remain anonymous, but most of the French literary world knew that he was, in fact, a minor aristocrat from Bordeaux, known previously only for his scientific and legal concerns, named Charles-Louis de Secondat, baron de Montesquieu. *The Persian Letters* talked at length of sex and offered its avid and numerous readers a glittering image of difference. Here, it seemed, was a place in which the somewhat constrained and constraining imagination of European fashion could expand without limit. It raised troubling questions about the origins, and the possible future, of the West itself. And it held a darkly distorting mirror up to European convention, European morals, and European complacency. *The Persian Letters* was a runaway success, going through more than ten printings in one year. Montesquieu later observed that publishers did everything in their power to get a sequel out of him or anyone else. "They went about pulling the sleeves of everyone they met," he wrote. " 'Sir,' they would say, 'do some Persian Letters for me, I beg you.' "³ Some, but not Montesquieu, obliged. In 1739, *The Persian Letters* was followed by the (infinitely less amusing) *Chinese Letters* of the Marquis d'Argens, the author of the erotic philosophical novel *Thérèse philosophe,* and in 1762 by Oliver Goldsmith's *Letters from a Citizen of the World to His Friends in the East.* Both were widely read, although neither could match either the brilliance or the popularity of the original.

Montesquieu's East, like that of Argens, Goldsmith, and a host of other lesser figures, is self-consciously the creature of his own political passions. He made no pretense to ethnographic accuracy. Although he went on in 1748 to write the first sustained treatise of comparative sociology—*The Spirit of the Laws*—which drew on examples taken from a wide range of cultures, Persian among them, his vision of Persian life, and of Islam, was largely dependent upon what he had read in Chardin and Tavernier. The image of the harem he creates through Usbek's correspondence with his chief eunuch is meant to be a portrait of France under the monarchy of Louis XIV, and his portrayal of Islam is as hostile to religion in general as is his depiction of the Christian Church as seen through Persian eyes. His Persians are fictions, mouthpieces for his own scathing criticisms of the institutions and customs of prerevolutionary France, the hypocrisy of European sexual mores, and the stultifying emptiness of conventional religion, which he looked upon as the main source of human fear, mistrust, and sexual misery—and which merely mirrored the prejudiced view of the world held by the believer.

Montesquieu was no Orientalist, but he was the beneficiary of a new attitude toward, and substantial new information about, the various nations of Asia. Neither are Usbek and Rica total fantasies. However artificial they may be, they come from a place with a recognized history, and they belong to a world that was becoming increasingly familiar to European readers. True, the French, on being told that Rica is a Persian, cry "Oh! Oh! What an extraordinary thing! How can one be a Persian?"[4] Yet it is Rica who is allowed time and again to see through the fabric of European convention. It is Rica who stoically concludes of the pretensions of all religions that "When I see men creeping over an atom, that is, the earth, which is merely a point in the universe, and immediately setting themselves up as models of Providence, I am unable to reconcile such extravagance with such smallness." It is Rica who makes the famous—and at the time scandalous—observation that if triangles had a god they would give him three sides.[5] To allow an "Oriental" and a Muslim, however fanciful, to criticize in such language what were clearly identifiable as the cherished habits and beliefs of Christians marked a shift away from the religious fervor, the dogmatic certainties that so many of the earlier accounts of Asia had been made to carry.

By the time Montesquieu came to write his novel, attitudes toward the cultures of the East had begun, perceptibly, to change from what they had been even a century earlier. Now there was no longer one voice crying out about the "pagans," the "infidels," the "Saracens," but several. And now those voices were by no means unanimously hostile. Europe grew ever closer to Asia. What now drove the two worlds together was not, however, a lessening of religious conviction in Europe and the rise of a secular enlightenment, although these clearly played a part. It was the great eighteenth-century machine of enlightened human progress: commerce. Commerce, or, as Montesquieu famously called it, "sweet commerce," as it was understood in the eighteenth century, went far beyond simple trade.[6] Exchanging goods meant also exchanging views. It meant, as the likes of Chardin, Tavernier, and Rycaut had, in their modest ways, demonstrated, coming to terms with peoples who at first meeting might seem alien and threatening. "Interest," declared the great Orientalist Sir William Jones—by which he meant economic interest—"was the magic wand which brought them [East and West] within one circle."

Commerce also had more immediate practical consequences. In order to trade with distant peoples, you had to know something about their customs and you had to be able to converse with them. Commerce, as Jones phrased it, was "the charm which gave the languages of the East a real and solid importance."[7] In 1453, there had been few people in Europe who had had a mastery of even Arabic and fewer still who knew Persian, Sanskrit, or Turkish—let alone Chinese or Japanese. Some two centuries later there were chairs in Arabic in Leiden in the Netherlands, at Cambridge, at Oxford, and at the Collège de France.

It has now become a commonplace to ascribe most of this increased interest in the East, in Eastern languages, culture, history, literature, and religion to a growing desire on the part of the European colonizing powers, and in particular France and Britain, to create an image of the Orient as essentially weak, monochrome, and subservient. By this reckoning the "Orient" is a fabrication, an imagined culture reaching all the way from the Bosporus to the China Sea, cobbled together from an ill-assorted collection of travelers' tales and a certain amount of disingenuous scholarship, so as to allow the colonial powers of Europe to assert their political power over the very diverse peoples living there. "European ideas about the Orient," claimed the polemicist and literary theorist Edward Said, served only to reiterate "European superiority over Oriental backwardness, usually overriding the possibility that a more independent, or more skeptical, thinking might have different views on the matter."[8]

This is a very crude description of a very complex picture.[9] It is certainly true that some European accounts of the Orient were patently false and patent attempts to show how cowered and inferior such peoples were (whether that made subjugating them any easier, however, is doubtful). It is also true that many self-styled Orientalists, and later many anthropologists, in particular those concerned with India, were closely involved with the imperial administration. The first Europeans to take an active interest in the lives and societies of non-European peoples from Asia to Africa and the Americas had been soldiers, merchants, imperial administrators, missionaries—all people with vested interests and often with quite obvious axes to grind. Before tourism, of one kind or another, got under way in the nineteenth century, nobody much else had the opportunity, the incentive, or the resources to leave home for long enough.

But soldiers often have a very shrewd understanding of their potential opponents, merchants are frequently able to see beyond their immediate economic interests, missionaries (sometimes) set aside their mission, and being involved in the imperial administration did not necessarily mean being subservient to it. Sir William Jones, one of the greatest linguists of the eighteenth century, is best known today for having suggested that the linguistic affinities between Sanskrit—a language he declared to be "more perfect than the Greek, more copious than the Latin, and more exquisitely refined than either"—and most of what are now called the Indo-European languages implied that they must all share a common ancestry.[10] (This idea later gave rise to the far less innocuous—and far less plausible—claim that all their speakers must, therefore, also share a common racial origin.) Jones was not only a linguist but also a highly respected jurist, who in 1783 became judge of the Supreme Court of Judicature at Calcutta, which made him, in effect, an employee of the East India Company. Like Jones, the legislator and philologist Nathaniel Halhed was also an employee of the company; so, too, at one point in his career, was the great nineteenth-century jurist Henry Sumner Maine, who claimed a common Indo-European origin for the Greek city-state and the Indian village, thus locating the origins of European democracy firmly in Asia.

None of these men was the simple lackey of a colonial ideology or the mere instrument of imperial propaganda. Jones was not, it is true, a disinterested polymath. His prime concern was the legal administration of British India, and the main question he set out to answer was: What set of laws would allow the Indians to live productive and peaceful lives under British rule?[11] Far from representing "Orientals" as in any way inferior to Westerners, Jones lamented that "Europeans for the most part treat Orientals as ignorant savages." It was, he complained, only "our prejudices which all derive from the same sources: self-love and ignorance . . . which lead us to believe that whatever is ours is superior to all else in the world."[12] He had even graver doubts about the wisdom of the European process of "civilization." Civilization, he wrote, can be described in many terms, "each measuring it by the habits and prejudices of his own country; but, if courtesy and urbanity, a love of poetry and eloquence and the practice of exalted virtue be a juster measure of a perfect society," then the Arabs were "eminently civilized for many ages before their conquest of Persia"— and thus long before the modern Europeans.[13] He ranked the sages of ancient India—Valmiki, Vyasa, and Kalidasa—as equal to Plato and the Greek poet Pindar. And, he told his friend Richard Johnson in 1784, "*Jûdishtêir, 'Arjun, Corneo* and the other warriors of the *M'hab'harat* appear greater in my eyes than Agamemnon, Ajax and Achilles when I first read the *Iliad.*"[14] The Persians were also, he insisted time and again, a "nation equally distinguished in ancient history" as the Greeks or Romans; and the Persian Hâfez was no lesser a poet than Horace.[15] It was only unfamiliarity that made us praise the one and ridicule the other.

It is true, however, that Jones's scholarship was not entirely disinterested—no great or compelling scholarship ever is. Nor was he hostile to the final objectives of the administration he served in India. He wished to see it improved, not replaced. He firmly believed that thanks to our "beautiful and wise laws or rather perhaps to our holy religion, on which they are based, we will never be as despotic as the kings of the Orient," and this at least made

British rule morally tolerable.[16] But he also believed, as did many subsequent British officials in India, that European law could not simply be imposed upon the Hindu and Muslim populations of India. "A system of *liberty*," he once argued, if "forced upon a people invincibly attached to opposite *habits*, would be a system of cruel *tyranny*."[17] And in the interest of harmonizing the various legal systems that prevailed in India, he undertook, in 1788, the colossal task of compiling a digest of Hindu and Muslim law. This, he told the governor-general, Lord Cornwallis, would give to the people of India "security for the due administration of justice among them, similar to that which Justinian gave to his Greek and Roman subjects."[18] Equipped with Jones's code, Cornwallis would have become the "Justinian of India." Jones died, however, before he could finish the task.

Jones's vast admiration of Indian, Persian, and Arabic literature was not driven solely by a desire to rehabilitate a number of rich but unjustly despised cultures.[19] What was also at stake, what had inspired this new enlightened concern with Asia, was ultimately the desire to find the source of European and finally all human civilization, to explain why they differed from one another, how they had evolved, and, crucially, where they were likely to end up. It was this that prompted Dr. Johnson to dub him "Harmonious Jones."[20] Men such as Jones, or Charles Wilkins, who in 1784 completed the first English translation of the *Bhagavad-Gita*, or Nathaniel Halhed, who compiled the first grammar of Bengali, were all driven by a desire to know how the "West" had come to be the way it was. To do this, they knew, they had to look to the East, for, as Jones told the members of the Asiatic Society of Calcutta (of which he was one of the founders):

> It will be sufficient in this dissertation to assume, what might be proved beyond controversy, that we [in India] now live among the doers of those very deities, who were worshiped under different names in old Greece and Italy,

and among the professors of those philosophical tenets, which the Ionic and Attic writers illustrated with all the beauties of their melodious language.[21]

Here, in the farthest reaches of India, lay the ultimate source of what had, in time, become the great civilizations of the ancient world, to which the enlightened Europe of Jones's own day owed its existence.

Jones ended the treatise he wrote (in French) on "Oriental literature" with a plea to the "princes of Europe" to encourage the study of the Asiatic languages, to

> set before the world those precious treasures of which you are only the repositories, and which are not treasures until they become useful, bring to light those admirable manuscripts which adorn your cabinets without enriching your minds, like the Chinese characters on porcelain vases, whose beautiful shapes we admire without ever understanding their meaning.[22]

For the great nineteenth-century German Orientalist Max Müller, the discovery of the association between the cultures of East and West had been one of the greatest in the history of mankind. The study of the origins of the Indo-European languages had now shown beyond any doubt that there had been a time when "the first ancestors of the Indians, the Persians, the Greeks, the Romans, the Slavs, the Celts, and the Germans" had all lived together "within the same enclosures, nay, under the same roof."[23] There might, he recognized, "be still some troglodytes left" who refused to recognize any affinity between Greek and Indian culture, as if Greek mythology were "a lotus swimming on water without any stem, without any roots." But they could only be compared to those who still persisted in believing that the world was flat.[24]

All these Orientalists were engaged in a highly ambitious project to write a complex, interlocking history not of Europe, but of the Indo-European world. It would persist through Müller and Maine, whose *Ancient Law* claimed to have discovered a formal association between the Indian village, the Greek *polis,* and the Scandinavian *mark*—all the sources of the Indo-European experiment in democracy—to the great French Indo-Europeanist Georges Dumézil in the mid–twentieth century. Thereafter, mired by the Nazi abuse of the "Aryan" myth, the project fell into disrepute. But if today it has become a simple fact that the history of Europe cannot be detached from that of significant parts of Asia, this is due, in great part, to the Orientalists of the Enlightenment.

## II

One man who paradoxically came to loathe and be loathed by William Jones, who took the message of the new Orientalism most to heart, was the French Sanskritist Abraham-Hyacinthe Anquetil-Duperron.[25] His life, and the story of how he came to be at the center of one of the eighteenth century's great intellectual controversies, says a great deal about the extent to which the Western image of the "Orient" was transformed during the Enlightenment and how, intellectually, it set the scene for the final conflict between East and West that would take place in the nine- teenth and twentieth centuries.

After Jones, Anquetil-Duperron became perhaps the most celebrated Orientalist of his day. Garrulous and self-referential, a man who loved nothing so much as to talk about himself, Anquetil-Duperron was, in most respects, the typical antiquarian. As such he made few friends among the learned, literary world of Paris, which had first been fascinated by his endeavors and then revolted by their results. He was, said the *philosophe* and Neapolitan ambassador to France Ferdinando Galiani, "all that a traveler should be, exact, precise, incapable of creating any system, incapable of seeing what is useful and what not."[26] This is not quite fair. Anquetil-Duperron's writings are certainly long-winded and overburdened with often irrelevant detail (most of it about himself). But for all his tiresome narcissism, he was convinced that what he had done was to construct a powerful and compelling system for understanding the "East" and through the East ultimately what he described as "Man, the center, somehow, of nature, the being who is of most interest to us."[27] He came to take a vivid interest not only in ancient but also in modern Asia (and even, toward the end of his life, in America)[28] and wrote a passionate defense of both the Turkish and Arab modes of government. The study of the East, he enthused, would "perfect the knowledge of mankind, and above all assure us of the inalienable rights of humanity."[29]

Anquetil-Duperron was born in Paris on December 7, 1731, the fourth child of a spice merchant of modest means.[30] He began his career by studying theology and Hebrew at the Sorbonne, and later Persian and Arabic at a seminary at Rhynwijk in Holland. In 1752, he went back to Paris with a position as attaché at the department of Oriental manuscripts of the Royal Library. Here, in 1754, he was shown a facsimile of four leaves from a manuscript of the Vendîdâd, which had been acquired by an English agent from the Parsees of Surat and given by him to the Bodleian Library in Oxford. Staring at what was then an unintelligible ancient manuscript, the young Abraham-Hyacinthe discovered his vocation. "There and then," he later recalled with his usual modesty, "I decided to enrich my country with the remarkable work. I dared to form a plan to translate the work, and with that in mind to travel to Gujarat or Kirman to learn ancient Persian." What little he knew of the Avesta had shown him that here, in the "ancient Orient," he might find "the enlightenment which had been sought for in vain among the Latins and the Greeks."[31]

What Anquetil-Duperron had seen that day in the Royal Library was a fragment from the twenty-two sections of the "Law Against Demons," or the rules concerning the purity of the faith of the Avesta. The Avesta, or Zend Avesta—both words in fact describe the forms of ancient Persian in which the work is written— is the sacred book of the Parsees, which dates from the Sassanid period of Iranian history, shortly before the Islamic conquests in 634. It is composed of 21 books divided into 815 chapters, of which only 348 now survive. It is claimed to be all that remains of the writings of the Persian sage and prophet Zoroaster.[32]

In the eighteenth century, nothing much was known about either Zoroaster or the religion he was widely believed to have founded, except for some stray, and somewhat unhelpful, references in a number of Greek texts, where he is treated as both a religious leader—in effect a creator of gods—and a lawgiver. He was widely believed, however, to have been the creator of a body of archaic wisdom, pre-Christian and possibly even pre-Mosaic, something that, it was thought, if only more were known about it, would demonstrate the links that connected the ancient Greek world to its Asian roots. If Anquetil-Duperron could find a faithful and complete original text of the Avesta and translate that into a modern European language, he would be able to transform everything then known about the most formative period in Western history.

He now had his grand idea. All he needed was patronage. He got in touch with some of the more prominent members of the various academic institutions of Paris. They listened. They were enthusiastic. They showed him—from a distance—the Académie des Belles Lettres, then the most distinguished academic body in France, and promised to make him a member if he succeeded. They agreed to speak to the minister on his behalf and to seek the support of the Compagnie des Indes, which controlled parts of southern India, for the project. As a result of all this activity, he had "the honour of being entertained several times" by Étienne de Silhouette, who was not only the king's commissioner for the Compagnie and comptroller general of finances of France, but also the author of a treatise on Chinese morality and government and well-known to be "a protector of young talent," in particular of those with an interest in the East. But still nothing happened. Anquetil-Duperron was an impatient man, and toward the end of 1754, he decided to take matters into his own hands. On November 7, carrying nothing with him except two pairs of briefs, two handkerchiefs, a pair of socks, a mathematics manual, and copies of the Hebrew Bible, the *Essays* of Montaigne, and *Treaty on Wisdom* by the Jansenist Pierre Charon, he enlisted as a foot soldier in the army of the Compagnie.

On February 24, 1755, he set sail for India aboard the *Duc d'Aquitaine,* a "moving fortress" bound for French India. By then, however, news of his enlistment had reached Silhouette, who secured him a cabin, a salary of 500 livres—not much, as Anquetil-Duperron ungraciously pointed out, but enough to survive on—and a place at the captain's table. In the end he traveled to India in some style. But this could not diminish the horrors of the crossing. Disease carried off some hundred persons; half the crew lay prostrate in their hammocks. Day in, day out, he lay in his cabin listening to the groans of the ship's timbers, punctuated every so often by cannon shots announcing that some poor corpse had been jettisoned overboard. "Everywhere," he wrote in his diary, "the stench of the dying was suffocating."

On August 9, this floating charnel house reached Pondicherry, the French settlement in southern India. There Anquetil-Duperron politely took his leave of the Compagnie des Indes and began an arduous journey, most of it spent, by his own account, in a state of semiconsciousness as a result of one or another kind of fever, toward Benares to study Sanskrit. No sooner had he set out than his ever-fragile health forced him to take refuge in a brothel at Bernagor. Two of the whores watched over him for five hours, ministering to him with sage tea. Enchanted, as he put it, "by the humanity of these poor victims of debauchery," he "rewarded them amply"—although what form the reward took, he prudishly does not say.[33]

By now, however, the French and British were at war. In March, Anquetil-Duperron reached the French factory at Chandernagore just as it fell to the British. Disguised as an Indian Muslim, he was forced to make a hasty retreat back to Pondicherry and then made his way to Calicut, Cochin, Mangalore, and the Portuguese settlement at Goa.[34] On the way he spoke endlessly with just about anyone who was prepared to listen. He reflected on the caste system, on the migration of souls, on the origins of the Christians of Malabar, and, at length, on the relationship between the Marathas and the Spartans. Finally, at five in the afternoon on April 30, 1758, "extremely weakened by dysentery," he arrived in Surat, the source of the fragments that had gripped his imagination four years before and where he would spend the next three years.

After some initial difficulties, he succeeded in making contact with a Parsee priest named Darab Sorabji Coomana and his cousin Kaos. From them he was able to secure a copy of the Vendîdâd, or at least of a document they claimed to be the Vendîdâd. This initial success, however, soon ran into difficulties. Delay followed delay. Finally, Anquetil-Duperron came to suspect that Coomana and Kaos had seen in him a steady source of income and had become alarmed by the progress he was making. He also discovered that the copy of the Vendîdâd he had been given was corrupt. Finally, Coomana seems to have come clean, to have provided him with a correct transcription of the Vendîdâd and a Persian Pahlavi grammar, "together with a number of other manuscripts in both ancient and modern Persian and a small history in verse of the retreat of the Parsees into India."[35] By the end of March 1759, he was able to begin work on the translation. By June the text of the Vendîdâd was finished. Then another disaster struck. As if a seemingly permanent state of dysentery were not enough, on September 26, Anquetil-Duperron was assaulted in broad daylight and in front of four hundred people by a French merchant named Jean Biquant. He retired bleeding but still alive from three thrusts from an epée and two blows from a saber—M. Biquant had evidently been taking no chances. Just why Biquant had wanted to kill him is not clear. Anquetil-Duperron describes the incident as if it were largely unprovoked, although clearly the two men knew each other. Other sources suggest that the conflict was over a woman whom Anquetil-Duperron had tried to seduce. The fact that after the event he was forced to seek asylum with the British certainly implies that he was not as innocent as he claimed. Beneath his prim, censorious exterior, and despite his delicate health, Anquetil-Duperron seems to have enjoyed a picaresque sexual life while in India. His convalescence in the brothel in Bernagor was probably not as innocent as he claimed.

A month later, and now under the protection of the British factory at Surat, he was back at work on translations of the other parts of the Avesta. He began collecting manuscripts in Persian, Zend, Pahlavi, and Sanskrit and sampling the pleasures, physical as well as intellectual, of the region, and he made plans to return to Benares and then to travel on to China and to learn Chinese. His main task, however, was now complete, and this alone, the abbé Jean-Jacques Barthélemy, keeper of the cabinet of medals at the Royal Library, assured him, would be sufficient "to illuminate his name throughout Europe, and make him famous."[36]

The political situation in India, however, was worsening for the French, and this, despite his close personal friendship with the governor of the English factory at Surat, made him increasingly uneasy. His persistent poor health seemed to be deteriorating even further. Finally, he decided to abandon any idea of further travel in the East and return to Europe. On March 15, 1761, he left Surat for Bombay, and two months later, and despite the fact that France was still at war with Britain, took passage on an East Indiaman, the *Bristol,* bound for Portsmouth. He reached England the following year and at once set out for Oxford, where he was also allowed to see, for the first time, the original of the manuscript that had affected him so powerfully eight years earlier. Somewhat to his surprise, it was attached to the library wall by a chain (many rare books in the Bodleian Library still are). The room in which it lay was also, he recalled, "very cold," and he was somewhat put out to discover that he was not allowed to take the volume back to his hotel to compare it with his own copies. Having taken a long look at the reading room, he concluded haughtily that "in general it does not look like our own public libraries" and made preparations to depart.[37]

By 1762, he was safely back in Paris, poorer than he had been when he set out in 1754 but "rich in rare and ancient manuscripts and in knowledge, which my youth (I was barely thirty years old) would give me time to develop at my leisure, and that was the sole fortune I had sought in India."[38] He now began preparing the various manuscripts, and the lengthy account of his travels, for the press. It took him another nine years. In 1771, the final version of Anquetil-Duperron's translation of the Avesta, together with the complete account of his travels and working methods, an essay on the customs and religious practices of the Parsees, and a Zend-French and Persian-French dictionary, was published in Paris in three hefty tomes. Here, he claimed, were laid before the learned European public, for the very first time, the authentic words of the prophet Zoroaster, "one of the first legislators of the ancient world."

It turned out to be a huge delusion. It was, Friedrich Melchior von Grimm, a prolific writer of letters to everyone of any literary or cultural significance, said unkindly, a work that "has not sold and no one can read."[39] Unreadable or not, however, it unleashed a controversy that would last for the rest of the century and involve at one time or another all the prominent Orientalists of Europe.

Immediately the book appeared, the then twenty-five-year-old William Jones published, anonymously and in French, an open letter to Anquetil-Duperron. Jones, who later accused his rival, not unjustly, of being "foul-mouthed and arrogant," loathed the long description of his journey with which he had prefaced his translation. "Some five hundred pages filled with puerile details," he said of it, "disgusting descriptions, barbarous words, and satires as unjust as they are gross."[40] He loathed the rudeness with which Anquetil-Duperron described his subsequent visit to Oxford. ("What punishment," he asked, "would your Zoroaster have ordered for such ingratitude? How much ox's urine would you have been obliged to drink?")[41] Above all, he loathed the scathing comments that Anquetil-Duperron had made about the work of

Thomas Hyde, an Oxford Orientalist who, in 1700, in a similar attempt to recover the bones of Zoroastrianism, had combined what he could find of the Muslim accounts of pre-Islamic Iran with "the true and genuine monuments of ancient Persians."[42]

But this was more than a squabble over scholarly methods or the wounded national sensibilities in the never-ending tussle between scholars in France and England—unfaded to this day—for intellectual preeminence. The Zoroaster whom educated Europe had been expecting was a sage. What Anquetil-Duperron had given it was what Voltaire called an "abominable hodgepodge." How, Voltaire asked, could a collection of such silly tales, grotesque gods and demons, and outlandish laws possibly be the work of someone as reputedly wise as Zoroaster? The "platitudes, blunders and contradictions" that Voltaire found so offensive had been offered by Anquetil-Duperron precisely as proof of the great antiquity of the texts he had used. "For that very reason," retorted Jones, "we have come to the conclusion that it is very modern or that it cannot have been composed by a man of intelligence and a philosopher as Zoroaster is represented by the historians." Had this collection of gibberish really been the laws and religion of the ancient Persians, "would it have been worth going so far to be instructed in them?"[43] Better stay at home, said Jones, taking a dig at Anquetil-Duperron's toadying dependence upon royal patronage, and be content "with your fine feudal laws and the Roman religion, which you seem to cherish." The writings Anquetil-Duperron had gone all the way to Surat to find were "barbarous in themselves and have not gained anything from your barbarous translation."[44] The work, said Jones, was obviously the outcome of a collaboration between one man—who knew very little modern Persian and, despite all his boasting, almost no Pahlavi—and another—the maligned Dr. Darab Sorabji Coomana—who barely understood what he was reading. Either that or it was an outright forgery. Jean Chardin agreed with him; so did the German scholar Christoph Meiners; so did the English linguist John Richardson, who tried to show that the text of the Avesta was filled with Arabic loanwords and that the harsh texture of the language in which it was written contrasted sharply with anything one might have expected from an ancient Persian magus.[45] Most of the circle around the *Encyclopédie* came to the same conclusion. "If these," said Grimm, "are the original books of Zoroaster, this legislator of the ancient Persians was nothing more than an illustrious driveling fool who, following the example of his confreres, mixed a modicum of absurd and superstitious opinions with a little of the common morality that may be found in all the laws on earth." Clearly, poor Anquetil-Duperron had thrown away his life, "uselessly and laboriously," by "traveling to the furthest ends of the earth to find a collection of inanities."[46]

As it turned out they were all wrong. Vain and self-indulgent Anquetil-Duperron certainly was. But he was not quite the linguistic farceur Jones made him out to be. It took until 1826, however, before the Danish linguist Rasmus Rask demonstrated that the Avestan that Anquetil-Duperron had read was not corrupted Sanskrit and had indeed been written at least before 334 B.C.E.[47] As Max Müller remarked, all Jones and the others had really done was to show that "the authors of the Avesta had not read the *Encyclopédie*."[48] That, however, had been the point. What was at stake was more than a philologist's squabble over sources; it was a conflict about the origins of European civilization and the degree to which East and West could be distinguished from each other. Zoroaster had been an Indo-European. As such, although an "Oriental," his works belonged to the East of Herodotus, the East that the Greeks had shared with the Persians, the East that had been the cradle of what would later become Western civilization. All that had emerged from Anquetil-Duperron's rendering of the Avesta was a farrago of superstitious nonsense, quite as absurd, some would say, as anything to be found in the Bible or the Qur'an.

## III

Anquetil-Duperron's troubled attempts to win fame, if not fortune, by offering the writings of Zoroaster to the learned public of enlightened Europe was but one part of the eighteenth century's rediscovery of Asia. Another was the work of Voltaire. As we have seen, Voltaire was one of those involved in the chorus of disbelief that greeted the publication of Anquetil-Duperron's translation. He had good reason to be dismayed. For, like most of those who reviled Anquetil-Duperron's work, he had invested heavily in the image of an ancient Asia whose civilization had been equal to that of the Greeks and the Romans. Here were the semi-mythical lands of the gymnosophists to whom "the Greeks before Pythagoras had gone for instruction," the lands that, together with Egypt, were the acknowledged source of both ancient Persian and Greek science and philosophy.

Dazzled by what he saw as the hitherto disregarded continuities between the Western and Eastern worlds, Voltaire had set out in 1740 to write a new kind of history, one that would look at the civilization of Europe and all its achievements from a truly cosmopolitan perspective. All previous world histories had viewed the West from the perspective of the Jews or the Christians, something that, in his opinion, was as absurd as writing a history of the Roman Empire from the perspective of Wales.[49] To counter this he would describe, and compare, all the civilized peoples of the planet and trace the trajectory of civilization from East to West. He called this great work *Essay on the Manners and Spirit of Nations* (*Essai sur les moeurs*), "manners" or customs being the thing that most distinguished one people from another and that determined character, behavior, and belief. The *Essay*'s secular, irreligious, and cosmopolitan viewpoint would give the Europeans, "so proud as they are of their civilization, a precious lesson in humility."

His work would also, Voltaire hoped, provide an answer to a question that, in one form or another, Europeans had been asking themselves for centuries: If the East had, as Voltaire put it, been "the nursery of all the arts, to which the Western world owes everything it now enjoys," why was it that we, the nations of Europe, of the West, which "seemed to have been born only yesterday . . . now go further than any other people in more than one field?"[50] Plato had asked himself much the same question, phrased somewhat differently, it is true, but essentially, as Voltaire was surely aware, the same. It would continue to be asked by Europeans until the collapse of the Ottoman Empire in the early twentieth century made the triumph of the West seem, at least for a while, inevitable. And it has continued to be asked to this day by increasingly puzzled and humiliated Muslims. Upon its answer turned the whole web of relations that had, since Herodotus's days, constituted the perpetual enmity between the East and the West and has, in only lightly disguised ways, continued to determine it to this day.

One answer could be found where Montesquieu believed he had found it, developing a theory that had, in fact, been around since Aristotle. Peoples were, he argued, the product of the environments, and in particular the climates, they inhabited. Asia has no temperate zone, and its peoples are consequently locked into a constant struggle between extremes in which the weak (from the torrid south) face the strong (from the frozen north). "Therefore," Montesquieu concluded, "one must be the conquered, the other the conqueror." In Europe, however, the strong had always faced the strong, which provided a situation of balance achieved through almost constant struggle. In Europe liberty was a matter of human will. In Asia, by contrast, it would seem that no matter how hard its peoples might battle their tyrannical rulers, the weak were always destined to submit to the strong, never to become strong themselves. "Which is why," Montesquieu concluded, "liberty never increases in Asia, whereas in Europe it increases or decreases according to circumstance." (David Hume's characteristically skeptical observation on this was that if it were true that northerners had always plundered southerners, this had nothing to do with climate and everything to do with poverty: the north was poor, the south rich.)

Although Montesquieu claimed proudly that this reason for the division between the East and the West "had never before been observed," it did not win many adherents.[51] Climate may have been a factor in determining "national character," but there were also the populations of modern Greece and modern Egypt to remind one, in Voltaire's words, that "if there was irrefutably proof that climate influenced the character of men, government always had a far greater influence." David Hume took the same view. "I believe no-one," he remarked caustically, "attributes the difference of manners in Wapping and St James's [poor and rich districts of London] to a difference of air or climate."[52]

No, the Europeans, although they might have started later than the Arabs or the Persians, had now overtaken both, not because of climate or because of some peculiar property of the "Oriental" mind that inclined it to slothfulness and imitation. If human nature was, in Hume's celebrated phrase, "so much the same in all times and places, that history informs us of nothing new or strange in this particular," the reason for the differences between the various peoples of the world had to be sought elsewhere than in supposed natural dispositions. Nations might indeed have characters, but those were acquired, not innate.

If the peoples of Asia from Constantinople to Delhi dwelt in a state of torpor, unable to free themselves from tyrannical overlords and incapable of benefiting even from the great achievements of their ancestors, this had to have something to do with the cultures they shared, the religions they embraced, and, most significantly of all, the governments by which they were ruled.

By the late seventeenth century, the political system that supposedly prevailed throughout the whole of Asia had come to be known as "Oriental despotism."[53] The term, or at least its popularity, we owe largely to the French philosopher and doctor François Bernier (1620–1688), and it was based on his experience of Mughal India. But the principles would do for all three of the great Muslim empires of the period. And in a somewhat different register, it would do also for China.[54]

Bernier had spent twelve years as physician to the Mughal emperor Aurangzeb, and in 1684 he published what must count as one of the earliest works of racism: *A New Division of the Earth According to the Different Species or Races Which Inhabit It* (the last chapter, however, was a treatise on female beauty). It was, however, his voluminous writings on Turkey, Persia, and Mughal India that would have a long and persistent influence. They clearly impressed the radical utilitarian James Mill, the author of the earliest, and certainly one of the most devastating, critiques of the British occupation of India and father of the great liberal philosopher John Stuart Mill. They also provided Karl Marx with much of the evidence he used to build his celebrated notion of an "Asiatic mode of production."

Bernier's argument was simple, and it was sustained by what was said to be firsthand experience. Under Muslim and Hindu rule, there was no law as it was understood in Europe. Legislation depended upon the whim of the prince or, in case of the Shari'a, on the collected whims of a long-dead ruler masquerading as divine scripture. The Oriental despot did not rule over but actually owned the state. In the West the status and identity of an individual were to a great extent determined by his or her ability to own property. In the East, however, everything was owned by the sovereign, and it was this that linked together the three great Islamic civilizations: the Ottoman Empire, Persia, and Mughal India. For all of these, claimed Bernier, had abolished

> the right of possession that is the basis of all that is beautiful and good in this world. They cannot, therefore, fail to resemble each other closely. And since they all have the same faults, they must, sooner or later, suffer the same fate that must inevitably follow: tyranny, ruin, and desolation.[55]

In the Orient, chattel slavery was a condition for many, as indeed it was in many parts of Europe and, with ever increasing brutality, in Europe's overseas colonies. But only in the Orient did there exist what Montesquieu termed "political slavery," the absence of any freedom to act or express oneself independently of the sovereign's will. And the sovereign's will was enforced not through honour, as occurs in monarchies, nor through virtue, as is the case in republics, but through fear, which is why in despotic states, in particular those in Asia, religion is so important, for all religion is always "fear added to fear."[56] For this reason, claimed Rhedi, one of the correspondents in *The Persian Letters*, with the exception of a few towns in Asia Minor and the perhaps more troubling exception of Carthage, republics were unknown in Asia and in Africa, which had "always been crushed under despotism."[57]

Such laws as do exist in Oriental despotisms are few and unchanging, since "when you instruct a beast, you take care not to let him change masters, training, or gait; you stamp his brain with two or three impulses and no more."[58] For Montesquieu, despotic societies resembled not states but large families. "Everything," he wrote, "comes down to reconciling political and civil government with domestic government, the officers of the state with those of the seraglio."

In this new image of the East, Muhammad had now succeeded in escaping from his place in Christian demonology, but only to be thrust into a perhaps even less appealing, if rather more plausible, role as the archtyrant, a skillful, shrewd, if also disreputable (but what prophet was not?), leader of men. This is how he is depicted by Voltaire in his tragedy *Fanaticism, or Mahomet the Prophet* of 1741. Voltaire's Muhammad is still a scheming despot with strong sexual appetites. But he is also a brilliant tactician and passionately devoted to the future of the Arabs, whom he calls "this generous people, too long unknown."

As he tells Zopire, the sharif of Mecca:

> *I will speak only through the God who inspires me*
> *The sword and the Koran, in my bloody hand*
> *Will silence all the rest of humankind.*

*My voice will be to them like a thunderclap*
*And I will see their forehead bent to the ground.*

*I am ambitious. Every man is, no doubt.*
*But never has king, pope, chief, or citizen*
*Conceived a project quite as great as mine.*
*Every people has its moment of glory on this earth,*
*By its laws, its arts, and above all by its wars.*
*The time of Arabia has at last come around.*[59]

Condorcet, Gibbon, Hume, and Rousseau all cast Muhammad in much the same role. "In order to give a leader to a nation which until then had been ungovernable," Condorcet had written in cautious admiration, "he began by raising up on the debris of their ancient cults a more refined religion. Legislator, prophet, pontiff, judge, general, every means of subjugating men were in his hands, and he knew how to use them with ability, with grandeur."[60] It was not a particularly original claim. It was not even an exclusively Western one. Four centuries earlier, Ibn Khaldûn had said something similar—although, of course, with the difference that for him Muhammad's status as "Seal of the Prophets" was uncontested. "Bedouins," he wrote,

> can acquire royal authority only by making use of religious coloring, such as prophethood or sainthood, or some great religious event in general. The reason is that because of their savagery, the Bedouins are the least willing of nations to subordinate themselves to each other. . . . But when there is religion [among them], through prophethood or sainthood, they have some restraining influence upon themselves.[61]

But original or not, the image of Muhammad as an armed prophet made of him the perfect embodiment of a society that, in almost every important respect, was a transposition to the Muslim world of an image formed by the ancient Greeks of their would-be Achaemenid conquerors. This made it comfortingly familiar. In the triumphal history of the ancient world, at Marathon and again at Salamis, the small, independent, liberty-loving, law-abiding Greeks had successfully defended themselves against a massive despotic state. Salamis had not only been, in effect, the end of the Persian menace; as we have seen, it had also been the beginning of the Greek Empire, which would continue to grow until Alexander invaded Persia, burned its capital, and set about uniting the world, as he understood it, into one culture. The success story of Europe, which every educated person in eighteenth-century Europe carried in his or her head, had begun at Salamis, had been consolidated by Alexander, and had then been transformed into a worldwide civilization by Rome, of which the modern European states and their overseas settler populations were the heirs. On the far side of the Bosporus, the Achaemenids had been replaced by the Parthians, who had been replaced by the Sassanids, who had been replaced by the Arabs, who had finally been overcome by the Safavids, the Ottomans, and the Mughals.

The trouble with this was that on closer and more sympathetic examination, the Ottomans, Safavids, and Mughals turned out to be nothing like as barbarous as so many Europeans supposed. A ruler such as Mehmed II, the conqueror of Constantinople, on whom Voltaire heaped unstinting praise, was widely perceived as a man of great culture and was in some sense the true bearer of values of the ancient world—values that the Christians, epitomized by the squabbling rapacious Crusaders, who were the real cause of the collapse of the Byzantine Empire, or the corrupt, decadent Greeks, had long since abandoned. It was Mehmed, Voltaire pointed out, who set up an academy in Istanbul, where ancient Greek, which the modern Greeks had all but forgotten, was taught, together with "the philosophy of Aristotle, theology, and medicine."[62]

On this point, at least, Anquetil-Duperron and Voltaire were of one mind. Exasperated by the simplicity of the traditional image of the Oriental state, in 1778 Anquetil-Duperron wrote a treatise entitled *Oriental Legislation* to demonstrate, as he put it, that "the manner in which despotism has been represented until now" in Turkey, Persia, or India, "cannot fail to give anything but an absolutely false image of the government of these places." So much of what had been written about the East, he complained, derived not merely from a misunderstanding but from a tendency to describe ills that were common enough the world over, or even the outcome of natural causes, to the will of men or, as he put it, to "government." "All that is wrong in Asia," he wrote,

> is always as a result of the government. Locusts have dev-
> astated one canton; war has depopulated another; lack of
> rain has created a famine that forces a father to sell his
> child in order to live. (I saw that in 1755 in Bengal.) Once
> again it is the government. The traveler writes his work in
> Paris or London or Amsterdam, where one can say any-
> thing against the Orient. The same troubles in their own
> country they attribute to the sky or the malice of men.[63]

Voltaire agreed. What, he asked time and again, would a Persian or Turkish chronicler have made of the European feudal system? Does that seem anything less like the possession of the subjects by their sovereign than what actually prevails among Muslims? On any closer examination, not only did all Oriental societies differ quite markedly from one another—as France, say, differed from Venice—but even the most extreme of them—that is, as most were agreed, the Ottoman Empire—could not really be described as "despotic." It was, declared Voltaire, absurd to suppose

> that the people are the slaves of the sultan, that they own
> nothing of their own, that their goods and they them-
> selves are the chattels of their master. Such an administra-

tion would destroy itself. It would be bizarre if the Greeks who have been conquered were not slaves, yet their conquerors were.[64]

One of the strengths (as well as, ultimately, one of the weaknesses) of the Ottoman Empire, was, in Voltaire's view, that neither the Turks nor the Arabs, unlike the Romans, had attempted to transform the world into a single nation. Nor had they, as the Mongol hordes in China had done, simply allowed themselves to be absorbed by the cultures they had conquered. What they had done was to create societies that, far from being the grim satanic places that appeared in most Christian accounts of them, were in fact welcoming, tolerant, and ecumenical. All you had to do to rise to the highest positions of power was to convert to Islam. The Ottoman sultan was indisputably a supreme ruler, but he ruled through local potentates. One visitor to the Sublime Porte, the Italian adventurer Luigi Ferdinand Marsili, who had spent some time in Istanbul in the early eighteenth century and written a treatise, *The Military State of the Ottoman Empire*, had concluded that the sultanate was indeed despotic but that it was in effect a democracy, rather than a monarchy, since as far as he could see, it was the Janissary Corps that was the ultimate source of power and each region was governed by its own local ruler.[65] Once the conqueror had triumphed, the government was restored to the conquered.[66] As for Persia, which, as the direct heir of the Achaemenid Empire, was in all respects more civilized than either Turkey or India, "there is no monarchy," declared Voltaire, "where there are greater human rights" (*droits de l'humanité*).[67] On this account the Ottomans and the Safavids come out looking much more like the heirs of Alexander, "the monarch of each nation and the first citizen of each town," than that of Xerxes.[68]

Many of these claims were merely provocative, aimed at European rulers whom men like Voltaire held to be quite as despotic as their Eastern counterparts. Yet even if the great empires of the East were not truly despotic—or at least no more despotic than most European monarchies under the *ancien régime*—even if they had until very recently been more than a match both militarily and culturally for most of the societies of Europe, they had also by the eighteenth century begun visibly and seemingly inexorably to wither away. The force that had driven the Arabs out of the Arabian Peninsula and across the world as far as southern France, that had led the Ottoman Turks from the mountains of Anatolia through the centers of Asian and European civilization, until by the seventeenth century Süleyman the Magnificent had outpaced even the ambitions of Xerxes and come to a halt just beyond the Danube, seemed now to be all but spent. The Arabs had borrowed from the Sassanids something of their political organization and from them and the Greeks most of their military technologies. The Turks seemed to have learned a great deal from the Persian and later Byzantine societies they had overrun. They had built up a centralized government machine second to none, with archival resources unheard of in Europe at the time. They had adopted Greek architectural techniques and Greek decorations. Yet, in the eyes of most European observers, neither people had apparently learned enough to move far beyond the raid-and-plunder stage of human history. For all their immense initial success, they were stagnant. Their military achievements, awe-inspiring though they had been, had not been accompanied, as had the conquests of the Greeks and Romans, by any corresponding development in the arts and the sciences. All they had ever done was to acquire more territory. The Ottoman state had fulfilled the ambitions of its Achaemenid predecessors by much the same means as they. The modern Turks knew nothing, Voltaire pointed out, about modern economics, extraordinary taxation, or advanced loans. They were not haunted by the specter of public debt or state banking. "These potentates," he said of the sultans, "know only how to accumulate gold and precious stones, as they have done since the day of Cyrus."[69] Safavid Persia might in part be an exception to this rule but even there the sciences that had once been equal to the Greeks' had perished "by the changes in the state."[70] To read Jean

Chardin's *Travels* was, said Voltaire, to "imagine an account from the time of Xerxes."[71]

But why had this happened? If in the sixteenth century the Ottomans, like the caliphate and the Safavids, had been more advanced than the Christian states of Europe, why had they failed to maintain their advantage? The simplest, most compelling answer was and would remain: religion.

Europe had succeed in resisting the Church's attempts to encroach upon the government of mere mortals. In the sixteenth century, as we have seen, internal struggles within the Church had stripped it of most of its authority in any realms other than the purely spiritual. Secularization was only partial, and it remains only partial to this day. But it was sufficient to guarantee the development of an independent scientific culture. In the Islamic world, however, things were very different. For Islam was the basis of the civil law. It had been the supreme instrument employed by Muhammad, in his role as despot and skilled military leader, to keep an unruly and belligerent people in check. To continue to perform its function, it had to be exempt from interpretations of the kind to which the civil law in the West was subject. Such interpretations of the Qur'an as did exist came down to being, as Voltaire said, "little more than a recommendation not to dispute with the wise." Even the word "Islam" itself, as he pointed out, signified resignation and acceptance of the word of God.[72]

The history of the westward march of civilization that Voltaire had set out to record in *Essay on the Manners and Spirit of Nations* was now complete. All of Asia, a vaguely conceived classical term given merely to the lands beyond the Hellespont, had acquired something of a common identity. Its peoples were immensely varied—which every European with even the most scanty understanding of the place would agree. But they had one thing in common: they were all, in their different ways, ruled by despots, enthralled to systems of government upheld by religions whose objective was to persuade the masses that neither nature nor any of their gods offered them any other way of living. They were societies composed of hordes, not individuals. And as long as they remained immured behind their self-imposed walls of ignorance and apathy, nothing could help them. For them, time and progress had little meaning; the truth of everything, including what Europeans looked upon as science, could be established only by reference to the past.

Much the same was also said about the other great civilization of the "East," China. China was in many respects thought to be wholly unlike its western Asian neighbors. But to European eyes it displayed at least some of the same properties: immobility, stagnation, an unwillingness to live by anything other than ancient inherited laws. It was, too, a despotism—a benign despotism in the eyes of some, it is true, but a despotism nonetheless. And without freedom of choice and freedom of expression, the kinds of scientific and economic achievements that had, by the mid–eighteenth century, allowed the West to dominate so much of the planet were unthinkable. "Behind its mountains," wrote Johann Gottfried von Herder, "orbicular China is a uniform and secluded empire; all its provinces, however different its peoples, governed by the principles of an ancient constitution, are in a state not of rivalry with one another, but of the profoundest obedience."[73] For the great twentieth-century social theorist Max Weber, this was the reason that China, like India and the Muslim East, had failed to achieve the economic takeoff experienced by northern Europe in the sixteenth century or to devise their own version of the industrial revolution in the nineteenth.[74]

The whole of Asia, from the Bosporus to the South China Sea, had its face turned resolutely toward the past. As Henry Sumner Maine reflected in 1881, in all "those great and unexplored regions we vaguely term the East . . . the distinction between the Present and the Past disappears."[75]

## IV

By the early nineteenth century, it had become clear to all but the most pessimistic observers in Europe that sooner or later, the various societies of Asia would have to perish or surrender themselves to the West. In the winter of 1822–1823, the great German philosopher Georg Friedrich Hegel gave a series of lectures on the philosophy of history at the University of Berlin. His purpose was to set before his students a picture of the forward march of "reasoning," of the ever onward progress of the human spirit, which, he said, like the sun itself, moved inexorably from "the *Orient* quarter of the globe—the region of origination," to the West, "where in the ages that lie before us, the burden of the World's History shall reveal itself."[76] In Asia, the world was static. The Muslim East had made no progress since the fall of the caliphate, while "the Indian nation, which we classify as belonging to the beginning of the world, is a static nation, like the Chinese. That which is now is as it has always been." In none of these places was there, nor could there ever be, what he called "progress to something else." Because of this, in India "the English, or rather the East-India Company are masters of the land; for it is the necessary fate of Asiatic Empires to be subjugated to the Europeans; and China will, some day or other, be obliged to submit to this fate."[77]

Hegel could afford to be confident—not merely because he was Hegel but also because long before he had come to these damning conclusions, the balance of power between Europe and the Islamic societies of Asia had begun to change dramatically in favor of the West. The Ottoman Empire had been in retreat since the 1780s. The Hasharid dynasty in Iran had fallen into evident and bewildering disarray after the assassination of Nadir Shah in 1747. The great Mughal Empire had effectively been in the hands of the British East India Company since the battle of Plassey in 1757, when the forces of the "heaven-born general" Robert Clive had defeated the nawab of Bengal and seized control of most of northwest India. Even "orbicular China" had, under the Qing emperors, been obliged to open up to European merchants—and European missionaries—so that by the end of the nineteenth century much of its trade was in European hands. To many, the collapse of the ancient orders of Asia and their absorption by the "West" seemed imminent. It would be the final realization of Alexander's ambition, thwarted over centuries. All it required now was someone with the vision, and the power, required to bring it about.

It was in this atmosphere of lingering possibility that late in 1782, Constantin-François Chasseboeuf, a restless twenty-six-year-old from Mayenne in northwest France, set out on what would be a highly significant journey to Egypt and Syria. Chasseboeuf had spent the past few years in Paris studying medicine and Arabic and frequenting the coterie of the notorious atheist and materialist Baron d'Holbach. At some point he had decided that his family name (which translates roughly as "beef prodder" or "beef hunter") was inappropriate for the literary career he was planning for himself and in honor of Voltaire had adopted the sobriquet "Volney," made up of the first three letters of his hero's own *nom de plume* and the last three of "Ferney," the name of Voltaire's château in Switzerland. Volney would go on to become a major, if somewhat transitory, figure on the postrevolutionary literary stage in France.

In December 1782, the now renamed Constantin-François Volney left from Marseille, bound for Egypt. In January of the following year, he reached Alexandria. His first encounter with the East was quite literally shocking. It was, he later recalled, wholly unlike anything he could ever have imagined. No amount of reading, no attempt at imagined reconstruction of what the "appearance of the land, the order in the towns, the dress and customs of the inhabitants" might be like could possibly prepare the European traveler for what he would find. "Everything he could have thought dissolves and vanishes, and he is left only with feelings of surprise and wonder"—and, he would later discover, horror and disgust.[78]

After a few weeks Volney left Alexandria for Cairo, a city that looked to him less like the European capitals to which it was commonly compared than something from the tenth century, a low, disordered place, dust-ridden and with unpaved streets where a mass of camels, donkeys, dogs, and humans thronged together.[79]

Here he stayed until September, studying the peoples, the agriculture, the effects of wind and water, the nature of the various religious and social groups, and the diseases that afflicted them. In September, after having paid an obligatory visit to the Pyramids—which impressed him less by their size and grandeur than they "pained him to think that in order to build a useless tomb, it had been necessary to torture an entire nation for twenty years"—he departed for Syria.[80] He then traveled to Jaffa and Acre, Tyre and Beirut, Aleppo, and Tripoli. He spent some time in Damascus and from there made his way to Jerusalem, Bethlehem, Jericho, and the Dead Sea before leaving once again for Alexandria and then for France.

On his return in 1787, Volney published an account of his journey, *Travels Through Syria and Egypt in 1783, 1784, and 1785*. If the title was prosaic and matter-of-fact, the content, with its harrowing descriptions of the misery of life in Egypt and its soaring accounts of the ruins of the once-great civilization of the Near East, was anything but. It was also quite unlike most travel accounts of the time in being a purposefully scientific description of the region and its peoples, driven, or so Volney imagined, by an "impartial love of the truth."

One moment in Volney's journey would prove crucial both to Volney's own future intellectual development and more sweepingly to the entire history of Western perceptions of the East. In 1784, after three arduous days' march "in arid solitude where all about I saw nothing but brigandage and devastation, tyranny and misery," he had come across the ruins of the city of Palmyra. There he sat down on a column, much as Gibbon had done two decades earlier in the Capitol in Rome, rested his head in his hands, fixed his gaze upon the desert, and meditated on the rise and fall of civilizations. Volney's thoughts, like Gibbon's, dwelt on contrasts—in his case the contrast between the former grandeur of Palmyra, capital of the warrior queen Zenobia, and the desert, "grayish and monotone," that now surrounded it. He was conscious that what lay about him was all that remained of a once-great civilization that had been Roman, Parthian, and Sassanid before falling into Ottoman hands and subsequently into ruins. He wanted to know not only why empires rise and fall but the ultimate cause of the prosperity of nations and "on which principles the peace between men and happiness of societies might be established." In pursuit of these elusive goals his mind wandered off to the sources of all the civilizations of the ancient world, both East and West—to Nineveh and Balbeck, Babylon and Persepolis, Jerusalem, Sidon, and Tyr. All now were in ruins.

Thus caught up in his meditations, a thought struck him, "which threw my heart into trouble and uncertainty." Each of these places, "when they enjoyed all that of which the glory and the well-being of man is made up," had once been inhabited by "infidels," peoples who, although like the Phoenicians they venerated the homicidal Moloch or prostrated themselves before serpents or worshiped fire, did not imagine their deities to be the only ones or that peoples who knew nothing of them should be bound by the same beliefs or laws as they. These infidels had been great empire builders. In time, however, they had all been replaced by the avatars of one or another of the world's great monotheistic religions—Christianity, Islam, or Judaism—and in time all that they had made had crumbled away. Once, under its "infidel" masters, the city of Palmyra and the oasis on which it had been built had been rich and fertile. But now, Volney observed sarcastically, "that believers and saints occupy these lands, there is nothing but sterility and solitude."[81] Yet did not these saints and believers claim to be "the elect of heaven, covered in grace and miracles?" Why, then, had such privileged peoples proved unable to enjoy the same blessings as the damned and despised infidels?

This unanswered, unanswerable question led him to reflect that it was clear that what he called the "scepter of the world" had now passed from ancient Asia to modern Europe. It was a reflection that made him tremble. For although "it pleased me to recover the past splendor of Asia in Europe," the vision of so much destitution also made him wonder if one day future travelers might not find, beside the banks of the Seine or the Thames or the Zuider Zee, just such "mute ruins" as now lay around him, and they, too, would "weep alone amid the ashes of the peoples and the memories of the greatness," as he did now.

Could this dismal process, this constant rise and fall of civilizations, be arrested? Volney asked. Could the scepter of civilization remain where it now was, among what, in his opinion, was clearly the most advanced people on Earth? Could History, which Aristides had once thought had come to its conclusion in Rome, now be brought to an end in the modern West? Volney thought it could. Modern civilization, he believed, could not now fail to draw men closer together. Soon "the entire species will become one great society, one and the same family governed by the same spirit, by common laws, and all enjoying the same happiness of which the human race is capable." And then, at last,

> The earth will give rise to a supreme power. The earth awaits a people to legislate for it. . . . A cry of "liberty" raised over distant riverbanks has echoed across the ancient continent. . . . A new century is about to dawn, a century that will astonish the masses, surprise and affront the tyrants, that will emancipate a great people and bring hope to the entire earth.[82]

And from whence would this come? For Napoleon Bonaparte, who declared Volney's book to be the only one ever written about the East "that has never lied," the answer was obvious.[83]

# 10

## *THE MUHAMMAD OF THE WEST*

### I

For Volney, musing amid the ruins of Palmyra, all Asia seemed to be in a state of ruin. It had a past but neither a present nor, as yet, a future. For more than two centuries the simplicity and ferocity of the Ottomans had driven all before them. By 1780, however, it was clear that the seemingly invincible empire of the Sublime Porte was in precipitous and irreversible decline. Nothing could save it now from ultimate disintegration. Like all empires, that of the Turks had been created by forcing together diverse peoples. In Europe it was also widely believed that, in the Ottoman case, they had been held together through fear and repression. The servitude to which the Greeks, the Egyptians, the Bulgarians, the Croatians, the Serbs, some Hungarians, and the heirs of the older, grander Arab caliphates in Syria and what is today Iraq had been subjected was the cause of their decline. Free them, the enlightened voices of Europe cried, and they might again reconstitute the great nations they once had been. "The empire of the Crescent," Volney predicted, would soon suffer the ultimate fate of all despotic states, and "the peoples of the empire, released from the yoke that bound them together, will recover their former identities." All that was now needed was, Volney declared, "a virtuous leader" and "a people powerful and just" to carry out the task.[1] But who would that leader be, and which the people?

By the mid–eighteenth century the armies of the Ottoman Empire, despite attempts at modernization, had suffered severe setbacks, in 1718 and again in 1730, at the hands of the Iranians. From the West its old enemies, Austria and Russia, hungry Nordic wolves, began to close in on the stumbling giant. In 1768 the sultan, egged on by France to defend the Polish Commonwealth, declared war on Russia. The consequences for the Ottomans were disastrous. The victorious Russian armies marched eastward through the Balkans, crushing the Turks at Khotyn in the Ukraine and then again in 1770 at Kagul on the Prut in modern Romania. In 1770, a Russian fleet sailed into the Mediterranean to help their Orthodox coreligionists rise against their Ottoman rulers. Revolts broke out in Montenegro, Bosnia, Herzegovina, and Albania. On July 5, virtually the entire Ottoman navy was set ablaze in the waters off the port of Cesme, near Izmir, and some 5,000 Turkish seamen lost their lives. It was a disaster that jubilant Christians compared to Lepanto, and for a while it even seemed as if the Russians might get all the way to Istanbul itself. Later that same year a Russian army invaded the Crimea. Then on July 21, 1774, after two disastrous battles at Suvorovo and Shumen, south of the Danube, the Sultan was forced to sign a treaty with the czar at Küçük Kaynarca whose terms were even more humiliating than those of the Treaty of Carlowitz seventy-five years previously. The Russians were given free access to the Black Sea and the Mediterranean and the right to intervene directly in the affairs of the Ottoman state on behalf of Russian Orthodox Christian communities in the empire. The treaty also obliged the Ottomans to pay a war indemnity spread over the next three years.

One of the more curious consequences of the treaty was the re-creation of the Ottoman sultan as "sultan-caliph." Although all the Ottoman sultans since Süleyman the Magnificent had consented to be addressed as caliph, none had formally adopted or made very much use of the title, and there had never been any attempt to redescribe the Ottoman Empire as a new caliphate.

At the suggestion of the French ambassador to the Porte, François Emmanuel Guignard, comte de Saint-Priest, the treaty included a clause claiming for the sultan spiritual jurisdiction over all Muslims both outside and within the borders of the Ottoman Empire, who were now to "conduct themselves toward . . . the Supreme Mohammedan Caliph . . . as is prescribed in the rules of their religion." No such rules, of course, existed. But by means of a wholly misleading analogy between caliph and pope, the sultan had been made the formal guardian of the entire Islamic world—and by his Christian enemies. It was a position that subsequent sultans accepted happily, and versions of Saint-Priest's clause were inserted into several later treaties. In 1808, Mahmud II inaugurated the custom that the sultan, at his accession, was ceremonially girded with the sword of the Caliph Umar; and the Ottoman constitution of 1876 declared that "His majesty, the Sultan, as Supreme Caliph, is the Protector of the Muslim Religion." All of this led to the creation of a powerful political movement known as "pan Islamicism," by which not only the Ottoman sultans but a number of Arab and even Indian rulers aspired to reunite the scattered Muslim world in the hope of reversing the crushing defeats it had suffered at the hands of its Western enemies.[2] The assumption that the caliph was a kind of Muslim pope was, however, as we shall see, to have grave consequences for the future of the sultan-caliph's relationship with his Arab subjects after the outbreak of the First World War in 1914.[3]

The treaty of Küçük Kaynarca had also stipulated that the sultan recognize the "independence" of the khans of the Crimea. It soon became clear, however, that this had been intended only as a preliminary to full annexation. This duly followed in 1783. For the Turks, as Volney commented later, the Russian occupation had "introduced their enemy into the heart of their empire; established him at the gates of the capital." It had inflicted on the Ottomans "the misery of the humbling of an ancient grandeur."[4]

To the other European powers, which waited anxiously to see what the consequences might be for them of the end of the Russian-Turkish conflict, the Treaty of Küçük Kaynarca and the subsequent loss of the Crimea looked very much like the beginning of the Ottoman Empire's final days. From Istanbul, Saint-Priest, urged his king to take an active part in dismembering the empire, before the Russians, the Austrians, and, worst of all, the British got there first. His eyes fell on the richest, most exposed, and most western of the Turkish provinces: Egypt.

For centuries Egypt had lain at the frontier between East and West. The ancient civilization of the pharaohs had provided the basis for much Greek science, and Egyptian deities and architectural motifs had found their way into every aspect of Western civilization. "Egyptomania" had been a recurring, if persistently suspect, theme in Roman relationships with the East, and the image of Egypt as a land of ancient, arcane, and sometimes terrifying power has lasted to this day. The country, however, rather than the semi-mythical land of the pharaohs, had, since the fifth century B.C.E., been a land of conquests, washed over by wave after wave of immigrants from both East and West. It had been ruled first by the Achaemenids, then briefly by Alexander, and after him by his Ptolemaic heirs; then by the Romans, the Arabs, and finally the Turks. By the 1770s, it was under the sway of the Mamluks, Turkish and Circassian slave soldiers who had been imported into Egypt under the Ayyubid caliphs in the twelfth century. By 1250, however, the Mamluks had overthrown their former masters and created a military oligarchy, and it had been they who, in 1291, had finally driven the Crusaders from Acre and thus from the entire *dar al-Islam*. When Egypt was conquered by the Ottomans in 1517, they had been reconfirmed in power, but now under the suzerainty of a pasha appointed from Istanbul. By the mid–eighteenth century, however, Ottoman authority had so far diminished that the pasha was kept a virtual prisoner in Cairo and the tribute the sultan was due arrived only fitfully.

The country was divided into twenty-four provinces, commanded by a number of beys (Ottoman regional governors), who fought constantly among themselves. In 1776, two of them, Ibrahim, who had charge of the administration, and Murad, who commanded the army, formed an alliance and succeeded in ousting all the others. In 1786–1787 the government in Istanbul made an attempt to bring the beys to heel, but with little lasting effect. It was clear to all the waiting Europeans that the sultan now no longer possessed the power to control the quasi-independent feudal armies through which many of the outlying provinces of the Ottoman Empire had traditionally been ruled.

In 1797 the Directory—the five-man executive council that had ruled France since October 1795—sent an infantry captain, Joseph-Félix Lazowski, on a fact-finding mission to Turkey. When he returned in January the following year, he submitted a report that urged the French government to seize both Egypt and the Greek islands. The excuse, he suggested, should be the (supposed) Mamluk harassments of the local French merchants who had been operating in Egypt and throughout the entire Ottoman Empire since the early seventeenth century. It was a lame and self-serving excuse, but, as Charles-Maurice de Talleyrand-Périgord, the minister of foreign affairs of the republic, pointed out, at this stage the French hardly needed an excuse. The mission to spread the values of the revolution—and unasked-for assistance—to their Ottoman allies was sufficient.

Into all this mounting interest in the possible fate of the Ottoman Empire came Napoleon Bonaparte. Ever since he was a boy, Napoleon had had his eyes fixed on the East. He was, as he said of himself, a fervent "Orientalist." On the eve of the revolution, while biding his time in Corsica, he had made detailed notes on the Abbé de Marigny's *History of the Arabs* and Baron de Tott's *Memoirs of the Turks and the Tartars*. He had also written a brief "Arabic tale," "The Masked Prophet," a story of a revolt by an impostor named Hakim against the rule of the early Abbasids. It is hardly great literature, and most of it is cribbed directly from Marigny, but with hindsight some of it sounds almost prophetic.

Napoleon describes Hakim in terms very like those of the persona he would himself later adopt in Egypt. He was a man "who said he was the envoy of God, who preached a pure morality that was pleasing to the multitude: the equality of all ranks and fortunes was the normal content of his sermons."[5] In 1795, while Napoleon/Hakim was an officer without a posting, impatient and angry at the apparent refusal of the French government to promote him, Napoleon had seriously contemplated going alone to Turkey and had written to Volney for advice. ("If circumstances assist him only a little," Volney had remarked on meeting the young officer "this will be the head of Caesar on the shoulders of Alexander.")[6]

There was a more serious side to this fascination. The Orient was a large and still largely unknown field in which all of Bonaparte's vast ambitions might be fulfilled. "Europe is too small for me," he is once said to have exclaimed. "I must go East." And the route to the East lay through Egypt. The abbé Guillaume Raynal's *Philosophical and Political History of the Two Indies,* one of the fiercest condemnations of European colonialism to appear during the eighteenth century, was one of Napoleon's favorite books, and in one of the many extensive notes he kept of his reading he had copied the following passage:

> Egypt is situated between two seas, in reality between the East and the West. Alexander the Great had intended to make his capital there and make of Egypt the center of the commerce of the whole world. This enlightened conqueror had understood that only through Egypt, which linked Africa, Asia, and Europe, would he be able to unite all his conquests into one state.[7]

Napoleon had always fancied himself a new Alexander. He could also see both the symbolic and strategic importance of Egypt as a base from which to harass the English in the eastern Mediterranean. It might also be a bridge to a new empire in Asia.

Napoleon could imagine a French expedition that would follow Alexander's route from Egypt to Syria, then Iran and Afghanistan, and finally drive the hated English out of India. It would be a fitting recompense for the humiliations inflicted on France at the end of the Seven Years' War in 1763. In August 1797, four months after negotiating the terms of the treaty of Campo Formio with Austria, which gave him control of most of northern Italy, the Austrian Netherlands, and the Ionian islands and, in effect, brought to an end the war in continental Europe, Napoleon wrote to the Directory, "The Turkish empire crumbles day by day. . . . The time is not far off when, if we really wish to destroy England, we will have to take possession of Egypt."[8]

The answer to Volney's question was now, at least, clear. It was Napoleon and the French who would lead the scattered and demoralized peoples of the Ottoman Empire out of their servitude to constitute new nations (under, of course, French tutelage).

On February 9 of the following year, Charles Magallon, the French consul general in Cairo for more than thirty years, sent a memoir to Talleyrand. His long experience in Egypt, he wrote, had taught him that a French conquest would be enormously advantageous and "would present no inconveniences." The previous French attempt to invade the Muslim East—the Crusades—he said, had really been prompted by Christian designs on the commerce of Asia: "Religion served as a pretext for the politician." Their failure had been due to the Crusaders' incompetence. "With some prudently managed successes," Magallon concluded, "one would have seen European colonies formed on the shores of Egypt and Syria."[9] That opportunity had been squandered. This new one should not be.

Talleyrand agreed with him and five days later drew up a full invasion plan. France, he announced, was now going to reverse history in the East as it had done in the West. "Egypt," he told the Directory, "had been a province of the Roman Republic; it should now become a province of the French Republic." The Roman conquest, he added somewhat inconsistently, had resulted in a period of decadence for that beautiful country; "the French conquest will lead to one of prosperity."[10]

On March 5, 1798, Napoleon was given command of the Army of the Orient, whose task was to invade Egypt and, in the name of the Ottoman sultan, to depose its Mamluk rulers and—somewhat vaguely, since this would seem to deny the French pretense of being allied with the sultan—to establish a French colony. Once this had been achieved, Napoleon was instructed to chase the British out of the Red Sea and seize control of the Isthmus of Suez and then, circumstances permitting, to march eastward and drive the British out of India.

Strategic concerns were not, however, Napoleon's only ones. Ostensibly at least, the Army of the Orient was bound not on an invasion but on a rescue mission. Napoleon had been instructed by the Directors "to improve by all the means at his disposition the lot of the natives of Egypt." Historical analogies are always hazardous, but if any Western military and cultural "mission" to the East foreshadowed the difficulties and the disasters that most subsequent ones would encounter, from the British "occupation" of Egypt from 1883 to 1956 to the American-led invasion of Iraq in 2003, it was this one. It was, as the First Lord of the Admiralty, Lord Spencer, observed sardonically in June 1789, a plan "so chimerical and romantic that little credit was given to it."[11] Most of its successors have been decidedly less romantic, but none have been less chimerical.

Unlike some later Western conquerors of the Muslim East, from General Sir Garnet Wolseley in Alexandria in 1883 and George W. Bush—or the architects of his foreign policy—in Baghdad in 2003, Napoleon in 1798 set out to achieve his strategic political objectives not merely by military force but by the manipulation, as he saw it, of culture and religion. He went to Egypt in order to secure what, had he succeeded, would have been a vital French colony, one that would, in all likelihood, have broken Britain's ties with India and with them Britain's growing maritime empire. But Napoleon, and those who traveled with him,

also went to bring to the peoples of Egypt the liberating message that the revolution had brought to France and that Bonaparte was in the process of exporting to the rest of a somewhat
reluctant Europe.

Napoleon's objective, declared the philosopher and mathematician Jean-Baptiste Fourier, who accompanied the expedition,
had been to "render the condition of the inhabitants [of Egypt]
more gentle and to bring to them all the advantages of a perfect
civilization." Such an objective, however, could not have been
achieved merely by arms or persuasion, or even by legislation; it
also required the "constant application of the sciences and the
arts."[12] To this end, the Army of the Orient would carry with it not
only arms and munitions but a thousand-book library containing
copies of the classics of Western literature, as well as, in Napoleon's
own words, "the elite of our moralists and our novelists": copies of
Montesquieu and Rousseau, Voltaire and Montaigne, as well as
the Qur'an (in both Arabic and French) and—just in case Napoleon
made it all the way to India—the Vedas. Traveling with the army
was also an entire scientific academy, the Institut d'Égypte, an institution "dedicated to the progress of all useful knowledge"
whose activities would, in the long run, prove the most enduring
legacy of the entire venture.

Napoleon went to Egypt to finish, in his own inimitable way,
what Alexander had begun. Happily, he was at the time twenty-
nine years old, the same age Alexander himself had been when he
had conquered Egypt and established his capital in the city that
bore his name and his tomb. Later, in a dialogue that Napoleon
claimed took place at the Great Pyramid of Cheops between himself and "several muftis and imams," he had the chief mufti of
Cairo address him as "noble successor of Alexander."[13] And long
after the great adventure was over and he was in exile on the island
of Saint Helena, he told Madame Claire de Rémusat:

> In Egypt I created a religion. I saw myself on the high road
> of Asia, mounted on an elephant, a turban on my head,

and in my hand a new Qur'an that I would have written as
I pleased. I would have united in my undertakings the experiences of two worlds, searching for my ends through all
the histories of the world.[14]

If this was not merely the whimsical recollections of a conqueror
whose glories were well behind him, we may be fairly certain that
he also had something like this in mind at the beginning of 1789.

From February to April, an army of more than 30,000 men,
and the fleet needed to carry it, was assembled along the Mediterranean coasts of France and Italy and then transported, together
with munitions and supplies, to the port of Toulon. Meanwhile,
in Paris, Napoleon's agents recruited a formidable collection of interpreters, artists, poets, architects, economists, astronomers, anti-
quarians, draftsmen, mineralogists, botanists, zoologists, chemists,
engineers, a sculptor, twenty-two printers, a balloonist, and "one
ex baritone from the Paris Opera"—167 in all. Most of them were
members of the Institut National. The Institut had been created
in 1795 to replace the Académie Française and the Académie des
Inscriptions, both of which had been royal foundations. In 1797,
Napoleon had himself been made a member, although on quite
what grounds it is not clear—certainly not for his services to literature. It was something of an empty gesture, but Napoleon was
immensely flattered and never failed to mention "Membre de l'Institut" among his titles, even after he had appropriated the far
grander one of "Emperor." Academic institutions in the eighteenth century still carried great cachet, and Napoleon had always
wanted to be remembered as something more than a brilliant military commander.

These "savants"—the wise men, as they would come to be
called—were enticed, persuaded, or bullied into forming the Institut d'Égypte.[15] A few, such as the linguist Louis-Mathieu Langlès,
who "knew every language spoken on the Tower of Babel," held
out, never to be forgiven by Napoleon; but the majority acqui-

esced willingly enough. The great German scientist and explorer Alexander von Humboldt, however, declined.

The objectives of the Institut were, admittedly, somewhat vague. It was there, clearly, to collect information and, where possible, things. The great museums of Europe and the United States are filled with ancient objects carried away from various parts of the Ottoman Empire in the eighteenth and nineteenth centuries by eager, and often amateur, travelers. For the Ottomans, in particular the local governors in Greece, Iraq, and Syria, short both of cash and concern for the fate of these unwelcome reminders of the non-Muslim past, such trafficking was a valuable source of additional income. But that was by no means the Institut's sole objective. For archaeology was not simple looting, nor was it the disinterested academic study it tends—or at least tries—to be today. In the eighteenth century, it had a clear political and cultural purpose. Just as William Jones, Anquetil-Duperron, and the other Orientalists had hoped to find the sources of European civilization in India, the members of the Institut went looking for them in Egypt. This was the quest on which Volney had embarked on his voyage into Syria and Egypt between 1783 and 1785. He had gone to the East, he told the readers of his *Travels Through Syria and Egypt,* because these were the places where

> most of the opinions by which we govern our lives were born. It is there that the religious ideas that have had such a powerful influence on our private and public conduct, on our laws, and on our entire social condition originated. It is interesting, therefore, to know the places where these ideas had their birth, the customs and habits of which they were made, the mind and the character of the peoples who established them. It is interesting to examine to what degree that mind and those habits and customs have been changed or preserved; to investigate what might have been the influence of climate, the effects of government, or the

consequences of habit upon them—in short to judge by
their present condition what they might have been like in
the past.[16]

Volney himself played no direct role in the expedition, but it was
as much his inspiration as it was Napoleon's project.

## II

On the morning of May 19, 1798, the fleet of nearly three hundred
heavily laden vessels weighed anchor and slowly moved out into
the Mediterranean, bound for the westernmost rim of Asia. The
flagship, aptly named *L'Orient* and the largest ship afloat at the
time, brought up the rear. On the quarterdeck stood Napoleon,
one foot on the gunwale, staring into his future.

On June 10, the fleet stopped off at Malta and, after a brief
skirmish, seized the island. Ever since 1530, Malta had been ruled
by an international group of armed clergy calling themselves the
Knights of Saint John. For more than two centuries they had
made a rich living by privateering raids against the Turks and
their allies in North Africa. Although by 1798 they had become
corrupt and decadent and now posed very little threat to the Ottoman Empire, they were still the sworn enemies of Islam and a
living relic of the Crusades, something Napoleon hoped to turn to
his advantage in Egypt. The French made off with the treasuries
of most of the churches on the island, together with some seven
hundred Muslim slaves from Tripoli, Algeria, Tunisia, Morocco,
Syria, and Istanbul itself. These unfortunates were formally given
their freedom, fed and clothed, and ushered aboard the flagship,
to serve as interpreters and spokesmen for the new, enlightened
regime that was now about to descend upon the Egyptian people.

Two days later, Admiral Horatio Nelson, who was then at
Messina, received news of the French occupation. Reckoning that
Napoleon must now be bound for Egypt, he set off in pursuit. But
in the days before radar or satellite positioning, naval encounters
were frequently hit and miss. In this case it was miss: the slowness

with which Napoleon's huge fleet was compelled to move worked
to his advantage. The British overtook the French during the
night of June 22–23, at a distance of twenty-two leagues, beyond
the range of any telescope, without even catching sight of them.
Five days later, making good speed in light seas, Nelson was in
Alexandria. Having no idea where the French might now be, he
sailed down the Syrian coast and then to Cyprus. When nothing
showed up there, he returned to Naples to take on food and water
and wait.

Meanwhile, on the morning of June 28 and wholly unaware of
what had happened, Napoleon reached Alexandria. "My Lord,"
wrote Koraim, the sheikh of Alexandria, to Murad Bey, "the fleet
that has just appeared is so vast that one cannot see where it begins or ends. For the love of God and the Prophet, send us fighters."[17] Murad promised help, but none came.

On the night of July 1, the French began to disembark.
Napoleon had already abandoned his first pious hope that the
Egyptians would welcome him as a liberator. But their resistance
was even fiercer than he had expected. So, too, was the heat. As the
Army of the Orient had chosen the hottest period of the year to
make its attack, the troops had to battle not only with the elusive
and deadly Egyptians in their floating robes but also with flies and
mosquitoes the likes of which they had never seen before and with
temperatures that in July can reach well into the hundreds
Fahrenheit, even at night. "The natives pressed around us fear-
lessly," recalled one dragoon, "our copper helmets shone in the
sun, our heavy cloth jackets and our leather trousers stuffed into
high boots formed a stark contrast with their light floating wool
garments so much better suited than ours to the boiling climate."[18] By the time they had waded ashore through the pound-
ing surf, most of the troops, drenched in salt water, were already
half dead of thirst. "The army had reached such a point," wrote
one of them, "that we had to find water or perish."[19] In the end,
however, sheer force prevailed against even the pitiless Egyptian
sun. By nightfall of the following day, Alexandria was in French

hands. The Islamic world's long isolation from its western neighbors had come brusquely to an end.

Napoleon himself disembarked. He may have abandoned any hope of a peaceful occupation, but, like many later European invaders of this part of Asia, he expected that sooner rather than later the civilian population would realize that he had come, as had Alexander, not to conquer but to liberate. Like most of his successors, he recognized that military success could be achieved—and, more important sustained—only once the conquered had come to accept the political, ideological, cultural vision in whose name the invader had come. He had carefully marked a passage in Volney's *Travels Through Syria and Egypt.* "Anyone who wished to seize control of Egypt," Volney had written, would have to fight three wars: one against England, one against the Ottoman Empire, "but the third, the most difficult of all, would be against those who make up the population of the country. This latter will result in so many losses that it should, perhaps, be considered an insurmountable obstacle."

Napoleon did not think it would be insurmountable. But in order to win over the people, he knew that the French had to avoid becoming the targets of what he called the "anathemas of the Prophet" and being cast as the enemies of Islam. With the help of the Orientalists he had brought with him, he set about adapting the principles of the revolution in such a way, he hoped, as to "win over the muftis, the ulema, the sharifs, the imams, so that they will interpret the Qur'an in favor of the army."[20]

While at sea, during the passage from Malta to Alexandria, Venture de Paradis, "Secretary-Interpreter of Oriental Languages for the French Republic," had sat in a cabin on *L'Orient* and drafted, at Napoleon's dictation, a "Proclamation to the Egyptians" both in Arabic and (in a further attempt to persuade the sultan that the French were there on his behalf) in Turkish. It is worth taking a closer look at this document, for it summarizes not only the French hopes for the "Orient" but also the ultimate failure of both sides to come to any approximate understanding of each other.

It began with a familiar Muslim invocation: "In the name of God the Merciful, the Compassionate. There is no God but God. He has no son nor has he any associate in His Dominion," which was intended to indicate clearly that the French were not Christians. It then went on to assure the Egyptian people that Napoleon Bonaparte, commander of the French army, "on behalf of the French Republic, which is based upon the foundations of liberty and equality," had not come to Egypt, as the Mamluks had put it about, "like the Crusaders," in order to destroy the power of Islam. Nothing, Napoleon assured his readers, could be further from the truth.

> Tell the slanderers that I have not come to you except for the purpose of restoring to you your rights from the oppressors, that I, more than the Mamluks, serve God—may He be praised and Exalted—and revere his prophet Muhammad and the glorious Qur'an. . . . And tell them also that all people are equal in the eyes of God and that the circumstances which distinguish one from other are reason, virtue and knowledge.[21]

Having thus done his best to conflate the principle of human rights—in a language in which there exists no obvious translation for the world "right"[22]—with what the Orientalists had persuaded him were the basic tenants of Islam, the man whom Victor Hugo would later describe as the "Muhammad of the West" continued:

> O ye Qadis [judges], Shaykhs and Imams; O ye Sharba-jiyya [cavalry officers] and men of circumstance, tell the nation that the French are also faithful Muslims and in confirmation of this they invaded Rome and destroyed there the Holy See, which was always exhorting the Chris-

tians to make war on Islam. And then they went to the island of Malta from where they expelled the knights who
claimed that God the Exalted required them to fight the
Muslims. [23]

(Napoleon, as Sir Walter Scott remarked contemptuously, never
tired of using, and abusing, "the inflated language of the East.")[24]

On July 2, each of the slaves Napoleon had rescued from
Malta was given a copy of the declaration and urged to spread the
good news.

It is hard to say how much Napoleon believed in all this. One
of his generals later told a friend of his in Toulouse that "we
tricked the Egyptians with our feigned love of their religion, in
which Bonaparte and we no more believe than we do in that of the
late pope."[25] But Napoleon's personal beliefs were largely beside
the point. The point was policy. Napoleon had always practiced
religious toleration because he knew that religious faiths could
make deadly enemies. Toleration, however, was one thing; credence, even respect, was another. It is indeed highly unlikely that
Napoleon had read much of the Qur'an he claimed to venerate. As
he told Madame de Rémusat, the only holy book that would have
been of any interest to him would have been one he had written
himself. He had, however, read Rousseau's *Social Contract,* the text
that more than any other had provided the ideological inspiration
for the revolution. And from the final chapter, entitled appropriately "On Civil Religion," Napoleon would have learned that
"no state has ever been founded without religion serving as its
base." Religion, even if it were merely the reflection of a childish
longing for security, an "exhausted error," in Rousseau's words,
had proved, over the centuries, to have one enduring value of
which Napoleon was acutely aware: a single faith, based, as
Rousseau would have it, upon "a belief in the happiness of the
just, the punishment of the wicked, the sanctity of the social contract and of the laws," was the only cement capable of holding a
society together. Whether it was true or not was immaterial. What

mattered was that it should be one and indivisible. For, claimed Rousseau, "Wherever theological intolerance is allowed, it is impossible for it not to have some civil effect; and once it does, the sovereign no longer is sovereign, not even in temporal affairs. Thenceforth priests are the true masters; kings are simply their officers."[26] It was in accordance with this general premise that Napoleon had opened the ghettoes in every Italian city he entered. "It was by making myself a Muslim," he said later, "that I established myself in Egypt, in making myself Ultramontane [a devotee of the papacy] that I won men's hearts in Italy. If I were to govern a Jewish people, I would re-establish Solomon's temple."[27]

In time the principles of the revolution, which had been enshrined in the civic education of the French people, would come to replace the "anathemas of the Prophet" in the hearts of the Egyptian people. By then, of course, the Egyptians would have been fully civilized. To make that possible, however, they had first to be shown how the two creeds—Islam and the doctrines of the revolution—could be made to resemble each other.

Just as most Muslims today have failed to be persuaded that Western social values can be made compatible with the Shari'a, so, too, were the Egyptians who confronted Napoleon. We know something of how they reacted to Napoleon's profession of love for Islam from the account of the first seven months of the occupation written by a member of the Dîvân—or Imperial Council—of Cairo named 'Abd al-Rahman al-Jabarti. Al-Jabarti was a well-read, perceptive man who was not unimpressed by French skills and technology (he was particularly taken by the wheelbarrow) and ungrudgingly admired French courage and discipline on the battlefield, which he compared, glowingly, to that of the mujahideen, the Muslim warriors of the jihad.[28] But for all that, he was a firm Muslim who could conceive of no good, no truth, that did not emanate from the word of God as conveyed by the Prophet.

He excoriated Napoleon's declaration for its language, its poor style, grammatical errors, and the "incoherent words and vulgar constructions" with which it was strewn and that often made nonsense of what Napoleon had intended to convey—all of which was no tribute to the skills of Venture de Paradis or those of the French Arabists in the expedition. But al-Jabarti reserved his most searing criticism for what he repeatedly described as French hypocrisy. The opening phrase of the declaration suggested to him not, as Napoleon had meant it to, a preference on the part of a tolerant nation for Islam but rather that the French gave equal credence to all three religions—Islam, Christianity, and Judaism—which in effect meant that they had no belief in any. Toleration, for a Muslim such as al-Jabarti, was as meaningless as it would have been for any sincere believer. It was merely a way of condoning error. The years when some kind of rapprochement between Judaism and its two major heresies might have been possible were long since past. There could now be only one true faith and any number of false ones. Napoleon could not claim to "revere" the Prophet without also believing in his message. The same applied to the Qur'an. You could not merely "respect" the literal word of God. You had to accept it as the only law, not one among many. "This is a lie," thundered al-Jabarti. "To respect the Qur'an means to glorify it, and one glorifies only by believing in what it contains."

Napoleon was clearly a liar. Worse, he was also the agent of a society that was obviously committed to the elimination not only of Islam but of all belief, all religion. The invocation of the "republic," al-Jabarti explained to his Muslim readers, was a reference to the godless state the French had set up for themselves after they had betrayed and then murdered their "sultan." By killing Louis XVI, the French had turned against the man they had taken—wrongly because their understanding of God was erroneous, but sincerely nevertheless—to be God's representative on Earth. In his place they had raised an abstraction, this "republic," in whose name Napoleon, who had come not in peace as he claimed but at the head of a conquering army, now professed to speak. Since for a Muslim there can be no secular state, no law that is not also God's law, the French insistence that it was only "reason, virtue and knowledge" that separated one man from another was clearly an absurdity. For "God," declared al-Jabarti, "has made some superior to others as is testified by the dwellers in the Heavens and on the Earth."

There are few things a believer, especially a believer in the fundamental sacredness of a script, dislikes more than a nonbeliever. To al-Jabarti the French seemed to be not would-be Muslims but atheists. The Muslims may have fought the Christians for centuries, but Christianity was still one of the accepted religions. Christians were one of the "Peoples of the Book" and Christ was still a prophet, not the last and greatest but a genuine emissary of God nonetheless. For a Christian to renounce his religion without becoming at the same time a Muslim was the worst of all crimes. The boast that the French had destroyed the Holy See and abolished the Knights of Malta was met with horror, not relief.

What al-Jabarti had rightly seen was that underneath all Napoleon's protestations was a profound distrust of any of the supposed words of God. The French were, he had experienced at first hand, indeed materialists, in that they sought to understand the world and to control it only through their own experience, something that, as he could see, they managed with alarming success. Despite his admiration for French technology and French courage, al-Jabarti could not conceive of a science that did not begin, and very largely end, with a divine revelation. In his view, the French were, as the pagan Arabs had been, atheists, "materialists," and as atheism and materialism were among the errors that Muhammad had been sent to the earth to correct, sooner or later, they would, like all those who had preceded them, either be converted to Islam or destroyed. The Egyptians had only to wait and remain steadfast in their faith, for all the technological novelties, the knickknacks, and the wheelbarrows the invaders might tempt them with.

Al-Jabarti was not alone in his indignation, nor was he the only one to recognize the ideological threat Napoleon's presence now posed for Islam. In a proclamation, composed in both Arabic and Turkish, the Ottoman sultan warned his subjects in Egypt of the new terrors the Europeans now carried with them:

> The French nation (may God devastate their dwellings and abase their banners for they are tyrannical infidels and dissident evildoers) do not believe in the Oneness of the Lord of Heaven and Earth nor in the mission of the intercessor on the Day of Judgment but have abandoned all religion and deny the afterworld and its penalties. . . . They assert that the books which the Prophets brought are manifest error and that the Qur'an, the Torah and the Gospels are nothing but lies and idle talk and that those who claim to be Prophets lied to ignorant people . . . that all men are equal in humanity and alike in being men, none has any superiority of merit over any other and everyone disposes of his soul and arranges his own livelihood in this life. And in this vain and preposterous opinion they have erected new principles and set laws and established what Satan whispered to them and destroyed the basis of religions and made lawful to themselves forbidden things, and permitted themselves whatever their passion desires, and have enticed into their iniquity the common people who have become as raving madmen and have sown sedition among religions and thrown mischief between kings and states.[29]

For the first time in its history, Islam was facing a quite unprecedented and largely unimaginable and unintelligible challenge. For centuries it had been an article of faith in the Muslim world that sooner rather than later the Shari'a would be proclaimed in every city from London to Vienna; the great cathedrals of Europe would, like those of Constantinople, be washed clean of their idolatrous images, their bell towers replaced by minarets, and the faithful around the globe would turn five times a day in prayer toward Mecca.

Now, however, not only did it look as if things were going in reverse, but the old enemy seemed suddenly to have shed his former identity. Instead of insisting that his religion was the only true one, he was now speaking of something entirely absent from any true religion, something to which both al-Jabarti and the author of the sultan's proclamation nervously alluded: the vision of a life that did not require obeisance before any god or his self-appointed representatives on Earth. Worse still, the whispered blandishments of Satan were insinuating that in this new, godless world, the masses, no less despised and powerless under Muslim rule than they had been under Christian, might indeed be able to achieve if not exactly "whatever their passion desires," then at least a life of some dignity and security and, above all, freedom to choose.

In every encounter, of which this was arguably the first, between a secularized European West and an Islamic East, from Egypt to India, what has stood in the way of any possible understanding has been much the same. Both sides hold that their values and, more fundamentally, their understanding of how the universe operates, applies equally to all humankind. But whereas in the West, that understanding is assumed to be something humans have arrived at for themselves by the application of reason and without any direct assistance from any deity, in Islam, the only universal truths—the only truths of any kind—have to come from the word of God. Christians did, and some Christians still do, hold similar views. But they had always had to struggle against an increasing tendency within Western society to restrict the role played by the deity and limit the authority of its self-appointed intermediaries. In Islam, however, history had moved in the opposite direction—or rather, in the opinion of Napoleon and his companions, had not moved at all.

Whenever anyone in Napoleon's entourage came up with a development that appeared to derive entirely from secular reason, the Egyptians would respond by asserting that all such things could be found in sacred scripture if only one knew how to look for them. In December 1789, a true dialogue of the deaf took place. One day, Napoleon was dining with members of the Ulema at the home of Sheikh al-Sadat. He told the sheikhs that under the caliphs the Arabs had cultivated the arts and sciences but that "today they lived in the most profound ignorance and that there now remained nothing of the knowledge their ancestors had once had." Al-Sadat replied to this indignantly by saying that they still had the Qur'an, which contained within it all knowledge. Napoleon then asked if the Qur'an could tell one how to cast cannons. "All the sheiks replied emphatically, 'Yes.' "[30] (Later, in 1883, an exiled Egyptian intellectual called Sayyid Jamâl ad-Dîn, known as al-Afghâni, suggested that such things as railroads, the modern principles of economics and taxation, and germ theory had all been foreseen in the Qur'an.)[31]


MOST OF THIS, however, seems to have been lost on Napoleon, as it has been lost on many subsequent generations of Western civilizers, down to and including the most recent attempt to introduce democracy, the formal modern equivalent of the principles of the French Revolution, by force or persuasion into various areas of the Arab world. Like so many of his successors, Napoleon also seems to have believed that the secret was perseverance. When the Egyptians proved to be reluctant to welcome the French, this was inevitably attributed to the overwhelming power of a perverse minority and to the inescapable ignorance of the masses. Immediately after the French invasion a resistance—an "insurgency," it would be called today—was formed. This, much like the one operating in Iraq in the wake of the American-led invasion in 2003, was a loose alliance of local peoples united less by any common interest or conviction than by a common dislike of a foreign, non-Muslim presence on their soil.

A few "people of good sense," wrote another member of the Institut, the painter and engraver Dominque-Vivant Denon, had grasped what France was offering them and had done their best to persuade their fellows. But "the masses of the nation, those who had nothing to lose, accustomed to belonging to cruel masters, assumed that the equality we showed them was a sign of weakness, and continued to be seduced by their beys," who, by "exploiting the prejudices of religion," maintained a constant opposition to all attempts to civilize them.[32] In the fourteenth century the Mamluks might have been, as Ibn Khaldûn said of them, "true believers," possessed of "nomadic virtues unsullied by debased nature, unadulterated with the filth of pleasure, undefiled by the ways of civilized living," but by the eighteenth they had become cruel, inefficient, and corrupt, even in the eyes of their coreligionists.

Every soldier, every savant in Napoleon's army knew that the Egyptians had been made the way they were not because they were "Orientals," not because of race—that interpretation would arrive slightly later—but because of the brutal tyranny of the Mamluks.[33] Such despotism, the French knew, created slavery, and slavery made men into animals and brought once-great civilizations to ruin. It was a lesson the well-read among them had learned from the ancients and most recently from Montesquieu, Voltaire, Rousseau, Condorcet, and Volney. Yet it was also a truth that was widely recognized even by the rank and file. It had, after all, been one of the key propositions of the revolution, one that Napoleon had dinned into his troops time and again during his campaigns in Europe. Looking on the ruins of Alexander's city and on what he had been told was "Cleopatra's palace," François Bernoyer, a young tailor who had been placed in charge of outfitting the army, wrote to his wife that he could only wonder at the brutal contrast between the centuries past, "which had given birth to those whose love of their country had led them to create such extraordinary things," and the fate of "the Egyptians of today, who are born on the same soil, in the same climate as their ancestors, yet their houses are miserable, for the most part made of mud mixed with the excrement of cows. . . . Thus, my dear," he concluded, "are the inevitable effects of a government based upon despotism."[34]

Bernoyer was also not alone in his conviction that, given time and the defeat of Egypt's present rulers, Egypt's cowered population would come to understand the self-evident benefits to be had from modern, secular society. Napoleon was, after all, the bearer of a new way of life that had already triumphed over just such despotic masters within Europe itself. As he had told his troops shortly before their departure from Toulon: "The genius of Liberty, which has made the republic the arbiter of Europe ever since its creation, now wishes that it shall become the arbiter for the most distant lands and seas." Or, as he expressed it more robustly elsewhere, *"Ce qui est bon pour les français est bon pour tout le monde."*[35]

## III

Today the Alexandria on whose beaches the French landed is a dismal coastal port, much of it dating from the 1950s and '60s. Little now remains of the cosmopolitan city, the "great wine-press of love" that so enthralled the novelist Lawrence Durrell before the Second World War. Even by the early 1950s it had been reduced, in Durrell's departing words, to a "thousand dust-tormented streets."[36] It was certainly even less inspiring at the end of the eighteenth century. The Arab city, said the Danish naval captain Friderik Ludvig Norden, who had passed through in 1737 on his way to the Sudan, was "not a Phoenix rising from the ashes; it is rather a vermin, emerging from the mud and the dust, in which the Koran has infected the whole place."[37] The masses who huddled together into the narrow houses and thronged the dusty, raucous alleyways were as alien as anything the French had ever seen. "A mass of unknown objects assails the senses," recorded Volney when he arrived in 1783. A bizarre assortment of persons dressed in long soiled robes and crowds of famished dogs thronged the streets from which a babble of noises "assaults the ear."

Beyond all this lay the ancient city of Alexander the Great, of the Ptolemies, and of Antony, then a steadily diminishing collection of ruins. François-René de Chateaubriand recalled in 1806, five years after Napoleon's final departure, how he had tried to catch, in the shadows of the ruins that surrounded him, the image of a city that once "had been the rival of Thebes and Memphis, which had had three thousand inhabitants and had been the sanctuary of the muses" or hear a distant echo of "the burning orgies of Antony and Cleopatra." But all to no avail. The shades of the Hellenistic and Roman past had quit the place for good. Instead there was only an oppressive reminder, once again, of the ravaging effects of despotic rule:

> A fatal spell had plunged the people of the new Alexandria into silence. That spell was despotism, which extinguishes all joy, and I could not suppress a cry of pain. Ah! What sounds could one hope to hear from a city one third of which at least was abandoned, in which another third had become a graveyard, and the inhabited third, caught between these two deathly extremes, was a sort of palpitating trunk, powerless, amongst the ruins and the graves, to shake off its chains?[38]

Farther out still, beyond the limits of any kind of settlement, were the Bedouin. The image that, at the turn of the nineteenth century, the British would create of the desert Arabs as "magnates of the wilderness," imbued with a "nobility, dignity, manliness, gracefulness, and virility," reminiscent of Rudolph Valentino's 1921 film *The Sheikh,* still lay some time in the future.[39] The peoples whom the French encountered in 1798, said one of the savants, the engineer Gilbert-Joseph Volvic de Chabrol, were merely rootless predators. "They know nothing of agriculture or commerce," he wrote. "Brigands by choice, they have become assassins though cupidity."[40]

One day, six of the members of the Institut, carried away by their enthusiasm, found themselves outside the French lines. There they were seized by a group of armed horsemen and carried back to their camp. After much debate the Bedouins decided to return them. Napoleon, never one to miss an opportunity for showmanship, was there to greet them and to thank the Bedouin leader for his "humanity." "So long as you behave towards me with rectitude," Napoleon assured him, "so long will you find in me a protector and a friend." The chieftain refused the offer of money for his troubles but gratefully accepted Bonaparte's gift of a gold watch. "They had superb horses," noted François Bernoyer, who had witnessed this little scene, "and magnificent weapons decorated with silver. But, he added with his tailor's eye, their clothes were "lamentable, nothing short of miserable."[41]

For the French, reared on the sentimental vision of the "noble savages" that the eighteenth century had pursued from the South Pacific to Africa, here was tangible evidence of what it really meant to live outside civilization. "They are the most horrible savages," wrote Napoleon's brother Louis Bonaparte, another devotee of Rousseau. "Oh, Jean-Jacques [Rousseau]! If only you could see these men you call 'natural,' you would tremble with shame and surprise at having admired them." Even Napoleon himself, who by the time he had reached Cairo had abandoned his lingering hope of discovering in these nomads the nobility that any man on a horse ought to inspire, wrote to the Directory, "Their ferocity is equaled by the miserable life they lead, exposed for days to the burning sand beneath the ferocity of the sun without water to cool them. They are without pity and without faith. It is the most hideous spectacle of the savage one could possibly imagine."[42]

ONCE ALEXANDRIA HAD been secured, the Army of the Orient moved across the desert toward Cairo, led by Napoleon, his head swathed in a mosquito net. The desert crossing was a far greater ordeal than the invasion of Alexandria. The overloaded troops, staggering under the burning sun, often threw away their food and water supplies in a desperate attempt to escape from exhaustion, only to collapse from thirst and hunger. Even those who could find enough water and food became afflicted, Napoleon recalled later, "by a vague melancholy that nothing seemed able to overcome . . . several soldiers tried to drown themselves in the Nile. . . . 'Why have we come here?' they asked. 'The Directory has deported us.' "43

Undeterred, Napoleon pushed on toward Cairo. On July 21, on a plain outside what is now Imbâbah, on the west bank of the Nile, he came face-to-face with Murad Bey at the head of an army of some 12,000 cavalry and 40,000 infantry. Napoleon drew up his army into squares, a novel tactic that the Austrians and the Russians had used to good effect against the Ottoman armies in the early years of the century. "Go and remember that forty centuries are looking down upon you," he told his soldiers and prepared to face the vastly superior Mamluk forces. The Mamluk cavalry were renowned for their ferocity and their courage. But they were also undisciplined and had only primitive handguns and no artillery. Within two hours, the ordered squares, regulated musket fire, and cannon of the French had broken them. When it was all over only 29 Frenchmen had been killed, but 10,000 Egyptians lay dead or dying under the pitiless sun. Murad Bey, accompanied by the 3,000 cavalry who had survived the battle, fled south through the desert, to Upper Egypt. The Mamluk residents of Cairo hastily followed him, taking with them all they could salvage from their houses. The "Battle of the Pyramids" as it came to be called—because it sounded good and because in the far distance it was just possible to glimpse the pyramids of Giza—left the French in possession of the Egyptian capital. Once again Napoleon assured the Egyptian people that they had nothing to fear for their families, their houses, or their property, "and above all for the religion of the Prophet, which I love."44

Once installed in the former palace of Alfi Bey, overlooking the vast square of Ezbekiyya, said by one French observer to be larger than the Place de la Concorde in Paris, Napoleon set about restructuring the fiscal system and administration of his new dependency. As he told Sheikh al-Sarqaui, the president of the *dîwân,* his objective was to "establish a single regime, based on the principles of the Qur'an, which are the only true ones and which alone can bring happiness to mankind."[45]

Quite how this was to be achieved was not altogether clear, since, as Napoleon was certainly aware, any attempt to have passed into legislative practice his claims to be able to reconcile the Rights of Man with the Shari'a would immediately have revealed the glaring contradictions between the two. The outcome somehow depended on his own personal destiny—"for who would be so incredulous as to doubt that everything in this vast universe is subject to the empire of destiny"—which in his mind had now become one with that of the entire Orient.

On December 21, 1798, he addressed the people of Cairo in a language that hovered somewhere between the Bible and *The Masked Prophet:*

> Let the people know that it was written that after having destroyed the enemies of Islam, having destroyed the crosses, I would come from the West to fulfill the task that has been given to me. Let the people know that in the holy book of the Qur'an, in more than twenty passages, that which has already happened has been foretold, and that which will happen also.

The day would come, he went on, warming to his theme, when "all the world will see the evidence that I have been guided by orders from on high and that no human efforts can prevail against me. Happy will they be who, in good faith, are the first to join themselves to me."[46]

He was, he now told the astonished Ulema, the new Mahdi. He even tried dressing up in the stylized "Oriental" garb many Europeans took to be Turkish dress, but he looked so gauche and uncomfortable in his turban and flowing Oriental robes that when his generals saw him, they burst out laughing.[47] It was all, he later admitted from exile, "charlatanism, but of the very highest kind."[48] The only people who seemed to have given any credence to Napoleon's frequently repeated claims that the French were now really Muslims—and the Muslims, therefore, French—were the British. Sir Sidney Smith, who would eventually defeat Napoleon at Acre and who evidently saw himself as some kind of latter-day Richard the Lion-heart, spoke of the need for "the absolute eradication of this French Mahometan colony from Africa." Napoleon's apparent "Orientalization" and rumored conversion to Islam—together with his Corsican origins—gave him a wildly exotic image that, together with his reputation for cruelty and tyranny, would endure until his death.[49]

One of Napoleon's generals, however, does seem to have taken at least some of his master's protestations to heart. Jacques Menou, not the most charismatic of men, was small, round, balding, and over fifty. But he proved to be an able administrator, and later, in September 1799, after Napoleon himself had left the scene, and his deputy, Jean-Baptiste Kléber, had been killed, Menou found himself in charge of what was left of the French army and the French administration in Cairo. Earlier in the campaign, in order to marry a Muslim woman named Zobaidah and, as he admitted, "for political reasons," he had converted to Islam and taken the name "Abdallah Jacques Menou," an act that was greeted by many, including al-Jabarti, as further evidence of the self-serving hypocrisy of the French.[50] Menou himself, who clearly had no more regard for Islam than he did for Christianity, would probably have agreed. But he evidently took some pride in the fact that his wife was a descendant of the Prophet. "I am, consequently," he boasted, "a cousin of Muhammad and all the green turbans of the world are my relatives." Napoleon disapproved, saying that it was contrary to French custom, made Menou look ridiculous in the eyes of the army, and was likely to lead to trouble. Menou, however, really does seem to have hoped to create some kind of parity between the Arabs and their French invaders. "Be generous with the Egyptians," he urged his soldiers on one occasion. "But what am I saying? The Egyptians today are Frenchmen; they are your brothers."[51]

Napoleon's own attempts to transform Egypt into an Islamic département of France, however, took the unpromising form of a *dîwân* composed of compliant members of the Cairo elite that was to govern Cairo in his name, and crippling fines on the wives of Mamluks who had been unwise enough to remain behind in the city after their husbands had fled. A new tax structure, which the population soon came to loathe, was set up along French lines with French administrators. Forced loans were extracted from likely victims. The merchant community of Alexandria alone was taken for 300,000 francs. Agricultural lands formerly held by the Mamluks were confiscated and redescribed as "national domains." Over all this, Napoleon, playing the not unfamiliar role of Oriental despot, dispensed summary justice. "Every day," he boasted to Menou, "I cut off five or six heads in the streets of Cairo." He had, he explained, to make these people obey. "And for them to obey is to fear."

On October 21, 1798, the people of Cairo, outraged by the French attempt to conduct a census that involved entering homes without their owners' consent, rose in revolt. The chief *qadi*, or judge, who had been placed in charge of the census was killed, as were a number of French officers caught on their own in the streets. The muftis then intervened, transforming a street protest into a jihad against both the French and their Muslim collaborators. Napoleon's response was rapid, brutal, and effective. He bombarded the rebellious district of the town until more than three thousand Egyptians lay dead and the ringleaders came to him begging for mercy. He then sent in a contingent to sack the al-Azhar quarter and the mosque in retaliation. French cavalry rode their horses into the holiest site in all Egypt, an act of desecration that revealed, as nothing else had, just how little regard the French had for the realties of Islam. Little wonder perhaps that al-Jabarti insisted that "most of the Egyptian people, and in particular the peasants, resented the government of the French." In his characteristic ruthlessness, Napoleon had made the same mistake that thirty years later the French would face again in Algeria as they struggled against the forces led by the emir 'Abd al-Qâdir. If the French behaved like barbarians, Alexis de Tocqueville warned the government in Paris on that occasion, the Turks "would always have the advantage over us of being Muslim barbarians."[52]

## IV

Meanwhile the savants were conducting another kind of assault on the Egyptian consciousness.

Some two kilometers to the south of the center of Cairo, in the suburb of Nasrihe, Napoleon formally installed the Institut d'Égypte in four palaces abandoned by their Mamluk owners. Into these lavish buildings the savants moved their library, their mathematical and physical instruments, a chemical laboratory, an observatory, a printing press, a menagerie, a botanical garden, a cabinet of natural history and one of mineralogy, and a collection of archaeological artifacts, together with a number of workshops.

On 5 Fructidor of the sixth year of the revolution—or August 22, 1798—the Institut was formally opened with the mathematician Gaspard Monge as president; Napoleon himself, with uncharacteristic modesty, as vice president; and Jean-Baptiste Fourier as permanent secretary. A charter was drawn up setting out in detail what the Institut's objectives were to be. It was to conduct research into every aspect—natural, human, historical, and political—of Egypt, and of Asia more generally. It was to give advice to the new administration on any matter on which it might be consulted. Above all, however, its general purpose was to ensure the "progress and propagation of enlightenment in Egypt."[53] It was provided with an official journal named *Décade Égyptienne*, the first number of which ended on a suitably triumphal note.

"Learned Europe," it declared, "cannot look with indifference upon the power of the applied sciences in a country to which they have been brought by armed wisdom and the love of humanity, after having been exiled for so long by barbarism and religious fury."[54] Napoleon himself attended most of the meetings, which, it was said, were the only gatherings where he would allow his policies to be criticized and contradicted. So keen on the place was he that the army took to calling it "the general in chief's favorite mistress."[55]

The activities of the savants greatly puzzled the Egyptians. Who were these men and what exactly were they doing? They were not soldiers or jurists (although some of their activities were legislative); they were not administrators; nor clearly, since all Europeans were atheists, did they have anything to do with religion. Given the variety, not to say peculiarity, of the kinds of activities the Institut sponsored, their bewilderment is hardly surprising. In the halls that had once housed the harem of Hassan Bey Kachef, papers were read on the formation of mirages, on the creation of salt of ammonia, on the fabrication of indigo in Egypt. Fourier offered a new solution for algebraic equations, and François Parseval, one of the Institut's poets, read from his verse translation of—appropriately for the occasion—Tasso's great poem on the Christian capture of Jerusalem, *Gerusalemme liberata*. Geoffroy Saint-Hilaire, one of the most distinguished of the savants, read a paper on the ostrich wing in which he attempted to prove that the bird was incapable of flight. (François Bernoyer was present at this session. It lasted, he said, for nearly three hours, at the end of which no conclusion had been reached as to whether the ostrich was meant to run or to fly. "I have never listened to anything so foolish," he told his wife. "The most ignorant person could have provided an answer by simply pointing out that as nature had given the ostrich great big legs, it was clearly intended to run. If it had been meant to fly, it would have been provided with great big wings.")[56]

After listening to accounts of all these activities, the Egyptians came to the general conclusion that they were all a subterfuge. The real work was being carried on in the chemical laboratory: there the savants were obviously trying to manufacture gold.

The cultural activities of the Institut were one way of promoting "armed wisdom and the love of humanity." There were, however, other more direct and explicit means. The French Revolution had already perfected the use of large-scale public celebrations to propagate its political message, and Napoleon adopted the same strategy in Egypt. Every Muslim feast was hijacked to make the same—for the Muslims—blasphemous and incoherent claims that Napoleon had put forward in his first declaration. On September 21, 1798, in an attempt to merge the Muslim and revolutionary calendars, the French staged a festival to celebrate the anniversary of the founding of the republic. With the help of the members of the Institut, an obelisk of wood and cloth some six feet in height was set up bearing the names of the French soldiers who had died in the campaign. On the perimeter of the Ezbekiyya Square a colonnade was built and over it a triumphal arch decorated with scenes from the Battle of the Pyramids.

On the day of the festival, the army, the members of the Ulema and the *dîwân* of Cairo, all the sheiks, the leader of the Janissary Corps, and the representative of the Pasha all took part in a parade through the city. This was followed by a grand lunch for 150 dignitaries on the ground floor of Napoleon's residence, in a room hung with French and Turkish flags, Phrygian bonnets and crescents, and quotations from the Declaration of the Rights of Man and the Qur'an. A number of toasts were made, culminating in that of Gaspard Monge, who raised his glass to "the perfection of the human spirit and the progress of Enlightenment." The day ended with the traditional fireworks display.

This, at least, was the account reported in the army's official publication, the *Courier d'Égypte*. The reality was somewhat less spectacular. As with so many of the other French attempts to display simultaneously power and goodwill, the Egyptians cast a cold, skeptical eye over most of the proceedings. Many of the invited guests failed to turn up or did so only reluctantly. Many of the fireworks failed to ignite. The cantata was cacophonous to Arab ears. The famous obelisk was impressive only from a considerable distance. Hastily assembled from bits and pieces, none of which was long enough to support its weight, it soon began to sag. Once the festivities were over, the soldiers made a hole in the base and transformed the interior into a makeshift brothel. Niqula al-Turk, a Greek Catholic poet who wrote in Arabic and who left us the other local eyewitness account of the French occupation, saw the comings and goings at the base of the obelisk. "The French," he remarked, "claimed that this column was the tree of liberty; but the Egyptians replied that it was rather the stake on which they had been impaled and the symbol of the conquest of their country."[57]

French attempts to overawe the Egyptians with the wonders of French science suffered a similar fate. Napoleon had brought with him a hot-air balloon—a "Montgolfier"—and a chief balloonist, Nicolas-Jacques Conté, a colorful figure with tightly curled hair whose left eye, lost in an explosion in 1795, was covered by a bandanna. On August 21, 1798, Conté and his men organized an elaborately staged launching of the balloon. The balloon, a long cloth sack painted in red, white, and blue, was suspended from a pole. With great ceremony, and to the blare of military trumpets, Condé ignited the burner. "Its smoke," recalled al-Jabarti, "rose into the cloth and filled it." The balloon then lifted slowly off the ground and "began to sail with the wind for a very little while and then its bowl [burner] fell as the wind fell and the cloth followed suit." As it hit the ground, it burst into flames, and many of those present, assuming this to be some new weapon the French were preparing to use against them, fled the square in panic. The French, wrote al-Jabarti, had boasted that "this apparatus is like a vessel in which people sit and travel to other countries." Clearly this was yet another materialist lie. In his opinion, the flimsy, brightly colored cloth was nothing more than an elaborate toy, "like the kites household servants build for festivals and other happy occasions." Undeterred by this disaster, another launch was attempted on January 16, 1799. This time the balloon got farther than it had previously but still fell to Earth in full view of the spectators. Had it succeeded in disappearing from sight, al-Jabarti commented sarcastically, "the French would have claimed that it had departed for distant lands."[58]

## V

By the time this fiasco took place, however, the French hold over Egypt had already begun to weaken. On August 1, 1798, Nelson had returned to Alexandria and caught the French fleet off guard in the Bay of Aboukir. Most of the French artillery had been moved onto land, where it could not reach the British ships, and the French ships huddled in the bay, which was shallow and poorly protected, were unable to maneuver with ease. Nelson's warships met with only limited opposition and hour after hour pounded away at the beleaguered French fleet. Shortly after ten in the evening, *L'Orient* caught fire. For about an hour it smoldered, sending sparks and bursts of white flame into the night air; then it exploded, taking with it the admiral of the fleet, Brueys, and those of its crew who had stayed behind in a desperate attempt to fight the blaze. The detonation was so loud that it could be heard across the entire bay, and for a few minutes the guns of both sides fell eerily silent. All that you could hear, recalled Bernoyer, was the sound of the debris from the ship, which had been propelled "to a prodigious height" by the explosion, falling back into the water.[59] "After that," wrote Saint-Hilaire, "tumult seized hold of our navy."[60]

Three quarters of an hour later, the guns began again. The battle continued all night, but at about noon the following day Admiral Pierre Villeneuve (who would face, and be defeated by,

Nelson again at Trafalgar) decided to make for Europe with all
that now remained of the fleet. The French lost 700 men killed or
drowned, 1,500 wounded, and 3,000 taken prisoner. Denon, who
arrived at Aboukir when it was all over, recalled, with "lowered
head and heavy heart," the sight of the Bedouin camped along the
shore and, by the ghastly flickering light of their fires, picking over
the litter of the broken ships and the bodies of the dead.[61]

The Army of the Orient was now stranded. When the news
reached Istanbul, the sultan declared war on France. In an attempt to forestall a wholesale invasion of Egypt, Napoleon sent a
delegation to the sultan's emissary at Acre, a ruthless Bosnian
named Ahmad Pasha al-Jazzar, reassuring him, once again, that
the French had not, as rumor now had it, come to reconquer
Jerusalem. This was not, he insisted, a new crusade. It was an attempt to recover Egypt from the Mamluks for the sultan, in
whose name Napoleon now claimed to be ruling. Al-Jazzar refused even to receive the delegation, which was forced to beat a
hasty retreat back to Cairo.[62] On September 9, Selim III declared a
jihad against the French. "It is the duty of every Muslim to go to
war against France," he said, since it was clear from all their actions that the French had "no other intention but to disturb the
order and harmony of the entire world and to sever the links
which bind together all peoples and nations."[63]

In Paris, however, hopes for some happy outcome to the expedition were still high. On November 21, 1798, Volney wrote an article in the official revolutionary journal *Le Moniteur*. "As everyone
writes his own novel about the army of Egypt," he told his readers,
"here is mine." Napoleon had exploited the divisions among
Copts, Bedouin, and peasants to win the people over to his side.
By "adopting many of their customs so that they should adopt
ours," he had flattered their amour propre. When he found them,
they had been "somber, quick to anger, and quarrelsome, as a consequence of tyranny." Now he made them "gay, amiable, good,"
with amusements, music, and public works. He repaired the

bridges, roads, and canals. He found the peasants in the condition of mere serfs and gave them property. He changed the inheritance laws so that not only would children enjoy equal shares, but women, too, could inherit. He forbade the marriage of under-age children and gently opposed the practice of polygamy. He founded a new civil code in Asia, which Volney predicted would change it for good. He transformed the economy, created schools in which Arabs, Copts, and Frenchmen would learn side by side, in both Arabic and French, all the natural sciences. He recalled the Arabs to the glory of their ancestors. "In a word, he created a nation."

Volney went on to imagine that after the loss of the French fleet at Aboukir, the Ottoman declaration of war against France, and the entry of the Russian fleet into the Mediterranean, Napoleon would turn his back on India. "Why go to the end of the universe," he made Napoleon say, "to an obscure and barbarous theater to spend all my efforts on so little glory and no benefit?" No. It is now toward Europe that he must look. And as "the imprudent Turk has raised the standard [against me] it is in Constantinople that I will take it from him." Having seized the Ottoman capital, he would call to arm the Kurds, the Armenians, the Persians, the Turcomans, the Bedouin to finish off "our common enemy." And having thus re-created a "new Byzantine Empire," he would sweep through the Mediterranean and across central Europe. Prussia would resume its old alliance with France. Moscow would break free of Saint Petersburg (thus putting an end to France's other enemy, the Russian Empire), and liberty would return to the English in their island fastness. Then all the governments of the world would be capable of living in peace with one another. "I can see that day," Volney cried, "the only one worthy of glory." And on the base of the great obelisk at Istanbul, where the three intertwined bronze serpents cast to commemorate the final defeat of the Persians had once stood, an inscription would be placed in gratitude. It would read:

> *To the French army, victorious*
> IN ITALY
> IN AFRICA
> AND IN ASIA
> *To Bonaparte, Member of the Institut National*
> PACIFIER OF EUROPE[64]

Thus would the East, once again, be united to the West and the civilizing mission—Alexander's mission—finally completed after more than two millennia.

---

AT THE TIME, however, Napoleon's own objectives were rather less imposing. He was concerned primarily with securing whatever he could of his diminishing hold over Egypt and then preparing for his return to France, with as much of the Army of the Orient as possible. On February 6, 1799, with a force of 13,000, he left Katia, east of Damietta, and marched into Syria before the Turks had time to collect their forces. He succeeded in taking Gaza and then Jaffa on March 3. There he conducted what appears to have been a repeat performance of the massacres of the Mamluk warlord Muhammad 'Abu Dahab in 1776, which Volney had described in horrified detail. Terror, it would seem, had now become, as much for Napoleon as it had been for his Mamluk predecessors, the main means of conquering Palestine.[65]

More than 2,500 victims died that day, so many that the firing squads ran out of ammunition and had to resort to bayonets to finish the job. Once the slaughter was done, Napoleon, still maintaining, as best he could, his role as the Muhammad of the West, addressed the people of Palestine. He had, he announced, come to make war not on them but on al-Jazzar. He promised them, as always, freedom of worship and full possession of all their property. He cautioned them against any resistance to one whose destiny had chosen him to rule over them. "You should know that all human efforts are useless against me," he told them. "For I succeed in all that I undertake. Those who declare themselves to be my friends prosper. Those who declare themselves to be my enemies shall perish."

In fact, however, it was his own soldiers who were perishing. No sooner had the French entered the city than they were struck by the plague. As more and more of his soldiers fell ill and news of the spread of the disease could no longer be contained, Napoleon, in another attempt to demonstrate that he possessed near-immortal powers, paid a visit to the plague hospital. The painting by Antoine-Jean Gros of Napoleon amid the sick, a cloth held to his nose, a dying soldier reaching out to touch his side as if he were Christ, would, ironically, become the most abiding image of the entire campaign.

On March 14, 1799, Napoleon took what now remained of his army out of Jaffa and marched toward al-Jazzar's capital at Acre. Here he met his first but finally decisive defeat. A British squadron under the command of Sir Sidney Smith had arrived at Haifa shortly before the French, providing al-Jazzar with munitions and the supply lines needed to resist a siege. The British had also captured the siege artillery that Napoleon had had transported by sea from Alexandria, which now passed into al-Jazzar's hands. By May 20 it was clear that Acre could be neither stormed nor forced into submission. The Army of the Orient struck camp during the night and began the long and hazardous retreat south. On June 14, Napoleon made a carefully staged triumphal entry into Cairo. But it was clear to all that it was only a matter of time before he would be forced to leave Egypt for good.

Meanwhile in Europe, things were going badly for the French. In June 1799, the short-lived "Parthenopean Republic," which had been installed in Naples with French help, collapsed. The following month, the French lost most of what they had gained in northern Italy, and an Anglo-Russian army was making its way toward Holland. The Vendée in northwest France, which had always harbored royalist preferences and had already rebelled once against the revolution, began to look as if it might be preparing to do so again. In general the Directory was held responsible for these disasters, for having, as one critic put it, "exiled to perish in the deserts of Arabia the elite of the army of Italy, its most celebrated general and the most gifted of our military leaders."

On September 10, the Directory decided to evacuate the Army of the Orient on almost whatever conditions it could secure and began negotiations with the sultan. Napoleon had no intention of perishing in any desert or of staying around to negotiate a humiliating peace. France, he informed Menou, was hovering between foreign invasion and civil war thanks to the incompetence of the Directory. Egypt was now secure and no longer needed his presence. Jean-Baptiste Kléber could hold it without further assistance. By the end of the month he was back in Paris, and Kléber was left with the thankless task of extricating what remained of the French forces as best he could. "He has left us," he remarked bitterly of his commander in chief, "with his trousers full of shit." And he promised to return to France "and wipe his face in it." But in June 1800, Kléber was assassinated. His place was taken by Menou, who by July of the following year was driven, with all that remained of his forces, out of Cairo. Menou retreated to Alexandria and surrendered to General John Hely-Hutchinson in September, and between September and October 1801, all that now remained of the Army of the Orient, together with the members of the Institut d'Égypte, left Alexandria, and Egypt, for France. The expedition was at an end, and Egypt returned, nominally at least, to Ottoman rule until 1883.

As a military operation the expedition had been an unqualified failure. Napoleon had failed to create a French colony in the Near East, failed to win over the Egyptians, failed to persuade all but a tiny handful of Muslims of the virtues of the rights of man and the principles of equality, and failed to reach India. But its long-term and frequently unintended consequences were many.

## VI

One of these, and in some respects the most enduring, was the work of the Institut. Once back in Paris, the savants, under the auspices of Dominique-Vivant Denon, began the work of making their findings public. If France had not been able to colonize Egypt, it would henceforth colonize whatever Europe came to think about Egypt and through Egypt of the whole wide world of the "Orient."

Denon proved to be an able publicist. A former minor aristo-crat and diplomat and an able draftsman (he once sketched Sir William Hamilton and his celebrated wife, Emma, at Naples), he eventually became the director of the Louvre (renamed the Musée Napoleon) and the chief architect of Napoleon's transformation of himself from gifted military commander to a Renaissance high prince of culture. The fruit of Denon's editorial labors, the mas-sive *Description de l'Égypte*, was published between 1809 and 1828 in twenty-three huge folio volumes. This was the first time a de-tailed scientific inquiry had been undertaken into the customs of the peoples of the modern East based upon long-term personal observation, and the image it conveyed of the Egyptians, and of Muslims more generally, would make a lasting impression of all future perceptions of the "East."

The appearance of *Description de l'Égypte* has often been de-scribed as a significant stage in the fabrication of a common, and highly derogatory, identity for all the peoples of an area that reached all the way from the eastern Mediterranean to the borders of China.[66] In part this is justified. With few exceptions, the view that most Europeans had acquired of the Muslim East by the early nineteenth century was largely negative. With the decline of the Ottoman Empire, there had declined, too, the often favorable view of the Sublime Porte that Voltaire and others had taken. The Napoleonic encounter with the Egyptian East had also been deeply marked by expectations. For the scholars, even for the edu-cated soldiery, who accompanied Napoleon, Egypt was the land of the pharaohs, of Alexander, and of the Ptolemies, a place of ancient and mysterious wisdom.

Even when these myths had finally been exploded, it was still widely accepted that it was the ancient Egyptians who had given to the Greeks the fundamental principles of the natural and mathematical sciences, on which Greek and subsequently European civilization had been based. It would have been easy, and entirely reasonable, for European visitors to suppose that the modern Egyptians had no more connection to the cultures that had once flourished on the banks of the Nile than the modern Italians did to the ancient Romans or the modern Greeks to the ancient. But the European historical imagination, in particular in the eighteenth century, could never quite accept that there might exist no association between places and the peoples who inhabit them. What, they asked themselves, could have become of the civilization of ancient Egypt? Why did the modern Egyptians live in such ignorance amid the ruins of their former glory? Why did they know nothing, and seemingly take pride in knowing nothing, of the arts and sciences their own ancestors had created? Why, for instance, were these descendants of the pharaohs now seemingly unable to perform even the simplest mathematical tasks? One of the members of the engineering corps of the Army of the Orient, Louis Thurman, told Chabrol how one day he was talking to "one of their principal architects":

He took out his prayer beads and began to make a calculation with them, probably in order to impress me with his learning. The calculation took a long time, and he began to scratch his ear. As I was sitting beside him I glanced over at his work. I saw that he was multiplying 250 and 30 (or some similar sum involving zeros) by writing out the number 250, 30 times and then adding the total. I offered to help him, which he readily accepted, certain that the outcome could only cause me embarrassment. You will, of course, understand that the correct answer was reached in the stroke of a pen. Seeing such marvelous rapidity, he exclaimed, "Allah! You French are sorcerers as certainly as Muhammad is the messenger of God. You know the secret of the highest angels. Is it they who have taught you?" I replied that as an officer I had to know such things and more, that there was not even a sergeant major among us who could not do as well. He was stupefied, and I heard him murmuring the Muslim profession of faith, apparently to ward off the evil spirit.[67]

A similar anecdote could just as well have been said about a contemporary European peasant. But the modern Egyptian failure to master basic multiplication struck Captain Thurman—as it did Chabrol—so forcibly not only because, in the history he had learned, mathematics had been "invented" in Egypt, but also because it had been the Arabs who had devised the numerical system he was himself employing.

The Egyptians also seemed to be similarly incapable of, or unwilling to, contemplate work of any but the most limited kind. They never repaired anything, recalled Denon. If a wall collapsed in a house, they simply had one less room. If the whole building fell down, they set up home next to the ruins.[68] All this sitting around was not simply idleness. There was, the French concluded, a certain pride in their indolence. Everything about Egyptian society conspired to the horizontal and the unmoving. The sofas on which, Denon observed, "they lie rather than sit"; the clothes they wore, with their long dangling sleeves that covered their hands, thus making most manual activities impossible; even their turbans, which he thought had been designed so as to keep their heads erect without effort—"all this discourages any activity, any imagination. They dream without any purpose, without pleasure, every day the same thing." One day a Turk "as arrogant as he was ignorant" explained to him that for Muslims, work was something done only by slaves and the vanquished. The only reply a French artist received when he tried to persuade an Egyptian of "the superiority of the Europeans over the Arabs in the arts and in industry" was to be told, "I believe that you infidels are condemned to work, while we disciples of Muhammad were born for rest and the contemplation of the glorious Koran."[69]

Idle and incurious, the modern Egyptians also greeted any new event, however remarkable, however tragic, with a passivity bordering on indifference. This struck the French more forcibly even than the squalor and the abandon by which they were surrounded. "Their sangfroid is astonishing," wrote one. "Nothing seems to trouble them. Death is for them what a voyage to America is for the English."[70] This was one of the reasons—or so it was comfortingly suggested—that they had failed to take much interest in the fate of the hot-air balloons.

It was not simply the wonders of French technology that left these people indifferent. "Whether they be consumed by anxiety or by remorse," noted Chabrol, "drunk with happiness, struck down by an unforeseen reversal of fortune, tormented by jealousy or hatred, boiling with anger, or tormented by vengeance, their demeanor always maintains the same impassivity."[71] They had patiently and uncomplainingly borne all the miseries the Mamluks had inflicted on them, wrote Bernoyer, miseries that no "educated and enlightened people would have tolerated." This observation led him to the melancholy conclusion that perhaps after all Rousseau had been right in arguing that the arts and the sciences, "enlightenment" in general, was nothing short of an infliction if all they had ever done was demonstrate to humans how miserable they really were. Perhaps, after all, it really was true, he told his long-suffering and distant wife, that man "cannot be truly happy except in his natural state, that is to say, his savage condition."[72]

The Egyptians—with the dubious exception of the Bedouin—were not, however, savages, certainly not the kind of savages Rousseau had had in mind. They were the remnants of a succession of once great civilizations. What, then, had brought them to this condition?

The immediate answer was, of course, once again, despotism.

Individual Egyptians, as Denon had observed, were, when given the opportunity, industrious and adroit. And since "like savages they lack any kind of tools, what they managed with their hands was remarkable." Potentially, at least, they were also ideal soldiers; eminently sober and disciplined, "they ride like centaurs and swim like tritons." Yet "a population of more than a million people, all of whom possessed these qualities, could be held subject by four thousand isolated Frenchmen controlling two hundred localities." Such, he concluded, was the force of habitual obedience.[73] Chabrol shared much the same view. The modern Egyptians, humble and passive, "vegetated in their uncertainty and never reflected upon their deplorable condition." Once they had had a vibrant cultural and intellectual life, but now their pleasures were few and simple: singing, telling tales, stroking their beards, "and the delights of the harem."[74]

Despotism, as both Denon and Chabrol knew, created such habits. It frustrated the imagination, deprived men of their natural capacity for creation. Yet Chabrol, at least, also thought that he could see that beneath these impassive exteriors "an ardent imagination is hidden." Dissimulation and lack of foresight had become for the Egyptians, "as for all Orientals in general, a refuge against violence." Yet under their unreflecting exterior their sensitivity had, if anything, gained in intensity, and "these men whom we take to be plunged into the most absolute apathy" had, in fact, brought their capacity for concentration and their memory "to the very highest pitch."[75] They waited now for the European imagination, European individualism to release them—if only, that is, they could be brought to understand all the benefits that these things would convey.

A little more than half a century later John Stuart Mill extended this observation to cover the entire "East":

The greatest part of the world has, properly speaking, no history because the despotism of custom is complete. This is the case over the whole East. Custom is there, in all things, the final appeal: justice and right mean conformity to custom; the argument of custom no-one, unless some tyrant intoxicated with power, thinks of resisting.

Where now, he asked, were "the greatest and most powerful nations of the world"? They have become, he answered, "the subjects or dependents of tribes whose forefathers wandered in the forests when theirs had magnificent palaces and gorgeous temples." Because barbarous though these new overlords might once have been, among them "custom exercised only a divided rule with liberty and progress."[76]

Here, however, there was another problem, as much for Denon as it would be for Mill. It seemed to most French observers highly unlikely that despotism alone could have made of men the kind of supine beings the Egyptians had apparently become under their Mamluk masters. There had to be something else, something that had prepared them to accept their masters' demands with so little resistance. As always, the most likely candidate was religion. *"God wishes it. God is great. God is merciful,"* wrote Chabrol. "Those are the only words which ever escape their mouths on hearing of an unexpected success or of the most terrible misfortune."[77] Islam, based as it was upon "the dogma of fatalism," had taught the Egyptians that they could achieve nothing in this world. Some years later, Edward Lane, an Arabist and former printer who lived in Cairo as an Egyptian from 1825 to 1828 and then again from 1833 to 1835, observed the same phenomenon. Lane, who took a far more positive view of Islam than did most Europeans at the time, interpreted the Muslim "resignation and fortitude nearly approaching to apathy" as proof not of indifference or sloth but of an "exemplary patience."[78] In the end, however, it made very little difference whether you chose to call it patience or indifference: the consequences were the same. By successfully crushing all resistance from his followers, Muhammad had ultimately made them incapable of change or self-improvement. Muslims accepted predestination without even understanding what it was they were accepting. They believed that nothing could ever be altered by human agency. Even such things as contagious disease could not be avoided. They had merely to be accepted as part of the ever-mysterious will of Allah.[79] As a consequence, Chabrol concluded, they had fallen into a "limitless resignation that sets them apart from all other peoples."

FOR THE FRENCH, one of the clearest indications of the malevolent impact of Islam on the minds of its adherents was the way in which the Egyptians treated women. For centuries Europeans had looked upon the sexual customs of their eastern neighbors with a mixture of prurience and contempt. Polygamy and concubinage appalled and fascinated them. Turks in particular became synonymous with unbridled sexual appetites, something perhaps to be ashamed of but also envied. The harem or seraglio, in particular, was a source of apparently limitless curiosity. Here was literally a storehouse of lasciviousness. Travelers to Turkey, Persia, and Mughal India, although none of them ever penetrated into the inner sanctum where the women were housed, were intrigued not only by its very existence but by all its exotic details, its odalisques, its eunuchs, its deaf-mutes and dwarfs. For Montesquieu it made the perfect model of the despotic society, based on fear and reverence, filled with subjects who looked upon their sovereign with awe and adoration only because they knew no better. In Islam women were to their husbands or—since so many of them were literally slaves—their owners as his subjects were to the sultan: mere chattels, somewhat less valuable than camels, if more pleasant to have around.

The French invaders claimed to be scandalized by these things. In actual practice, however, there was also often very little to choose between the behavior of the Egyptians and the French, in particular in their treatment of unmarried women. The young men of which the Army of the Orient was largely composed accepted the presence of available women as a matter of convention.

In November 1798, Bernoyer sent his cousin—confessing that this was hardly a tale he could tell his wife—a detailed and intimate account of how, after "so long a privation," he had been driven to take a concubine. One of the army of procurers who thronged the French quarters of Cairo provided him with twelve young women, each one veiled and dressed from head to ankle in a "long shirt of blue cloth" so that all he could see of them were their feet. "They aroused my pity more than my passion," he confessed. Until one of them, "with a single gesture threw off her clothing and stood naked before me with a shout of laughter as if mocking my timidity." He was instantly enchanted. "Her huge eyes, black and ravishing, and above all her fourteen years of age, captivated me, and I proclaimed her my sultana."[80]

Not content with his "beautiful child," sometime later, Bernoyer and a Captain Lunel from Avignon attempted to buy for themselves two women slaves, both of them black—apparently oblivious of, or indifferent to, the fact that one of the items on the civilizing agenda of the Army of the Orient had been precisely the abolition of slavery, described by the Convention Nationale in 1793 as "the greatest affront to our nature."[81] If Bernoyer is to be believed, when he asked the slave trader if he had any white women for sale, he was told that there was only one but that Napoleon himself had ordered that she be set aside until he had an opportunity to take a look at her. Faced, as Bernoyer put it, with "such a weighty argument," he shrugged his shoulders and departed. Whether this was true or not, Bernoyer seems to have accepted without question, and showed not the slightest surprise, that the champion of the rights of man—and woman—should be thinking of buying himself a concubine.[82]

To the French, the Eastern women they encountered seemed to be, in the full sense of that hackneyed phrase, "sex objects" to be bartered, bought, enjoyed, traded, and discarded. They existed to give pleasure or to produce heirs, and their lives were cheap. Just how cheap startled even the insouciant Bernoyer. When the mistress of another French soldier threw a stone at his "sultana" and seriously wounded her, Bernoyer had her assailant taken by two French fusiliers to what he calls the "Turkish commissar" and demanded that she be put in jail for two weeks to cool off. Immediately the woman was seized, bound hand and foot, and trussed up in a sack. "That's a strange way to put a person into prison," Bernoyer remarked. "What," replied the commissar, "don't you want her thrown into the Nile?" Bernoyer was horrified and immediately ordered the poor woman to be released. The shock she had received was, he reflected, sufficient punishment.

Here was the image of the "Orient" captured in countless nineteenth-century paintings, of seraglios and slave auctions, images of young, half-clothed women (and occasionally boys) with downcast but slyly lascivious eyes, submissive, available, everything in fact that the starched and corseted maidens of contemporary Europe were not. Several years later, in 1849, the same imagery would lure the great French novelist and pioneer sex tourist Gustave Flaubert to the same city. There he met, and slept with, Kuchuk Hanem, an Egyptian dancer, entranced and repelled (Flaubert had a hard time distinguishing between the two) by "her shaven cunt, dry though fatty," which "gave the effect of a plague victim or a leper-house."

The Egyptians, for their part, had an equally lurid vision of the sexuality of the Europeans. For the first time since the Crusades, Napoleon's invasion had exposed a Muslim population to Western cultural behavior for a prolonged period. Hitherto, all any Muslim had ever seen of the infidel outsiders were travelers and ambassadors, isolated individuals who were, on the whole, discreet and circumspect, eager, if only for the sake of their own safety, to conform to local conventions—in particular where sex was concerned. With the arrival of Napoleon, the Egyptians were brought face-to-face with a people who were at best indifferent to their sensibilities and who, for the time being at least, had the upper hand. To make matters worse, the officers and some of the savants had also brought their wives along with them. No Muslim, certainly no Egyptian, had ever seen a European woman at close quarters before, or if they had they had been as sheltered and secluded as any Muslim woman would have been. The French women in Napoleon's Egypt, by contrast, felt free to behave as if they had never left France.

The Egyptian men did not like what they saw. These creatures who strutted about in public with exposed hair and exposed faces, who spoke freely and ate in the presence of their men, were an affront to God. The obvious materialism of the French, reasoned al-Jabarti, had clearly poisoned their understanding of the universe and of the natural relationship between the sexes. "Their women do not veil themselves and have no modesty," he began a passage that, like so many European accounts of alien and alarming "others," combines simple observation with equally simple fantasy and describes what are certainly isolated or bizarre incidents as if they were age-old customs. "They do not care whether they uncover their private parts," he declared. "They have intercourse with any women who please them and vice versa. Sometimes one of the women goes into a barber's shop and invites him to shave her pubic hair. If he wishes he can take his fee in kind."[83]

The Egyptian women, when they had the opportunity, seemed to have reacted somewhat differently to the scandalous freedom allowed the French. One at least, no less a person than the daughter of Shaykh al-Bakri, the greatest of the religious dignitaries of Cairo, fell so deeply under the spell of the outrageous customs of the French that she took to dressing in European clothes and going about in public unveiled. After the French left and the Turks returned, she was publicly executed for the moral good of the people.

Mutual incomprehension, distrust, and disgust had colored the East-West perception of sex and of women since antiquity. For centuries this had largely focused, as with al-Jabarti's fantasy of pubic barbershops, on supposedly abnormal or at least unusual sex. Westerners were filthy, indiscreet, and immodest, and their women were an affront to the divinely ordered hierarchy of the sexes. The Muslims were lascivious, cruel, hypocritical, frequently bisexual, and also, curiously at the same time, effete and effeminate. Those stereotypes persisted—and some still do. But by the time Napoleon arrived in Egypt, Europeans had begun to be less concerned with how women were treated in Muslim societies as sexual beings than with how they were treated as persons.

Until the early eighteenth century there had been relatively little to choose between the position of women under Islamic and Christian rule. But with the retreat of the authority of the Church, the slow advance of the European Enlightenment, and increasing prosperity across Europe, the status and condition of women, and above all the esteem and respect that a woman could demand and confidently expect to receive—at least from some sectors of society—had improved immeasurably. By 1869, John Stuart Mill could condemn the continuing legal subjection of women to men as the last remaining instance of a form of bondage that belonged to an earlier age, that had become "one of the chief hindrances to human improvement." It was, admittedly, a radical statement, which all too many beyond the enlightened educated liberal circles in which he moved might well have mocked. But it reflected a condition that was far removed from the attitude with which Muslim males looked—and in most quarters still look—upon their women.[84]

Bernoyer, whose standards, in common with most Europeans' of his day, were certainly not merely double but multiple, while he could happily accept the existence of slave women, was horrified to discover that mothers and daughters in Muslim households were also looked upon, as far as he could see, as little more than chattels. It was, he concluded, but another demoralizing aspect of despotism, and, as with most forms of despotism, this one had succeeded in co-opting its victims. For Chabrol the victims' compliance was further indication of the capacity of all forms of despotism to crush the power to resist. In Muslim society, every male became in his own household what the sultan was to the society at large. "The women," he wrote with outrage, "isolated from society and condemned to absolute nothingness, are, in the eyes of the Mohammedans, barely worthy of being considered beings with intelligence and the privilege of reason."[85]

The isolation of Muslim women, symbolized by the various forms of dress with which they were shrouded, had struck Volney with equal force on his visit to Egypt. Nothing about Alexandria had appalled him more than the vision of a "species of ambulating specters from out of whose enveloping drapery the only glimmer of humanity is a pair of female eyes."[86] A society from which women were excluded, he said, "cannot possess that mixture of sweetness and politeness which distinguishes the nations of Europe." Women were for him the creators of the modern nations of the world. Without their presence the social conventions that in the eighteenth century went under the general name of "politeness" could never have come into being, and without them the primitive tribal people that in Europe had more than once brought flourishing and stable civilizations to an end could never hope to become true nations.

THE FAULT, OF course, lay with what was widely believed to be the image of women in the Qur'an and the laws that governed their conduct and their exclusion from the Muslim vision of the afterlife. To exclude women from Paradise, as Chabrol believed the Qur'an to do—although in fact it does not—was to make of them nonbeings, and thus to banish them and their necessary "civilizing influence from the world of the living.[87] Doubtless, Chabrol noted sarcastically, "in order to sustain the monstrous scaffolding of his supposed paradise," with its "seventy-two wives of the girls of Paradise" and the perpetual orgasms to which each believer was entitled, Muhammad had had to exclude true women from it. But could he not, he wondered, "have found a more equitable way of reconciling the marvelous with reason and justice?"[88]

## VII

An image of the Muslim East as a land rotting in despotic lethargy, constrained by a simple and savage religion that denied half of its peoples their humanity and in so doing prevented any possibility of progress and enlightenment, was not the expedition's only legacy. Napoleon's presence in Egypt had also disturbed the unstable balance of the Ottoman world. Throughout the Balkans, and in Greece in particular, the French seemed to offer a promise of liberation from what the Greek revolutionary hero Rigas Velestinlis called "the most abominable Ottoman despotism" of "that tyrant called the Sultan, wholly given over to his filthy woman-obsessed appetites."[89] Napoleon's invasion also opened the way for a French presence in the Middle East that would lead first to the invasion of Algeria in 1830 (in which many of the veterans of the expedition participated) and after 1918 to the French involvement in Syria, Palestine, Libya, and, once again, in Egypt itself. Every one of these carried with them the memories of both the successes and the failures of the Napoleonic campaign in the same region. France, declared the foreign minister, Barthélemy-Saint-Hilaire, in 1881,

> has always maintained in this country [Egypt], as in this whole region of the Orient, secular traditions that have given it a prestige and an authority that we cannot allow to diminish. At the end of the last century, our expedition, half military, half scientific, resurrected Egypt, which has, since then, never ceased to be the object of our solicitude.[90]

Saint-Hilaire's claim that Napoleon had "resurrected" and brought "modernity" to Egypt, still popular in the late twentieth century, now seems like the most obvious piece of Orientalist fantasy. Yet in one indirect sense it has some truth to it. Napoleon himself, and in particular his attempt to blend revolutionary ideologies with his own understanding of the Qur'an, may have passed into oblivion. But his presence and the disruption it created in the Ottoman Empire were certainly responsible for the beginning of a slow process of reform that by the end of the nineteenth century would transform Egypt into, if not a modern European state, certainly into a modern nation.

In 1805, the Ulema of Cairo, who were looked upon as the spokesmen of both God and the people, asked Muhammad 'Ali, an Albanian Turk who was the leader of Albanian troops who had landed with the British in 1801, to replace the ineffectual Ottoman governor, Khurshid Pasha. The sultan, Selim III, rightly suspicious of 'Ali's obvious ambitions, was reluctant, but in the end had no alternative but to accept or face full-scale revolt, which he had neither the forces nor the will to suppress. 'Ali was therefore made *wali* of Egypt and initiated a dynasty that would rule, almost without interruption, although for much of the time under British supervision, until 1952. He was, by any account, a remarkable man. Within a few years he had routed the remaining Mamluks who had escaped Napoleon—and butchered many of them—and made himself master not only of all of Egypt but also of the Sudan. He soon extended his control over the Nile Valley and the Red Sea and over most of the eastern Mediterranean as well. His declared ambition was to create an Arabic (unrealistically, perhaps, since he himself spoke no Arabic) Muslim empire on the ruins of the Ottoman with himself as the new caliph. He also inaugurated an ambitious program of modernization. He created a powerful central administration, reformed agriculture, and created a range of manufacturing industries. Crude and primitive though these were, they were the first in Egypt and among the earliest in any part of the Ottoman world. And behind all these ambitious projects stood the French. Young Egyptians were sent to France to study industry, engineering, and agriculture. French advisers helped create a system of state education, and French doctors were brought in to set up hospitals and a rudimentary system of public health.

One of those who benefited from Muhammad 'Ali's policy of

sending promising young men to France was a writer named Rifa'ah Rafi' at-Tahtâwî. In 1834, he published in Arabic (and in Turkish in 1839) a defense of "civilization" that provided an ideological justification for Muhammad 'Ali's reforming regime and also one of the earliest attempts to reconcile Islam with the values of the European Enlightenment. At-Tahtâwî's vision of a modernized form of Islam owed much to the five years he spent in Paris from 1826 until 1831 and to his reading of the French literature of the eighteenth century, in particular Montesquieu, Voltaire, and Rousseau. But it had been formed, while he was a student, by Sheikh Hasan al-'Attar, one of the great Egyptian scholars of his age who, twenty years earlier, had visited the Institut in Cairo and seen for himself, and come to appreciate, something of modern European science.

At-Tahtâwî—who became the director of Egypt's first national museum—was the first Egyptian to demand a form of nationalism that, in imitation of the European concepts of national rebuilding, attempted to link an Islamic present and future back to the past of the pharaohs.[91] His great history of Egypt has some harsh words for Napoleon the invader. But the French presence, and the work of the Institut in particular, is credited with having aroused the Egyptian national consciousness and with having shown the Egyptian people the glories of both their pharaonic and their Islamic pasts. (At-Tahtâwî was also responsible for translating into Arabic both the Code Napoléon and the French commercial code.) Muhammad 'Ali, who never thought of himself as Egyptian and was less interested in the pharaohs than he was in securing French assistance for his modernizing program, was nonetheless persuaded to reopen the Institut d'Égypte in 1859. One of the savants, Edmé-François Jomard, who had been a patron of at-Tahtâwî during his years in Paris, was young enough to have been a member of the first version of the Institut and old enough to visit the second.

In the subsequent history of Egyptian nationalism the same theme recurs. The French occupation had been a fiasco. But the

French presence had stimulated the Egyptians to seize control of their own national heritage, first from the Ottomans and then, after 1882, from the British. The French occupation had also kindled a form of nationalism that would eventually spread from Egypt throughout the entire Arab world. In 1962, Gamal Abdel Nasser, the nationalist leader who in 1952 forced Muhammad 'Ali's great-great-grandson King Farouk to abdicate and transformed Egypt into a republic, grappling somewhat uncertainly with at-Tahtâwî's thesis, described Napoleon's expedition as having given "a new assistance to the revolutionary energy of the Egyptian people." The French, he claimed, had brought with them "certain aspects of modern science that had been perfected by European civilization after having been initiated elsewhere, particularly in the two civilizations of the Pharaohs and the Arabs."[92]

By this account, Napoleon had launched Egypt on its subsequent career as the preeminent Arab nation. By the end of the nineteenth century, it had more sophisticated cities, a stronger economy, and a far richer literary and intellectual life than any other state of the Arab world.[93] It had taken all that it required from the West but at the same time had, as at-Tahtâwî had tried to demonstrate, remained true to its Islamic roots. For men like the historian Jamal Hamdan, even the caliphates, when compared to the role played by Egypt, had been marginal to the onward march of Arab-Muslim history. The Umayyads had collapsed, then fled to Spain, and the Abbasids had been destroyed by the Mongols in the thirteenth century. It had been Egypt that had provided Saladin with the base and the forces necessary to destroy the Crusader kingdoms of the Levant. Earlier still there had been the Romans, the Ptolemies, Alexander the Great, all the way back to the pharaohs. And although these could not be said to have played any direct role in Arab history, their passage through the land had given it a standing, a place in the cycles of human history, unrivaled by any other place.

Such are often the unintended consequences of imperialism.

## VIII

The Napoleonic invasion also had another lingering heritage. On May 22, 1799, this brief, elliptic notice appeared in the *Moniteur:*

> CONSTANTINOPLE: News of a proclamation by General Bonaparte to the Jews, in which he invites them to rally to his standards, to go to rebuild the walls of Jerusalem.[94]

As far as we know, no such proclamation was ever issued. A year earlier, however, and before the departure of the expedition, an article, signed simply "L.B.," had been published in the *Décade Philosophique, Littéraire et Politique,* the house journal of the intellectual group to which Volney belonged, known as the "Ideologues," with whom Napoleon was, at that time, on close terms. This discussed in detail the possibility of a future Jewish state of Palestine, to be created with French arms.[95] Such a state, the author claimed, would not only right a "persecution that has lasted for more than eighteen centuries," it would also place in the heart of the Ottoman Empire a group of well-funded settlers who would bring with them all the benefits they had acquired through their participation in "the enlightenment of Europe." These they would then pass on to the miserable populations of Syria. This, of course, corresponds very closely to Napoleon's own declared "civilizing" intentions—and, as we shall see, to later British projects for Palestine—and Napoleon may well have read the article. ("L.B." may also have stood for Louis Bonaparte, Napoleon's brother.) There are, too, scattered remarks in Napoleon's correspondence and his later reminiscences from Saint Helena that suggest that, from time to time, he entertained the idea of creating a Jewish state, with himself as a new Solomon, just as he had entertained the idea of conquering Constantinople as a new Mehmed II or of ousting the British from India and installing himself as a new Akbar.

Napoleon's attitude toward Judaism (if not the Jews themselves) was very similar to his attitude—real, not declared—toward

Islam. In both cases, he looked upon religion as having usurped the civil law, something that always occurred during the "infancy of nations"; and in both cases he hoped that exposure to European ways would gradually eliminate what he described, in the Jewish case, as "the tendency on the part of the Jewish people to a very great number of practices contrary to civilization and the good order of society." Once the Jews had been assimilated, they would, he believed, "cease to have Jewish interests and sentiments but would adopt French sentiments and interests." All that would remain of their Judaism would be their religion, and that, like Christianity in Europe and eventually Islam in Egypt and Syria, would be a purely private, not a civil, affair.[96] In fact, Napoleon never came near the city of Jerusalem itself, whose population kept the French pinned down on the coast. It was only with the siege of Acre in March 1799 that Napoleon entered Palestine, but his military operations were limited to Galilee and by then he was in no position to attempt to create, populate, and defend an independent state.

Nevertheless, the rumor of Napoleon's supposed desire to restore the Jews to their ancestral home seems to have circulated rapidly through the communities of the Diaspora. Napoleon had broken open the ghettoes of Ancona, and the grateful Jews had planted "trees of liberty" and been given tricolor cockades to wear. Might he have been planning something even grander had not the forces of the British driven him first from Palestine and then from Egypt? Certainly by the late nineteenth century it seems to have become an accepted fact in Zionist circles that this had, indeed, been Napoleon's intention. In 1898, Theodor Herzl, the intellectual founder of the Zionist movement, told the German kaiser, Wilhelm II, that the creation of a Jewish state in Palestine, the "idea that has seized me, had already conquered the heart of that great ruler Bonaparte. What could not be realized under his reign, another emperor might undertake today—you!" In 1915, Israel Zangwill, the president of the Jewish Historical Society of England, wrote to David Lloyd George, at that time minister of munitions, that Britain had a historical duty to "follow Napoleon's example" and permit the Jews to "recover their ancient homeland." Ten years later, when Palestine had a sizable Jewish population but was still a British mandate, Zangwill rose during a meeting of the society, at which Lloyd George was present, and, appropriating Napoleon's declaration before the "Battle of the Pyramids," declared that "Napoleon under the spell of forty centuries that regarded him from the Pyramids announced his desire to restore the Jews to their land." Would England, he went on, "with Egypt equally at her feet, carry out the plans in which she had foiled Napoleon ?"97

At the end of the conference he got his answer. Lloyd George—who was said to "talk about Jerusalem with the same enthusiasm as about his native hills"—by then out of office but still a formidable political presence—made a speech in which he compared the situation in 1799 to that at the end of the First World War.98 The Allies, he claimed, had taken up Napoleon's ideas, but whereas Napoleon, a Frenchman, could not be trusted to maintain his promises, as indeed had turned out to be the case, the English certainly could. "The Jews," he bellowed, "knew the signature of Napoleon was not of much use, but they also knew that the British signature is invariably honored." All present then stood and, with Lloyd George leading the way, sang the Zionist anthem, the "Haitkvah."

Whatever the real impact of Napoleon's expedition may finally have been, here at least were the two things that would forever change the nature of the Muslim world and of all its future dealing with the West: the roots of nationalism and, like a spear thrust into its side, the state of Israel.

# 11

## THE EASTWARD COURSE OF EMPIRE

**I**

In January 1853, Czar Nicholas I of Russia, during a dinner at the palace of the Grand Duchess Helena in Moscow, told the British ambassador, Sir Hamilton Seymour, that it was time their two countries carved up the failing Ottoman Empire. "We have a sick man on our hands," the czar is reported to have said, "a man gravely ill. It will be a great misfortune if one of these days he slips through our hands." And, he might have added, into the waiting hands of the French, the Austrians, the Germans. The phrase stuck. For the next sixty years, until it finally disappeared, the Ottoman Empire was known as "The Sick Man of Europe."[1]

The sickening had, in fact, been going on for a long time. The defeat at Vienna in 1683; the humiliations imposed upon the empire by the Treaty of Carlowitz, when the sultan had been forced to accept the Russian czar, an unbeliever and the ruler of a horde of seminomadic barbarians, as his equal; the loss of the Crimea; the Treaty of Küçük Kaynarca; Napoleon's invasion of Egypt and Syria—these had been only the first signs of the disease. Throughout the nineteenth century, they continued unchecked. In 1804, a Serbian force under the command of George Petrović, known as

Kara (Black) George, successfully besieged the fortress of Belgrade, and after another eleven years of intermittent fighting and intermittent Russian intervention, the Serbs were able to secure quasi-autonomous status for themselves. By 1812, Russia had also seized parts of Moldavia before the Napoleonic invasion of Russia itself forced the czar to accept a settlement.

Many of the Sick Man's ailments had been inflicted on him by his foreign, Christian enemies, most of them by the Russians. But he was also being dismembered from within. In the middle of the eighteenth century in the Nejd, in the center of the Arabian peninsula, a religious leader named Muhammad ibn 'Abd al-Wahhâb began preaching an extreme form of Islamic puritanism, known as the doctrine of *tawhid*, "uniqueness of God." Like most puritans, al-Wahhâb claimed that the current misfortunes of the world were due to backsliding on the part of the faithful, who were much less faithful than they should have been, aided, or at least tacitly abetted, by a lax and impure regime in Istanbul. Like Luther, al-Wahhâb would have been ignored had it not been for the political support he was given, in this case by a powerful tribal chieftain named 'Abd al-'Aziz ibn Muhammad al-Sa'ud, who, with al-Wahhâb's help, founded a dynasty that rules what is now named Saudi Arabia to this day. By the time of his death in 1792, Ibn Saud had created a state that had all but severed its allegiance to the corrupt and impious Ottoman sultanate. Wahhâbi warriors then began spreading northward to the Persian Gulf and then into Iraq, where, in 1802, they sacked the holy places of the heretical Shiites at Kerbala and Nejaf. The following year they seized the city of Mecca itself and stripped the Ka'ba in the Great Mosque of all its ornaments. Although they were driven out again in 1805, by the sharif of Mecca, they returned the following year and then went on to seize Medina as well. Two years later, Ibn Sa'ud's successor, Sa'ud ibn 'Abd al-Aziz, substituted his own name for that of the sultan in the Friday prayers, thus formally, and ostentatiously, renouncing all his ties with Istanbul. Now the whole of the Arabian Peninsula and the holiest sites of Islam were in Saudi hands. Strategically and economically, Arabia counted for very little; but it was of immense religious, and thus political, significance. The sultan's hold over his subject peoples relied heavily on his undisputed role as the supreme leader in the Islamic world. By usurping the sultan's place in the Friday prayers, by driving his representatives out of Mecca and Medina, Sa'ud ibn 'Abd al-Aziz had, in effect, destroyed Selim III's claim to be the "Commander of the Faithful," much less caliph.

The next, more serious, bout of the disease arrived, appropriately, from the West, and not merely from the West but from Greece, the former heartland of all Western culture. The Ottoman Empire was a vast multiethnic state. But it was administered on the basis not of ethnicity or of nationality, but of religion. It included not only Muslims (of various persuasions) but also diverse kinds of Jews and several different denominations of Christians—Orthodox, Roman, Assyrian, Armenian, and Nestorian. Life for these non-Muslim subjects of the sultan was rarely quite as dire as it was often portrayed in the West. Under Islamic law, as we have seen, the so-called protected peoples (*dhimmah*) were entitled to a limited freedom of worship and administrative autonomy so long as they paid the appropriate taxes. They were second-class beings, but for all, that they could live prosperous and untroubled lives and even, on occasion, rise quite high in the various state hierarchies.[2] The Ottomans had extended these rulings into the *millet* system, which created semiautonomous territories, provided for separate courts for cases involving non-Muslims, and placed each minority—of which there were three main ones, Greek Orthodox, Jewish, and Armenian—under a national, and usually also religious, leader (known in Greek as the *ethnarchis* or *millet bashi*), who was responsible directly and personally to the sultan for the conduct of the members of his *millet*. Throughout most of the history of the empire, Christians and Jews had more dealings with their own religious authorities than they ever did with the Ottoman civil power.

The *millet* system brought a degree of independence and also of what might be called benign neglect to the non-Muslim subjects of the empire; but the degree to which it was benign and not merely neglectful should not be exaggerated. It is not obvious that most minorities under Ottoman rule would have endorsed Edward Said's claim that "what they had then seems a lot more humane than what we have now"—although, of course, it rather depends on who the "they" and the "we" are.[3]

Of all the sultan's non-Muslim subjects, the most significant in terms of sheer numbers and power were the Greeks. The capital of the empire was still in many respects a Greek city. (In 1918, it was possible for those who wished to make Istanbul into an international mandate to argue that it was not, and never had been, Muslim, since only 458,000 of its population were Muslim as against 685,000 non-Muslims, and that only a small proportion of the Muslims were actually Turks.)[4] The surviving Greek aristocracy, known as the Phanariot Greeks, since they all lived in the Fener quarter of Istanbul on the Golden Horn, were drawn from a tightly knit group of eleven families, all claiming descent from the ancient Byzantine nobility. They constituted a wealthy and powerful community who collaborated fully and enthusiastically with their Ottoman overlords. Some of them even became the sultan's envoys to the West and regional governors in the Balkans. "We conform," said one of the most illustrious of them, Prince Alexandros Mavrokordatos, later to be named the first president of the independent republic of Greece, "to the prescription of the Gospel, 'Render unto Caesar, the things which are Caesar's.' It is not the custom of us Christians to confuse what is temporal and corruptible with what is divine and eternal."[5]

Such admirable pragmatism often struck outsiders as mere opportunism. It was, said Sir William Eton, a disdainful English observer in 1791, strange to witness the infatuation with which these people sought the political favor of the sultan. "Though styling themselves noble and affecting superiority over the other Greeks," he wrote, "they are the only part of their nation who have totally relinquished the ancient Greek spirit; they seem not anxious, as the islanders are, for liberty, but delight in the false magnificence, and in the petty intrigues of the seraglio."[6]

The *millet* system also conferred upon the Greek Orthodox patriarchs and their metropolitans and bishops a degree of power their Byzantine predecessors had never enjoyed. Despite Mavrokordatos's adroit exploitation of Christ's injunction not to confuse God and emperor, in the Christian communities "in infidel lands" (*in partibus infidelis*), as they were called, the *millet* system had finally broken the long-respected division between Church and state in favor of the Church. In order to protect this authority—and their own lives—many churchmen urged complete submission upon their flock even to the point of representing the Ottoman conquest of the Byzantine Empire as a divine punishment for the Orthodox Church's many sins. "This powerful empire of the Ottomans," the patriarch Anthimus of Jerusalem told his flock in 1798, had been raised up "higher than any other kingdom" to keep the poor Orthodox Greeks from slipping into the heretical clutches of the Latin Western Church. The sultan had thus bestowed on "us the people of the East a means of salvation." He had also thoughtfully provided his Christian subjects protection against another creeping evil, disseminated recently by the Devil and the French Republic, namely "the much vaunted system of liberty," which was nothing other than a trap to "drive the people headlong into corruption and confusion."[7]

But if the churchmen and the Phanariots had flourished through, in Eton's words, "treachery, ingratitude, cruelty and intrigue which stops at no means," both the Greek peasantry and the merchant communities suffered increasingly during the eighteenth and nineteenth centuries as the sultan gradually lost control of his provincial governors. Corruption, banditry, and piracy at sea drove more and more wealthy Greeks overseas. Trade in Greece, as in other parts of the Ottoman Empire, fell into the hands of western Europeans prepared, and equipped, to take the high risks involved. In exile in France, Britain, Germany, and even Russia, the Greek intelligentsia and the Greek merchant class had

witnessed the ever-widening contrast between the polish and re-
finement, the respect for life, order, and property of the European
nations, and the deepening chaos and misrule that prevailed
within the frontiers of the Ottoman state, where the sultan, "slaying, drowning, hanging with his will as the only law," provided the
only source of political authority.[8] "Rise up," called one of them,
Ioannis Pringos, from Amsterdam in 1768, "another Alexander,
who as he drove the Persians out of Greece, may expel this tyrant,
that Christianity will once again shine in the Greek lands as of
old."[9]


ENLIGHTENMENT AND THE prosperity and relative political
stability of late-eighteenth-century Europe had provided educated Greeks with one incentive to throw over their troubled and
decaying Ottoman rulers. The French Revolution had provided
another. In 1797, Rigas Velestinlis (or Rigas Feraios), a hellenized
Vlach from Thessaly, poet, pamphleteer, and the protomartyr of
Greek independence, called for the creation of a multicultural
Greek republic—freedom-loving Turks would also be welcome—
modeled on the French constitutions of 1793 and 1795. Velestinlis's new Greece was to be, in effect, a restored version of the
Byzantine Empire with a republican rather than a monarchical
constitution. Nothing came of this or of any of his other revolutionary writings. In 1798, Velestinlis set out for Venice to make
contact with Napoleon and offer him the services of a shadowy
Greek revolutionary society that may have existed only in his
imagination. Before he arrived, however, he was betrayed and
handed over to the pasha of Belgrade, who ordered him to be
strangled and his body thrown into the Danube. Just before his
death he is said to have told his captors, "This is how brave men
die. I have sown; the time will come when my country will gather
the harvest."[10] They turned out to be prophetic words, for if Velestinlis's political vision of an enlightened, republican Greece
that would rekindle the splendors of democratic Athens and the

grandeur of Byzantium remained largely unfulfilled until the late twentieth century, his martyrdom became a powerful inspiration for the leaders of the uprising that began some twenty years later—as a tribute to which his face now appears on the Greek ten-euro-cent coin.

The first serious Greek move to expel the new Persians from Greece itself came in September 1814 with the creation by three Greek expatriates—the freemason Emmanuel Xanthos, Nikolaos Skouphas, and Athanasios Tsakalov—in the Russian port of Odessa of the Philiki Etairia, or "Friendly Society." Other societies with specifically Hellenic aims had been created before. The Ellinoglosson Xenodocheion, or "Greek Hostel," had been set up in Paris in 1807, and in 1814, the Philomousos Etairia, or "Society of the Friends of the Muses," was founded in Athens and Vienna. But these had been intended to remind the Greeks of their unique cultural heritage through education, archaeology, and philology and had had no overt political objectives. The Philiki Etairia, by contrast, was dedicated to the cause of national autonomy and the liberation of the "Motherland" by violent means, and it was duly anathematized by the Holy Synod in 1821 as an unwarranted, "evil and insubordinate" assault on "our common, generous, solicitous, powerful and invincible Empire."[11] Despite, or perhaps because of, ecclesiastical censure, over the next few years membership of the Philiki Etairia grew considerably. Since the society aimed only at the overthrow of the Ottoman government and espoused no ideology, nor offered any vision of what kind of Greek society would replace the empire, it found support in all ranks of society, even among some of the lesser clergy, although more than half of its initiates seem to have been merchants of one kind or another, mostly poor and marginalized ones. Although the Philiki Etairia, like most secret societies, was strong on rhetoric, ritual, and madcap schemes and poor in strategies and objectives; and although by itself it achieved almost nothing, it did succeed in creating a network of dedicated patriots who pro-

vided much of the organizational groundwork for the coming revolt.

The spark that lit the flame came on March 25, 1821. Germanos, the metropolitan of Old Patras, in defiance of the Muslim
prohibition against any display of religious symbols by "protected
peoples," raised the cross at Kalvryta in the northern Peloponnese. This, or so legend has it, was the beginning of the Greek War
of Independence. A month later, on Easter Saturday, April 22, the
Orthodox patriarch, Gregory V, who as *ethnarchis* of the Greek *millet* the sultan held responsible for the behavior of all his Christian
Greek subjects, was hanged publicly from the lintel of the gate of
the patriarchate in Istanbul, together with two of his chaplains.
His body, one English witness recalled, "was suffered to remain
suspended at the doorway, so that everyone who went in and out
was compelled to push it to one side." Three days later, it was
taken down, thrown to a Jewish mob assembled for the purpose
by the Ottoman authorities, and dragged by the neck through the
streets, "where offals of all kind were lying about in foul masses"
and then dumped into the harbor, "where the waters closed over
it."[12] After Gregory's death, the gate of the patriarchate was closed
and it remains closed to this day. A pact that had been in existence
since the days of Mehmed the Conqueror had been broken.

Of all the wars for independence by all the various peoples
who made up the Ottoman Empire in all its long history, from the
Bosporus to the shores of the Caspian, this, for the West, was the
most emotive. The uprising of March 21, however, had begun not
in the name of ancient Greece but in that of the Greek Orthodox
Church and as such aroused little immediate sympathy in western
Europe. Quite apart from what was looked upon, even by many
Greeks, as the submissive, sycophantic behavior of the Orthodox
hierarchy, the view of most western Europeans toward the Greek
Church had changed little since the fifteenth century. Nothing,
wrote one English traveler on the eve of the uprising, revolted him
as much as "the inane ceremonies and disgusting superstitions"

of the Greek brand of Christianity.[13] Neither were the modern
Greeks themselves an altogether inspiring people: ignorant, factious, and rude, they seemed, as they had to Voltaire, to be living
proof of the belief that despotisms can transform even the most
liberty-loving of peoples into mere slaves. In 1820, the Russian
poet Aleksandr Pushkin had written an "Ode to Freedom" in defense of the Greek cause (for which he had been rewarded by the
less-than-liberty-loving czar with imprisonment in Bessarabia)
and called for the Turks to withdraw from Hellas, "the legal heir
of Homer and Themistocles." But when later he came across some
Greek merchants in Odessa and Kishinev, he was horrified. These
"new Leonidases," he wrote in disgust, were not the heroes he had
been expecting but a "nasty people of bandits and shopkeepers."

Despite the unpromising nature of the modern Greeks, it did
not take much to transform the struggle for the independence of
modern Greece into a struggle for the liberation of the entire heritage of the ancient world. Here was the "cradle of Western civilization," pitted against the most tyrannical, most ferocious of all
the despotates of the East. Already by February 1821, one month
before Germanos raised the cross in the Peloponnese, the Phanariot Prince Alexandros Ypsilantis, who had taken a commission
in the Russian army and risen to become the czar's aide-de-camp,
crossed the River Pruth at the head of a motley Russian-backed
army. On the twenty-fourth, in the Moldavian capital of Jassy, he
issued a proclamation calling upon his fellow Greeks to "invite
Liberty to the classical land of Greece." Let us, he went on, "do
battle between Marathon and Thermopylae. Let us fight on the
tombs of our fathers, who, so as to leave us free, fought and died
here." And he called up the shades of all the Greeks who had died
in the cause of liberty, ending with "Leonidas and the Three Hundred, who cut down the innumerable armies of the barbarous Persians, whose most barbarous and inhuman descendants we today,
with very little effort, are about to annihilate completely."

The task of cutting down the Turks, however, turned out to be
rather less effortless than he had supposed. After several ineffec-

tual skirmishes, Ypsilantis's army and the "Sacred Battalion" of Greek students who had joined it, was destroyed in June at the battle of Dragatsani. Ypsilantis, however, had made his point—at least as far as the rest of western Europe was concerned. The fight for Greece was to be a fight for the virtues of the entire Western world, which had been born in Greece and for which, or so Ypsilantis professed to believe, the "enlightened peoples of Europe" were "full of gratitude."[14]

On April 9, 1821, shortly before the killing of Gregory V, the revolutionary leader Petros Mavromichalis, known as Petrobey, proclaimed leader of the senate of Messenia, "commander in chief of the Spartan forces," in the southern Peloponnese, addressed the rulers of Europe. "Greece, our mother, was the lamp that illuminated you," he told them. "On these grounds she reckons on your active philanthropy. Arms, money and counsel are what she expects of you."[15] Not much, however, of any of those things arrived. The call for the independence of the Greek world initially aroused only suspicion in the minds of most of the leaders of the West, ever wary of the possible consequences that the collapse of the Ottoman Empire might have for the balance of power within post-Napoleonic Europe. Prince Klemens von Metternich, the Austrian chancellor, dismissed Greece, as he would later dismiss Italy, as merely a geographical expression, with no real claim to a national identity. After the killing of the Greek patriarch, Czar Alexander I withdrew his ambassador from Istanbul. But like most such gestures today, this had no significant impact.

The Greek rebellion found, however, a ready response among the educated, liberal-minded, mostly middle-class elites of both Europe and the United States. "We are all Greeks," wrote an enthusiastic Percy Bysshe Shelley in the autumn of 1821, after he had read about the initial uprising.

> The apathy of the rulers of the civilized world to the astonishing circumstance of the descendants of that nation to which they owe their civilization . . . is something perfectly

inexplicable to a mere spectator of the shews of this mortal scene. . . . Our laws, our literature, our religion, our arts have their root in Greece. But for Greece . . . we might still have been savages and idolaters; or what is worse, might have arrived at a stagnant and miserable state of social institutions as China and Japan possess.

*Another Athens shall arise*
*And to remoter time*
*Bequeath, like sunset to the skies,*
*The splendour of its prime;*
*And leave, if naught so bright may live,*
*All earth can take or Heaven give.*

Some months later, from exile in Paris, one of the earliest prophets of the revolution, the classical scholar Adamantios Koraïs, wrote an appeal to the "citizens of the United States," as the new heirs of the ancient Athenians. "It is in your lands that liberty has fixed its abode," he told them. "Free and prosperous yourselves, you are desirous that all men share the same blessings; that all should enjoy those rights to which all are by nature equally entitled." And it was now up to the Americans to restore Greece to its rightful place in the world.[16] Despite Koraïs's embarrassing belief that the new abode of liberty had also been responsible for the abolition of slavery, the response was rapid. In July 1821, a dinner for Americans was held in Paris at which both Washington Irving and Lafayette were present and at which a toast was proposed to "The land of Minerva, the birthplace of the Arts, Poetry and Freedom—civilizing her conquerors in her decline, regenerating Europe in her fall. May her sons rebuild her clime the home of Liberty."[17] Edward Everett, author of the (first) Gettysburg Address, elected professor of Greek at Harvard in 1815 at the age of twenty-one, and editor of the *North American Review*, a journal with an immense circulation, used his influence to have Koraïs's letter printed in newspapers up and down the country. Donations poured in. In the winter of 1821–1822, private donors from Charleston, South Carolina, alone sent fifty barrels of dried meat to feed the insurgents.

On the final day of 1821, a National Assembly was convened near Epidaurus that declared Greece to be an independent republic, adopted a constitution loosely modeled on the French constitution of 1799, elected as its president Prince Alexandros Mavrokordatos, and stated that "the Greek nation calls Heaven and Earth to witness that in spite of the dreadful yoke of the Ottomans, which threatened it with destruction, it still exists." Greeks now took to slaughtering Turks and Turks, in retaliation, Greeks. In the end it was, of course, the Greeks who prevailed. More than twenty thousand Turkish men, women, and children, some of whom had lived in Greece for generations, were hunted down by bands armed with clubs and scythes, often led by priests who months before had protested their unwavering loyalty to the "blessed sultan." Within weeks of the outbreak of the revolution, the Turkish and Albanian Muslim population of the Peloponnese had ceased altogether to exist as a community. What little remained of it huddled for protection in the remaining Turkish enclaves on the coast, but these, too, were under siege. The moon, as the Greeks said, had devoured them all. It was a slaughter that Greece and Europe, swept along by talk of the heroic liberators of ancient liberty, has chosen to forget.[18]

Despite these events and all the noise now being generated at home, the governments of Europe still remained aloof. In April 1822, however, a Turkish expeditionary force sacked the island of Chios and butchered and enslaved its inhabitants. The reaction throughout Europe was one of horror and disgust.[19] In March 1823, a London Greek Committee was formed and began collecting funds on behalf of the cause. The same month the foreign secretary, George Canning, officially recognized the Greeks as belligerents, which, among other things, gave them the right to search neutral ships for war supplies.

Slowly, haltingly, the governments of Europe moved toward

recognizing, if not actually assisting, the Greek cause. Official reluctance, however, was offset by personal enthusiasm. Volunteers flooded in from all over the Western world to reclaim the cradle of civilization from the hands of "the monstrous Ottoman dynasty."[20] They called themselves the "Philhellenes"—the "Greece Lovers"—and they came from Britain, France, Italy, the German states, Spain, Portugal, Hungary, Poland, Switzerland, Sweden, Denmark, the United States—even one from Cuba. Most were idealists. Some were former soldiers looking for employment, some were speculators and spies, some simple opportunists. There were even missionaries from England and the United States trying—with singular lack of success—to turn Greek Orthodox Christians into any one of several kinds of Protestants. And, as with all such irregular conflicts, there was the usual handful of simple eccentrics: a Bavarian china manufacturer hoping to set up a factory in Greece, an unemployed French actor, a bogus Danish count who traveled around with a lithographic press on his back, a dancing master from Rostock, a Spanish girl disguised as a man.[21] There was also—or rather there was not—Edgar Allen Poe, who put about the story that he had set out "without a dollar on a quixotic expedition to join the Greeks then struggling for liberty." In fact, he got no farther than Boston.[22]

The most famous of these, however, although he never saw a shot fired in anger, was George Gordon, Lord Byron.

On April 7, 1823, Edward Blanquiere, a former Irish sea captain and a member of the London Greek Committee, together with a Greek named Andreas Louriotis, whose task was to raise money for the war, called on Byron, who was then living in Albaro, outside Genoa. Byron, who had recently had a brief and unsatisfactory fling with the Carbonari, the Italian partisans seeking independence from Austria, and a slightly longer and more satisfactory one with Countess Teresa Guiccioli, was restless and undecided whether to spend, as he put it, a "simple but useful life" in South America or to go to Greece. Blanquiere and Louriotis made up his mind for him. "I am determined," he wrote two months later, "to go to Greece; it is the only place I was ever contented in."[23] He gathered together about 9,000 pounds sterling—an immense sum in those days. "Cash is the sinew of war," Byron once wrote, "as indeed of most things—love excepted and occasionally of that too." He sold his sailing ship, aptly named *Bolivar* in honor of his South American aspirations, and chartered a three-masted, round-bottomed, bluff-bowed clipper of 120 tons, called—also aptly, in anticipation of his new Greek ambitions—*Hercules*.[24] On July 16, she set sail with Byron, Blanquiere, Louriotis, and an assortment of friends and retainers on board, and made her way down the coast to Livorno and then to Kephalonia. There Byron went ashore, took a house at Metaxata, on the coast south of Argostoli, and pondered what to do next. After four months of fending off incessant demands from the several factions now fighting for control of Greece, he decided to go to Missolonghi, which, despite lying on the edge of a huge and in those days mosquito-infested lagoon, was then the most prosperous town in western Greece. In 1821 its population had massacred or driven out all the Turks, and it was rapidly becoming the epicenter of the uprising.

The move would prove to be a fatal mistake. Byron was rich, he was famous, he was "Milord." He arrived to a twenty-gun salute and cheering crowds, like, it was said at the time, the Messiah—a Messiah in a gaudy red military uniform of his own devising and with a number of gold helmets. (In the end he never wore any of them, although he had himself painted with them piled at his feet.)

Before long Byron had recruited his own private army, the "Byron Brigade," composed mostly of Albanian refugees and mercenaries posing as the victims of Turkish atrocities. Each morning, Milord would go riding at the head of this motley gang of brigands, and in the afternoon he held inconclusive conferences in a house near the shore. Byron also did something that almost no other Greek commander did, or could: he paid his men. Before long, recruits began to flood in from all over Europe. All roads, it seemed, now led to Missolonghi. As Count Petro Gamba, a younger brother of Teresa Guiccioli, who had accompanied Byron from Genoa, tellingly said of the Philhellenes, "we had them of all nations—English, Scotch, Irish, Americans, Germans, Swiss, Belgians, Russians, Swedes, Danes, Hungarians and Italians. We were a sort of crusade in miniature."[25] But this diverse assembly of characters did nothing except fight among themselves while waiting for a conflict that never came. The Philhellenes squabbled endlessly over rank and status. The Albanians mutinied for more pay, and both they and the Greeks attempted to pillage the arsenal. Then, in February 1824, after riding through a rainstorm, Byron contracted some kind of fever—probably tick fever, a disease transmitted by dog fleas—and took to his bed. The doctors then finished him off. They not only bled him almost dry (it has been calculated that the amount of blood that was taken from him would alone have proved fatal, even if he had been in perfect health), they also applied a desperate collection of patent remedies, lunar caustic, purgatives, and a solution of cream of tartar known as "imperial lemonade," all of which only lowered the resistance of his already wasted body. He died around six in the evening of April 19, 1824. After his death his personal physician, Julius Millingen, who was responsible for most of the damage, sent in a bill of 200 guineas with the comment "Lords do not die every day."[26]

While Byron had been conducting his own war from Missolonghi, the situation in the rest of Greece had deteriorated into a condition of virtual civil war. In December 1821, Demitrios Ypsilantis, Prince Alexandros's brother, had succeeded in convening a national assembly at Argos. This drafted a constitution for a new independent Greece, modeled, as Rigas Velestinlis's had been, on the French constitutions of 1793 and 1795. Another assembly was held two years later in Astros. Although both initiatives laid the grounds for some kind of future Greek state, they did very little to ease the hostility among the various factions. By the summer of 1823, the Peloponnese and the eastern and western parts of the Greek mainland had been liberated from Ottoman rule. Who now governed them, however, was by no means certain. By the end of 1823, there were two rival governments claiming to represent the Greek insurgency, neither of which exercised undisputed control over any single area, and it was not long before war broke out between them. One of the difficulties faced by all the factions was that although everyone knew, or imagined that they knew, what ancient Greece and Byzantine Greece had been, no one had any clear idea of what modern Greece might, or should, look like. For most Greeks, in particular most of those doing the fighting, the term "motherland," *patrida,* used by Rigas Velestinlis, Ypsilantis, and other cosmopolitan patriots to describe the new Greece, referred not to a nation but to a region, sometimes even a village. The Ottomans—and Metternich—were in some sense right in claiming that there was no such place as Greece. Like so many of the nations freed from long-dominant imperial overlords, it had to be invented. The trouble was that, in this case, the process of invention turned out to be a protracted, costly, and bloody affair.

In 1824, the Sultan Mahmud II persuaded Muhammad 'Ali, the semi-independent *wali* of Egypt, whom we met in the last chapter, to put an end to the insurgency while the insurgents were still busy fighting one another. In exchange he was promised the *pashalik* (governorship) of Crete for himself and that of the Peloponnese for his son Ibrahim. Muhammad 'Ali commanded a far more imposing army than the demoralized Ottoman forces and, perhaps more significantly, had a modern navy equipped with French-built ships commanded by former French naval officers. In June, he took first Crete and then Cyprus. The following year, Ibrahim landed on the Morea. In April 1826, he sacked and destroyed Missolonghi, leaving only three buildings standing, one of which was Byron's former residence. This, said the picaresque adventurer Edward Trelawny, who had been a member of Milord's entourage and returned briefly to Missolonghi after the sack, "loomed like a lonely column in the midst of a desert."[27]

By August, Ibrahim was in Athens. The population fled to the Acropolis, as it had done in 480 B.C.E., and from there the women,

children, and aged took refuge at Salamis. Alas, this time there
was no Greek fleet waiting for them. And Themistocles, in the
shape of the British admiral and sailor of fortune Lord Cochrane,
failed to arrive until the spring of the following year. As he sailed
past the still-besieged Acropolis, Cochrane, summoning all his
meager powers of description, wrote in his diary, "There was the
seat of science, of literature. At this instant the barbarian Turk is
actually demolishing by the shells that are flying through the air
the scanty remains of the once magnificent temples of the Acropolis."[28]

Ibrahim's ravages through the Morea forced the European
Great Powers to act. On July 6, 1827, France, Britain, and Russia
signed the Treaty of London. Greece was declared to be an autonomous, although not sovereign, state, and the three powers
undertook to impose some kind of mediation on the various belligerents, both Greek and Turk. The Greeks of all factions agreed,
at least in principle. The Ottoman sultan, not surprisingly, did
not. But it had become clear to Ibrahim, and his father, that the
entry of the allies into the war had altered their situation dramatically for the worse. The Prophet, Ibrahim told his representative
in Istanbul, had promised the faithful that Islam would one day
triumph over the entire world: but that might just as well be a reason for caution as for desperate measures. Better to retreat now,
and return at some later and more propitious time, than risk losing the entire fleet, and the 40,000 Muslims who sailed in it, to a
now far more powerful adversary.

His canny prudence, however, came too late to save him. On
October 20, the combined Egyptian and Turkish fleets, 89 ships
and 2,240 guns, were lying in the Bay of Navarino. Late in the
morning and with a light following wind, the allied fleet entered
the bay led by the British admiral Lord Codington in a ship somewhat inauspiciously named *Asia*. By six in the evening, the battle
was over. The Ottoman fleet lost 81 ships of the line and between
4,000 and 6,000 men. The Allies lost only 174 men and not a single ship. The battle of Navarino in effect secured the future inde-

pendence of Greece. It did not, however, secure the kind of liberal republican constitution that the founders of the revolution had had in mind. With the Great Powers now firmly in charge, what the Greeks got was yet another monarchy. After a certain amount of squabbling about who should be the new king, in May 1832, Britain, Russia, France, and Bavaria conferred upon Prince Frederick Otto of Wittelsbach, the seventeen-year-old son of King Ludwig of Bavaria, the "hereditary sovereignty" of the new "monarchical and independent state" of Greece. His heirs clung precariously to the throne through the next century and a half of civil war, invasion, coup, and countercoup until the last king of the Hellenes, ironically named, like the last Byzantine Emperor, Constantine, departed into exile in 1967.

THE CESSATION OF Greece, the first of its provinces to win recognition as an independent state, was the worst blow the Ottoman Empire had to suffer, or would suffer again until 1918. But it was also only the beginning. In 1814, the British had occupied Corfu. In 1882, they returned to Cairo and, despite repeated protestations that they were about to leave, remained until Egypt became formally independent in 1922. Until the outbreak of the First World War, Egypt was nominally a quasi-independent state ruled by a khedive—all descendants of Muhammad 'Ali—under the sovereignty of the Ottoman sultan. In reality, however, Egyptian independence was, in the words of the prime minister, Lord Salisbury, who was largely responsible for the situation, "a screaming farce."[29] British rule was just about as indirect as it could possibly be. As Ronald Storrs, Oriental secretary (or specialist in "Eastern affairs") wrote in 1914 to the governor-general of Egypt and the Sudan, Lord Kitchener, in nicely calibrated Foreign Office prose, "We deprecated the Imperative, preferring the Subjunctive, even the wistful Optative mood."[30] But no matter in which mood the British chose to govern, govern they did, and although the sultan retained his claim to sovereignty, Egypt be-

came, in all but name, a British protectorate (a "veiled protectorate," Lord Milner, another imperial proconsul, called it), in
which the khedive and his cabinet issued the commands that were
written for them by their British advisers. Looking back on
Britain's successful seizure of this most important part of the
Arab world from the Ottoman Empire, Evelyn Baring, Lord
Cromer, who had been Salisbury's "agent"—in every sense of the
term—in Egypt for twenty-four years, noted with satisfaction, "All
history was there to prove that when once a civilized Power lays its
hands on a weak State in a barbarous semi-civilized condition, it
rarely relaxes its grip."[31]

## II

The impact of these humiliations over the course of more than
two centuries upon the standing of the Ottoman sultan with all
Muslims had been immense. If the Western infidel enemy had triumphed so decisively and incontrovertibly over God's peoples,
clearly something must be wrong. But what? As in nearly all such
situations, the possible replies were limited to two: defeat was due
to either the skill or virtues of the victor or it could be blamed on
some internal weakness of the vanquished. If the former were the
case, the obvious thing to do was to find what it was that had
made the enemy successful and then imitate it. If the latter, then
the only course open was to discover what had gone wrong and to
try to put it right. The latter invariably came down to assuming
that either society was corrupt and decadent or some kind of sin
had been committed—or, most often, a combination of the two.
Remedy, therefore, demanded that either society should be purged
or—which often came to the same thing—an irate god should be
placated by enforcing fierce and often literalist interpretations of
his imagined wishes. Ever since the late eighteenth century, the
story of the response of the Muslim world to the steady and inexorable encroachment of the West oscillated, at times violently, between these two extremes.

The first response from the great powers of the Islamic world was to attempt to discover just what it was that had suddenly made the hitherto chaotic and ineffectual Franks so powerful. As early as the late fifteenth century, Ottoman sultans had begun to adopt European artillery, handguns, and mines and had also employed a number of Europeans to show them how to make and deploy them. Turks and Greeks had cooperated in one way or another on the frontier for centuries. They had influenced each other in dress, in diet, in speech, even in popular religious observances. (There are more than a few saints, holy places, and festivals common to both Greek Christians and Turkish Muslims.) It was not difficult to extend these exchanges into the fields of science and technology. "No nation in the world," wrote the emperor Ferdinand's ambassador to Istanbul, Baron Ogier Ghiselin de Busbecq, in 1560, "has shown greater readiness than the Turks to avail themselves of the useful inventions of foreigners, as is proved by their employment of cannons and mortars, and many other things invented by Christians."[32] The Ottomans also copied European shipbuilding, initially from the Venetians and later from the English and the French, so that by 1682, three-decked square-rigged sailing ships capable of mounting large numbers of guns, of the kind the English and the Dutch had been using for more than a century, were being produced by Ottoman shipyards.

These moves, however, were all limited, and they certainly did not include any greater interest in the culture that had been able to produce, in this way, better weapons, larger and faster ships, and more accurate maps. "They cannot," noted Busbecq of the Ottomans, "be induced as yet to use printing, or to establish public clocks, because they think that the Scriptures, that is, their sacred books—would no longer be *scriptures* if they were printed, and that if public clocks were introduced, the authority of their muezzins and their ancient rites would be thereby impaired."[33] Things that were evidently useful and with apparently no religious connotations might be borrowed, although the more devout had their doubts about even those; but anything that posed even the most tenuous threat to the status of Islam and the purity of the Qur'an was unacceptable.

Print technology and clocks were not the only European innovations that were cautiously examined and then rejected as potentially subversive. Very little knowledge of European medicine, for instance, ever made its way into the Islamic world, not that it would have been of much use to anyone if it had. (It was, after all, the Ottomans who were indirectly responsible for introducing inoculation, perhaps the most significant medical discovery of the eighteenth century, into the West.)[34] But the same distaste for anything Western, anything apparently Christian, that had shut out European medicine also shut out all the other European sciences. To the Islamic mind, the works of Descartes, Kepler, and Galileo were as false and irrelevant as the writings of Luther and Calvin.[35]

After Carlowitz, however, and a similarly humiliating treaty signed with the Austrians at Passarowitz in 1718, which effectively restored the frontiers of Hungary and Croatia to where they had been before the conquests of Süleyman the Magnificent, the Ottoman authorities began to give serious consideration to the notion that the by now undeniable technological and organizational superiority of the Christians might have something to do with the ways in which their societies were organized and governed. In 1719, the grand vizier, Damad Ibrahim Pasha, sent an ambassador, Mehmed Said Effendi, to Paris with instructions to "make a thorough study of the means of civilization and education, and report on those capable of application."[36] It was a promising beginning, and in the next few decades a number of Europeans, most of them French, traveled to Istanbul to help modernize the archaic, creaking Ottoman state. In July 1727, the sultan, Ahmed III, issued an imperial *ferman* authorizing the establishment of a printing press in the "High-God-Guarded City of Constantinople" that was licensed to print all manner of books in

Turkish—except of course the Qur'an itself. They also began to import European clocks and clock makers, one of whom was Isaac Rousseau, the father of Jean-Jacques, who, if his son is to be believed, became the official "watchmaker to the Seraglio."[37]

In 1731, a renegade French nobleman, Count Claude-Alexandre de Bonneval, was commissioned to reform—in effect create—the Bombardier Corps, for which he was made a pasha by the grateful sultan. In 1720, another Frenchman, named David—who converted to Islam and took the name Gerçek—successfully organized the first fire brigade in Istanbul. More significantly, in 1734, a new training center and school of geometry, the Hendese-hane, was opened in Üsküdar, whose purpose was to teach the basic principles of Western mathematics to Ottomans. One of its instructors, Mehmed Said, designed a "two-arch quadrant" to be used for sighting long-range guns and wrote a treatise on trigonometry. In 1773, a school of mathematics was opened for the use of the navy under the auspices of the Baron de Tott, a French officer of Hungarian origin who helped train the sultan's army, managed his arsenal, and was the author of *Memoirs of the Turks and the Tartars,* from which Napoleon had taken much of what he knew about the Ottoman Empire. For a while, most things European, and in particular French, were cautiously accepted by the wealthier citizens of Istanbul. Even French gardens and French furniture enjoyed a brief popularity among the members of the court, and for the first time since the reign of Mehmed II, an Ottoman sultan allowed himself to be painted by a European painter.

This was one side of the coin. The other is provided by a poem Sultan Mustafa III wrote in 1774, shortly before his death, on the condition of his realms. It is filled with self-pity and despair:

> *The world is in decay, do not think it will be right with us;*
> *The state has declined into meanness and vulgarity,*
> *Everyone at the court is concerned with pleasure;*
> *Nothing remains for us but divine mercy.*[38]

In Mustafa's opinion it was the world that had declined, not the Ottoman state; it was the frivolity of the sultan's subjects that had been responsible for the loss of the Crimea and the crippling defeats at the hands of the Russian infidels. The only remedy was to surrender to the will of Allah and hope for his mercy—that, and reform the morals of the court, reinforce the power of the Shari'a—reaffirm, in other words, all that was old in the forlorn hope that the old, once made strong enough, would be able, ultimately, to resist the new. As always in such responses, the question asked was inevitably the wrong one—not what have they done that makes them strong, but what have *we* done that has made us weak?

Mustafa did not live long enough to do any of the things he proposed. It is unlikely that, even if he had, the reforms, uncertain though they were, could have been reversed. The presence of the Russians was always there to remind the sultan of the technical shortcomings of his military machine. But the seesaw between an always limited and hesitant bid for change and a call, loudly backed by the religious establishment, for moral and spiritual retrenchment—which invariably meant a rejection of anything from the West—continued until the empire itself was no more.

Mustafa did not survive to sign the treaty of Küçük Kaynarca in 1774. This unhappy task fell to his brother, Abdülhamid I, who watched as the empress Catherine marched through her newly acquired territories in the Crimea, seemingly bent on restoring the Byzantine Empire with herself as the new basileia. In 1779, to commemorate the birth of her provocatively named grandson Constantine Pavlovich, she had a coin struck with the image of the Hagia Sophia. In 1789, she met the emperor Joseph II in a camp erected outside Kherson in the Ukraine. In an elaborately stage-managed triumphal entry, intended to evoke those of the Roman emperors, the two rulers passed into the city beneath an arch inscribed in Greek (a language spoken by neither ruler) with the legend "The Way to Byzantium." Whether or not all this was mere charade, Abdülhamid took it seriously enough, and Russia and the Ottoman Empire went back to war again. Again the outcome was a humiliating peace, signed at Jassy in January 1792.

The need for reform grew ever more pressing. Already in 1791, Selim III, fully aware of the impending outcome of the war with Russia, had issued a command to a group of twenty-two soldiers, bureaucrats, and clerics to come up with some quick and effective way of stemming the tide of destruction that was threatening to sweep the empire into oblivion. Among them were two Christians: a French officer named Bertrand; and Mouradgea d'Ohsson, the Armenian dragoman (translator) at the Swedish Embassy in Istanbul. They presented their replies in the form of memorials—*lâyiha*—by analogy, perhaps, with the French *cahiers* of 1789 that had helped bring about the revolution.[39] Their responses were predictably mixed. Some suggested retrenchment, a reassertion of the old values, a return to former military glories by former methods that, they somehow imagined, had merely been corrupted, rather than overtaken by superior Western military technologies. Some suggested the (cautious) adoption of Western techniques, training, and weapons, if not the Western ideas that lay behind those methods. Some, however, went so far as to suggest that the army should be entirely replaced with a new one, trained, equipped, and armed along European lines.[40] Selim himself, more audacious than any of his predecessors, tended to agree with the last of these suggestions, and between 1792 and 1793, he put into place a series of reforms, known collectively and significantly as the Nizam-i Cedid, the "New Order." New regulations were imposed on provincial governors, taxation, and control of the grain trade in an attempt to modernize the economy of the empire and reverse the slowly disintegrating hold the sultan had over the provinces. (It is no coincidence that some thirty years earlier, Charles III of Spain, had attempted similar reforms in a similar, and equally doomed, bid to prevent the disintegration of another archaic empire. The force of reform generated by the European Enlightenment, moderate though it was, proved to be irresistible.)

Most radical of all was the creation of new military and naval schools teaching not only the rudiments of Western techniques of gunnery, fortification, and navigation but also the Western science required to master them. These schools were staffed by French officers, and French was made a compulsory language for all students. In defiance of the general Islamic injunction against all forms of non-Islamic learning and non-Islamic writings, these schools were provided with a library of some four thousand European books, which included a copy of the *Encyclopédie*—a work the conservative Muslims found quite as offensive as conservative Christians did.

All of this was brusquely interrupted in 1798 by Napoleon's invasion of Egypt. But Selim was determined, despite his appeal to Islamic universalism in his attempt to regain control of Egypt, to rule as a European monarch. For a while, as we have seen, he railed against Western atheism, the evils of the doctrine of equality, and the potentially destructive force of Western rationalism. But once the French were out of Egypt, the policies of westernization began again. (He even forgave Napoleon for the Egyptian expedition, officially recognized him as emperor in 1806, and presented him with a portrait of himself.)

The "New Order" proved, however, to be more than the old order, in particular the Janissary Corps, could tolerate. In the summer of 1806, a series of uprisings against Selim's reforms spread to Istanbul. Angry and devout Muslims rose up, in the words of the imperial, and deeply conservative, chronicler Ahmed Asim Effendi, to resist "the malicious crew and abominable band" of the Franks, who "by incitement and seduction to their ways of thought" had sought to undermine "the principles of the Holy Law."[41] Within a year the forces of reaction had won, and on May 28, 1807, Selim III was deposed. The "New Order" was extinct.

But not for long. Selim's successor, Mustafa IV, brief and ineffectual, was assassinated the year after he came to power, and his successor, Mahmud III, initiated a far more radical process of reform than even Selim had imagined—although wisely, he chose not to give it a suggestive and menacing name, in particular one that included the word "new," against which his enemies could focus their resistance. Provincial nobles were steadily stripped of their power. A new ministerial system was created and the finances of the empire were overhauled. A new system of taxation was introduced, and a census—the first of its kind in the empire—was taken, to establish who could pay what. In the early nineteenth century the Ottoman sultan was doing what Philip II of Spain and Louis XIV of France had done in the sixteenth and seventeenth.

On June 14, 1826, after a brief but bloody struggle, in which large numbers were burned to death in their barracks and hunted down as they tried to flee, the Janissaries were suppressed. A new army, equipped and trained by Europeans, took their place. "The Victorious Soldiers of Mahmud," as they were called, were given uniforms and a new headgear, the fez, which replaced the traditional turban. The sultan himself began to adopt a modified version, known as *istanbulin*, of the traditional frock coat and trousers worn by the monarchs of Europe.

Mahmud had been careful to introduce all of his reforms with the active, if often forced, participation of the clergy and to pass all of them off as the expression of religious duty. He had also been scrupulous to avoid any suggestion that his reforms might be the advance guard of a more systematic process of westernization. But for all his assurances, it was clear to outside observers that these were the first, uncertain steps in the transformation of a tribal Islamic empire into a modern absolute monarchy.

It would be another ten years, however, and another change of sultan, before the process began in earnest. In 1839, Sultan Abdülmecid I inaugurated what came to be called the Tanzimat, or "Reordering." Cautiously the word suggested that all Abdülmecid was attempting was a rearranging, a reinvigoration of the old ways. In fact, however, the Tanzimat would go a long way beyond either Mahmud's reforms or Selim's aborted "New Order." These had confined themselves to changing the way the existing institutions operated. But they had left most of those institutions, with the notable exception of the armed forces, largely unaltered. The Tanzimat was aimed at the very heart of Ottoman Islamic society, for at its core lay a wide-ranging, radical attempt to replace the Islamic system of law, which resulted in a significant, if still only partial, secularization of the entire society.

The "Noble Rescript of the Rose Chamber," which Abdülmecid issued on November 3, 1839, contained a number of concepts hitherto unknown in Ottoman law: the principles—if not quite "rights"—of life, dignity, and property of the subject and of a fair and public trial for all. The following year a new penal code was introduced, which claimed to offer to "the shepherd in the mountains and the minister" (although not yet the sultan) equality before the law.[42] Most controversial of all, it established the principle of the equality of all religions, thus, in effect, abolishing the privileged fiscal and legal position to which all Muslims believed themselves entitled. For their part, the peoples of the "protected religions" were no longer required to pay special taxes, and—all reforms have unfortunate consequences for some—the minority *millet*s lost their right to be governed according to their own laws. The culmination was the creation, for the very first time in any Islamic state, of a code of civil law, the Medjelle, although it would take nearly forty more years for it to be completed.

The Noble Rescript was also radical in another way. It openly acknowledged these principles as innovations, and "innovation" (*bid'a*) in traditional Muslim usage was the converse of Sunna—the way of the Prophet. "The worse things are those that are novelties," runs one of the Hadith. "Every novelty is an innovation, every innovation is an error and every error leads to hellfire." Most previous reforms, with the exception of Selim III's short-lived "New Order," had been disguised, as best as possible, as a return to older, now corrupted ways. Not the Rescript. For many Muslims it was a direct affront to religion: an act of heresy.[43]

The Tanzimat implied a cleavage between justice, and thus also politics, and religion. It was, in effect, a violation of the most sacred tenet of Islam: that the law is, and can only be, the commands of God as transmitted to mankind by the last of his prophets. In 1841, the minister Mustafa Resid Pasha introduced a new commercial code, derived almost entirely from French models, to the High Council. He was asked by members of the Ulema if it was in conformity with Holy Law. "The Holy Law has nothing to do with the matter," he replied. "Blasphemy!" shouted the Ulema. The ensuing uproar forced the sultan to dismiss Resid Pasha—although he was later reinstated and became minister of foreign affairs and subsequently grand vizier. The process of reform was arrested briefly, but it could not be halted. In 1847, mixed civil and criminal courts were created, with an equal number of European and Ottoman judges, and bound by the rules of evidence and procedure drawn from European—again predominantly French—not Islamic practice.

The administration of the law was not all that was slowly being pried away from the Ulema. So, too, was education. A number of schools were created in 1846, and although they provided for instruction in "the duties and obligations that religion imposes on man," it was clear that they were preparing the way for the eventual introduction of a wholly secular system of education. In 1868, the Imperial Ottoman Lycée was opened at Galatasaray. Classes were taught in French, the curriculum was Western, and Muslim and Christian pupils sat in the same classrooms side by side. The Mulkiye, or civil service training college, and the Harbiye, or war college, were greatly expanded and modernized. The long-delayed project for a University of Istanbul, destined to be the first true university in the Muslim world, was established, although it did not admit its first students until 1900.

In 1871, another attempt was made by the Ulema to halt the process of reform. Once again, the sultan bowed to pressure. Once again, however, the forces of reaction had only a temporary success. Ottoman Christians were dismissed from their posts (and later reinstated) and the Medjelle was set aside, but only for a while. The empire was now too far along the road of reform to turn back. Modern Turkey is the most westernized, most modern of all Muslim states, the most secular, and the most progressive. Many Turks today claim, with good reason, that their country is a natural part of Europe, and on this basis, Turkey will certainly before very long be admitted to the European Union. That the heartland of what had for centuries been Europe's most bitter enemy can now seek to merge itself with Europe is due in great part to the reform program begun by Selim III, Mahmud II, and Abdülmecid I.

Modernization was not, however, wholly beneficial for all. The sultan's prime concerns were to strengthen his armies and preserve his empire. As means to those ends, Abdülmecid went a long way toward divorcing religion from politics and creating a measure of legal equality for all the subjects of the empire. Yet while the Tanzimat indubitably paved the way for a modern Western society, it did so by strengthening the central power of the state. In the process some of the older, informal liberties the people had enjoyed disappeared.

Adolphus Slade, a British naval officer who visited Turkey many times in the 1830s, had this to say of the impact of Mahmud II's reforms:

> Hitherto the Osmanley has enjoyed by custom some of the dearest privileges of freemen, for which Christian nations have so long struggled. He paid nothing to the government beyond a moderate land-tax, although liable it is true, to extortions which might be classed with assessed taxes. He paid no tithes, the vacouf [lands let on lease] sufficing for the ministers of Islamism. He traveled where he pleased without passports; no custom-house officer intruded his eyes and dirty fingers among his baggage; no police watched his motions, or listened for his words. His sons were never taken from his side by soldiers, unless war called them. His views of ambition were not restricted by barriers of birth or wealth: from the lowest origin he

might aspire without presumption to the rank of pasha; if he could read, to that of grand vezir; and this consciousness, instilled and supported by numberless precedents, ennobled his mind, and enabled him to enter on the duties of high office without embarrassment. Is not this the advantage so prized by free nations?[44]

It was a somewhat rosy view of social conditions in the early Ottoman world, but it was not entirely false. Islam had been born of a tribal society, and Islamic law and practice had preserved some of the egalitarian features of their tribal origins, although it would be a mistake to describe these as "democratic" in any way, as some modern scholars have tried to do. In addition, the Ottomans themselves had never wholly forgotten their own beginnings as Ghazi warriors. The sultan was, as the Prophet and the caliphs had been, a sole unchallenged ruler—that, after all, is what the word "sultan" means—none of whose subjects ever enjoyed anything like "rights," in the Western sense of the term. But while it may not have been quite true that those "from the lowest origin" could aspire to become grand vizier, an ever-proliferating Ottoman bureaucracy did provide a means of advancement for many who would have had no such opportunity in the societies of old-regime Europe.

What the reforms had achieved was in some respects not unlike the changes brought about in western Europe in the seventeenth century. When the earlier, quasi-egalitarian political traditions of the old Germanic tribes, which had survived into the Middle Ages—and been revived during the French Wars of Religion—were steadily snuffed out, replaced by the unquestioned power of the state under an absolute monarchy. In most European countries, with the partial exception of England and parts of Scandinavia, codified Roman law, the law of the "unfettered legislator," of the Emperor Justinian, replaced that of the older Germanic or customary law, just as the Medjelle replaced not only the Shari'a but also the earlier customary practices of the Turkic warrior band. It is some irony that it took reforms borrowed from European models to transform the Ottoman sultanate into what its harshest European critics had claimed it had been all along. As Slade observed, Mahmud had made "European subservient to Asiatic, rather than Asiatic to European manners."[45] Reform had given him the instruments that he had previously lacked. "The sovereign," reflected Slade, "who before found his power (despotic in name) circumscribed, because with all the will, he had not the real art of oppressing, by the aid of science finds himself a giant—his mace exchanged for a sword."[46]

In the first decades of the nineteenth century, the Ottomans had embraced one kind of westernization—absolute monarchy—at the very moment that it was being slowly superseded in the West itself by another form of government: liberal democracy. Even when, in 1876, the sultan Abdülhamid did reluctantly grant his people a constitution, it was not, as all European constitutions had been, wrested from the sovereign, the expression of a contract between ruler and ruled. It was instead, as one modern observer has put it, "a benevolent 'grant' by an absolute sovereign to his subjects, the parody rather than the analogue of what obtained in Europe."[47] The shift from benevolent grant to contract the Ottomans would succeed in making only by ceasing to be Ottomans and becoming Turks, and the Arabs have still to make this shift.

Modernization also had a further and, from the point of view of the sultan, far more unfortunate consequence. The creation of new, largely secular schools soon created a new educated middle class, which rapidly came to resent the sultan's unrestrained hold on power. Toward the end of the nineteenth century a revolutionary group calling themselves the Young Ottomans demanded both a return to the Islamic roots of Ottoman society and at the same time a more liberal form of government. On December 23, 1876, in a desperate bid to keep the peace, Abdülhamid II agreed to most of their demands. For the first time the Ottoman state was given a European-style constitution and a parliament, al-

though it had hardly any powers and was chosen by means of indirect elections. But Abdülhamid was no "enlightened despot."
Known in the West as "Abdül the Damned" and the "Red Sultan"
for the string of atrocities over which he presided during his
thirty-three-year reign, he was deeply conservative and suspicious
to the point of paranoia. His aim was not to create the sort of liberal Western society the Young Ottomans had demanded, but to
return to the autocratic pan-Islamicism of his ancestors. In February 1877, he suspended the constitution and a month later pro-
rogued the parliament. Thereafter, he retreated to his "Islamic
Vatican," as the French called it, and attempted to preserve as
much of his dwindling empire as he could by using as much force,
and as little tact, as he could muster.

But, as with the Tanzimat, the move to constitutional government could not be so easily reversed. It took another forty-one
years, but then, in 1908 in Salonika, the members of another revolutionary group, formed in 1894 to promote broadly liberal ideals
of unity and equality for all the races and creeds within the empire
and calling itself the Committee of Union and Progress (CUP),
or "Young Turks," as it was popularly known, led a revolt against
the sultan, demanding the restoration of constitutional government.[48] "Henceforth," proclaimed the leader of the revolt, an army
major named Enver Bey, from the steps of the government building, "we are all brothers. There are no longer Bulgars, Greeks, Ro-
manians, Jews, Muslims; under the same blue sky we are all equal,
we glory in being Ottomans." Abdülhamid gave way. On July 24,
he issued a decree ordering the restitution of the constitution and
setting a date for elections. Crowds lined the streets of Istanbul,
chanting "Long live the sultan. Long live the constitution."[49]

Abdülhamid's time as a constitutional monarch, however, was
brief. Capitulating to the CUP had put an end to his credibility,
and the CUP wanted not a share in power but power itself. In
April 1909, the sultan was deposed and sent into exile, along with
two of his sons and a number of concubines, to live out the rest of
his life in a private villa in, ironically, Salonika, the very city from

which his nemesis had come. He was replaced by his compliant, mild-mannered, sixty-four-year-old brother who ascended to the throne as Mehmed V. In the summer the constitution was revised. The powers of the sultan were reduced below those of most contemporary European monarchs. He now neither reigned nor ruled, and all he could really do was confirm decisions already made by his parliament.

A new and quite unexpected version of the "Turkish menace" now presented itself to the still-autocratic governments of western Europe. Subversive radicalism, uncomfortably reminiscent of the French Revolution, seemed suddenly to have replaced "Oriental despotism"—and of the two the powers of western Europe, on the whole, much preferred despotism. The Young Turks, who saw themselves as "the vanguard of awakened Asia," wrote Sir Gerard Lowther, the British ambassador to Istanbul, on May 29, 1910, were now bent on "imitating the French revolution and its godless and leveling methods."[50] (The expression "Young Turk" is still used in English to describe any hotheaded disrupter of the public peace.) Austria-Hungary, terrified that the Young Turks might spread principles of parliamentary government to Bosnia-Herzegovina, over which it had exercised a measure of control since 1878, and from there possibly even as far as Budapest or Vienna itself, quickly annexed the province outright.[51] Thus began a series of Balkan crises that would not only contribute to the final demise of the Austro-Hungarian Empire but with whose consequences we are still living today.

Once in power, however, the Young Turks turned out to be much less liberal—in any sense of that troubled word—than they had been in opposition. Their ideals of religious and ethnic unification, which Enver had so loudly proclaimed at Salonika, very rapidly became the far less ecumenical objective of "Turkification." They outlawed any groups with ethnic or national aims. They imposed the use of Turkish in all secondary schools and in the law courts, which hitherto had employed the languages of the regions where they were located. This aroused the anger in particular of the Arabs, the largest non-Turkic minority within the empire and, because of their special relationship to Islam, the most fiercely protective of their language and their culture. Islam, which had once served to unite all Muslims under a single ruler, was slowly unraveling with what would prove to be disastrous consequences, in the long run, for the entire Middle East.

In March 1912, Serbia and Bulgaria, joined by Greece and Montenegro, formed the Balkan League, which first demanded far-reaching reforms for those areas still under Ottoman rule and then, when these were not met, declared war. Within months the armies of the league had stripped the Ottoman Empire of what little of it still remained within Europe. By the end of the year, the Ottoman army had been forced back to within fifty kilometers of Istanbul itself. On January 23, 1913, Enver Bey and a group of other officers broke into the cabinet chamber, shot dead the minister for war, Nazim Pasha, forced the resignation of the grand vizier Kamil, and dissolved the government. The Ottoman experiment with liberalism was at an end, and with it any hope for the future of the Ottoman Empire.

## III

When war broke out in the summer of 1914, the Allies initially did their utmost to persuade the Ottomans to remain neutral. On August 18, 1914, a little more than a month after the first shots had been fired, the British foreign secretary, Sir Edward Grey, assured the Ottoman ambassador to London that, if the Sublime Porte stayed out of the war, the empire's territorial integrity "would be preserved in any condition of peace which affected the Near East."[52] But the triumvirate of Young Turks—Enver, Jemal, and Talaat Pashas—who now ruled the empire had been convinced by Enver Pasha that Germany would win the war and that this was an opportunity for the Ottomans to regain if not all, then certainly part, of what they had lost in Europe.

Enver's conviction of the invincibility of German arms was the sad outcome of a German, Prussian presence in Turkey since the mid-1830s. By the late nineteenth century, German military advisers had replaced the French, and—in an unprecedented reversal of the customary direction of travel—Ottoman officers paid regular visits to Germany. The grand vizier Mahmud Sevket Pasha, who had been instrumental in the overthrow of Abdülhamid, had even spent ten years there as a young man. When in 1885 the Liberal prime minister of Britain, William Gladstone, withdrew British support from the sultan, the Germany of the "Iron Chancellor," Otto von Bismarck, was only too happy to take its place. From 1888, when Queen Victoria's nephew Wilhelm II acceded to the throne, Bismarck created a policy known as the "Drive to the East," whose objective was to transform the Ottoman Empire—with the unwitting assistance of the Sultan Abdülhamid, who saw in Germany the only successful Western state not apparently indebted to dangerous liberal and constitutional ideals—into a willing client of the rising German empire.

Buoyed up by an overriding confidence in the supremacy of German efficiency and German manufacturing, on November 11, 1914, the sultan Mehmed V, as titular head of state, declared war on the Allies. Two days later, in the Topkapi palace, surrounded by relics of the Prophet, the coming war was declared to be a jihad, to which the sultan-caliph called all Muslims. "Our participation in the world war represents the vindication of our national ideal, the ideal of our nation and people lead us . . . to obtain a natural frontier to our empire, which should include and unite all branches of our race," declared the Triumvirate, employing the newly minted language of racism, another unfortunate Western import.[53] The Sick Man was about to commit suicide.

The Ottoman involvement in the First World War brought about the final collapse of the empire and its subsequent transformation into the modern Republic of Turkey. It resulted in the creation all across the Middle East of a number of largely artificial, Western-backed satrapies and divided the Islamic world against itself in ways that would prove irreversible. It is no small irony that the modern struggle between Islam and the West began with an alliance between a Muslim state, with one infidel power in a struggle to the death with three others. The Ottoman entry into the First World War would also divide the various Muslim groups of which the empire had been composed for centuries in ways that were unprecedented. Far from becoming the jihad the sultan and his ministers and their German patrons had hoped for, the war became the first of many subsequent conflicts between a westernized Islamic state and a number of purist Islamic leaders—now called somewhat misleadingly "fundamentalists"—bent on returning Islam to what they conceived to be the true way of the Prophet.

The first split between the sultan and his Muslim subjects came in February 1915. With British and French fleets blockading his ports, Cemal Pasha, commander of the Ottoman Fourth Army and absolute ruler of Syria, became convinced that the Arabs under his command were plotting an uprising. To forestall this, he executed a number of leading Arabs and deported their families to Anatolia, and the program of "Turkification," which had been allowed to lapse, was vigorously reinforced. All of these moves had the predictable effect of turning paranoid fear into something resembling a reality.

Britain, which had hitherto ignored the Arabs, now began to see in them a possible ally. Might they not have, or be persuaded to acquire, the nationalist sentiments that had been the undoing of the Ottomans in the Balkans? One of those who became convinced that they might was the British war minister, Lord Kitchener of Khartoum, whose square jaw, fierce blue eyes, and bushy mustachio stared out from the familiar recruiting poster for the war, above the ominous and threatening legend YOUR COUNTRY NEEDS YOU. Kitchener had spent much of his career in the East. It had been he who, at the famous battle of Omdurman on September 2, 1898 (in which an enthusiastic young Winston Churchill had participated), had finally destroyed the regime of the successor of Muhammad ibn Abdullah, the self-declared mahdi whose forces had killed the national hero General Charles George Gordon on the steps of the government house in Khartoum in 1885. In 1911, Kitchener had become the British governor-general of Egypt and the Sudan. During all this time, in common with most British army officers in "the East," Kitchener had acquired very little understanding of the peoples he had fought against and ruled over. His understanding of Islam was about as profound and accurate (although rather more sympathetic) as his understanding of Catholicism, and he believed, as his Russian predecessors had believed in their dealings with Abdülhamid I in 1774, that the caliph occupied a position analogous to that of the pope, or rather to what a stout British Anglican imagined the position of the pope to be.

Like most British officers, Kitchener was therefore unduly alarmed by the possible effect that the sultan-caliph's declaration of jihad against Britain and its allies might have on Egypt, the Sudan, and, most crucial of all, India, which were home to more than half of the world's Muslims, all of whom were ruled, directly or indirectly, by Britain. The experience with the mahdi may not have taught the British very much about the intricacies of Islam, but it had certainly taught them the very real potential of Islamic religious fervor. Kitchener believed that religious conflagration of the kind that had erupted in the Sudan in 1885 could perhaps be avoided if only the caliphate could be detached from the Ottoman sultan and transferred to someone more sympathetic to Britain.

The idea was not, in fact, a new one. As early as 1877, the only plausible alternative caliph, the sharif of Mecca, the Hashemite ruler of the Hijaz and direct descendant of the prophet himself, had already made promising, but inconclusive, overtures to the British administration in Cairo. At the time, the British had had no wish to see the Ottoman Empire disintegrate still further and had politely discouraged the move—but had kept the possibility in mind just in case. Now, however, the reigning sultan-caliph had become the pawn of an infidel power, and there was a new sharif, Hussein ibn Ali, at Mecca who was eager to revive his predecessors' ambitions.[54]

Hussein's willingness to advance the Allied cause, however, seems to have derived less from religious purity than from the clash between his own brand of "Oriental despotism" and the reformed Ottoman government's new absolutism. "The Grand Shereef," wrote the acting British consul in Mecca, Abdurrahman, "is naturally opposed to any reform and wants everything to run in the same rut."[55] It was also an open secret at the time that the Porte was making plans to depose him. In February and again in April 1914, Hussein's favorite son, Abdullah, had gone to Cairo in an attempt to persuade the British that the rival Arab emirs of Arabia, the Idrisi of Asir (who in the end remained pro-Turkish), Ibn Sa'ud of the Nejd (who was the sharif's sworn enemy, "feeling for him," remarked Storrs, "as an Ebenezer Chapel might for Rome"), and possibly Ibn Rashid, the ruler of Hayil, were ready to lay their differences aside and to unite behind Hussein to fight for an "Arabia for the Arabs."[56]

On October 31, Lord Kitchener wrote to Abdullah, repeating yet again that the Allies had done their best to keep the Ottoman Empire out of the war, but since "Germany has bought the Turkish Government with gold" and was, therefore, by implication an apostate, their hands had been forced. Now, he went on, Great Britain, which had until then "defended and befriended Islam on the person of the Turks; henceforward it shall be that of the noble Arab." He ended his letter with the salutation "The good tidings of the freedom of the Arabs, and the rising of the sun over Arabia."[57]

By November 1915, Hussein had persuaded the British high commissioner in Egypt, Sir Arthur McMahon, that he now represented "the whole of the Arab nation without any exception." He did not. Not only was there at the time no such thing as an Arab "nation," but the vast majority of the 8 million to 10 million Arabic speakers of which it might be thought to be composed, even most of the inhabitants of Mecca itself, remained loyal to the sultan-caliph in Istanbul. As T. E. Lawrence wrote in a memorandum of 1915:

Between town and town, village and village, family and family, creed and creed, exist intimate jealousies, sedulously fostered by the Turks to render a spontaneous union impossible. The largest indigenous political entity settled in Syria is only the village under its sheikh, and in patriarchal Syria the tribe under its chief. . . . All the constitution above them is the artificial bureaucracy of the Turk.[58]

But McMahon had not read Lawrence's memorandum and doubtless would not have paid much attention to it if he had. After some deliberation, he wrote to Hussein on October 24, 1915, promising that "Great Britain is prepared to recognize and uphold the independence of the Arabs in all the regions lying within the frontiers proposed by the Sharif of Mecca," with the significant exception of those "portions of Syria laying to the west of the districts of Damascus, Homs, Hama and Aleppo"—which meant not only much of inland Syria but also the entire western coastal region—and the "Vilayets [administrative districts] of Baghdad and Basra," on the grounds that they "cannot be said to be truly Arab and must on that account be exempted from the proposed delimitation."[59] Hussein for the moment demurred but wrote back ominously that as all these regions, in particular Baghdad and Basra, had once formed a part of "the pure Arab kingdom," it would be "impossible to persuade or compel the Arab nation to renounce the honorable association."

Hussein was clearly happy to employ the language of the nation-state when speaking to the British, but he must also have been aware not only that no such thing as an Arab nation then existed but that a national caliphate could only be an oxymoron. The caliphate had to embrace all Muslims and, if the injunctions and the prophecies of the Prophet were to be made real, had, one day, to embrace all mankind. Even Hussein's initial territorial demands for a united Arabia from Aleppo to Aden included not only Arabs but Turks, Armenians, Kurds, Assyrians, Chechens, and

Circassians; and it included not only Muslims but also various kinds of Christians and Jews. But Hussein's caliphate was never intended to be limited by narrow European conceptions of nationality. His ambition was to create not the simulacrum of a modern nation-state, which, after the war, the Allies did their best to force upon him and all the other Arab warlords; it was to restore the caliphate in all its former imperial glory. As Abdullah told T. E. Lawrence in May 1917, "it was . . . up to the British government to see that the Arab kingdom is such as will make it a substitute for the Ottoman Empire."[60] As a first step, on November 2, 1916, Hussein had himself proclaimed by his followers "King of the Arab Countries." Britain and France promptly refused to accept this. "We could not conceal from ourselves (and with difficulty from him)," observed Ronald Storrs laconically, "that his pretensions bordered on the tragi-comic"; and in January he was persuaded to make do with the rather more modest, and more plausible, "King of the Hijaz."[61]

In the meantime, the British MP Sir Mark Sykes, who fancied himself an expert on "Middle Eastern" affairs and would play a crucial role in the development of the future Arab Revolt, had begun a series of negotiations with the French over the possible postwar division of the lands of the Ottoman Empire. After months of wrangling between November 23, 1915, and January 3 of the following year, Sykes and the French negotiator, François Georges Picot, finally came to an agreement. This partitioned Syria and the former Ottoman *velayets* of Basra, Baghdad, and Mosul—a vast swath of territory from Russia to the Persian Gulf, then called Mesopotamia, together with a large part of southern Turkey—into spheres of direct, or indirect, French and British influence. France's share corresponded to what are now the states of Syria and Lebanon, while Britain got what is now Iraq and Transjordan (later Jordan). The interior would be under the indirect control of both powers, but they were prepared to "recognize and uphold an independent Arab state or confederation of Arab states, under the suzerainty of an Arab chief" stretching from

Aleppo to Rawandaz and from the Egyptian-Ottoman border to Kuwait.

The subsequently infamous Sykes-Picot agreement has been persistently represented, in particular in the Arab world, as an act of barefaced European colonialism, a legacy of the Crusades, and a portent of the American imperialism to come. The Palestinian leader Yasser Arafat was merely expressing an accepted commonplace of Arab history when, in August 1968, he claimed that "our ancestors fought the crusaders for a hundred years, and later Ottoman imperialism, then British and French imperialism for years and years."[62] In fact, although the Allies were not prepared to indulge Hussein in his wilder flights of fantasy about a new caliphate to replace the Ottoman Empire, the territory finally allotted to the "Arab chief" under the agreement was not much less than what he had originally asked for. Certainly, Sykes seems to have believed until his death from influenza in the Hôtel Lotti in Paris in 1919 that he had succeeded in keeping the promises the British had made to Hussein.[63]

In the meantime, Hussein himself was busy trying to remain neutral for as long as possible so as to collect as many bribes from both sides as he could. By June 1916, he had gathered more than 50,000 pounds in gold from Istanbul to support an entirely illusory struggle against the British and a substantial sum from the British to help him equip a rebellion against the Turks. "All is fair," as Lawrence later remarked, "in love, war and alliance."[64] The Porte, however, was growing suspicious of his activities. When, in April 1916, he was informed that an Ottoman army large enough to destroy him was planning to march through his domains, he hastily decided that he now had no option but to act. Sometime between June 5 and 10, he raised his revolt against the sultan-caliph, and on June 16 he seized Mecca from its small Turkish garrison. He now waited for the 100,000 or so Arab troops whom he and his son Faisal expected to flock to their standard. None came except for a few thousand Bedouin, lured by the promise of British gold. The revolt was a damp squib. The subsequent military successes attributed to the uprising—the seizure of the ports of Jeddah, Rabegh, and Yanbo on the Red Sea—were all achieved by British ships and contingents of the British-controlled Egyptian army. But the British could go no farther since Hussein refused to contemplate the idea of British, Christian troops stationed anywhere on the land that contained the holiest sites of Islam. To have done so, he believed, would have compromised his standing with the entire Muslim world. If the problems his Saudi successors have had with non-Muslim troops—American this time—stationed in Arabia are anything to go by, he was probably right.

Meanwhile, Hussein continued as best he could to hedge his bets by suggesting to the CUP leadership in Istanbul that he might, if the proper incentives were offered, be prepared to shift back to the Ottoman side. It seems likely that the emir of Nejd, Abdül Aziz Ibn Sa'ud, was about right in claiming that his hated rival's immediate political objective was "to play off the British against the Turks and thus get the Turks to grant him independence guaranteed by Germany."[65]

No matter what Hussein was up to, from the view of the Arab Bureau in Cairo, which had been charged with overseeing the revolt, it was clear that the sharif's small, ill-disciplined, and ill-equipped forces were of relatively little use against the Turks. "Their preference," wrote one member of the Bureau of the Bedouin, "is for the showy side of warfare and it will be difficult to hold them together for any length of time, unless the pay and rations are attractive." Pay and rations—in particular pay—would prove crucial.[66]

At this point enter T. E. Lawrence. Small—"my little genius" Ronald Storrs called him—and unduly sensitive, with "piercing gentian-blue eyes," made miserable by being illegitimate and more miserable still by being gay, Lawrence was a potentially brilliant archaeologist, an indifferent poet, and a gifted linguist.[67] He was also a mythomaniac and an accomplished self-publicist. With the help of a wandering newspaperman from Ohio named Thomas

Lowell, who wrote dispatches that greatly exaggerated his exploits and sent heroic photographs back from the front (at that time a fairly novel form of journalism), Lawrence was transformed from the son of a disgraced Irish baronet into "Lawrence of Arabia." Here was just what the increasingly dispirited British public most needed: a British hero in a remote and highly romantic theater of war, far removed from the sorry, sordid trench warfare in Europe in which too many had already died anonymous deaths—that "comfortless bloody business," as the novelist John Buchan called it. After the war Lowell presented his version of Lawrence of Arabia's revolt against the Ottomans in the form of a lecture with photographs, which opened to a packed house at the Century Theatre in New York in 1919 and was later taken to London. It was called, predictably, "The Last Crusade."

Lawrence sentimentalized the Bedouin. His vision of the "East" had been shaped by a generation of Romantics—men like Charles Doughty, for whose *Travels in Arabia Deserta* (1888) he wrote a preface in 1921, and Sir Richard Burton, celebrated as a libertine, fencer, explorer, diplomat, translator of the *Kama Sutra* and the *Thousand and One Nights,* and, in 1853, the second (the first was Ludovico di Barthema in 1503) infidel to enter the great mosque at Mecca (disguised as a Pashtun). These men had come to see the Bedouin not as the merciless brigands Napoleon's troops had encountered but as the upholders of an ancient warrior creed and an ordered, stratified society of the type that was fast disappearing in Europe itself beneath the weight of democracy and industrialization: stout English yeomen, in other words, "translated into another idiom."[68]

Like Doughty's and Burton's, however, Lawrence's attitude toward the Arabs was also not a little patronizing. These were "magnificent men," loyal, fierce, proud, and true. But they were also unreliable, irresolute, and more interested in booty than in the destiny of their nation. "Our fighting," he later told his friend the poet Robert Graves, who wrote his biography, "was a luxury we indulged in only to save the Arab's self respect."[69] (When I met

Graves in 1971 or 1972, he recalled that Lawrence had never seemed to be quite so enraptured of the desert Arabs as he claimed. As with all such imaginings, the brute reality never quite lived up to the image. And there was a world of difference between a man like Faisal and a common herdsman.)

In Lawrence's view the Arabs would never be anything but desert tribesmen, living by plundering one another, unless a European—preferably Lawrence himself—were prepared to set them on the stony path to nationhood. "My object with the Arabs," he told Graves in 1926 in the tones of a petulant schoolmaster, had been "to make them stand on their own feet."[70] For although at the beginning of the war Lawrence seems to have believed that no Arab nation did, or could, exist, by November 1916, he had clearly changed his mind. In the *Arab Bulletin,* a publication of the Arab Bureau, which he had helped set up, he reported from the Hijaz that tribal opinion "struck me as immensely national and more sophisticated than the appearance of the tribesmen led one to expect." This he attributed to the Germans, who, having tried preaching jihad, " 'til they saw that that idea had fallen flat," had then taken up nationalism in an attempt to "awaken in the provinces, the (in their opinion) dormant Ottoman sensibility." The effect of this, however, had been to stir up not Ottoman but Arab nationalism. "Whatever the cause," Lawrence concluded, "the Arab feeling in the Hejaz runs from complete patriotism amongst the educated Sheriffs down to racial fanaticism in the ignorant."[71]

In his own version of events, it had been he who had drawn on these feelings to make of the Arabs a fighting force capable of transforming themselves from a collection of scattered tribes into a nation. It had been he, as he wrote in the poem that begins his own account of the events, *The Seven Pillars of Wisdom,* a book meant to make him the Herodotus of the early twentieth century, who had finally gathered this "tide of men" into his able hands and allowed him to "write his will" across the sky of the new Arabia.[72]

In December 1916, he became the liaison officer to Prince Faisal, charged with salvaging the revolt by making use of the Bedouins' indubitable qualities as guerrilla fighters and their command of the desert. Intoxicated by this position, in which he operated on his own, and entranced by Faisal, whom he described as "very like the monument of Richard I at Fontevraud" and "an absolute ripper," Lawrence, mounted on a camel and dressed in flowing white robes said to be more costly than Faisal's own, "went native."[73] The only substantial military achievement, however, in which he had any part was the capture of the port of Aqaba on the southern tip of Palestine. Aqaba lay at the entrance to a channel of the Red Sea, and its shore batteries prevented the Royal Navy from transporting Arabian tribesmen into Palestine, where most of the Turkish forces were concentrated. The overland route was blocked by the Turkish garrison at Medina, whose population, significantly, had shown no interest whatsoever in joining Hussein's revolt and were more than a match for Faisal's Bedouin troops. The seizure of Aqaba might, therefore, transform the nature of the entire revolt.

In the spring of 1917, Lawrence disappeared into the desert. Ten thousand pounds in gold sovereigns had secured him the allegiance of Auda abu Tayi, sheikh of the eastern Howeitat, whom he now persuaded to launch an attack on Aqaba. Behind Aqaba lay the desert of the Hijaz, one of the most inhospitable places on Earth. Since the Turks could not possibly imagine any threat from that quarter, they had placed all their artillery facing toward the sea. Auda proposed to do precisely what the Turks, but not the Bedouin, considered to be impossible: cross the desert and take the city from the rear. In Lawrence's stirring retelling of this event, it is he who had the idea, although it has all the markings of the kind of Bedouin tactics that twelve centuries earlier had given them victory over the Sassanid emperor Yezdegird at al-Qadisiyya. Lawrence also contrives to give the impression, veiled and somewhat elusive, that it had been he who had actually commanded the raid. The decision of Faisal, "hitherto the public leader," to remain behind in Wejh, he wrote, "threw the ungrateful primacy of this northern expedition upon myself."[74] Only in Lawrence's imagination would a man such as Auda have consented to being "gathered into the hands" of a lone British liaison officer, no matter how much gold he carried in his saddlebags. Lawrence may have spoken fluent, although by all accounts inaccurate, Arabic. He may have known long passages of the Qur'an by heart. He may even have won over Auda and the other Bedouin leaders with all the male bonding that goes on in *The Seven Pillars of Wisdom*. But he was still an infidel, and no Bedouin sheikh would have consented to being led into battle by an infidel. As with all Bedouin raids, this one involved a number of clans, each under its own leader, all operating largely on their own under the general leadership of Auda. It is highly unlikely that, apart perhaps from assisting in the negotiations with the Turks, Lawrence played any significant military role in the operation.

On July 6, Auda's forces emerged from the desert, to the astonishment of Aqaba's small and unprepared Turkish garrison. Within a few hours, after widespread looting but—thanks, perhaps, to Lawrence's diplomatic skills—relatively little killing, Aqaba was in Allied hands. (With perhaps unintended irony, in June 2002 it was chosen as the site of a summit meeting among George W. Bush, Ariel Sharon, and the Palestinian leader Mahmoud Abbas.)

Lawrence set off immediately for Suez, virtually on his own, across enemy-controlled territory, to report the capture, presumably with the intention of forestalling any other, more accurate, version of the story reaching British headquarters before his did. So compelling, indeed, was his own account of the raid, and so willing were his commanding officers to believe that any measure of success must have been due to one of their own, that there was even talk of his being awarded the Victoria Cross, the highest award for bravery in the face of the enemy. (In the end, however, Lawrence was denied the honor because the terms stipulate that a British officer has to have observed the act of heroism for which it is being made. And no one, British or otherwise, ever seems to have witnessed any of Lawrence's self-described feats of heroism.) After the war and the publication of *The Seven Pillars of Wisdom,* it became a historical fact, played to audiences across the globe, from Lowell Thomas's *The Last Crusade* to David Lean's block-buster movie of 1962, *Lawrence of Arabia,* that it had been Lawrence, dressed in his flowing white robes and mounted on a camel, who had led the skeptical, fatalistic, and deferential Howei-tat to victory.

Lawrence's most important task, however, apart from playing the war hero to a disheartened British public, had been to supply the Arabs with British gold—immense quantities of it carried across the desert by camels in sovereigns packed into cartridge boxes. After the war was over, Ronald Storrs calculated that Hussein's largely ineffectual revolt had cost the British some 11 million pounds sterling, some $400 million in today's currency. Nearly half a century later, when the war was over and Lawrence dead from a motorcycle accident in the Oxfordshire countryside, a Bedouin sheikh, when asked if he remembered him, replied laconically that yes, "he was the man with the gold."[75]


ON MARCH 11, 1917, Baghdad fell, virtually unopposed, to the Anglo-Indian Army of the Tigris. It was a symbolic rather than strategic victory, but it offered Sir Mark Sykes the opportunity to make a declaration to the Arabs, setting out, in suitably grand but guarded terms, the British intentions for their future. He spoke to them of the tyrannies and indignities they had suffered. "Since the days of Halaga [the Mongol khan Hülegü]," he announced, "your palaces have fallen into ruins; your gardens have sunken in desolation, and your forefathers and yourselves have groaned in bondage." Now, he promised them, "in due time you may be in a position to unite with your kinsmen in north, east, south, and west in realizing the aspirations of your race." In 1965, long after the war was over and Sykes dead, the aged Sir Ronald Wingate,

Kitchener's successor in the Sudan, still remembered this declaration with irritation as a "farrago of nonsense composed by presumably some 'ebullient Orientalist' such as Sir Mark Sykes, based on the *Arabian Nights,* [and] the play of *Kismet* which had been a great success before the War."[76]

Rather more modestly and more realistically, Sykes also urged the Iraqis to prepare the way so that the British "when the time comes [may] give freedom to those who have proved themselves worthy to enjoy their own wealth and substance under their own institutions and laws," and pointed vaguely toward a future Middle Eastern federation under the rule of Hussein now promoted from sharif and emir to king.[77]

Despite its cautious Foreign Office undertones—the use of the "wistful Optative mood" and the refusal to set even an approximate date as to when the Arabs might be ready to rule themselves or realize their racial aspirations (assuming that they had any)—the declaration entirely disregarded the fact that the populations of the provinces of Baghdad and Basra were then, as they are now, predominantly Shiite. Hussein was, of course, a Sunni. Indeed, as sharif of Mecca and now "king of Arabia," he was the leader of all Sunnis, and the enmities between him and the Shiites—and between both and the Jews, who were economically the most powerful minority in Baghdad at the time—had been simmering for centuries. To the south, in the mountain valleys, where the border of modern Turkey meets Iraq, Syria, Iran, and Russia, were the Kurds, transhumant shepherds who were Sunni but not Arabs and who were, and have remained, at odds with just about all their neighbors, although in 1917 they were neither as numerous nor politically as significant as they are today.

Although there was no uprising against the British at this stage, the invasion, with its promises of new freedoms imported from the West (constitutional and monarchical rather than democratic, but almost as alien) under the benevolent hand of a conquering infidel power, and its complete disregard for centuries-old religious and ethnic divisions, which former Muslim imperial powers, first the Abbasids and then the Ottomans, had succeeded in containing, was an ominous foretaste of the American (and British) invasions of the same region in 2003. Little wonder that the commander in chief of the Army of the Tigris, General Stanley Maude, who knew only too well what the vague promises contained in Sykes's declaration might bring in their wake, did his best to suppress the document. "Before an Arab façade can be applied to the edifice," he told his superiors in Britain, "it seems essential that a foundation of law and order should be well and truly laid."[78]

Sir Mark, in the meantime, busied himself with designing a flag for Hussein's new Arab "nation." It was a combination of black, white, green, and red, all associated with past Muslim, although not exclusively Arab, glories. It is no small irony that today (2007) substantially the same flag flies over the headquarters of the Hamas-controlled Palestinian Government building in Ramallah.

On December 11, a combined British-Arab force, led by General Sir Edmund Allenby (known as "the Bull"), entered Jerusalem by the Jaffa Gate. He went on foot for fear of being mistaken for a latter-day Tancred. Major Vivian Gilbert, who accompanied him and later wrote a popular account of the campaign, reflected, as he walked through the almost deserted streets, that it was "a strange irony which has wrought such a beautiful city out of the love and hate of three creeds! Mosque, church, long-roofed convent, synagogue, dome, jostling each other; all white and glistening with a show of peace in the clear light of mid-day." A new crusader army had now taken Jerusalem not in the name of creed or glory but in order to liberate an oppressed people. Nothing, thought Gilbert, could be more revealing of "England's intentions in Palestine, than the order of the commander-in-chief that no British flag should fly over the conquered city, and no flag was flown save the Red Cross flag, emblem of succor to the distressed, and this proud pennon flew from the American Hospital."[79]

Two of the most symbolically important cities in the Islamic world were now, in effect, in infidel hands. On October 1, 1918, Allenby took Damascus, completing the British seizure of the Arab-speaking lands of the Ottoman Empire, with the exception of the holy places of Islam, Mecca and Medina, which the British had no wish to occupy. Damascus had been the ancient capital of the Umayyad Caliphate, and although it was no longer of any great strategic importance, it was, like Jerusalem, of immense political significance. Because of this, the city was handed over to Faisal who, with Lawrence in tow (and, if one were to believe the account in *The Seven Pillars of Wisdom,* very much in charge), staged a triumphal entry and took formal possession of the city in the name of his majesty the "King of the Arabs."[80]

The surrender of Damascus to Faisal may have salvaged Arab sensibilities, but it also promised to undermine the terms of the Sykes-Picot agreement, by which France had been given the status of "protecting power" in Syria and direct control over Beirut, both of which were now, nominally at least, in Faisal's hands. There were many in Britain, including the ever-present Lawrence, who, in defiance of the agreement that they had originally endorsed, now backed Faisal's inflated ambitions in Syria and Lebanon.[81]

Another part of the subsequent Lawrence myth casts him as a sort of postcolonial warrior struggling for the creation of a free and independent Arab world against cold, steely imperialists, bent on carving it, and its vast oil reserves, up among themselves. It is true that Lawrence was critical of British policy in Mesopotamia, but largely because it was not tough enough on the French. In fact, Lawrence's ambition for the Arab world was to create not a new "Arab nation," much less the new caliphate Hussein had hankered after, but rather a free Arab Dominion—a sort of Arab Australia—which would form a part not of the British Empire but of the British Commonwealth. In a much-quoted phrase, he once wrote, "My own ambition is that the Arabs should be our first brown dominion, and not our last brown colony."[82] To achieve this end, the French, who saw Syria as yet another Algeria, had first to be removed from the picture.

Most of the power brokers in London tended to agree with Lawrence, at least as far as ridding themselves of the commitments made by Sykes to the French was concerned. The formidable Lord Nathaniel Curzon, who was then chairman of the Eastern Committee, declared the Sykes-Picot agreement to be "absolutely impracticable" and hoped to expel France from Syria altogether.[83]

To make complicated matters more complex still, the United States now entered the fray, in the person of President Woodrow Wilson, who came brandishing a new and unsettling doctrine, to which the British had subscribed in theory for some time but had never seriously considered implementing, and called by Wilson "self-determination." Wilson forced upon Britain and France the recognition that in the future all Arab territories liberated by the Arabs (an admittedly somewhat ambiguous definition) would be governed solely with the "consent of the governed." In his famous "Fourteen Points" speech of January 8, 1918, to the U.S. Congress, Wilson declared that in all subsequent discussions over the future of formerly colonized peoples, no matter by whom they had been colonized, "the interest of the populations concerned should be paramount."[84] This also became one of Wilson's "Four Ends of Peace," which he outlined six months later, on July 4, 1918, and which was subsequently incorporated into the charter of the League of Nations.

On November 7, 1918, Britain and France signed an accord by which they agreed to uphold the complete liberation of all peoples "oppressed" by the Ottomans and the creation of national governments that would express the will of their populations. The Middle East was not to become yet another case of the "translation of empire." Its peoples were to be free to pursue their own ends, however they might choose to define them. Both the Allies and the Americans, however, were extremely wary about how this was to be done. It might, they thought, take some time—more, certainly, than Wilson had in mind. In an imaginary dialogue between an Egyptian and an Englishman written at the turn of the century, Lord Milner had summed up the ambiguities of the British stance on self-determination nicely. "We could not let you continue in your old paths," he makes his Englishman explain to the Egyptian, "because you were a proved failure. . . . But on the other hand, we English don't want to stay in your country forever. We don't despair of your learning to manage your own affairs. . . . You need to be shown what to do, but you also need to practice doing it. You need energy, initiative, self-reliance. How could you develop them if we were to keep you absolutely in leading strings?"[85]


WHERE THE MIDDLE East led, the rest of the world would soon follow—or so Wilson hoped. Self-determination was a doctrine to be shared by all the colonized world (with, however, the significant exception of Africa). Things did not, of course, turn out quite like that. But the dice had been cast, and although it would be some forty years before they finally stopped rolling, by 1945 and the end of what the great Spanish statesman Salvador de Madariaga once called Europe's thirty-year civil war, it had become obvious to all but an obdurate few that imperialism was now a thing of the past, even in the hands-off version known as "indirect rule" favored by the British.[86] Of course it never really occurred to Wilson, any more than it did to subsequent American presidents, that the populations now being allowed to pursue their own intentions might choose something unpalatable to modern, democratic interests. Democracy, which for most of the rulers of Europe was still an uncomfortable compromise with the kind of autocratic rule they preferred, was for Wilson, and has been ever since, a new universalism, and he was confident that it would prove to be irresistible.

On November 13, 1918, a British army, followed soon after by the French and Italians, entered the city of Istanbul. Far more than the capture of Jerusalem and Damascus, which had, at least nominally, been carried out by Arabs, the capture of Istanbul by forces that were wholly infidel was seen by many Muslims as the return of the Crusaders to the lands of Islam.

Istanbul had been the greatest city in the Islamic world. For more than three centuries it had been the Belde-i Tayyibe ("The Pure City"), the Der–i Saadet ("The Abode of Happiness"), and, above all, the Darül'l Islam ("The Abode of Islam"). Of all the Muslim cities taken by the Allies, it was by far the most significant. "Think what Constantinople is to the East," Churchill had written in 1915, during the disastrous Gallipoli campaign. "It is more than London, Paris and Berlin all rolled into one are to the West. Think how it has dominated the East. Think what its fall will mean."

He was not exaggerating. Constantinople had been the greatest single prize ever seized by a Muslim power from the West. Its recapture after nearly five hundred years signaled the end of an empire that, though it may have become, in the eyes of many of its former subjects, a decayed and corrupted state, was still the seat of the caliph-sultan, the "Commander of the Faithful." It was an end, too, of the independence of the *dar al-Islam* and of thirteen hundred years of Muslim imperialism.

The mighty terror of the world had now been reduced to a humble supplicant of the peoples it had once terrorized. The "one great anti-human specimen of humanity," as William Gladstone had angrily called the Ottomans in 1876, were now cowered and destitute.[87] For many, the time had come for the righting of past wrongs. There were those—such as Virginia Woolf's generally pragmatic and liberal-minded husband, Leonard—who argued that as the Turks had entered Constantinople in 1453 as alien conquerors, the city should be placed permanently under an international administration, awaiting, perhaps, the arrival of a new Byzantine emperor to reclaim it.[88] The British, Curzon in particular, were keen to drive the Turks out of their capital, in order, if nothing else, to deprive the Ottomans of their still-lingering status as a great empire and with it their hold on the imagination of the entire Islamic world.[89] Under one scheme, dubbed the "Vatican proposal," the caliph-sultan would remain the "spiritual head of all Mohammedans" but the city itself would be governed by a commission made up of delegates from any number of non-Muslim nations, including the United States and in one suggestion even Brazil and Japan—just to show that this was meant to be a truly "international" zone, not merely a western European one. To this the French prime minister, Georges Clemenceau, tartly remarked that it was "quite bad enough to have one pope in the West" without creating another in the East.[90]

The occupation of Istanbul, although relatively brief (the Allies left again in October 1923), was but one stage in a process of dismemberment, absorption, recovery, and further dismemberment that has continued, as war follows war, until today. Looked at another way, it was the culmination of a long series of humiliations that had brought to its knees the last great Islamic empire—the last great Eastern empire now that both the Mughals and the various dynasties that had ruled over what was once ancient Persia were long gone. It was also the beginning of a new kind of nation building on the ruins of the Sublime Porte that would shape the future not only of the Islamic world but also of the West for the foreseeable future.

All of those caught up in this process of imperial disintegration, greater powers and lesser, now slouched toward the final reckoning, which was due to take place in Paris at the beginning of 1919.

The Paris Peace Conference opened on January 18. Nothing quite like it had ever happened before. As the British prime minister, David Lloyd George, told the House of Commons, the Congress of Vienna, which had brought the Napoleonic Wars to an end in 1814, had lasted eleven months. But Vienna had resolved only the affairs of Europe. At Paris, "it is not one continent that is engaged—every continent is affected."[91] It was meant to bring about a new world order, to ensure that the First World War really was the "war to end all wars." Sadly, as it turned out, it only laid the grounds for another series of conflicts that, on several fronts, have continued and continue to this day. As Field Marshal Earl Wavell remarked bitterly when it was all done, "After the 'war to end war,' they seem to have been pretty successful in Paris at making the 'Peace to end Peace.' "[92]

When it came to settling the future of the Middle East, the main point of disagreement between the Allies was not so much over the wishes of the Arabs for self-determination, as Wilson had wished and which the British claimed to support, as over the obvious British attempt to detach Syria from French control and give it to Faisal under British control. Finally, the exasperated Georges Clemenceau, who had to endure not only Lloyd George's machinations but also indignant criticism from a nationalistic French press eager to acquire another Muslim colony in Syria, exploded. "I won't budge," he told the former prime minister Raymond Poincaré in March. "I won't give way on anything anymore. Lloyd George is a cheat."[93]

Woodrow Wilson now surprised both parties by suggesting that the best way to resolve the question was to ask the Arabs what they wanted. Both the French and the British greeted this as an example of characteristic American political naiveté and refused to have anything to do with it. Public opinion was a feature of states with parliamentary governments. It simply did not exist among Arabs. But Wilson ignored them and insisted on sending a commission to Syria and Palestine to discover whom, if anyone, the Arabs wished to rule over them. The commission consisted of two of Wilson's cronies, Dr. Henry King, president of Oberlin College in Ohio and director of religious work for the YMCA, and C. R. Crane, a Chicago millionaire. The mission was a farce. Both King and Crane were confirmed Francophobes who had persuaded themselves early on that among the Arabs, "the American training and Anglo-Saxon literature and civilization are regarded as morally superior to the French." They came away from their carefully managed tour filled with admiration for the new spirit of democratic enthusiasm among the Arabs. On the delicate issue of the status and treatment of women, they had been persuaded that "the Moslems recognize that the time has come for the education of their women." The Arabs, they reported, did not, however, much like the way the French had handled the issue, since "they say that those [women] who receive French education tend to become uncontrollable"—memories perhaps of the Napoleonic occupation. Better to entrust the whole affair to the Americans. They were also assured by Faisal that, if the Americans would provide the aid, he would open an "American college for women at Mecca."

On their return, King and Crane recommended that Syria be handed over to the Americans, but if Wilson would not have it, it should be ceded to the British. The members of the American peace delegation, far less naive then their president, were so embarrassed by this nonsense that they refused even to show the report to the Allies.[94]

In the end, and after a lot more wrangling, the British abandoned their claims to Syria on Faisal's behalf. In July, the unhappy Faisal, who only four months earlier had had himself crowned King Faisal I of Syria, was driven out of Damascus. He took refuge at Haifa, where he was received with full honors by the British high commissioner. An article in the London *Times* hailed him as a modern Saladin, although quite how being driven *out* of Damascus by a new group of Franks qualified him for that was left unsaid.

Britain emerged from the conference with Iraq, Egypt, Persia (as an informal protectorate), Palestine, and Transjordan, plus control of the sheikhdoms of the Persian Gulf. Lloyd George fiercely rejected any suggestion that he might have cheated not only the French but also the Arabs out of what they had been promised. "No peace settlement," he later wrote in his memoirs, "has ever emancipated as many subject nationalities from the grip of foreign tyranny as did that of 1919. . . . No race has done better out of the fidelity with which the Allies redeemed their promises to the oppressed races than the Arabs."[95] Needless to say, successive generations of Arabs have not seen it that way.

## IV

The Ottoman Empire was not the only victim of the relentless en-croachment of Western, European, and now increasingly Ameri-can technologies and Western imperial ambitions. Farther east lay the other great power of the Islamic world: Persia.

By the second decade of the eighteenth century the fortunes of the Safavid dynasty, which had held both the Russians and the Ottomans in check for most of the previous two centuries, was in steady, if erratic, decline. In 1722, Peter the Great seized part of the country, to be followed immediately by the Ottomans. The two powers then divided up northern and western Persia between them, leaving the Safavids isolated in the eastern heartlands of the empire. In 1736, however, the last Safavid shah, Abbas II, was deposed by Nadir Qoli Beg, of the Turcoman Afshar tribe, who in-stalled himself as Nadir I. In 1739, having driven out the Ot-tomans and the Russians, he crossed the famous Khyber Pass into India, defeated the Mughal emperor, Muhammad Shâh, and in March entered Delhi in triumph. "The accumulated wealth of 348 years," wrote one Indian historian, "changed owners in a mo-ment." One of the spoils of this victory was the Peacock Throne, created in 1635 by Shâh Jahân, who also built the Taj Mahal. Jean-Baptiste Tavernier who saw it in 1676, described it as being en-tirely covered with precious stones, while the peacock itself, perched up above the canopy, had an "elevated tail made of blue sapphires, and other colored stones, the body being gold inlaid with precious stones, having a large ruby front of the breast, from whence hangs a pear-shaped pearl of fifty carats."[96] Nadir carried this back to Isfahan in triumph, and although the throne itself was apparently destroyed in 1747, various imitations were com-missioned by later shahs and the term "Peacock Throne" was used to describe the Iranian monarchy until its final overthrow in 1979.

Nadir then returned to Persia and restored Muhammad Shâh, whom as a descendant of Timur Nadir claimed as a relative, to his now much-impoverished empire, retaining only the southern banks of the Indus on the grounds that these had once belonged to the empire of Darius the Great. Nadir was a brilliant general who by 1740 had once again established Persia as a great military power. But he was also a negligent and inhumane ruler who was famous for piling up his victims' skulls into pyramids, said to be second in size only to those of Tamerlane some 350 years earlier. In 1747, his embittered and increasingly rebellious subjects could stand it no longer, and he was murdered by a group of his own officers.

He was succeeded in 1794 by Aghâ Muhammad Khan, the founder of the Qajar dynasty, which lasted until 1925. Persia now became a central piece in what was called the "Great Game," a phase immortalized by Rudyard Kipling in 1901 in his novel *Kim*, to describe the struggle among France, Britain, Russia, and later Germany for control of the chain of Islamic states that ran from the eastern Mediterranean to the Indus.[97] In the course of the game, Persia was buffeted back and forth between the competing ambitions of the British in India to the east and the Russians and Ottomans to the west, with occasional interference from the French under Napoleon.[98] Throughout much of this time, it remained a backward, autocratic society, in which the shah ruled over a population of wealthy and largely absentee landlords and impoverished and illiterate peasants, who, although legally free, were, in effect, like their equivalents in Russia, serfs. The Shiite religious hierarchy, made up of mullahs and mujtahids, scholars of Islamic law, was both larger and more powerful than the Ulema of Sunni Islam. But since the monarchy posed no threat to it, it offered no threat to the monarchy. Although Qajar Persia was engaged in almost constant warfare with one or another of its neighbors, its habitual defeats at the hand of technologically more advanced Europeans, in particular the Russians, did not result in any lasting desire for reform. Some attempts were made under the enterprising Mirzâ Taqî Khan, grand vizier to the shah Nasir ad-Din, to introduce a version of the Tanzimat. But the failure of that in Turkey and Abdülhamid's return to autocratic rule in 1870 put an end to the experiment, as well as, on the insistence of the shah's formidable mother, to the life of the unfortunate vizier.

The Persian court was costly, flamboyant, and corrupt. In order to sustain his extravagant lifestyle, and unable to develop anything in the way of modern manufacturing industries or even mechanized agriculture (carpets and textiles were virtually Persia's only exports) to subsidize it, Nasir ad-Din began to make ever more extensive concessions to foreign commercial interests. In 1873, a naturalized British subject, the Baron Paul Julius von Reuter, an adventurer and the creator of the famous news agency Reuters, was granted a seventy-year monopoly on all Persian railroads and streetcars—although at the time they barely existed—and on the exploitation of all mineral resources and government-owned forests, including all uncultivated lands, and sundry other concessions, including the right to collect all Persian customs dues for twenty-five years. In exchange Reuter was to pay the shah 20 percent of the railway profits and 15 percent from all other sources. It was, said Lord Curzon, "the most complete and extraordinary surrender of the entire industrial resources of a kingdom into foreign hands that has probably ever been dreamt of."[99]

No one in Persia itself seems to have much minded the alienation of so much of the nation's resources to a foreign, infidel interloper. In 1892, however, yet another massive concession, this time of the tobacco monopoly to a British company, led to widespread riots. (The Persians were very heavy smokers.) These, an omen of what would happen eighty-seven years later, were led by the clergy, who, at the head of an angry mob, marched on the royal palace. The terrified shah called out his Cossack Brigade, which had been created a decade earlier and was commanded by Russian officers, whom he believed to be unwaveringly loyal, only to discover that they had gone over to the clerics. He then capitulated "out of the love of his people," shut himself up in his harem, and spent the rest of his dismal reign doting on cats and marrying a long series of wives. The clerics rejoiced and returned home. On this occasion, their protestations had been short-lived and their actions, in the long run, largely ineffectual. But the outcome of the tobacco riots had demonstrated the power of the mullahs to defend their own interests and those of the people against any kind of innovation from above. It was a lesson they would not forget.[100]

The hemorrhaging of Persian resources continued. Persia, said Nasir ad-Din's brother, was like a "lump of sugar in a glass of water," slowly melting away. On May 1, 1896, the shah was assassinated and succeeded by the still more disastrous Muzaffar ad-Din, whose sole ambition seems to have been to go on a public tour of Europe, which he financed by mortgaging ever-larger sections of the Persian economy. Attempts to impose some kind of parliamentary control were finally forced upon him in 1906. A parliament, or Majilis, was created, which then enacted a Fundamental Law, known subsequently as the Constitutional Revolution. Although subsequent governments attempted to overturn or repeal it, and the last shah ignored it for most of his reign, it survived, and in some form survives even today in the institutions of the Islamic Republic of Iran.

The reform did not, however, put an end to the steady alienation of the country's resources. In 1901, this would take a still more dramatic turn, when an Englishman named William Knox D'Arcy was granted a sixty-year petroleum and gas concession for the entire Persian empire. By the end of the nineteenth century, it had already become clear that oil would be the fuel of the future. At that time, 90 percent of it came from the United States and Russia, and the world market was dominated by two companies: Standard Oil and the Royal Dutch Petroleum Company. The British, the Royal Navy in particular, urgently needed an independent source over which they could exercise direct control and in 1905 the Admiralty persuaded the British Burmah Oil Company to link up with D'Arcy to drill for oil in Persia. The only problem was that although he had been searching for five years, D'Arcy had as yet found no oil. Then, one day in 1908, just as he was on the point of abandoning the entire venture in despair, his drills bit into the Masjid-i-Sulaiman in southwest Persia, one of the largest oil fields in the world. The Anglo-Persian Oil Company was formed in London in April 1909 and immediately went public. In 1914, two months before the outbreak of war, Winston Churchill, then First Lord of the Admiralty, persuaded the British government to buy a controlling interest in Anglo-Persian Oil. Later this became the Anglo-Iranian Oil Company and, later still—the changes of name were all politically charged—British Petroleum.

On January 8, 1907, Muzaffar ad-Din died and was succeeded by Muhammad 'Ali Shah, who proved to be opposed to any kind of constitutional reform and did his (largely ineffectual) best to dispense with the Majlis and overturn the constitution. Meanwhile, now that Persia had been found to be potentially rich in oil, both Britain and Russia kept an anxious eye on what seemed like a process of ever-accelerating decay. "Persia was the danger point," Sir Edward Grey, the British foreign minister, recalled later. "The inefficiency of Persian governments, the state of their finances, the internal disorders, not only laid Persia open to foreign interference, but positively invited and attracted it."[101]

On August 31, 1907, the British and Russians reached a deal in Saint Petersburg. While this committed them to upholding the Persian Empire's "integrity and independence"—at least in principle—the two powers partitioned the country between them, with Russia in the north and Britain in the south. "The primary and cardinal object," as Lord Grey admitted later, as with so much British foreign policy in the region, had been not only the oil but also "the security of the Indian frontier."[102] And, for awhile at least, both objectives had been obtained.

For the moment, at least, the Persians had, like the Arabs and the Ottomans, been effectively swallowed up by the West. But in 1907, Sir Cecil Spring-Rice, the British ambassador to Tehran, had written a memorandum to Grey:

> I think that the European nations should be prepared to face in Persia, what they are beginning to experience else-

where, a national and religious movement, formless perhaps and misdirected, but of great vigor and intensity. And owing, perhaps, to the superior attainments of the Persian race, I think it not improbable that the leaders of the movement here . . . may occupy a prominent, perhaps a dominant position in the future development of the national and constitutional movement among the Mussulman peoples.[103]

It was, as it turned out, a remarkably prescient observation, although what finally emerged from Iran was far from being the kind of moderate, ultimately liberal, religious movement Spring-Rice seems to have had in mind—and the indisputably prominent position that Iran has come to occupy in the development of the "Mussulman peoples" has, alas, little to do with the "superior attainments of the Persian race."

## V

With the end of the Paris Peace Conference, the Allies began the process of reconstructing and reallocating all that now remained of the Ottoman Empire beyond the borders of modern Turkey. Technically the areas of the Middle East allocated to Britain and France had been given not as colonies but as mandates under the League of Nations. "Mandate" was a novel concept. Its purpose was to entrust to a developed Western power an undeveloped area of the globe, although in most cases these had been former colonies of somebody else's empire, with the charge that the mandatory power prepare that territory for independence and incorporation into the League as early as possible. The mandatory powers were required to submit annual reports on their trusteeship, and the League solicited petitions from the inhabitants of the territories in order to monitor the progress of the emancipation process. In London and Paris, however, a darkly skeptical eye was turned on this notion, American in inspiration and impossible of execution. As Lord Curzon, then British foreign secretary, told the House of Lords on June 25, 1920, "It is quite a mistake to suppose that . . . the gift of mandate rests with the League of Nations. . . . The mandate for Palestine and Mesopotamia was conferred upon us and accepted by us, and the mandate of Syria was conferred upon and accepted by France." And what had been conferred and accepted could not then be monitored or interfered with—not that the League of Nations, any more than its successor, the United Nations, actually possessed the operating authority to intervene in any significant way in what its mandated powers got up to.

The British thus went about their business of creating new and viable nations in its mandated territories as they saw fit. This was described by Gertrude Bell, a seasoned "Oriental" traveler who in 1918 became assistant to Arnold Wilson, the British civil commissioner for Mesopotamia, as the policy of "creating kings."[104] The British knew how to handle kings (unlike the French, who seemed to have lost the knack during the revolution). Autocratic and unaccountable to their own people, they sat on their thrones and, kept there by British arms, they were far easier to control than popularly elected parliaments. Hussein's second son, described by Winston Churchill as "a most polished and agreeable person," was therefore installed as king of Transjordan, now Jordan. He would, said Lawrence, make an ideal British agent because "he was not too powerful and . . . not an inhabitant of Trans-Jordania but . . . relied upon His Majesty's Government for the retention of his office." His descendants have continued to rely first upon His, then upon Her, Majesty's government and now upon the government of the United States for the retention of their offices.

In Mesopotamia, a series of Arab uprisings, intertribal feuds, and the repeated assassination of British officers led, in the summer of 1920, to a full-scale Arab revolt. In a what looks like a rehearsal for the present war in Iraq, Shiites fought Sunnis and both fought the British. The London *Times,* more outspoken than its modern, cautiously uncritical version, denounced "the foolish policy of the government in the Middle East." "How much longer," asked a leader of August 7, "are valuable lives to be sacrificed in the vain endeavor to impose upon the Arab population an elaborate and expensive administration which they never asked for and do not want?"[105]

The solution, proposed by Winston Churchill, who in 1921 was appointed colonial secretary with responsibility for British policies in the Middle East, was to install Faisal as the king of a newly created state of Iraq made up of the former Ottoman velayets of Basra, Baghdad, and Mosul. Faisal had, ever since his brief rule in Syria, been cultivating the image of himself as a modern national leader. "We are one people," he had declared in May 1919, "living in a region which is bounded by the sea to the east, the south and the west and by the Taurus mountains to the north." The obvious divisions, ethnic and above all religious, that had riven the entire area for centuries would now be washed away in the cleansing rivers of pan-Arabism. "We are Arabs," Faisal repeated, "before being Muslims, and Muhammad was an Arab before being a prophet," which may have been literally true but must also have been deeply repugnant to any true Muslim. But if Faisal endorsed, or claimed to endorse, secular Western conceptions of nationhood, he was also an Arab and from a family with irreproachable claims to tribal allegiance.[106] On July 11, 1921, the Council of Ministers of Baghdad declared Faisal to be the constitutional monarch of Iraq. The Iraqi "people"—although shortly before his death in 1932 Faisal admitted that there was no such entity—were then asked if they agreed.[107] On August 18, the Ministry of the Interior announced that the results of a yes/no plebiscite had demonstrated that they had done so, and overwhelmingly. Five days later Faisal was crowned, and the ancient term "Iraq"—meaning, ominously, as it turned out, "well-rooted country"—became the official name of the new kingdom. Faisal's coronation put an end, briefly, to the Arab uprisings, but Iraq, although indubitably fertile, has proved to be very far from "well rooted" in any political sense. The British attempt to build a nation where none had ever existed, which, as one American missionary wisely told Gertrude Bell in 1918, was a denial of four millennia of history, created a far more intractable problem than the one it was intended to resolve and for which no solution has yet been found.

To round out this process of kingmaking, Fû'ad I, a descendant of Muhammad 'Ali, was installed as the ruler of an independent Egypt in 1922. The three areas under British control—with the exception of Palestine—were now safely transformed into petty kingdoms, their monarchs bobbing and bowing to whatever tune His Majesty's government chose to transmit from London. It worked for a while, but not for long. As Lawrence had warned Churchill, these were a fierce people with a proud imperial past, which they were now bent on reviving. Faisal himself lasted until 1933 and the Hashemite kingdom of Iraq until 1958. The Kingdom of Egypt would vanish in the nationalist revolution of Gamal Abdel Nasser in 1952. The French, who had been given the mandates of Syria and Lebanon, ruled them as if they were dependencies of metropolitan France in accordance with a tradition that, in one way or another, went back to the seventeenth century. In 1930, however, Syria also broke away from its European puppet master and became a parliamentary republic independent in all respects except for security and foreign policy, both of which remained under French control.

THE OUTCOME OF the squabble between Britain and France over their respective mandates was not, however, the only, or even the most significant, outcome for the Middle East of the Paris Peace Conference. By far the most consequential was the decision to create, in the place the world called Palestine—the land of the Philistines—what all Jews called the Land of Israel, a Jewish national homeland. As we saw in the last chapter, as early as the late eighteenth century rumors had been circulating in the ghettoes of the Diaspora that Napoleon intended to found a Jewish state with its capital in the holy city of Jerusalem. In the 1830s and '40s, Lord Palmerston, the British prime minister, had supported the idea of a Jewish homeland. He had done so partly because he believed that a British-backed Jewish client state in Palestine would be a useful ally in Britain's constant tussle with France for control of the Middle East and the route to India, and partly because it revived a millenarian ambition that went all the way back to the seventeenth century and Oliver Cromwell's conviction that Puritan England was God's chosen instrument to bring the Jews back to the Holy Land.

In 1868 Britain also acquired, in the person of Benjamin Disraeli, known affectionately as "Dizzy," its first and to date only Jewish leader. In 1847, Disraeli had written a novel, *Tancred; or The New Crusade,* the story of a young aristocrat who travels to the Holy Land not to conquer but to find the spiritual enlightenment offered by all three of the great religions of Asia: Judaism, Christianity, and Islam. It was a long, somewhat inchoate expression of the conservative disenchantment with the "shipwrecked gaiety of Europe," and, like many such examples of "Occidentalism," it looked to Asia for redemption. "Asia's slumber," says Tancred at one point, "is more vital than the waking life of the rest of the globe."[108] *Tancred* is really about the present condition of England, but it nevertheless plays with the image of an East in which the great hostilities that had divided the three Abrahamic faiths over the centuries might finally be resolved and their resolution provide what the nostalgic Romantics believed that modern industrialized Europe lacked—a soul.

In the late nineteenth century, Zionism became a subject of wide concern in broadly liberal circles in Europe. In 1876 the novelist George Eliot (whose real name was Mary Ann Evans) published a long, complex novel entitled *Daniel Deronda,* in which a young, apparently English aristocrat discovers that he is, in fact, Jewish. Shattered but also deeply moved by this discovery, he immediately abandons his traditional English gentlemanly pursuits and the unfortunate woman who has seen in him the possibility of her own salvation and dedicates himself to the creation of a Jewish homeland in Palestine. The book caused a sensation (even in liberal circles Jewish heroes were not the accustomed staple of English fiction) and the disapproval of many of the literary establishment. But it also aroused a great deal of latent sympathy for the Jewish cause and was looked upon as a source of inspiration by many later Zionists. "One of our first visionaries," Abba Eban, one of the negotiators in the foundation of the state of Israel, called Eliot. Today there is a street named after her in every major city in Israel. Until the end of the nineteenth century, however, the ideal of a Jewish homeland—as yet no one spoke of a "state"—in Palestine remained confined to such prophetic fiction, as elusive as Cromwell's bid to hasten the Day of Judgment by returning the Jews to Judea or Napoleon's supposed intention to "rebuild the walls of Jerusalem."

In 1894, Theodor Herzl, the Paris correspondent of the Viennese newspaper *Neue Freie Presse,* was asked to cover the Dreyfus case. Alfred Dreyfus, an apparently loyal Jewish army officer, had been condemned to the infamous Devil's Island, a prison camp in French Guiana from which few ever returned, on charges of passing secrets to the Germans. When it emerged that the evidence against him was flimsy and circumstantial, the army, alarmed by the possibility of a scandal, pressed ahead with a court-martial that many, the hugely popular novelist Émile Zola among them, believed to have been rigged. What made the case sensational was not a dubious court-martial; it was the fact that Dreyfus was a Jew. It convulsed and divided French society. It also transformed Herzl's life. Hitherto, he had been a fully assimilated Jew for whom Judaism itself meant very little. Alarmed now by the widespread anti-Semitism revealed by the Dreyfus case, he had, he said, been taught "the emptiness and futility" of trying to "combat anti-Semitism." Two years later he published *The Jewish State* (*Der Judenstaat*). The plight of the Jews, he argued, was an international one, and it could be solved only internationally—by the creation of an independent Jewish state. Jews, he wrote, had dreamed the "kingly dream" of a separate Jewish state "for the whole night of their history." " 'Next year in Jerusalem' is our ancient watchword." It was now "a question of showing that the dream can be converted into a living reality."[109]

The book was an instant success, going through three editions in the year of publication alone, and it launched the movement called "political Zionism." In August 1897, this was given an institutional expression when Herzl convened the first Zionist Conference at Basel, which then created the World Zionist Organization and elected him its president.

Herzl's own vision of the Jewish homeland was very different from the "living reality" that finally emerged after 1948. His new state was to have embodied the egalitarian, transnational, multiracial, multicultural, transdenominational ideals to which enlightened Jewry had aspired since the great ecumenical German Jewish philosopher Moses Mendelssohn in the eighteenth century. It was to have no national language, certainly not Hebrew— "Who among us," asked Herzl, knows enough Hebrew to ask for a train ticket in this language?—but each of the several groups of which it was to be made up would preserve "his language which is the precious home of his thoughts." The army, which has since become an omnipresent part of Israeli life, would be confined to barracks except for emergencies, and "every man shall be free and undisturbed in his faith or his disbelief as he is in his nationality." A true cosmopolis, in other words, a Switzerland of the Middle East.[110]

The creation of a Jewish nation was to be a solution to the "Jewish problem," which nationalism itself had if not created then certainly exacerbated. Anti-Semitism, which had for centuries been Europe's vilest disease, had since the early nineteenth century acquired a new form. If there was now to be a "Europe of Nations," to which nation—if any—did the Jews belong? Were German Jews really German, with a recognized place in the new German nation Bismarck had created in 1871? Were the French Jews true citizens of the new society that had come into being after the revolution? The answers to these questions were unclear even to the Jews themselves. The creation of a Jewish nation would eliminate the need for them altogether. But the "Jewish home-land" was more than simply an expression, loaded with racial overtones, of the new nationalism. For ever since the Judeans, as they were then called, had been driven into exile after the destruc-tion of Jerusalem by the Roman emperor Titus in 70 C.E., the Jews had been dreaming of a return to the Promised Land. Each year at Passover, the international Jewish communities around the globe repeated the ritual prayer, "Next year in Jerusalem." For the first time in millennia, it now seemed as though it might soon be an-swered.

The uncomfortable fact that ever since the first century C.E. wave upon wave of immigrants, most of them Arabs, had settled in this same land was largely disregarded except by a few. When in 1897 the rabbis of Vienna sent a fact-finding mission to Palestine, they famously reported back that the bride "was beautiful but married to another man." But the implication of this wry remark—that the Zionists should attempt to marry someone else—was dis-regarded. In 1901, Herzl approached the sultan Abdülhamid with a request to lease part of Palestine for his project. At first the sul-tan was inclined to concede, attracted by the prospect of wealthy Jewish investors in the region, but in the end he declined when his ministers persuaded him that such a move would be disastrous for his pan-Islamic ambitions.

For Herzl himself, however, what mattered was not Palestine. Although it was indubitably "our unforgettable historical home," it was not exactly the most fertile or inviting part of the world. Al-most anywhere would do, as long as it was away from the clutches of anti-Semitic Europe. Argentina, to which there had been a con-siderable Jewish migration in the nineteenth century and which was clearly to be preferred to Palestine on grounds of climate, wealth, and fertility, had been another of Herzl's options; and when, in 1903, Joseph Chamberlain, at that time secretary of state for the colonies, offered him a site on the Uasin Gishu plateau near Nairobi, in what was then British East Africa, he was inclined to accept, at least as a temporary measure.[111] But the project was rejected the following year by the sixth Zionist Conference. The new Jewish state had to be where it had always been, in Judea itself, or nowhere.

The outbreak of the First World War provided a new impetus to the Zionist cause. If the Ottoman Empire was to be dismembered, as seemed very likely, the chances of Allied support for a Jewish state in Palestine would be good. The British had set their sights since the beginning on some kind of control over Palestine. To many in the Colonial Office, it seemed that by far the best way to achieve this would be by creating not an Arab state under British domination but a Jewish one under British tutelage. "In the Zionist movement," declared the *Sunday Chronicle* in April 1917, "we have a motive force which will make the extension of the British Empire into Palestine, otherwise a disagreeable necessity, a source of pride and a pillar of strength."

As the war progressed, the Allies moved closer and closer to some kind of understanding that they would support the creation of a Jewish homeland in Palestine, provided this could be achieved without threatening the lives of the existing population. The final declaration came in November 1917. It had previously been approved by France and the United States after consultation with Italy and even, albeit resentfully, with the Vatican. It took the form of a letter written and signed by the foreign secretary, Arthur Balfour, to Lord Rothschild, who, although no Zionist, was the most prominent member of the British Jewish community. This was published in the London *Times* on November 2, 1917. It read:

> Dear Lord Rothschild,
> I have much pleasure in conveying to you, on behalf of His Majesty's Government, the following declaration of sympathy with the Jewish Zionist aspirations which has been submitted to and approved by the Cabinet: "His Majesty's Government view with favor the establishment in Pales-

tine of a national home for the Jewish people, and will use
their best endeavours to facilitate the achievement of this
object, it being clearly understood that nothing shall be
done which may prejudice the civil and religious rights of
existing non-Jewish communities in Palestine, or the
rights and political status enjoyed by Jews in any other
country." I should be grateful you would bring this to the
knowledge of the Zionist Federation.

The Balfour Declaration, as this came to be known, was endorsed
by the French government on February 14, 1918; by the Italian
government on May 9; and by a joint resolution of the U.S. Congress on June 30, 1922. Mild though the language might seem—
far too mild in the opinion of Chaim Weizmann, president of the
British Zionist Federation—it was greeted as "the political charter
of the Jewish nation." Nothing so momentous had happened
since the proclamation of Cyrus the Great in 538 B.C.E.

The British motives for this act were complex and frequently
confused. At one level the Jewish homeland was seen as a possible
means of rallying support for the allied cause in the final years of
the war. "It was considered," explained Churchill in 1922, "that
the support which the Jews could give us all over the world and
particularly in the USA and Russia would be a definite palpable
advantage."[112] It was also feared at the time that the Germans
would establish themselves in Palestine and that a "Teutonized
Turkey," as Curzon put it, would constitute "an extreme and perpetual menace to the Empire."[113]

There was, however, another more ambitious long-term objective. A Jewish settlement at the heart of the fragmented world that
was bound to follow the inevitable collapse of the Ottoman Empire would be a force for order and stability. The educational and
technical superiority of the Jews made it seem that they might
succeed in transforming Palestine not only into a useful satrapy
of His Majesty's government but also into a prosperous modern
community, something that came close to echoing the French

suggestion made in 1798 that the Jews of the Diaspora would bring with them "the Enlightenment of Europe," which they had played a significant part in creating, and then pass it on to the miserable populations of Palestine and Syria.[114] Herzl, too, had seen it that way. "To Europe," he had written, "we would represent a part of the barrier against Asia; we would serve as the outpost of civilization against barbarism."[115] A part of the West, in other words, lodged securely in the flank of the East.

The East was once again to be westernized. Only this time the job was to be carried out not directly by Westerners but by means of a people who, though unmistakably "Eastern," had, over time, absorbed all the skills and attitudes of enlightened Europe. With the help of the industrious Jews a new Palestine would emerge, which the Arabs, "a handful of philosophic peoples," in Churchill's words, would never have been able to create for themselves.[116]

The Balfour Declaration linked together what turned out to be two incompatible objectives. But at the time the idea that Jews and Muslims (and a few Christians) might be persuaded to live together harmoniously in a Jewish-controlled land—under ultimate British sovereignty—did not seem so wildly improbable as it might today. Leopold Amery, who had helped draft the declaration, wrote later that he and others of his generation had believed that "the regeneration of the whole Middle Eastern region . . . would be far more effective, and, one hoped, more acceptable, if carried out by people who bring the knowledge and the energy of the West to bear, [yet] still regard the Middle East as their home. . . . Most of us younger men, like Mark Sykes, were pro-Arab as well as pro-Zionist and saw no essential incompatibility between the two ideals."[117]

Not everyone agreed. The only Jewish member of the British cabinet, Edwin Samuel Montagu, denounced the whole project as the reconstruction of the tower of Babel. "Palestine," he said, "would become the world's Ghetto." Curzon, too, denounced it as an act of "sentimental idealism" and remarked acidly that Britain had "a stronger claim to parts of France" than the Jews had to Palestine. But the opponents, whatever their motives, were overruled. As Balfour himself, in startling disregard for both international law and the principle of self-determination he was supposed to be upholding, told Lloyd George, there was no intention of going "through the form of consulting the wishes of the present inhabitants of the country." The Great Powers, he went on, were committed to Zionism, "be it right or wrong," because it was "rooted in age-long traditions, in present needs and future hopes" and that these were "of far profounder import than the desires and prejudices of the 700,000 Arabs who now inhabit that ancient land."[118]

At first the optimists seemed to be justified. Even Arab opposition to the Balfour Declaration and all that it implied was slow to mobilize. Initially the Arabs do not seem to have seen the steady influx of Jews who had already begun buying up land from absentee Arab landlords as a threat. In January 1919, Faisal, in his bid to secure Syria for himself, even signed an agreement with Chaim Weizmann, pledging his support for the declaration and for the adoption of "all necessary measures . . . to encourage and stimulate the immigration of Jews into Palestine on a large scale."[119] What he did not commit himself to was the creation of a Jewish state. And in the end it was this, of course, that would be the prime cause of the ensuing conflict.

Throughout the 1920s and '30s, Jews and Arabs continued to live in an uneasy relationship with one another. But the older illusion of a common multicultural society, in which Herzl had invested so heavily, rapidly faded. Jewish immigration steadily intensified, as did Arab hostility. Between 1922 and 1939, the Jewish colonies in Palestine increased from forty-seven to two hundred, and Jewish landholdings more than doubled. By the outbreak of the Second World War, the Palestinian Jews were in effect self-governing. They had had their own university on Mount Scopus in Jerusalem since 1925 and their own army, the Haganah, which although clandestine was officially tolerated and had even assisted British troops in putting down the so-called Arab Rebellion of 1936 to 1938.

The Holocaust changed all that. By 1945, the creation of a Jewish homeland could no longer be conceived by the victors of the Second World War as a somewhat quixotic attachment to the fulfillment of a biblical prophecy, nor as a means to smuggle the ideals of enlightened Europe into the Muslim East. It had become a moral duty. For the Jews it had acquired a new urgency as the only possible guarantee of their continuing existence. The flood of Jewish refugees, and the ferocity of the terrorist campaign the settlers began to wage against their former benefactors, soon destroyed any effective British control over the territory.[120] Gone now was any possibility that the state of Israel might form part of some larger British federation, some new "Commonwealth of Nations" under British tutelage that would replace the evidently archaic British Empire. Israel must now be a true nation, in a new postwar world composed of independent, self-determining nation-states. On May 14, 1948, standing beneath a portrait of Herzl, the Zionist leader, David Ben-Gurion proclaimed the existence of "the Jewish state to be called 'Israel.' "[121] Within hours it had received de facto recognition from President Truman and de jure recognition from the Soviet Union.

In many respects, Britain's reasons for supporting the state of Israel in the early twentieth century were not unlike those of the United States today. In both countries there was, and is, a wealthy and highly influential Jewish community. Both nations saw in Israel a potentially modern, stable, and enlightened—to which now add democratic—state in a backward, unstable, undemocratic region, one furthermore that would prove to be a valuable ally in any conflict that might subsequently arise with the Arabs. The British saw Israel as a buffer state inhabited by "an intensely patriotic race" favorable to British interests. The Americans have similarly looked upon Israel as a buffer state, first against the encroachments of the Soviet Union and now against the ever-elusive Islamic terrorists and their Arab and Iranian supporters.

## VI

The Paris Peace Conference had left virtually the whole of Asia west of the Himalayas under some kind of Western control, with one notable exception. In 1922, after a bitter and bloody war between the Ottoman Empire and Greece, a Young Turk named Mustafa Kemal seized power and immediately set out to transform what remained of the empire into a modern Western nation-state. On November 1, 1922, the sultanate was abolished, and after half a millennium the empire finally dissolved. Two years later the caliphate, which Kemal said later "could only have been a laughing-stock in the eyes of the really civilized and cultured people of the world," followed it into oblivion.[122] A legal code based on various European forms replaced the Shari'a, laicism was established as one of the six cardinal principles of the new state, and women were forbidden to veil themselves in public—all of which effectively destroyed the power of the Ulema. A form of parliamentary democracy was introduced, with a new constitution that guaranteed equality before the law and freedom of thought, speech, publication, and assembly and that placed sovereignty in the hands of an elected Grand National Assembly—although the real power remained, and some would say still remains, with the armed forces. The nation was now to be called Turkey, and its peoples were to acquire a new national identity—no matter to what ethnic group or religion they belonged—as "Turks." The capital was transferred, eleven days after the Allies evacuated it, from Istanbul to the then small, unpromising town of Ankara, where, ironically perhaps, in 1402 the Mongol Khan Tamerlane had all but annihilated the nascent Ottoman Empire.[123] On October 29, 1923, a theocratic empire formally became a modern secular republic. In 1928 the Arabic script was abandoned in favor of a modified version of the Latin alphabet. All Turks were required to have given and surnames, and Kemal himself took the name "Atatürk"—"Father of the Turks."

Turkey became an inspiration to other Muslim states. If the embrace of Western nationalism had allowed Atatürk to preserve his country from annexation and dismemberment, it might do the same for other Muslim states. The most obvious candidate for resurgence was Egypt. Egypt had been the first Muslim society to be "modernized"—if only fleetingly—by Napoleon, then by Muhammad 'Ali, and finally by the British. Now, it was by adopting Atatürk's Asian nationalism, it might be transformed into a modern nation.

The Egyptians had a number of very good reasons for hating both the French and the British as outsiders and usurpers. But there was no denying that, thanks to their interference, by the end of the First World War, Egypt had emerged as the richest, most powerful of all the Muslim societies. Its peoples were better educated, its armies better equipped and trained, its cities bigger and brighter, and its agriculture and manufacturing industries more productive than those of any other Middle Eastern society. It was no surprise, then, that it was in Egypt that the new Atatürk emerged. He was Gamal Abdel Nasser, the son of a postal clerk in a small provincial town, who like so many nationalist leaders from the so-called Third World, like Atatürk himself, had risen to power through the ranks of the army. On July 23, 1952, in what came to be called "the July Revolution" by analogy with the Bolshevik "October Revolution," Nasser toppled the highly compromised regime of King Farouk, and sent the obese monarch and his entourage off to disgrace themselves on the beaches of southern France.

Nasser was a nationalist, one of whose first acts was to bring a statue of the pharaoh Ramses II to Cairo. But he realized that without some kind of unity throughout the entire Arab world he would never be able to rid himself of the old colonial regimes, which, in his view, were largely responsible for the impotence and backwardness of the Middle East.[124] The time had come, he declared in his revolutionary manifesto of 1955, for the Arabs to cease their feeble lamentations that the hated West had betrayed and outsmarted them. Now was the time to strike back. "We are strong," he told them. "Strong not in the loudness of our voices when we wail or shout for help, but rather when we remain silent and measure the extent of our ability to act."[125] True to his word, the following year, he defied the British and French and nationalized the Suez Canal, the most potent symbol of Western colonialism west of Hong Kong. Britain and France, which had joint ownership of the canal, entered into secret negotiations with Israel in the town of Sèvres outside Paris. An agreement was reached under the terms of which the Israelis would invade Egypt, which, it was widely and, as it turned out, accurately, assumed they would be able to do with little serious opposition. At this point the British and the French would force an agreement upon Egypt whereby both armies would withdraw to within ten miles of the Canal Zone and the British and French would regain control of the canal itself. The whole thing was given the somewhat whimsical code name Operation Musketeer. On October 29, the Israeli army invaded the Gaza Strip and the Sinai Peninsula. The British and French duly offered to intervene to enforce the cease-fire demanded by the United Nations. Nasser, however, refused to accept their terms. A combined French and British force then seized the canal. All had, apparently, gone according to plan. At this point, however, the United States, fearful that the Soviet Union might now intervene on Nasser's side and embarrassed at condemning the Soviet invasion of Hungary while simultaneously tacitly condoning a similar operation in the Middle East, brought economic pressure to bear upon the allies, eventually forcing them to withdraw, leaving the canal once again in the hands of the jubilant Egyptians. The "Suez crisis," as it is called in the West, or the "War of the Tripartite Aggression," as it is commonly known in the Arab World, although a military defeat, had turned, unexpectedly, into an ideological triumph for Nasser.[126]

That the humiliation of the Franco-British-Israeli alliance had been achieved not through Egyptian arms but as a result of economic pressure from the United States went largely unnoticed in the Arab world. What mattered, as far as the Arabs were concerned, was that an Arab David had faced down the Western Goliath and won. Nasser's revolutionary nationalism now held out the hope of a new Arab, postimperial renaissance. "Our battle is against imperialism," he had proclaimed on ousting the British and French from the canal. "Our battle is against Israel, created by imperialism to destroy our Pan-Arabism just as it destroyed Palestine." The wording was highly significant. The enemy was imperialism, a word that Nasser had hijacked from the lexicon of Leninism but that now stood for Europe and the United States—"the West," in other words—and the objective was not only to destroy the satrapy it had established in the heart of the *dar al-Islam* but to create a united, and largely secular, Arab world—dominated by Nasser himself—"from the Atlantic Ocean to the Persian Gulf."[127]

Nasser became a hero to the Arab world, a beacon for the future, a demonstration of the ability of the united Arabs not only to destroy Israel but to confront the entire Western world on its own terms. More than a century before, Rifa'ah Rafi' at-Tahtâwî had come to the conclusion that the strength of the West lay less in its technological achievements or its secularized science, which could in his view easily be replicated by an Islamic society, than it did in Western patriotism. If only, he argued, Muslims could emulate that, they would acquire the "means to overcome the gap between the lands of Islam and Europe."[128] It was a devastatingly effective claim, even though it was also a denial of what had held Islam together, however unsteadily, for centuries. By abandoning the singularity of the Muslim community, the Umma, Rifa'ah Rafi' at-Tahtâwî and the new Arab leaders he inspired had also turned their back on the universalism and cosmopolitanism that had made of the caliphate a once-great civilization. It would be a tragic inspiration. Almost the last thing that the Muslim world needed in the mid–twentieth century was the already tawdry garb of European nationalism.

One by one, however, the major Arab states fell under its spell, aided and supplied, where expedient, by the Soviet Union. During the 1950s, Syria came increasingly under the control of the Ba'th

("Rebirth") party, founded in 1947 by two young Syrian intellectuals, Salâh ad-Din Bitar and a Paris-educated Christian, Michel
Aflaq, who rapidly became the party's chief ideologue and later its
secretary-general. Aflaq, like Nasser, preached a new pan-Arabic
nationalism, an "indivisible Arab nation," which asserted not only
the Muslim roots of all Arab society but, more emphatically still,
the Arab roots of Islam. "Muhammad," he once said, "was the
epitome of all Arabs . . . so let all Arabs today be Muhammad."
This was the Arabic Ba'th. And, as we shall see, it would stand in
stark contrast to the later "Islamic Ba'th," whose ideals would be
the inspiration of most of the Islamic revivalist movements after
the 1960s.[129] In pursuit of this ideal of Arab unity, the Ba'th party
pressed not merely for a closer alliance but for an actual union
with Egypt, and on February 2, 1958, Syria and Egypt merged to
form the United Arab Republic. (The UAR did not, however, last
very long. In February 1961, a more radical Ba'th party drove
Michel Aflaq into exile, seized power in a military coup, and has
held on to it ever since.)[130]

In July 1958, King Faisal II of Iraq, the last remaining sup-
porter of pro-Western policies of any significance, was ousted by a
group of officers led by a self-confessed Nasserite, Abdül al-Karim
Qasim. The most powerful states in the Arab world were now in
the hands of outwardly modernizing nationalist leaders, all
backed by the Soviet Union, all determined to drive the former
Western colonial powers from the region for good. Five years later,
however, Qasim was toppled from power and his place taken first
by 'Abd as-Salam Aref and then, when he died, by his brother 'Abd
ar-Rahman Aref. In 1968 the Ba'th party assumed control, and by
1979 this had passed into the frequently bloody hands of one
man and his family and clients: Saddam Hussein.

Saddam set about unifying what was in reality a far from unified state. He was a Sunni, and his main approach was to suppress,
as far and as brutally as he was able, Shiite and Kurdish opposition, not through any religious conviction but because the Sunnis
were the only group in whose loyalty he could trust. In the 1970s

and '80s he also began an ideological campaign aimed at persuading the peoples of Iraq that they could identify with something
older, and more compelling, than an unhappy coalition of Sunnis, Shiites, and Kurds fabricated by the British and now held together by a Sunni-led dictatorship. Like the shah, Saddam turned
to the pre-Islamic past. Ancient Babylonia, Akkadia, Assyria, and
Sumeria were all resurrected as "Iraqi," and their long-dead peoples provided with a "Semitic," and thus Arab, pedigree on the basis
of dubious nineteenth-century linguistic associations, and were
duly celebrated as "our grandfathers," and "our ancestors."[131]

All of these societies, with the partial exception of Turkey, had
pursued much the same goals. All had turned their backs on their
purely Islamic pasts, even if they had not all, as Atatürk had done,
utterly destroyed the power of the Ulema. All had seen modernization and westernization—albeit now under Soviet rather than
American or European guidance—as the means of resurgence. In
this they had unwittingly followed much the same path as the reforming Ottoman sultans in the eighteenth and nineteenth centuries. They, too, had adopted Western notions of reform. They,
too, had modernized and to some degree secularized, in the same
hope of being able to defeat their Western enemies. In the Ottoman case, as we have seen, the outcome had not been the creation of the kind of state the French and English liberals of the
nineteenth century had advocated. What had emerged instead
was precisely what the Europeans had shed in the late eighteenth
century: absolute monarchy, or not-so-enlightened despotism. A
similar fate now overtook the new-style nationalists. Instead of
creating the liberal democracies that had been the ultimate source
of the success of the Western powers, all the Arabs achieved was
new kinds of tyranny, also imported from the West. If Selim III
and Abdülmecid had turned out to be Oriental versions of Peter
the Great or Louis XIV, Nasser and the Nasserites most closely resembled Hitler and Stalin, even if they never quite indulged in the
genocidal fantasies of either. As the British constitutionalist
Thomas Erskine May had noted in 1877, "In the hands of Eastern

rulers, the civilization of the West is unfruitful; and instead of restoring a tottering state, appears to threaten it with speedier ruin"—a condescending Victorian observation, to be sure, but one that repeated experiments seem, all too depressingly, to have borne out.[132]

The same unhappy fate also overtook Persia. On February 21, 1921, the commander of the Persian Cossack Brigade, Reza Khan, with British encouragement, had deposed the last of the Qajar dynasty. Four years later, with the overwhelming support of the parliament, he elected himself shah, changed the name of the country from Persia to Iran, created a new dynasty named after an ancient language, and became Reza Shah Pahlavi.

Reza Shah, like Atatürk, whom he greatly admired, set out to modernize, westernize, and secularize the new Iran. He introduced a judicial system based on French models that, like its Turkish counterpart, replaced the Shari'a. The wearing of the veil was outlawed in 1936, and both sexes were obliged to discard traditional Persian dress in favor of European costumes. A Western system of education was introduced and a new university created. But the new shah, like his counterparts in the Arab world, rapidly became not a democratic leader but a dictator, lavish in his lifestyle and authoritarian in his government.

Both of these features of the new dynasty were continued, and if anything exacerbated, by his son, Mohammad Reza Pahlavi, who came to the throne in 1941, after Britain and the Soviet Union had forced Reza Shah to abdicate. The new shah's "White Revolution," financed by a massive oil boom and intended, as he once said, to make Iran into an "Asian Germany," brought significant improvements in education, health, and agriculture. But it benefited only a very few, while the cronyism, corruption, and self-indulgence that had marked the previous reign went on unchecked.

The ultimate focus of the ambitions of these new nationalistic Muslim dictators—with the exception of the shah (and he was neither an Arab nor, in the eyes of most Muslims, even a Muslim)—

was the final destruction of the state of Israel. By the end of the 1950s, Israel had come to be seen as a successor state to the Crusader kingdoms of the Levant. Jews and the Christian Crusaders, whom Muslims had for centuries treated as two quite distinct enemies, now became merged as one. For centuries Jews had been tolerated, even promoted, by the great Islamic states. Now that the Jews had become an instrument of the West, however, a series of injunctions to destroy them that had lain largely forgotten for centuries was revived and intensified. When Osama bin Laden was asked in October 2001 if he agreed with the notion, popularized by the American political scientist Samuel Huntingdon, of a "clash of civilizations," he replied:

> Absolutely. The [Holy] Book states it clearly. Jews and Americans invented the myth of peace on earth. That's a fairy tale. . . . The Prophet said: "The Hour will not come until the Muslims fight Jews and kill them." When a Jew hides behind a rock or a tree, it will say: "O Muslim! O servant of Allah! There is a Jew behind me, come and kill him!" Those who claim that there will be lasting peace between us and the Jews are unbelievers because they disagree with the Book and its contents.[133]

It need hardly be said that not all Muslims interpret the Hadith bin Laden was quoting, or rather misquoting, or the more ambiguous comment in the Qur'an that "They [the Jews] imitate the saying of those who disbelieved before [i.e., they are polytheists]; may Allah destroy them; how they are turned away!" (Q. 9:30) in this light. Not even all Islamic radicals have called for the destruction of all Jews, as opposed to all Israelis. Sheikh Qaradawi of the Muslim Brotherhood, for instance, has always insisted that a clear distinction should be made between Jews, as one of the "peoples of the Book," and the Israelis, as the conquerors of a sacred territory of the *dar al-Islam*. But for many Arabs, at least, as for bin Laden, Israelis, Jews, and Western "crusaders" have merged into one common enemy.[134] Before the broad-based terrorist group that bin Laden had helped create and fund became al-Qaeda ("The Base"), it had called itself the "World Front for Jihad Against Jews and Crusaders."

At first the Arabs had believed that they could easily defeat the new Crusader State. Some are said to have reasoned that as the Crusader States themselves had lasted for something less than a century, Israel could not be expected to last any longer. History, in the Muslim imagination, is always doomed to repeat itself.

The first Arab attack came in 1948, the very day after Israel had declared its existence. Early on the morning of May 15, a combined force of regular units from the Syrian, Jordanian, Iraqi, and Egyptian armies crossed the border into Palestine. So confident was King Farouk of Egypt of success that he even issued commemorative postage stamps to celebrate the inevitable victory. But the pan-Arab armies were chaotic and disorganized. (As one Arab remarked bitterly, they were fit only for staging military parades.) When the fighting ended in January 1949, the Israelis had occupied the Negev as far as the former Egypt-Palestine border, with the exception of the Gaza Strip, a narrow band of territory along the coast. Only 21 percent of Palestine now remained in Arab hands, and between 700,000 and 750,000 Palestinians had been driven into exile. About 1 percent of the entire Jewish population of Israel had also perished, but Israel's War of Independence, as it came to be called, had demonstrated the capacity of the new state to survive against overwhelming odds and the ever-mounting loathing and despair of its neighbors. It was the beginning of what in Arabic is called *al-Naqba,* the "catastrophe."

The second and still more catastrophic blow came in 1967. This time it was Israel that stuck first, after prolonged Syrian-backed terrorist attacks against its borders. The war lasted a mere six days, from June 5 to 10, but it left some 50,000 people, most of them Arabs, dead or injured. It resulted in the loss of all that then remained of Palestine and deprived some 300,000 Palestinians of their homes. Israel occupied and united the two parts of Jerusalem, which had hitherto been a divided city, and seized from Syria the Golan Heights and the west bank of the Jordan.

The Six-Day War, as it came to be called in the West—the Arabs dubbed it *al-Naksah*, "the setback"—also destroyed the military capabilities of Syria and shattered Nasser's self-image as the appointed leader of a new Arab world. It destroyed most hopes of pan-Arab unity, and it compromised the radical nationalism in whose name that unity was supposed to have been created. Later, conservative Saudis would call it a divine punishment for having forgotten religion.[135]

The defeat also forced upon the entire Muslim world a new vision of the struggle between Islam and the West. A new history began to emerge of which Osama bin Laden and his followers are the ultimate beneficiaries. The story goes something like this.

There have been three crusades. The first brought armed Christians, under the banner of the cross and their own version of jihad—vile but also intelligible—into the Holy Land. In the end, however, they were successfully repulsed by Saladin and his heirs. The second—the age of Western imperialism—began with Napoleon's invasion of Egypt in 1789. The third, which is really indistinct from the second, is the war not of arms (although that has continued unchecked) but of words and ideas. It is the attempt, also begun by Napoleon, to persuade the peoples of the East of the overwhelming supremacy of the West, of its technologies, its culture, its laws, and its political institutions. Before the Army of the Orient landed on the beaches at Alexandria, ordinary Muslims had had no previous contact with the "Franks" since the thirteenth century. They had, therefore, taken it for granted that the West was as backward and as poor as they. In 1798, they came face-to-face with an infidel people who were obviously wealthier, better armed, better trained, and more confident than they. Instead of being swept up into the all-powerful embrace of Islam, as they had been expected to, these infidels had triumphed and continued to triumph, each successive wave carrying them farther and farther into the *dar al-Islam*.

The response from the reformers in the late nineteenth century—men such as Jamâl ad-Dîn al-Afghânî; Muhammad 'Abduh, who had been with al-Afghânî in Paris, where both worked on the revolutionary Islamic journal *The Firmest Bond,* and in 1899 became chief mufti of Egypt; his disciple, Rashîd Ridâ, who became president of the Syrian Congress in 1919; Alal al-Fasi from Morocco; the Tunisian 'Abd al-'Aziz al-Thalabi; Algeria's 'Abd al-Hamid ibn Badis; and Muhammad Iqbâl from India—was to attempt to modernize Islam from within.[136] Their objective became known as *salafism,* which by revealing the roots of modernism that were thought to lie in the traditions of the ancestors (*salaf*) of Islam, sought to mix religion with the new imported European doctrine of nationalism in the hope, as al-Afghânî put it, "that Islam, while still being Islam, will succeed some day in breaking its bonds and marching resolutely in the path of civilization, after the manner of Western society."[137]

But their bright young nationalist vision had been betrayed and its visionaries had been turned into tormentors and murderers, who in turn had been tormented and murdered by others. The culmination of this descent into Hell had been the disastrous revolutions initiated by the Young Turks, who had betrayed the Ottoman sultan-caliph and, while still claiming to be Muslims, had in fact destroyed the traditional order of the Islamic world. The poor deluded Arabs, believing that the Western Allies would make good their promises to free them from a now corrupted and debased Ottoman sovereign, had found themselves, after the end of the First World War, "sold like slaves," as the infidels dismembered the Arab world and divided it up into a series of satrapies run in their own interests: Jews in Palestine, heterodox Alawites (who are related to the Shiites) in Syria, and Maronite Christians in Lebanon.[138]

Undeterred by this experience of what the West and all its

ways really offered, another generation of nationalist leaders had arisen, inspired this time not by liberalism or democracy but by various kinds of Marxist-Leninism. They, too, had gone the way of their predecessors. The Soviet Union, which in 1918 had turned its back on the imperialism of the Russian czars, had proved to be another false friend. Marxism was, for all its appeal to the dispossessed, still a Western creed. "Marx," wrote Muhammad Jalal Kishk, a prolific polemicist with connections to the Muslim Brotherhood, in 1969, "did not call for a new civilization; he is a faithful son of Western civilization. . . . Marx believed in the values and the history of Western civilization; he was proud of that history which he considered as a triumph for humanity on the way to its final victory."[139] He was, of course, broadly right.

The next and final phase in this history would be the rebirth of Islam, the only force that is capable of uniting the Muslim world and of resisting the undiminished onslaught of Western neocolonialism. That Islam, however, could not be the weakly anemic thing the reformers had made of it. It had to be pure, shorn of all the accretions that had clung to it over centuries of attempted accommodation with the godless West. It had to carry the full force of the Shari'a. For Hasan al-Banna', the founder of the Muslim Brotherhood, which for long had sustained a separate society within Egypt, providing many of the things the state could not or would not provide—hospitals, schools, factories, welfare societies—the Western concept of secular democratic government was a form of blasphemy. Although he had been a disciple of Muhammad 'Abduh, whose doctrine had been a denial of all literalism, al-Banna' insisted that only Islam had fully understood that God wished his creatures to be ruled by his dictates alone, not in accordance with the changing whims of mere humans. "It is the nature of Islam to dominate, not to be dominated," he declared, "to impose its law on all nations and to extend its power to the entire planet."[140]

This vision of the relationship between the Muslim East and first the Christian, then the secular, West, although necessarily overschematic, is far more than simple distortion. In desperation, as their worlds collapsed around them, Muslims turned, as their Ottoman predecessors had once done, to the past. Muslim societies had failed, because, unlike non-Muslim Asian societies, most obviously China and Japan, which had emerged triumphant from a brief period of Western domination, they had proved to be woefully ill equipped to turn Western norms and Western technologies to their own advantage. They had failed because they had abandoned and betrayed the old ways, which meant that they had turned aside from the path indicated to them by the Prophet. In attempting to become constitutional monarchs, secular nationalists, liberals, or Marxists, they had ceased to be Muslims. God, who by the late 1960s had been banished from the public sphere almost everywhere in the world, suddenly made a violent reappearance. But God, like all gods, is notoriously unable to help himself. It is his creatures who must, therefore, act for him. As al-Afghânî had warned the Egyptians in the 1880s:

> If someone says: "If the Islamic religion is as you say, then why are Muslims in such a sad condition?" I will answer: When they were truly Muslims, they were what they were and the world bears witness to their excellence. As for the present, I will content myself with the holy text: "Verily, God does not change the state of a people until they change themselves inwardly."[141]

The logic was not very good. But it is a message that later generations of dispossessed Muslims from Egypt to Iran have taken very much to heart. Only for them it is not so much inward change that is necessary as outward. The "national and religious movement" that Sir Cecil Spring-Rice had glimpsed in shadowy form in Iran in 1907 had in the intervening years, as one failed experiment in modernization followed another, become the last desperate bid by a once-great civilization to reverse the humiliation and despair that had engulfed it.

Appropriately, it was in Iran, seemingly the most secure of the new Muslim national states and the only one not to have been touched by the humiliation of the Six-Day War, where the blow fell first. The shah's ambition to create an "Asian Germany" had left the majority of his people disorientated and dissatisfied. He had done his best to limit the power and wealth of the religious community while failing to provide any alternative for the goods, charity, and assistance in times of need—welfare, in other words— that the mosque had traditionally offered. The extravagance of his personal lifestyle, followed in excruciating detail by the drooling Western gossip sheets, depressed and disgusted many even of his potential middle-class supporters. It also alienated many of his would-be Western allies, who increasingly came to see him as indistinguishable from the countless dictators who had brought ruin to many African and Latin American states. From the late 1960s, an opposition, which drew on both secular Marxist and Islamic fundamentalist ideals, gathered momentum. It was guided, improbably, by an elderly embittered ayatollah named Ruhollah Khomeini from exile in Paris.

The shah had seriously underestimated the degree of hostility that his White Revolution had created, in particular among the members of the still immensely influential Ulema. Once when a foreign visitor asked him what a protest taking place outside the royal palace in Tehran was all about, he replied laconically, "Just some mullahs pining for the eleventh century." It was a nice riposte, but in the end the insouciance it revealed would be his undoing. By 1978, an increasing number of Iranians were also pining for the eleventh century. Not only the peasantry and the disaffected proletariat who had nothing to lose and everything to gain from Khomeini's promise of an Islamic revolution, but increasingly also the middle classes and the military who had sustained the regime and benefited most from it, had joined the opposition. Neither they nor the European and American intellectuals on the left who saw the ayatollah as a new Lenin or Mao had much, if any, understanding of what Khomeini's heady mixture of Islam and popularism was intended to produce. Almost no one in Iran itself, much less in the West, had read Khomeini's lectures, published as *Islamic Government*. And even if they had, they would scarcely have believed that in the second half of the twentieth century, a popular and respected political leader, even if he was also the servant of an obscure and disregarded religious creed, could seriously be proposing to use his influence to resurrect the laws and customs of a primitive desert community. Yet that was precisely what he was intending to do.

On January 16, 1979, after a year of riots, the shah and his queen hastily departed the country—like the "Flying Dutchman," said Henry Kissinger—never to return. Khomeini emerged from exile to a jubilant welcome and immediately set about transforming the shah's new westernized Achaemenid Empire into a theocratic republic. The new Iran was to be based on the *velayat-e faqih,* or "government of the Islamic jurist" and administered by the clerical class—always more powerful in Iran than elsewhere in the Muslim world—as the Prophet's true heirs, in strict accordance with the Shari'a. It was, save for the oil and modern weaponry at its disposal, to resemble as closely as was possible in the second half of the twentieth century the Islamic community of Muhammad's own day. The mullahs needed to pine no more.

With the resurgence of religion came, too, a resurgence of the universal aspirations of Islam, which now took the form not of political uprising against a hated westernizing tyrant, but of a worldwide jihad on behalf of all Muslims, Shiite and Sunni, everywhere, against all the peoples of the West. The universal message of the new Islamic revolution now echoed around the globe. Every struggle, however local, from Bosnia and Chechnya to Somalia and the Philippines, became but another facet of the continuing Manichean struggle between the universal forces of Islam, backed by God, against what Khomeini called the "Great Satan." He was referring to the United States. But the United States, in his mind, was merely the latest manifestation of the Western crusader.

At the heart of the struggle there still lay the Palestinian-

Israeli conflict. This, however, had now ceased to be only a struggle over a piece of land. It had become a Holy War enjoined upon all Muslims no matter what their ethnic origins, a battle to regain a crucial piece of the *dar al-Islam* now in the hand of the infidels. As the constitution of Hamas, the Palestinian terrorist organization, puts it, "When our enemies usurp some part of the House of Islam, jihad becomes a duty binding on all Muslims."[142] What for the nationalists had been represented as a betrayal of the Arabs to be resolved by pan-Arab unity now became the betrayal of Muslims everywhere, the majority of whom are not Arab and many of whom do not even live within the "House of Islam." Palestine, Afghanistan, and Iraq are not separate theaters of war, and the struggle is no longer—if indeed it ever was—several but one. Now it has become possible for someone of Pakistani origin, born in Britain, speaking no language but English, to blow himself and a number of others to pieces on the bright London morning of June 8, 2005, in retaliation for the humiliations inflicted upon peoples he knew nothing of, in distant parts of the world he had never visited.

In 1979, as the shah left for what turned out to be a brief and uncomfortable exile, another "Western" power, the Soviet Union, invaded another part of the House of Islam, Afghanistan, in an attempt to bolster a debilitated and unpopular Communist government against Pakistan-sponsored Islamic insurgents. The Muslim world was outraged. The Soviet Union declared itself to be showing "respect for the religious feelings of the masses" and to be "holding out the hand of solidarity and friendship to all Muslims in their struggle against imperialist forces."[143] But the creaking language of Leninist anti-imperialism no longer impressed anyone, least of all the "Islamists"—as they have come to be called— who merely associated it with the failed nationalism of the hated secularizing regimes. The invasion confirmed, if confirmation were necessary, that the Soviets were repeating what the czarist Russians had done for centuries before them—making war on Muslims in order to increase their hold on central Asia.

The war in Afghanistan attracted militant Muslims from around the globe. It was a movement in which Arab states friendly to the West, such as Saudi Arabia—which offered would-be mujahideen a 70 percent discount on Saudi Airlines flights to Peshawar in Pakistan—and Egypt enthusiastically participated, because as the Egyptian president Sâdât declared, "They are our Muslim brothers and in trouble."[144] (These recruits, however, achieved very little. Their main contribution after the departure of the Red Army was chopping "atheist" Afghan prisoners into pieces in March 1981 and stuffing them into boxes, something that caused consternation among the Afghan mujahideen.)[145] The war was also backed by the United States and Israel: the first because it saw, rightly as it turned out, that this might be a means of debilitating the Soviet regime; the latter, partly to please the United States, and partly because it promised to channel some of the forces of the Palestinian opposition away from Palestine.

In February 1989, the Soviet forces gave up the struggle and withdrew. It seemed at the time to the Muslim world to be a magnificent victory. A loose association of untried and poorly trained (although, thanks to the CIA, very well equipped) irregulars had brought down one of the world's infidel superpowers. In 1994, after five years of internecine fighting among the various mujahideen forces, the Taliban, a group of battle-hardened mullahs and former students and bearers of a version of Wahhâbism more extreme than any that had preceded it, now swept to power. Many Muslims condemned the reign of terror that they immediately unleashed upon the unhappy Afghan population in the name of Islamic purity. But although the Taliban may not have been much to the liking of even many Islamic militants, their astonishing victory pointed the way forward to the day when the great Islamic state would encompass the world. For the militants, at least, it seemed as if the titanic struggle between good and evil, between the Islamic East and the crusader imperialist West was at least beginning to be resolved—and in their favor.

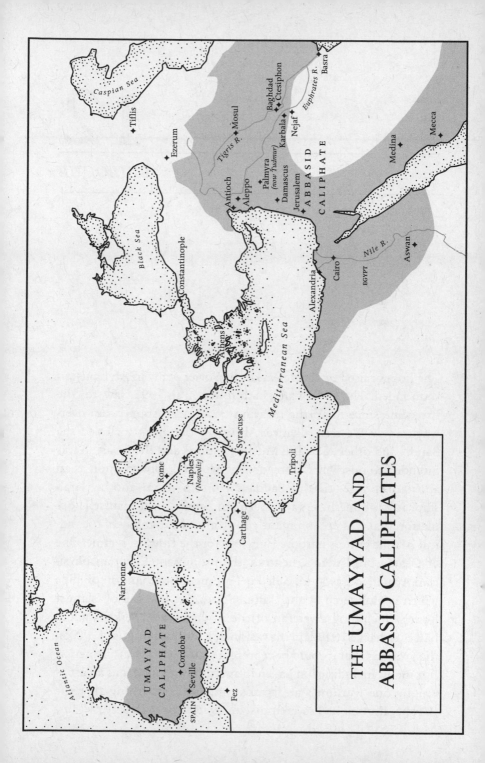

THE UMAYYAD AND
ABBASID CALIPHATES

# 12

## EPILOGUE

**I**

The hopes raised by the defeat of the Soviet Army in Afghanistan soon faded. The Gulf War, which followed in 1991, saw, for the first time since Suez, the invasion of an Arab state by an over-whelmingly Western alliance. Although the allies included the Saudis and other Arab and Muslim nations and the invasion was intended to oust what many devout Muslims looked upon as an apostate regime—that of Saddam Hussein—from another Arab state, it was nonetheless a Western war being waged against a pop-ulation that was predominantly Muslim. Not that any of the rad-ical Islamic groups harbored any love for the Kuwaiti regime. One of Osama bin Laden's mentors, the prominent Egyptian physi-cian Ayman al-Zawahiri, called it nothing more than "an oil pipe which is plundered by the United States."[1] A number of Islamist leaders, including a veteran of the Afghan war, Sheikh Tamini, had even backed Saddam's invasion of Kuwait in August 1990, on the grounds that it could be represented as a war against America. But many, including bin Laden himself, condemned it as an act of war by one Muslim state against another, something that the Prophet himself had severely discouraged. "Know that every Muslim is a Muslim's brother," he is reported to have said with almost his dying breath, "and that the Muslims are brethren; fighting between them should be avoided."[2] It was also a war that rapidly came to be seen as part of the wider American-Israeli plot to dominate the entire Middle East. "The best proof," declared bin Laden in a fatwa (a legal opinion) issued on February 23, 1998, of the fact that the ambition of the Americans and "crusader-Zionist alliance" in the Middle East was to "serve the Jews' petty state and divert attention from its occupation of Jerusalem and murder of Muslims there" was their apparent eagerness to "destroy Iraq, the strongest neighboring Arab state."[3]

Initially, Saddam had justified his invasion on grounds of national integrity, although he failed to mention that the emirate of Kuwait, which he was supposedly seeking to reintegrate into the "motherland" had come into being some two centuries before the British had dreamed up the modern state of Iraq. But as the war turned against him, he began to make increasingly inflated claims to be the champion of universal Islam. He represented himself as a devout Muslim, a fervent enemy of Zionism (this from a man who in 1985 had assured the Israelis that "no Arab leader looks forward to the destruction of Israel"), an ardent pan-Arabist, and, as we have seen, a new Saladin defending the holy lands of Islam from the infidels.

Few Muslims were taken in by any of this, and the more extreme elements among them were unlikely to have forgotten that in 1980 Saddam had executed the economist Baqir al-Sadr, an important ideologue for the Islamist Shiite resistance to Ba'thist nationalism. Most accepted Saddam's war for what it obviously was: a bid to lay his hands on an oil-rich territory at a time when his own economy was steadily collapsing through corruption, mismanagement, and the burden of a prolonged and inconclusive war with Iran.

The West was also able to employ Saddam's own nationalist arguments against him. Kuwait was no province of Iraq and never had been. Its invasion therefore constituted the violation of one sovereign state by another, something the United Nations and the "international community" more broadly could not simply stand by and ignore. (Although, as the American commander of Operation Desert Storm, General Norman Schwarzkopf, famously remarked, if all Kuwait had had was carrots instead of oil, he would never have been sent there.)

Few in the West, however, seem to be aware that although "Iraq" as a nation was largely a Western fiction, the territory it occupied had once been the heartland of the caliphate. For the Islamists, the invasion, although the Allied troops stopped within 150 miles of Baghdad and then began to withdraw, looked as if it were the beginning of yet another attempt to plant a Crusader State in the very heart of the *dar al-Islam,* something the subsequent American-led invasion of Iraq had finally brought to fruition. To make matters worse, once the Gulf War was over, the Saudis allowed American forces to be stationed close to Mecca and Medina, the holiest sites in the entire Muslim world. For those like bin Laden, himself a subject of the House of al-Sa'ud, it was the ultimate act of betrayal by a family that ever since it had formed an alliance with the religious leader Ibn 'Abd al-Wahhâb in the eighteenth century had by many Muslims been looked upon as the guardians of the purest form of their religion.

By 1991, the West seemed to have reclaimed the initiative. The once-confident Islamic militants now came to believe that something more was needed, something that would unite all Muslims and draw them away from their own petty squabbles and back to the one universal cause: the creation of an "Islamic state" that would one day claim the entire world as its own. If another Afghanistan could not easily be provided, then some sensational terrorist act aimed at the enemy's heartland might do almost as well. A number of such attacks were now planned, by various shadowy organizations, against Western targets or against those thought to be supporters of the "Crusader-Zionist Alliance." In February 1993, Ramzi Yousef, a young half-Palestinian, half-Pakistani engineer, planted a bomb in the parking garage of the

World Trade Center in New York, which tore a 150-foot-square crater in the floor of one building and killed six people. In 1996, a bomb destroyed the Khobar Towers, a U.S. military housing complex in Saudi Arabia, killing nineteen American servicemen. Two years later, the American embassies in Nairobi and Dar-es-Salaam were car-bombed within four minutes of each other. Then, on October 12, 2000, a U.S. Warship, the USS *Cole*, on a routine refueling stop in Aden harbor, was approached, apparently unchallenged, by a small boat loaded with explosives. The blast tore a hole nearly forty feet long in the half-inch-thick armored steel of the ship's hull and ripped though the galley, killing seventeen of the ship's crew who had been lining up for lunch.

The most devastating attack, however, came on September 11, 2001. With the destruction of the World Trade Center in New York, symbols to most Muslims of Western aggression and the staggering power of the Western economies—"those awesome symbolic towers," bin Laden called them, "that speak of liberty, human rights, and humanity"—were brought down.[4] "Wherever you are, death will overtake you in lofty towers" (4:78) the Qu'ran had promised. And so it had. A target deep within the lands of the "faraway enemy" had been successfully destroyed by a small handful of "martyrs," and thousands of infidel lives (as well as a number of Muslim ones) had been lost. Bin Laden and his associates were ecstatic.[5] He had, he was recorded as saying, never hoped that the mission would have achieved so much. Many in the Muslim world rejoiced along with him, even those who might at other times have had some misgivings about the taking of innocent lives—if that was what they were believed to be—which is, of course, as much forbidden by Islam as it is by all other religions. A university student sitting in a McDonald's in Cairo and reading *The Wall Street Journal* recalled that when the news broke, "Everyone celebrated. People honked in the streets, cheering that finally America had got what it deserved."[6] They were not alone.

While the Western world, at least, struggled to recover from the horror of the endlessly repeated images, broadcast around the globe on millions of television sets, of the flame-enveloped towers slowly cascading onto the humans beneath them, Ayman al-Zawahiri distributed on the Internet a text entitled "Knights Under the Prophet's Banner." The blow had been struck, al-Zawahiri explained, in order to mobilize the Muslim masses, who had grown dispirited and inactive after the end of the Gulf War, to show them that Islamic militants could successfully damage the United States, the most powerful of the enemies of Muslim societies across the globe and the succor of apostate regimes in the Middle East and North Africa. The attack had demonstrated that Western liberal democracy was weak and morally corrupt and that soon all the governments of the Western world would go the way of the twin towers. All that was now needed was a worldwide renewal of the jihad from Pakistan to the Philippines and by Muslims both outside the House of War and within. Terrorism was to be the favored method, because the felling of the twin towers had shown that this was "the only language the West understands."

On October 7, in the first video sent to the Arabic television station Aljazeera since the attacks, al-Zawahiri appeared, seated with bin Laden at the entrance to a cave somewhere in the mountains of Afghanistan, wearing a turban (unusual for an Egyptian pediatrician) and dressed in the costume favored by actors on Egyptian soap operas about the life and times of the Prophet. To make the point on which Islamists insist tirelessly, that the new terrorism is in effect but another phase in a war that reaches back to the Crusades, if not earlier, he called 9/11 "the true day, the sincere day, the day of challenge: your day of glory has arrived." "A new epoch in the history of Islam has begun," al-Zawahiri told the camera, "a new battle for the faith like the battles of Hattin, of Ain Jalut [the victory over the Mongols in 1260] and the conquest of Jerusalem. The epoch has begun, now hasten to defend the honour of Islam."[7]

Although 9/11 failed to bring about the awaited apocalypse, Western retaliations—the American-led destruction of the Taliban regime and the invasion and de facto occupation of Iraq—

have helped to spread the militants' view that the entire world is now convulsed by a struggle to the death between Islam and the West and all that the West stands for—secular government, democracy, rights, freedom of choice, and equality of the sexes. Moderate Muslim societies and more secular Muslim regimes, or at least regimes more friendly to the West—Yemen, Sudan, Saudi Arabia, Algeria, Egypt, Jordan, Pakistan, the Philippines, Indonesia—have all been convulsed by the growing influence of extreme forms of Islam. And the jihadists have continued to strike at targets "deep into Crusader Europe"—in Madrid in March 2004 and in London in July 2005.[8]

With each of these attacks the enemy has come to be conceived in broader and more general terms. Once the enemy was a religion—Christianity, Judaism—then it was a particular power— the British, the French, the Americans. Now it is merely the "West." The Western response to this has been mixed. With each successive attack hostility not merely to Islamic extremists but to Islam in general has grown. And that hostility has inevitably fueled the conviction of even more moderate Muslims that Western civilization, in whatever shape it might take, is bent upon their ultimate destruction. There are, however, many Western intellectuals who have continued to insist that the shared religious sentiments that once existed along the borders between Islam and Christendom, and the willingness of Muslim societies to tolerate Christians and Jews in their midst, proves the existence of a shared world, that the hardening of the demarcations between West and East is a recent phenomenon, largely the fault of the West and of Western colonialism.[9]

Many Muslims, even those living in the frontline states, share similar beliefs—or hopes—that the enmity between the secular (or Christian) West and the Islamic world can be resolved by dialogue, by mutual respect and understanding. But for others, and alas, often for those who are most intent on making themselves heard, all that this talk of dialogue and understanding amounts to is yet another form of westernization. Who says that tolerance, dialogue, and understanding are virtues? The answer is invariably: secular Westerners. The religion of the Prophet is not one of polite conversation. It is one of submission—that, after all, is what the words "Islam" and "Muslim" refer to. To believe otherwise is to fall into the same trap that Napoleon, with his blend of Islam and the Rights of Man, had hoped to set for the Egyptians. Coat your modern godless beliefs with enough Qur'anic sugar, and sooner or later, once the poor Muslims have been persuaded to embrace equality, individual freedom, self-expression, rights, and all those other shibboleths of Western society, they will come to see just how crude, brutal, and primitive their old ways always were. "All these people," wrote Mawlana Abdullah Mawdudi, one of the founders of the Jama'at-i-Islami (Islamic Society) in what was then India in 1941, and one of the revered teachers of the modern Islamists, of the reconcilers and appeasers, "in their misinformed and misguided zeal to serve what they hold to be the cause of Islam, are always at great pains to prove that Islam contains within itself the elements of all types of contemporary social and political thought and action." But this, he insisted, was an absurdity that derived not from conviction, but from an "inferiority complex" that had led his fellow Muslims to believe that "we . . . can earn no honor or respect unless we are able to show that our religion resembles modern creeds, and is in agreement with most of the contemporary ideologies."[10] In a more strident view the imam Kadhem al-Ebadi al-Nasseri told his congregation in Baghdad in May 2003, "The West wants to distract you with shiny slogans like freedom, democracy, culture and civil society." Do not listen to them, he cried, for "infidel corruption has entered our society through these concepts."[11]

This did not mean rejecting Western democracy as such, if all that was understood by the term was popular and elective government of the kind that was widely believed to have existed at Medina during the life of the Prophet. What it meant was rejecting the Western secularism that had accompanied—and in the eyes of most Muslims had been largely the consequence, rather than the condition of—the creation of the modern liberal democratic state. And that meant in effect rejecting what is, after all, the basic premise of all forms of democracy, modern and ancient: that sovereignty lies with the people and not, as it must in Islam, with God. The future Islamic state of Mawdudi's imagination was to be based upon a duly elected leadership, but it was to be bound not by the laws that its leadership might make, but only by the laws of God, that is, the Shari'a. He called this a "God-worshipping democratic caliphate"—which would seem to be an oxymoron— and a "theo-democracy."[12] What he had in mind was something not unlike what the modern Republic of Iran has become (not unlike, too, what many Christian fundamentalists might wish for the United States, were it not for the impossibility of fabricating anything much like the Shari'a out of the Gospels).

Mawdudi's views were shared by, among others, Sayyid Qutb, who is in many ways the patron saint of the modern jihadist movement. He is a thinker whose influence over radical Islam has been at least as great as that of Khomeini, and his most radical work, *Milestones,* written in the 1960s, was, and remains, a bestseller in the Muslim world. (Sayyid Qutb's brother, Muhammad, was also responsible for the instruction in Islamic studies that Osama bin Laden received at the Abfd al-Aziz university in Jeddah.)

For Qutb, the enemy was, in the first place, the nationalist Arab, and in particular Nasserite Egyptian, state. Islam, he argued, had replaced all former traditions, pagan, Christian, and Jewish. It had been "a spontaneous creation, a new birth and a reality apart."[13] The most important of the traditions that Islam had supplanted was the initially Christian but then more generally Western notion of the separation of religion and society. The failure of the West, as Qutb experienced it firsthand during the time he had spent in the United States in the late 1940s, had been its failure to take the form of a "system responsible for all of life and binding the kingdom of earth to the kingdom of heaven."[14] By adopting a similar distinction between the rule of man and the commands of God, the Arab nationalists had, in Qutb's view, plunged the Arab peoples back into the condition—known as *jahiliiyya*—that had existed before the coming of the Prophet and which was now acquainted with life in the thrall of "Western Civilization."

Nasser and his acolytes were the immediate agents of this process of retardation. They and other accommodating Muslims had "invented their own version of Islam, other than what God and His Messenger—peace be on him—had prescribed and explained, and call it, for example, 'progressive Islam.' "[15] But they were all really stooges of the West, which was the real enemy, because it was the West, and America in particular, that had succeeded in poisoning all of Islamic society, so that it too was now whoring after Western goods, Western pleasures, and Western morals—in particular sexual ones. (Qutb, who on his journey to the United States had had an unfortunate shipboard encounter with a half-drunk woman, was particularly agitated about sex. The American woman, he wrote in horrified fascination, "knows full well the beauties of her body, her face, her exciting eyes, her full lips, her bulging breasts, her full buttocks, and her smooth legs.")[16]

Qutb may have had his difficulties with Western sexual habits, but he was by no means hostile to Western science as such. No Islamist ever has been. "Europe's genius," he readily acknowledged, had "created . . . marvelous works in sciences, culture, law, and material production, due to which mankind has progressed to great heights of creativity and material comfort." It was not easy to find fault with "the inventors of such marvelous things." The problem was that they had all been based on "man-made traditions" and had driven out all "those vital values" that are necessary for "real progress," values that are to be found only in Islam. The values of the modern version of the "*jahiliiyyah* system," for all the material goods they have given to mankind, are, he argued, nevertheless "fundamentally at variance with Islam," and thus the system, "with the help of force and oppression, is keeping us from living the sort of life which is demanded by the Creator."[17] (In Iran while it was under the American-backed regime of the shah, this process of corruption was known not as Westernization but as "Westoxification"—*gharbzadegi*.)[18]

What was now required, Qutb argued, was a "new Qur'anic generation" to somehow replicate that of the companions of the Prophet, and rebuild Islam out of the ruins of Western national-ism, just as Muhammad had created the first Umma out of the ruins of Arab paganism.[19] To achieve that, it would be necessary to return to a pure understanding of the original sources, the Qur'an itself and the Hadith, and "the biography of the Apostle, and his exemplary actions." This is what Qutb and his followers understood the true meaning of *salaf* to be, not the liberal aposta-tizing, accommodating readings that had been proposed by Afghânî, Abduh, and Rida. Everything that was not a strict—although not always quite literal—understanding of the canonical sources had to be disregarded. For Qutb, this repudiation even reached back to the great Muslim writers of the Middle Ages, the "philosophers of Islam" Ibn Sind, Ibn Rushd, and al-Fârâbi who in his view taught "nothing but a shadow of Greek philosophy, alien in its spirit to the spirit of Islam."[20] The task of this new Qur'anic generation was to "declare the divinity of God alone," to "wrest the power from the hands of the human usurpers to return it to God alone," and to declare the "supremacy of divine law alone and the cancellation of human laws."[21] This was to be the basis of the "Islamic Resurgence" (*Ba'th*).

Not surprisingly, the largely secular and fully human authori-ties in Egypt did not much relish the prospect of having their purely human laws canceled, and just after dawn prayers on Au-gust 29, 1966, they hanged Qutb, who went to his death with the words "Thank God, I have performed jihad for fifteen years until I earned this martyrdom."[22]

For a true believer Qutb's views are neither extreme nor ab-surd. Just the same kind of argument could and has been made by many Christians against their own secular governments and against what they perceive as their own histories of apostasy and secularization.[23] Mawdudi and Qutb were in many respects literalists; but they were not the kind of militant Islamic literalists represented by the Taliban, the kind who believe that if the Prophet had a beard, every good Muslim must have one, too, that if the Prophet slept on his right side, every good Muslim must do so also, and so on. But they did believe that the only way forward for Islam was to return to its roots and rid itself altogether of Western influences, and they did believe that any modification or dilution of the Prophet's original message betrayed it.

Mawdudi, Qutb, and their yet more radical followers, who include most of the leaders of al-Qaeda, are all, in varying degrees, the heirs of two earlier jurists and theologians. One, Muhammad ibn 'Abd al-Wahhâb, we have already met. Behind Ibn 'Abd al-Wahhâb, however, stood the still more influential Taqi ad-Din ibn Taymîyah, a man whose writings were frequently cited in justification for the slaying of the Egyptian president Anwar Sâdât in 1981, by a group led by a young electrical engineer named Abdessalam Faraj, and used as a rallying call by extremists for the overthrow of the Saudi regime in the mid-1990s.[24]

Ibn Taymîyah was born in 1263 in Harran in what is now Turkey, not long after the Mongol destruction of Baghdad. The son and grandson of distinguished theologians, he was himself a brilliant jurist of the Hanbal school. He was a purist who held that only the strictest interpretation of the sacred texts was permissible, and he wrote a treatise condemning Greek logic as contrary to the teaching of Islam. He also took part in a number of campaigns against the Mongols and against the Armenian kingdom of Asia Minor, which was an ally of the Crusader States of the Levant. In time he became an intransigent puritan, convinced that the caliphate had become lax and departed from the true path of Islam and that only a purification of the lives of all those claiming to be Muslims could regain God's favor for the Umma. His views, angry and uncompromising like those of most puritans, landed him in trouble with the authorities, and he ended his days in a prison cell in Damascus, denied access to writing materials so as to stop the ceaseless flood of pamphlets that had come from his pen.

Today Ibn Taymîyah is best known among radical Muslims for having issued a famous fatwa against the Mongols. The Mongols had converted to Islam in 1295, but in Ibn Taymîyah's view they were "deviants," apostates, because of their uninhibited cruelty, their treatment of their Muslim subjects, their habitual drunkenness, and the facts that they did not properly observe the pillar of the faith and that they failed to live their lives by the Shari'a.[25] If the Mongols were, in effect, apostates, they were punishable by death. Ibn Taymîyah went one crucial step further by arguing that war against apostates also belonged to the jihad and, furthermore, that the category of apostate could be extended to all those, no matter how they might have lived their private lives, who had aided an apostate or infidel regime. If it turned out that they had done so unknowingly, then they were unfortunate, but God would redeem them in the afterlife.[26]

Ibn Taymîyah's hold over the minds of bin Laden and his followers is obviously considerable. In February 2001, Jamal Ahmad al-Fadl, a renegade member of al-Qaeda turned FBI informer, told his interrogator in broken English that "He [bin Laden] says ibn al Tamiyeh [sic] he make a fatwa. He said anybody around the tartar [the Mongols] he buy something from them and he sell them something you should kill him. . . . If you kill him you do not have to worry about that. If he's a good person, he go to Paradise and if he's a bad person he go to Hell."[27] This garbled recollection was taken to mean that although Ibn Taymîyah himself, in common with all Islamic jurists, forbade the slaying of genuine noncombatants, there had been no innocent victims on September 11. Even those Muslims who perished in the conflagration were, by their very presence in the towers, apostates. So, too, of course, are all those regimes, the Saudis in particular, who have assisted and continue to assist the West.

Like Ibn Taymîyah, like Ibn 'Abd al-Wahhâb, like Sayyid Qutb, like Mawlana Mawdudi, the modern "jihadists" believe that jihad against both infidels and their apostate Muslim accomplices is the duty of every true believer. There can be no reconciliation between the Muslim world and the West, Christian or secular, only the absorption, as the Prophet foretold, of the one by the other. All Islamic militants and radicals hold broadly similar views. Some of Mawdudi's followers even seem to have had a territorial sense of what constituted the "East" that closely resembled the Western conceptions of the late eighteenth century. All of Asia, not only the Muslim Middle East but also the Chinese and Japanese Far East, are gathered into one compatible, if not exactly homogeneous, cultural world. In support of this claim, in 1969, Muhammad Jalal Kishk asked his readers to consider the gathering antagonism between China and the Soviet Union. Here, too, he claimed, as in the Islamic world, a European ideology—in this case Marxism—had briefly masked the irreconcilable differences between a Western culture and an Eastern one. But in the end the old rivalries had proved too strong, and China had turned on its Western parent and resumed, in the Cultural Revolution, its old "Eastern" ways.[28]

Such views are obviously not typical. Ibn Taymîyah, and al-Wahhâb were both extremists condemned by the orthodox Ulema of their own day. The views of many of their followers, bin Laden and the spiritual leaders of the Taliban among them, are more extreme still. There are millions of Muslims who, like their Christian counterparts, have quietly abandoned the militant and millenarian aspects of their religion as best they can. Yet with a few notable exceptions, Turkey, Morocco, Tunisia, and perhaps some day Bosnia—all states that are at least nominally Muslim—only the Muslim world seems to be unable to reconcile traditional ways of life, traditional religious beliefs, with modern liberal forms of government. Why?

For many in the West the answer has always been much the same. What ultimately distinguishes the East from the West is not religion, history, or culture. It is simply government or, more broadly, politics. This was, as we have seen, also Herodotus's view. It was not race or climate—although such things may have played a role—that distinguished the Greeks from the Persians but a unique form of politics, *isonomia*, "the order of political equality." And it is the fundamental principles of *isonomia*, now recast as modern liberal democracy, that more than anything else define the "West" for both Muslims and non-Muslims. The problem with Muslim society, say those in the West, is not Islam, which many imagine to be an essentially peaceful religion much like Christianity is in principle, if rarely ever in practice. It is tyrannical and unrepresentative governments that are the problem, and few if any of these, with the notable exception of Iran, can make any claim to be authentically Islamic. Change those governments into democratic ones, and militant Islam will cease to have any attraction.

On this perhaps overly optimistic reading of Islamic history, there is no reason why Muslim societies should not, as Christian ones have, accommodate themselves to democratic forms of government. The advocates of the exportation of democratic institutions throughout the entire world assume that every human being—no matter what may be his or her race, creed, or past—desires, above all else, personal freedom. It is, and has been for centuries, an unchanging principle of European civilization. It can easily be defended. The assumptions that have followed from it, however, are often less easily defensible. One of these is that once the rituals and institutions of liberal democracy are in place—and often this has meant, as most recently in Iraq, nothing more than a single successful election—freedom will necessarily follow. And once a people has acquired a taste for freedom, it will never look back. The reasoning is not entirely false. There are many regions of the world today—China, for instance—where the demands for some kind of representative government and some measure of personal political freedom simply refuse to go away, no matter how successful the present system has proved to be in creating wealth and stability.

The problem with most attempts at democratic regime change, however, is that they have frequently, if not invariably, been based upon a very simplistic understanding of what democracy is. The key terms of democratic rhetoric can have very little meaning beyond the kinds of goods they can deliver. No one values "freedom" of anything as a merely abstract good. What freedom delivers in the West is a better, safer, richer life of largely unregulated (if by no means unlimited) choice. The German Democratic Republic was probably the most repressive of the Soviet satellites. But the East Germans had access to West German television, and many of them had relatives who lived in the West, with whom, despite severe restrictions, they were able to maintain regular contact. They, more than any other people within the Soviet block, could see just what modern liberal democracy had to offer. Because of this it was the East Germans who took the first steps that would eventually bring down the Communist regimes of eastern Europe.[29]

Just about all that the peoples of the Middle East have seen of the benefits to be had from modern liberal democracy are the traditional products of American consumerism. And desirable though these goods clearly are, there is no obvious link between them and the political and economic regimes that made them possible. They could, after all, just as easily have been produced by a Communist state such as China—as indeed many of them are. It is not insignificant that when the American troops first arrived in Baghdad in 2004, they were greeted by swarms of small boys running excitedly along beside the humvees shouting at the soldiers inside them not "Freedom, freedom" but "Whisky, whisky." Perhaps if the United States had listened and given them—not whisky certainly, but schools, hospitals, some basic security, and the rudiments of a welfare system, democracy might have seemed a more attractive option, something worth fighting for against the centuries-old struggle among Sunnis, Shiites, and Kurds.

The fundamental assumption in the West is that what the Chinese "paramount leader" Deng Xiaoping dismissed as "bourgeois liberal democracy," to which the students at Tiananmen Square had aspired, is a universal political system and thus instantly recognizable as the only way of achieving equity and freedom among human beings living together in society. But it is not. Despite certain general similarities with some aspects of early Islamic societies, it is the necessarily imperfect and incomplete expression of one political tradition—that of ancient Greece and its self-styled heirs. It is a product of what throughout this book I have been calling "the West."

This does not mean that the champions of democracy are not also right in claiming that it is also the best obtainable government there is and, with the failure of communism, the most equitable way of distributing power and goods—at least at present. The mistake is to assume that this fact must be simple and obvious, in particular to those who have had no prior experience of modern democracy, who inevitably equate it with Western imperialism and Western godlessness, and whose first encounter with it is often at the wrong end of a gun.

Islam, as Mawdudi and countless other liberal Muslims have pointed out, is perfectly compatible with some form of collective rule. The modern Republic of Iran may not be a democracy in the modern American sense, but it is not far from what the Athenians would have understood by the term, nor so far removed from the state created by the Jacobins in France in 1792. Iran, however, is also, in the Muslim world, something of a novelty. The "democratic" nature of early Islamic society, although it appeals as a concept to both Western and Muslim intellectuals, is unlikely to have much appeal in the slums of Baghdad. The actual experience of almost all Muslim societies, ever since the death of the Prophet, has been, as of course it has been in the West, one or another kind of autocratic rule. The West experienced a series of revolutions before ridding itself of Condorcet's "kings and priests" by which it had been plagued for so long. In most of the Muslim world they are still there. Iran has shed its kings, but it has still to get rid of its priests.

Otanes failed to persuade his fellow Persians of the value of *isonomia*, not because he could put his case persuasively but because the Persians had no understanding of what he was talking about. In the end, Darius's argument in favor of retaining the monarchy wins the day on the simple but overwhelmingly persuasive grounds that "we should refrain from changing ancient ways, which have served us well in the past. To do so would not profit us."[30]

Darius was certain of one thing: far from being a "natural" form of government, as is commonly supposed in the West, modern liberal democracy—any government based upon contract and consent—is just about the most unnatural there is. Most societies throughout history—including for centuries those in the West—have been hierarchical, patriarchal, kin-based, and often, in no small degree, theocratic. What sense does voting, and all that it implies, make in a world where the assumption that political authority derives from a contract between the ruler and the ruled is quite unknown and where the idea of the voluntary surrender of power is unimaginable? The trouble with the parliamentary system, as one British diplomat put it while he struggled to devise workable constitutions for former colonies, was the concept of the "loyal opposition." Just how, he asked, do you convince peoples inured for centuries to the idea that authority derives from on high, that once you have seized power you have to do your very best to hang on to it, and that you have a duty to use your position to favor family, friends, and clients—how do you persuade such people that justice and good government demand not only that you make no personal gain from your office but that you also exclude all those who might have some claims on you and that from time to time you voluntarily surrender power to your enemies?

Even in its place of origin, democratic government has had a long and troubled history. In its initial form in ancient Greece—which in fact bears little structural resemblance to what today we understand by the term—democracy was suppressed by an expansionist military power, that of Alexander, and it took the modern

West centuries and a number of sometimes violent revolutions to rebuild something like it. As late as the end of the nineteenth century, it was still regarded by the autocratic rulers of Europe with horror as the deadly child of the American and French Revolutions. Winston Churchill once remarked that of all the known forms of government it was the least bad option, and that was something of a concession. Even today in the most developed modern democracies there are many who visibly chafe against it, even if they dare not actually denounce it.

How, then, can such a concept simply be "introduced" into worlds that have no comparable history, without any prior preparation and usually by foreign arms? The view that all humans naturally desire freedom also ignores the fact they desire, often quite as strongly, order and direction in their lives. If they did not, religion, with its assumption that there exists some organizing principle—an "intelligent design"—amid the obvious chaos and that an ultimate impartial source of justice guides our lives, would have very little appeal. Many people, even in democratic societies, are far more concerned with finding a surrogate father, in Heaven or on Earth, than they are with abstract ideals of freedom. For centuries, Europeans took their kings to be their fathers—quite literally  and many Americans today take their president to be one.

The champions of democratic "regime change" also labor under another misapprehension: that the democratic process must necessarily lead to the creation of a modern bourgeois liberal democracy. In fact, however, it is perfectly possible to have an elective government without any attachment to liberal democratic principles. Iran has held regular elections since the success of the revolution in 1979, and although some of these have led to the appearance of moderate political actors, none has substantially changed the nature of Khomeini's original regime, as modern democratic theory demands that it should have done. A duly elected government, which is theocratic and very far from liberal, now (2007) rules in Palestine. And unless we think that this is all some aberration of the Islamic East, we need only remember that Hitler also came to power initially via the ballot box.

It is for this reason that most of the moderate pro-Western Muslim states have refused to be persuaded to hold popular elections. With the Islamists everywhere in the ascendance, they know that in many places free elections would inevitably result in the seizure of power by Islamist regimes, as indeed has occurred in Palestine and would have occurred in Algeria in 1991, had not the army decided to annul them after the Islamic Salvation Party won the first round. The same fate would almost certainly overtake Pakistan and Egypt—to name but two—if the military were ever to fully relinquish its hold over the country.

In the one predominantly Muslim state that does hold regular and entirely free elections, Turkey, an Islamic, although not openly Islamist, government has gained power. So far it has proved to be, if anything, more moderate and no less secularizing in the public sphere than many of its predecessors. It has also gone a long way toward making Turkey an acceptable candidate for entry into the European Union, which has meant abolishing torture and the death penalty (except in limited cases), introducing greater freedom of speech, and putting an end to the harassment of Kurds and other non-Turkic minorities. But here, as elsewhere in the Islamic world, the increasing polarization, real and perceived, between the "West" and the "East" has also strengthened the hand of Islam. Between 1999 and 2006, the number of Turks who identify themselves with Islam rose from 36 percent to 46 percent, a significant increase, however unreliable such statistics inevitably are in a country with a fiercely secular political culture.[31] There have also been some unsettling suggestions that demands for greater religious observance, always the not-so-thin edge of the Islamists' wedge, are beginning to make themselves heard. Judges who have upheld bans on the wearing of ostentatiously religious clothing have been repeatedly threatened, and one has been murdered. On one occasion, the state television station canceled the transmission of a "Winnie the Pooh" cartoon because one of the characters in it, Piglet, was a pig and thus offensive to Muslim sensibilities.

The only other Muslim state to have carried out successful elections recently is, of course, Western-occupied, post-Saddam Iraq. There the results were, as almost anyone not blinded by a faith in the evidently liberalizing influence of due democratic process could have foreseen, the creation of a Shiite government with strong ties to the government in Iran. As long as American and allied troops remain in Iraq, it is committed to behaving like a moderate Western-style government representing all the interests of "the Iraqi people" (an abstraction, in this case, even more amorphous than "the American people"). When, and if, they ever leave, Iraq will probably degenerate still further into chaos to emerge finally as another theocratic state like its now admired and powerful anti-Western neighbor. At the moment of writing—March 2007—Iraq is in a steadily deteriorating state of what few now trouble to deny is a civil war—not, in fact, unlike the condition that prevailed in Mesopotamia in the summer of 1920.

## II

Like all universalistic beliefs, democracy today is based on the assumption that nothing better will come along at some future date to replace it. There have been many such theories of the "end of history." As we have seen, the Greek orator Aristides believed that the Roman Empire, based upon a "mixed constitution" that blended together the supposedly perfect mix of democracy, aristocracy, and monarchy had finally brought history to a satisfactory conclusion. Others—Plato, Rousseau, Marx—have been more cautious and set their imagined ends of history in the future. All, however, assume that there is a point in time when humanity will cease to evolve, when the perfect state will come into being or have withered away altogether. None in the end, of course, has turned out to be the end of anything—except itself.

The stories told by the world's religious faiths have something in common with such secular tales as these. They are, however, rather less easy to discard. Political systems are all man-made, and thus, like all things made by man, even their most fervent admirers have to admit that they may in the end turn out simply to be wrong, or at least in need of serious modification. The same does not apply to religion. Religions are not, supposedly, made by man but by gods. The stories they tell cannot, therefore, turn out to be wrong, and they can have no alternative endings. The three great monotheisms, Judaism, Islam, and Christianity, all tell a slightly different version of much the same story. Each sets down strict laws of conduct for its adherents, and each denies the validity of the others. Each has strong messianic leanings, and both Christianity and Islam at least believe that one day they will triumph over all the other faiths in the world.

Christianity, however, has been compelled over time to be remarkably flexible and accommodating, in order to survive at all, in a world where its chief opponent is not another religion but a general indifference to all religions. It, too, has its share of fundamentalists. But they are relatively few and have, so far at least, in the long run, turned out to be largely ineffectual.

The same was also once true of Islam. But it is no longer so. Moderate Muslims may still make up the majority, but the "fundamentalists" are everywhere on the rise. This is in part a consequence of the current political state of the Muslim world; in part, as we have seen, an embittered reaction against Western neocolonialism. There is also a fundamental theological difference between Islam and Christianity that lies not in the ethical systems that underpin both, nor in their conception of God, nor much in their notion of what doing good and doing ill consist of. It lies instead, as all the militants have insisted, in the association between religion and the law. In a "Letter to America" distributed via the Internet in November 2002 and attributed (possibly falsely) to Osama bin Laden, the major crime of which the United States is accused is precisely the separation of the secular from the sacred:

You are the nation who, rather than ruling by the Shari'a of Allah in its Constitution and Laws, chose to invent your own laws as you will and desire. You separate religion from your politics, contradicting the pure nature which affirms Absolute Authority to the Lord your Creator. . . . You are the worst civilization witnessed by the history of mankind.[32]

Qutb, as we have seen, said much the same thing, and most Muslim theologians and jurists would have to agree with him. They might not choose to express themselves so crudely, but, in the end, this is what divides the West, and most of the actual societies in the Muslim world, from what the Islamists would make of them. Everything else follows from it. The society of Islam is ultimately based not upon human volition or upon contract but upon divine decree. In the societies of the West, by contrast, every aspect of life has been conceived as a question of human choice. And for the devout Muslim that is, and can only be, an offense against God. For "such a society," Qutb protested, "denies or suspends God's sovereignty on earth while God says plainly: 'It is He who is Sovereign in the heavens and Sovereign in [sic] the earth.' " (Q. 43:84)[33]

The division between Church and state, which has existed as long as the Church itself, was what made secularization in the West ultimately possible. It was the gate through which first reform and then the gradual minimization of religion were allowed to enter into daily life. And secularism in the West has shown that it can work. For all the evils of the last century (and this) committed by its adherents; for all the unjust wars fought in the name of modernization, liberalization, and democracy; for all the inequalities that still exist between the developed and the developing world; for all that globalization, in addition to its proven and obvious benefits, has also been responsible for some gross injustices and inequalities—for all that, the lives of the majority of the populations of the world are far better because of what the secularizing West has brought to them, directly or indirectly, than it was even a century ago. People live longer lives; they live freer lives, they may even live happier lives. To claim otherwise, to weep over the disappearance of tribe and community, is mere sentimentality. All, of course, is not perfect in the bright new secular modern world. Nor is it ever going to be. But that is simply a part of the human condition. What is wrong on this earth can never be remedied by appeals to God. There is no Eden at the end of history, neither here nor in some imagined afterlife.

This is something that convinced believers cannot accept. If the works of man have not turned out to be perfect, only perfectible, then, they argue, it is to God we must turn, for only if we are prepared to observe his commands will God deliver to his followers all that mere humans are unable to achieve. And since we can know what those commands are only through the scattered writings he has sent to his prophets, those writings must make up the laws we are to follow—no matter how disagreeable the consequences.

In June 2006, Iran's Holocaust-denying president, Mahmoud Ahmadinejad, wrote an open letter to George W. Bush as one believer to another. A true believer, he claimed, whether Muslim or Christian, could not now fail to see that "Liberalism and Western-style democracy have not been able to realize the ideals of humanity." Today, he went on, "these concepts have failed. Those with insight can already hear the sounds of the shattering and the fall of the ideology and thoughts of the liberal democratic system." It could not have turned out otherwise because, as Bush must surely recognize, such things were merely human and a disenchanted world was now "gravitating toward faith in the Almighty and justice, and the will of God will prevail over all things."[34] The letter was clearly intended to embarrass the American president by implying that although he spent a great deal of time talking about God, his actions were no different from those of any other Western infidel, and the claim that liberalism was on its last legs was merely whistling in the wind. But the message it conveyed was strictly in keeping with the belief not only of all Islamists but increasingly of many more accommodating Muslims.

Despite the resurgence in some Western states of Christian fundamentalists (on the political Right) and multiculturalists (on the political Left), both of whom, in their very different ways, seek to privilege religious belief and the ingrained and unexamined customs that John Stuart Mill once described as "the standing hindrance to human advancement"—broadly Enlightenment values still predominate throughout all the Western democracies and still dictate the actions of their governments.[35] Did they not, there would be no airlifts to poverty- and disease-stricken Africa, no aid programs, no Médecins sans Frontières, and Western governments would not expend vast resources and the lives of their own willing citizens to prevent genocide or bring an end to civil war in places to which they owe no special debt and with whose peoples they have no particular affinity.

Those Enlightenment values have been able to survive because the power of religious institutions and religious laws has been kept firmly in its place on the far side of the dividing line between politics and religion. This is not to suggest that "fundamentalist" Christianity—so far in its bitter resentfulness from anything to be found in the Gospels—does not represent a severe threat to the West. Yet even a self-confessed "born-again" Christian, such as George W. Bush, who when asked to name the political philosopher who had influenced him most, replied "Jesus Christ," would not consider attempting to replace modern representative government and the values associated with it by a rigid theocracy. (Presumably also if Bush were serious, he would have taken to heart the command to "render unto Caesar that which is Caesar's," which is Jesus' only identifiable political utterance.) True, some of the president's other remarks have been more disturbing. God, he told various evangelical groups in 2000, had called him to run for president—although quite how he did not say. "I know it won't be easy on me or my family," he acknowledged piously, "but God wants me to do it." The then attorney general and Pentecostal

Christian John Ashcroft made an even more direct attack on the founding principles of modern democracy by telling an audience at Bob Jones University that "we have no king but Jesus" and that the separation of Church and state was "a wall of religious oppression."[36] There have also been some unsettling indications—in stem cell research for instance—that the Christian opposition in the United States is capable of arresting the development of what, in this case, might yet turn out to be the most significant breakthrough in medical science since the discovery of antibiotics. God may well be dead in the Western world, but he has yet to lie down.

FOR A WHILE in the nineteenth century, as we have seen, a similar separation between the secular and the sacred seemed imaginable also for Islam. In 1883, Muhammad Abduh could argue that revelation was entirely compatible with human reason, that Islam condemned the blind following of tradition (*taqlid*) as contrary to the teaching of the Qur'an, and preached the need for a revision of the laws governing entire categories from the charging of interest to marriage and divorce.[37] He could insist that, while the laws concerned with religion were sacred and could not be altered in any way, those concerned with purely human conduct were not and could therefore be changed as circumstance demanded. What is most remarkable is that he was able to do all this while retaining his position as chief mufti of Egypt, one of the three highest religious dignitaries in the country.[38] But those days are long past. No Islamic theologian today is prepared to adopt such a position. Even if there were some kind of liberal Islam that was theologically viable, it is doubtful that it would have very much appeal beyond a small circle of already moderate, half-Westernized middle-class Muslims. For most others it would seem to be but yet another pandering to Western ideals.

When Muhammad Abduh was writing, Islam seemed to have a future as the belief system of a successful modern nation. By the 1960s, it became a means of resistance to imperialism wherever it was found and just about however it was conceived. As the American anthropologist Clifford Geertz put it in 1968, when the Algerian war was coming to its bloody end, Islam, despite its own imperialist past, could still fulfill the role of an anti-imperial creed simply because "the only thing the colonial elite was not, and a few ambiguous cases aside, could not become, was Muslim."[39] This is why it appealed at the time so strongly to African Americans. Today, in many parts of the world, it has become a religion of protest and resentment, most of it understandable, some of it justifiable, but all of it ultimately barren.

This is also why it is not only the impoverished and socially divided Muslim societies of the Middle East that have turned increasingly to militant forms of Islam. This is also the reason why immigrant Muslim societies in the West, and in Europe especially, have tried so hard to separate themselves from what they view as the threatening influence of the secular or "Christian" world around them. They live in alien lands, in the House of War, because the *dar al-Islam* is incapable of sustaining them economically. But they do so reluctantly, hating the fact that they must rely upon their self-described enemies for survival, always hoping that one day, somehow, the great revolution of Islam will sweep over the West.

Those who feel this way are inevitably those who have in the end gained far less from their relocation than they had expected. And this is far from true of all Muslims. There are thousands of well-integrated Muslims in most European societies. It is often forgotten that before the present wave of Muslim immigrants flooded into Europe, there was a smaller, silent immigration, mostly from Pakistan, India, and Bangladesh, for whom religion was wholly a private matter. When conflict broke out in Britain between unemployed whites and the more prosperous "Pakis," as they were insultingly called, in the 1960s and '70s, the issues were always ones of race—never religion.

But for those who are now confined to the dismal suburbs of the industrial cities of western Europe and North America, the

West has been a delusion. And it is a human failure to hate and resent what has disappointed us, which is not to say that the rioters (by no means all of whom, however, were Muslims) who set the Paris suburbs ablaze in 2005 did not have some real grievances against their adopted country and its government. The attraction of Islam to the newly dispossessed peoples of Europe is that it provides not only a cultural home and a set of unquestioned values but also a justification for hatred, a cause for which to fight that can all too easily be cast in terms of an apocalyptic struggle between good and evil. The appeal to certainty, even a certainty grounded on a premise as undemonstrable as divine revelation, has an overwhelming and entirely comprehensible attraction for the dispossessed. For many Muslims, living in conditions of great poverty and uncertainty, belief in an afterlife, even in the most extreme of cases one that can be rapidly obtained by suicide, seems more real than the remote, unexperienced benefits that the renunciation of God might have to offer.

For these peoples, and alas increasingly for many more moderate Muslims, the "clash of civilizations"—a crude but useful phrase—is the enduring reality of Islamic life, as it has always been a central feature of Islamic history. Other once-great civilizations of Asia, China, Japan, Korea, and India may all, in their own ways, have allowed themselves to reach an accommodation with the West. But the Muslim world, the Umma, can never do so. "O Prophet!" the angel Gabriel commanded Muhammad, "Urge the believers to war, if there are twenty patient ones of you they shall overcome two hundred, and if there are a hundred they shall overcome a thousand." (Q. 8:65) The patient ones are patient still.

It seems unlikely that the long struggle between East and West is going to end very soon. The battle lines drawn during the Persian Wars more than twenty-three centuries ago are still, in the selfsame corner of the world, very much where they were then.

# ACKNOWLEDGMENTS

I OWE A very great debt to my wife, Giulia Sissa. Not only did she provide me with the initial inspiration for this book and guide me through all the passages dealing with classical antiquity, she has, in extended conversations in many parts of the world, shaped my views on any number of other issues. To her wisdom, her learning, her keen intelligence, her generosity—and most of all her love—I owe more than I can ever repay.

Strobe Talbott took valuable time off from running the Brookings Institution and from writing a very important and related book of his own on world government to read almost the entire manuscript. Without his extensive and patient advice this book would have been a far lesser thing. My agent, Andrew Wylie, did much to shape the original idea for the book and has been unfailingly encouraging at every step of the way. My editor at Random House, Will Murphy, compelled me to prune a sometimes baggy and prolix manuscript, thus making this a far leaner and, I hope, more engaging work. To both I am greatly indebted. For all its shortcomings, however, no one is ultimately responsible but me alone.

Finally, a word on transliterations. With Arabic Persian and

Turkish I have made no attempt to be consistent. In general, where words are familiar in English I have used the conventional, if invariably imprecise, transliteration—Shiite, for instance, rather than Shi'īte. Elsewhere I have sometimes, where it seemed appropriate, resorted to more scholarly forms, but even these make no claim to be consistent. Those who can recognize the original script behind the transliteration will; those who cannot will not care.

# NOTES

## Preface

1. Although the first person to employ the term "Middle East"—in this case to describe the area around the Persian Gulf—seems to have been the American naval historian Alfred Mahan.

2. Quoted in J.G.A. Pocock, "Some Europes in their History," in *The Idea of Europe: From Antiquity to the European Union,* Anthony Pagden, ed. (Cambridge: Cambridge University Press, 2002), 58.

3. David Gress, *From Plato to NATO: The Idea of the West and Its Opponents* (New York: Free Press, 1998), 24–25.

4. The pope's quotation from the fourteenth-century Byzantine emperor Manuel II Paleologus that Islam contained nothing that was new that was not also "wicked (*schlechtes*) and inhumane (*inhuman*)" and that Muhammad had commanded his followers to "spread by the sword the faith" was, as many protested at the time, taken out of context. But since Manuel's comments have very little bearing on the context in question—the need for faith to be guided by reason and for the scientific community, which the pope was addressing, to accept faith as rational—it is hard not to see it, as most Muslims did, as a rebuke. *Faith, Reason and the University: Memories and Reflections* (Vatican City: Libreria Editrice Vaticana, 2006).

5. *De lingua latina,* VI, 3,1.

6. Herodotus, *Histories,* VII, 10–11. I have used, with occasional modifica-

tions, the translation of Aubrey de Sélincourt, revised by John Marincola (London: Penguin Press, 1996).

7. Nietzsche, "On the Uses and Disadvantages of History for Life," in *Untimely Meditations,* trans. R. J. Hollingdale (Cambridge: Cambridge University Press, 1983), 59.

8. The most recent account of the genocide is Taner Akçam, *A Shameful Act: The Armenian Genocide and the Question of Turkish Responsibility,* trans. Paul Bessemer (New York: Metropolitan Books, 2007).

## Chapter 1

1. Ovid, *Metamorphoses,* II, 862–4.

2. For a brilliant account and invocation of these myths, see Roberto Calasso, *The Marriage of Cadmus and Harmony* (New York: Vintage Books, 1993).

3. Paul Valéry, "Note, ou L'Européen," in *Varieté: Essais quasi politiques* (Paris: Gallimard, 1957).

4. On this and other versions of the myth, see Luisa Passerini, *Il mito d'Europa: radici antiche per nuovi simboli* (Florence: Giunti, 2002).

5. See J.A.S. Evans, "Father of History or Father of Lies? The Reputation of Herodotus," in *Classical Journal* 64 (1968): 11–17. Persian inscriptions speak only of the land (*bumi*) and peoples (*dahyu/dahyava*) of the great king. See Pierre Briant, *Histoire de l'Empire Perse: De Cyrus à Alexandre* (Paris: Fayard, 1996), 9.

6. Herodotus, *Histories,* VII, 104. Herodotus's understanding of the divisions of the world is, however, complex and sometimes contradictory. See Rosalind Thomas, *Herodotus in Context: Ethnography, Science and the Art of Persuasion* (Cambridge: Cambridge University Press, 2002), 80–86.

7. For a detailed and forceful account of the battle itself, see Victor Davis Hanson, *Carnage and Culture: Landmark Battles in the Rise of Western Power* (New York: Anchor Books, 2001), 27–59.

8. Herodotus, *Histories,* I, 209.

9. Aeschylus, *The Persians,* trans. Janet Lembke and C. J. Herington (New York and Oxford: Oxford University Press, 1981), 270–311. I have made some slight modifications to the translation.

10. For the early history of the Persian Empire, see Richard N. Frye, *The History of Ancient Iran* (Munich: C. H. Beck'sche Verlagsbuchhandlung, 1984), 91–96.

11. Isaiah 45, and also Ezra 1, who makes Cyrus declare, "The Lord God of Heaven hath given me all the kingdoms of the earth."

12. A. Kuhrt, "The Cyrus Cylinder and Achaemenid Imperial Policy," in *Journal for the Study of the Old Testament* 25 (1983): 83–94.

13. Herodotus, *Histories,* I, 205–14.

14. Plato, *Laws,* II, 694, claims that Cambyses lost his throne through drunkenness and debauchery.

15. Herodotus, *Histories,* III, 27–29.

16. This at least is Herodotus's version. Most scholars now discount the story of the slaying of the sacred bull. The bull appears to have died from natural causes in 524 B.C.E., during Cambyses' absence in Ethiopia. Following precedent, Cambyses had his name inscribed on the animal's sarcophagus with an accompanying limestone stele that depicts Cambyses in native royal costume, wearing the uraeus serpent and kneeling in reverence before the sacred beast. Whatever Cambyses' true behavior, we do know that he died at Agbatana near Mount Carmel and that on March 11, 522 B.C.E., his brother Bardiya, known to the Greeks as Smerdis, proclaimed himself king. On September 29, Bardiya was killed by Darius at Sikayauvatish in Medina Nisaya. See Albert Ten Eycle Olmstead, *History of the Persian Empire* (Chicago: University of Chicago Press, 1948), 107–18.

17. Herodotus, *Histories,* III, 79–83. The debate has been widely discussed. See, e.g., Norma Thompson, *Herodotus and the Origins of the Political Community* (New Haven, Conn.: Yale University Press, 1996), 52–78.

18. Pierre Briant suggests that what Herodotus's Persian sources may actually have described was not a debate over the best form of government but one over the problems of dynastic inheritance; *Histoire de l'Empire Perse,* 121.

19. Quoted in Olmstead, *History of the Persian Empire,* 107. For a more extended account of Darius's claims, see Briant, *Histoire de l'Empire Perse,* 121.

20. *Isonomia* should not, however, be taken to be identical with democracy (*demokratia*), although Herodotus does use both *demokratia* and *plethos,* "rule by the majority," in a later reference to the debate in V, 43. See Gregory Vlastos, "Isonomia politike," in *Platonic Studies* (Princeton, N.J.: Princeton University Press, 1981), 164–203.

21. See J. Peter Euben, "Political Equality and the Greek Polis," in *Liberalism and the Modern Polity,* ed. M. J. Gargas McGrath (New York: Marcel Dekker, 1959), 207–29.

22. Herodotus, *Histories,* I, 132.

23. On reverence and the Persian contempt for foreigners, see ibid., 134.

24. Ibid., 153.

25. Herodotus, *Histories,* IX, 16.

26. This interpretation of the "constitutional debate" is much indebted to Giulia Sissa, "The Irony of Travel: Democracy and Ethnocentrism in Herodotus," forthcoming in *Metis.*

27. Herodotus, *Histories,* III, 86. It has been suggested that Herodotus's story is a misunderstood and garbled version of a Persian ritual. There is also some evidence, not least of all from Plato (*Laws,* III 695c), that Darius may have been forced to share power with his five fellow conspirators. See Briant, *Histoire de l'Empire Perse,* 140–42.

28. This may have been an allusion to his economic policies. Darius seems to have been the first Persian ruler to standardize weights and measures throughout his domains. See Olmstead, *History of the Persian Empire,* 185–94.

29. On Mani, see pp. 154–56.

30. For a brilliant evocation of Persepolis, see Olmstead, *History of the Persian Empire,* 172–84.

31. Herodotus, *Histories,* V, 97; An echo of *Iliad* V, 62, and XI, 604.

32. Herodotus, *Histories,* VI, 43. See Pierre Briant, "La vengeance comme explication historique dans l'oeuvre d'Hérodote," *Revue des Études Grecques* 84 (1971): 319–35.

33. Herodotus, *Histories,* VI, 100–2.

34. Ibid., 106. The Spartans were celebrating the feast of Carneia, during which time they had to abstain from warfare.

35. See Nicole Loraux, *The Invention of Athens: The Funeral Oration in the Classical City* (Cambridge, Mass.: Harvard University Press, 1986), 162.

36. Herodotus, *Histories,* VI, 111–18.

37. Briant, *Histoire de l'Empire Perse,* 170–72.

38. At least according to the Akkadian sources; R. A. Parker and W. Dubberstein, *Babylonian Chronology* (Princeton, N.J.: Princeton University Press, 1956), 17.

39. Herodotus, *Histories,* VII, 8.

40. Herodotus, *Histories,* V, 78.

41. Gress, *From Plato to NATO,* 1.

42. Herodotus, *Histories,* VII, 10–11.

43. Aelius Aristides, "The Roman Oration," in James H. Oliver, *The Ruling Power: A Study of the Roman Empire in the Second Century After Christ*

*Through the Roman Oration of Aelius Aristides,* Transactions of the American Philosophical Society, New Series 23 (1953): 5.

44. Herodotus, *Histories,* VII, 42–44, and see Olmstead, *History of the Persian Empire,* 249–50.

45. Herodotus, *Histories,* VII, 33–36.

46. Lysias, *Funeral Oration,* 29. On Xerxes' bridges, see L. J. Roseman, "The Construction of Xerxes' Bridge over the Hellespont," *Journal of Hellenic Studies* 116 (1996): 88–108.

47. Herodotus, *Histories,* VII, 56–100.

48. Isocrates, *Panegyricus,* 150.

49. Herodotus, *Histories,* VII, 101–5.

50. Plutarch, *Parallel Stories,* 306.4.

51. Paul Cartledge, *Thermopylae: The Battle That Changed the World,* (London: Pan Books, 2006), 194–95, which is now the best account of the battle, its historical context, and its aftermath. See also Sarah B. Pomeroy, et al., *Ancient Greece: A Political, Social and Cultural History* (New York: Oxford University Press, 1999), 195–96.

52. Herodotus, *Histories,* VIII, 51–55.

53. Aeschylus, *Persians,* 630–42.

54. Ibid., 670–97.

55. Herodotus, *Histories,* VIII, 85–96.

56. Thucydides, *The Peloponnesian War,* 1. 138.

57. Plato, *Laws,* IV, 707 c2–8.

58. Georg Friedrich Hegel, *The Philosophy of History,* trans. J. Sibree (Dover Publications: New York, 1956), 257–58.

59. Xerxes may also have been forced to return to suppress a rebellion in Babylon that broke out in 479 B.C.E. See Pierre Briant, "La date des révoltes babyloniennes contra Xersès," *Studia Iranica,* 21 (1992): 12–13.

60. Aristides, "Roman Oration," 16.

61. Lysias, *Funeral Oration,* 47. Lysias, however, is here claiming that it was the Athenians alone who were ultimately responsible for driving the barbarians out of Europe. See Loraux, *The Invention of Athens,* 53–54.

## Chapter 2

1. Herodotus, *Histories,* IV, 177; IV, 183–84; V, 5.

2. Aristotle, *Politics,* 1252 b 4.

3. Plutarch, *On the Fortune of Alexander,* 329b.

4. Plato, *Statesman,* 262d. Plato's objection to this practice, however, was that there existed only one division of the human species, and that is gender. Human beings are divided into men and women. All other divisions are accidents of place and circumstance, significant certainly but in no way decisive.

5. Immanuel Kant, "Kant on the Metaphysics of Morals: Vigilantius's Lecture Notes," in *Lectures on Ethics,* eds. Peter Heath and J. B. Schneewind (Cambridge: Cambridge University Press, 1997), 406.

6. Claude Lévi-Strauss, *The Elementary Structures of Kinship,* trans. James Hare Bell (London: Eyre and Spottiswoode, 1968), 46.

7. Aeschylus, *Persians,* 331–35.

8. Ibid., 827.

9. See A.W.H. Adkins, *Moral Values and Political Behaviour in Ancient Greece* (New York: W. W. Norton, 1972), 100.

10. *The Peloponnesian War,* 1. 96.

11. Ibid., 74. 1.

12. Ibid., 75.

13. Diodorus Siculus, *Bibliotheca,* 15, 93.1. For the condition of Persia in these years, see Briant, *Histoire de l'Empire Perse.*

14. Isocrates, *To Philip,* 40.

15. Herodian, *History,* 3.2.8.

16. Book of Daniel: 7.

17. Isocrates, *Panegyricus,* 157–58.

18. My account of Philip's reign and of the life and campaigns of Alexander is heavily dependent upon A. B. Bosworth, *Conquest and Empire: The Reign of Alexander the Great* (Cambridge: Cambridge University Press, 1988), and the brilliant reconstructions in Peter Green, *Alexander of Macedon, 356–323 B.C.: A Historical Biography* (Berkeley: University of California Press, 1991), and Robin Lane Fox, *The Search for Alexander* (Boston: Little Brown, 1980).

19. Diodorus Siculus, *Bibliotheca,* 17, 17.2. See, however, the discussion of the meaning of this act in Bosworth, *Conquest and Empire,* 38.

20. For a detailed account of the battle, see Green, *Alexander of Macedon,* 172–81.

21. Cited in N.G.L. Hammond, "The Kingdom of Asia and the Persian Throne," in *Alexander the Great: A Reader,* ed. Ian Worthington (London: Routledge, 2003), 137.

22. See the descriptions in Green, *Alexander of Macedon,* 213–15, and Paul

Cartledge, *Alexander the Great: The Hunt for a New Past* (London: Macmillan, 2004), 114–15.

23. Green, *Alexander of Macedon,* 234–35.

24. Plutarch, *Life of Alexander,* 34. 1–4.

25. Green, *Alexander of Macedon,* 315–16.

26. The story has been told many times. My version is based on Green, *Alexander of Macedon,* 318–21, and Fox, *The Search for Alexander,* 244–54.

27. Quoted in Efraim Karsh, *Islamic Imperialism: A History* (New Haven, Conn.: Yale University Press, 2006), 198.

28. Herodotus, *Histories,* VII, 42–44.

29. Plutarch, *Life of Alexander,* 43. 3–4.

30. Johann Gustav Droysen, *Geschichte Alexanders des Grossen,* vol. 1 of *Geschichte des Hellenismus* (Basel: Schwabe, 1952), 83.

31. Arrian, *Campaigns of Alexander,* 519.4–5.

32. Strabo, *Geographia,* xv. 1. 6.

33. Fox, *The Search for Alexander,* 407–18.

34. Green, *Alexander of Macedon,* 483–84. Seneca, *Quaest, Nat.* VI, 23, and *Epistolae,* 91. 17.

35. Victor Davis Hanson, "Take Me to My Leader," *The Times Literary Supplement,* October 2, 2004, 11–27.

36. Seneca, *Florida,* VII.

37. W. W. Tarn, *Alexander the Great* (Cambridge: Cambridge University Press, 1948), 1: 145–48.

38. Cartledge, *Alexander the Great.*

39. Alexander's divinity is discussed in detail in Bosworth, *Conquest and Empire,* 278–90.

40. Isocrates, *Panegyricus,* 151.

41. Tarn, *Alexander the Great,* 1: 145–48.

42. Cartledge, *Alexander the Great,* ix.

43. Plutarch, *The Fortunes of Alexander,* 329.

44. On the importance of Stoicism, see pp. 119–20. Plutarch, *The Fortunes of Alexander,* 329.

45. Montesquieu, *L'Esprit des lois,* X, 14.

46. Tarn, *Alexander the Great,* 1: 145–48.

## Chapter 3

1. Aristides, "Roman Oration," 1.

2. Aristides' oration and its significance are brilliantly described in Aldo

Schiavone, *The End of the Past: Ancient Rome and the Modern West*, trans. Margaret J. Schneider (Cambridge, Mass.: Harvard University Press, 2000), 3–15.

3. Paul Veyne, *L'Empire gréco-romain* (Paris: Seuil, 2005), 245–47.

4. *De Consulatu Stilichonis,* III, 150–55.

5. The image is generally attributed to the sixteenth-century Italian poet Ludovico Ariosto, speaking of the empire of Charles V. Aristides' formulation is much less eloquent but carries the same meaning: "Your possession is equal to what the sun can pass, and the sun passes over your land." "Roman Oration," 10.

6. Aristides, "Roman Oration," 104.

7. Ibid., 90–91.

8. Gibbon, *Decline and Fall of the Roman Empire,* III.

9. Virgil, *Aeneid,* XII, 808–42. I would like to thank Maurizio Bettini for drawing this passage and its significance to my attention.

10. Horace, *Epistles,* 2.1. 156–57.

11. Ramsay MacMullen, *Romanization in the Time of Augustus* (New Haven, Conn.: Yale University Press, 2000), 2–3.

12. See in general, Fergus Millar, "Taking the Measure of the Ancient World," in *Rome, the Greek World and the East,* vol. 1: *The Roman Republic and the Augustan Revolution* (Chapel Hill: The University of North Carolina Press, 2002), 25–38.

13. Aristides, "Roman Oration," 96.

14. By Suetonius in *Lives of the Twelve Caesars,* V, 42. The translations are from Suetonius, *Lives of the Twelve Caesars,* trans. Robert Graves (New York: Welcome Rain, 2001).

15. Michael Grant, *The World of Rome* (London: Weidenfeld and Nicolson, 1960), 37–39.

16. Cicero, *Tusculanae Disputationes,* 4. 70. See also F.P.V.D. Balsdon, *Romans and Aliens* (London: Duckworth, 1979), 33, 225. And cf. Polybius 31. 25.3–5, who claims that homosexuality arrived in Rome following the victory over Perseus of Macedon in 167 B.C.E.

17. Livy, 9.17.16.

18. R. A. Gauthier, *Magnanimité: l'idéal de la grandeur dans la philosophie païenne et dans la théologie chrétienne* (Paris: Vrin, 1951), and Georges Dumézil, *Idées romaines* (Paris: Gallimard, 1969), 125–52.

19. Dumézil, *Idées romaines,* 48–59.

20. Cicero, *De Republica,* 3.35; Peter Garnsey, *Ideas of Slavery from Aristotle to Augustine* (Cambridge: Cambridge University Press, 1996), 40–43.

21. Petronius, *Satyricon,* 119. 19, 24–27.

22. Lucan, *Phars.* 7, 442; 8, 362; Balsdon, *Romans and Aliens*, 61.

23. Juvenal, *Sat* III, 60–85. See Mary Gordon, "The Nationality of Slaves Under the Early Roman Empire," in *Slavery in Classical Antiquity*, ed. M. I. Finley (Cambridge, U.K.: W. Heffer and Sons, 1960), 171–89.

24. Juneval, *Sat* III, 6–72.

25. *Catilinae coniuratio,* 11.5.

26. Homer, *Odyssey,* 13.271–86.

27. F. Mazza, "The Phoenicians as Seen by the Ancient World" in *The Phoenicians,* ed. Sabatino Moscati (London: I. B. Tauris, 2001), 548–67.

28. Arnaldo Momigliano, *Alien Wisdom: The Limits of Hellenization* (Cambridge: Cambridge University Press, 1975), 4.

29. Livy, 22.6I

30. For a brilliantly evocative account of the battle see Victor Davis Hanson, *Carnage and Culture: Landmark Battles in the Rise of Western Power* (New York: Anchor Books, 2001), 99–111.

31. Ibid., 110.

32. As told by Antisthenes of Rhodes. See Jean-Louis Ferrary, *Philhellénisme et imperialisme: aspects idéeologiques de la conquête du monde hellénistique* (Rome: Bibliothèque des Écoles Françaises d'Athènes et de Rome, 1988), 362, where the dream is attributed to Scipio Africanus.

33. Momigliano, *Alien Wisdom,* 4.

34. Quoted in Benjamin Isaac, *The Invention of Racism in Classical Antiquity* (Princeton, N.J.: Princeton University Press, 2004), 377.

35. Yves Albert Dauge, *Le barbare: recherches sur la conception de la barbarie et de la civilisation* (Bruxelles: Latomus Revue d'Études Latines, 1981), 99–261.

36. Seneca, *De Constantia,* 13.4.

37. The story is told in Aulus Gellius, *Noctes Atticae,* IV, 8.

38. Plutarch, *Julius Caesar,* 60.

39. Plutarch, *Mark Antony,* 26.

40. Ibid., 54.

41. Recorded by the second-century historian Cassius Dio, 50.24.6.

42. Virgil, *Aeneid,* VIII, 685–88.

43. Plutarch, *Mark Antony,* 66.

44. Suetonius, *Lives of the Twelve Caesars,* II, 17–18.

45. Claude Nicolet, *The World of the Citizen in Republican Rome,* trans. P. S. Falla (Berkeley: University of California Press, 1980), 21.

46. *Historia,* I, XVI.

47. Suetonius, *Lives of the Twelve Caesars,* V, 41.

48. Tacitus, *Annals,* XV, 41–42.

49. Aristides, "Roman Oration," 103.

50. Ibid., 15–26.

51. For this and other instances of imperial imagery, see Andrew Lintott, "What Was the *Imperium Romanum?*" *Greece and Rome* 28 (1981): 53–67.

52. Cicero, *De Republica*, 3.15.24.

53. Pliny, *Naturalis Historia*, 3.39.

54. See Clifford Ando's remarkable study *Imperial Ideology and Provincial Loyalty in the Roman Empire* (Berkeley: University of California Press, 2000), 67.

55. Quoted in P. A. Garnsey, "Laus Imperii," in *Imperialism in the Ancient World*, ed. P. A. Garnsey and C. R. Whittaker (Cambridge: Cambridge University Press, 1978), 168.

56. Quoted in Schiavone, *End of the Past*, 5.

57. Livy, 8.13.16.

58. Aristides, "Roman Oration," 22–23.

59. Ibid., 34.

60. Ibid., 11, 104; Schiavone, *End of the Past*, 7–8.

61. Veyne, *L'Empire gréco-romain*.

62. Described beautifully in Peter Brown, *The World of Late Antiquity* (New York: W. W. Norton, 1989), 11.

63. Millar, "Taking the Measure of the Ancient World," 31–33.

64. Modestinus, in *Digest*, 50.1.33.

65. Quoted in Brown, *The World of Late Antiquity*, 123.

66. James Wilson, "Lectures on Law: XI. Citizens and Aliens" [1790–91], *The Works of James Wilson*, ed. Robert Green McCloskey, 2 vols. (Cambridge, Mass.: Harvard University Press, 1967), II: 581.

67. M. I. Finley, *Ancient Slavery and Modern Ideology* (Harmondsworth, U.K.: Penguin Press, 1983), 107.

68. Ibid., 93.

69. Quoted in Nicolet, *World of the Citizen in Republican Rome*, 39.

70. The following clause, "provided that there remains . . . part of the *dediticius*," has created considerable controversy as to its meaning. The *dediticii* were provincials who had become subjects of Rome through formal surrender in war. See A. N. Sherwin White, *The Roman Citizenship* (Oxford: Oxford University Press, 1973), 380–86.

71. Quoted in White, *Roman Citizenship*, 435. As with most Christians, however, Tertullian's attitude toward the empire was not wholly unambiguous, and he was capable of imagining a unity of mankind that transcended the unity of the Roman world.

72. Tacitus, *Annals*, II, 23–24. For a detailed account of these events and their significance, see White, *Roman Citizenship*, 237–50. The struggle between the Gallic king Vercingetorix and Caesar—to which the senator alluded—had taken place in 52 B.C.E., nearly a century earlier, and it was Caesar who had besieged Vercingetorix, not vice versa.

73. Ando, *Imperial Ideology and Provincial Loyalty in the Roman Empire*, 41.

74. Cicero, *De Officiis*, II. 27.

75. Aristides, "Roman Oration," 59–60.

76. Nicolet, *World of the Citizen in Republican Rome*, 22.

77. Acts 21:37–39.

78. Acts 22:25–29.

79. Acts 25:10–12.

80. For a further discussion of these events, see Nicolet, *World of the Citizen in Republican Rome*, 18–20.

81. Petrus Baldus de Ubaldis, quoted in Luigi Prosdocimi, " 'Ex facto oritur ius': breve nota di diritti medievale," *Studi senesi* (1954–55), 66–67, 808–19.

82. Quoted in Donald R. Kelley, *Historians and the Law in Postrevolutionary France* (Princeton, N.J.: Princeton University Press, 1984), 45.

83. Gibbon, *Decline and Fall of the Roman Empire*, XLIV.

84. Cf. Cicero, *De Officiis*, I. 34–35.: "There are two types of conflict: the one proceeds by debate, the other by force. Since the former is the proper concern of man but the latter of beasts, one should resort to the latter only if one may not employ the former."

85. Cicero, *De Republica*, 3.34. See Jonathan Barnes, "Ciceron et la guerre juste," *Bulletin de la Société Française de Philosophie* 80 (1986): 41–80.

86. Cicero, *De Legibus*, I,x,29; I,xii,33.

87. Aristotle, *Rhetoric*, 1. 13 1373b, and see P. A. Garnsey, "Laus imperii," in *Imperialism in the Ancient World*, 159–91.

88. *De Finibus*, III 63.

89. *The Meditations of the Emperor Marcus Aurelius Antoninus*, vi. 50, 58.

90. Plutarch, *On the Fortune of Alexander*, 329.

91. Aristides, "Roman Oration," 102.

92. Ernest Barker, "The Conception of Empire," in *The Legacy of Rome*, ed. Cyril Bailey (Oxford: Oxford University Press, 1923), 53.

93. Aristides, "Roman Oration," 104, 99.

94. Ibid., 12–13.

95. Gibbon, *Decline and Fall of the Roman Empire*, X.

96. Veyne, *L'Empire gréco-romain*, 306–11.

## Chapter 4

1. Claudian, *De bello Getico,* 78.

2. Peter Brown, *Augustine of Hippo: A Biography* (London: Faber and Faber, 1976), 298.

3. Quoted in ibid., 289.

4. Augustine, *De Civ. Dei,* IV 7.

5. Daniel 7:14.

6. Samuel Purchas, *Hakluytus Posthumus or Purchas His Pilgrimes, Contayning a History of the World, in Sea Voyages and Lande-Travells by Englishmen & Others,* 5 vols. (London, 1625), I:45.

7. Quoted in Ando, *Imperial Ideology and Provincial Loyalty in the Roman Empire,* 63.

8. Quoted in Brown, *Augustine of Hippo,* 291.

9. Ramsay MacMullen, *Christianizing the Roman Empire, A.D. 100–400* (New Haven, Conn.: Yale Univeristy Press, 1984), 134, n. 14.

10. W.H.C. Frend, *Martyrdom and Persecution in the Early Church* (Oxford: Oxford University Press, 1965), 413, numbers the martyrs in hundreds rather than thousands.

11. Henry Chadwick, "Envoi: On Taking Leave of Antiquity," in *The Oxford History of the Classical World,* ed. John Boardman, Jasper Griffin, and Oswyn Murray (Oxford: Oxford Univeristy Press, 1986), 808.

12. Eusebius of Caesarea, *Life of Constantine,* trans. A. Cameron and S. Hall (Oxford: Oxford University Press, 1999), 1:28–32.

13. See Charles Freeman, *The Closing of the Western Mind: The Rise of Faith and the Fall of Reason* (New York: Alfred A. Knopf, 2003), 170–72.

14. Paul Veyne, *Quand notre monde est devenu chrétien (312–394)* (Paris: Albin Michel, 2007), 28.

15. The phrase is Peter Brown's in *The World of Late Antiquity,* 87.

16. A.H.M. Jones, *Constantine and the Conversion of Europe* (London: Hodder and Stoughton, 1948), 92–93.

17. F. E. Peters, *The Monotheists: Jews, Christians, and Muslims in Conflict and Competition* (Princeton, N.J.: Princeton University Press, 2003), I:248.

18. *De Civ. Dei,* V 15.

19. Jacques Heers, *Chute et mort de Constantinople, 1204–1453* (Paris: Perrin, 2005), 20.

20. The description is Steven Runciman's in *The Great Church in Captivity: A Study of the Patriarchate of Constantinople from the Eve of the Turkish Conquest to the Greek War of Independence* (Cambridge: Cambridge University Press, 1968), 7.

21. Quoted in Ibid., 59.

22. Carl Erdmann, *The Origin of the Idea of Crusade,* trans. Marshall W. Baldwin and Walter Goffart (Princeton, N.J.: Princeton University Press, 1977), 296–97.

23. Norman Davies, *Europe: A History* (Oxford: Oxford University Press, 1997), 341–42.

24. Niccolò Machiavelli, *The Prince,* ed. David Wootton (Indianapolis: Hackett, 1995), 5, [Cap. 18].

25. Quoted in Freeman, *The Closing of the Western Mind,* 176.

26. Gomes' Eanes de Zurara, *Crónica dos feitos na conquista de Guiné,* ed. Torquato de Sousa Soares (Lisbon Academia Portuguesa da Historia: 1961) 1:145–148. For an account of Henry's career, see Peter Russell, *Prince Henry "The Navigator": A Life* (New Haven, Conn.: Yale University Press, 2000).

27. Quoted in Karen Ordahl Kupperman, *Settling with the Indians: The Meeting of English and Indian Cultures in America, 1580–1640* (Totowa, N.J.: Rowman and Littlefield, 1980), 166.

28. Peters, *The Monotheists,* 2:138–39.

29. *De Civ. Dei,* XI 13.

30. See Brown, *Augustine of Hippo,* 58–59.

# Chapter 5

1. There are several excellent histories of the rise and fall of the empire of the Arabs. In the following narrative, I have drawn extensively on Michael Cook, *Muhammad* (Oxford: Oxford University Press, 1983); J.-Cl. Garcin, ed., *États, sociétés et cultures du monde musulman médiéval, Xe.–XVe. siècle* (3 vols.) (Paris: PUF, 1995–2000); Albert Hourani, *A History of the Arab Peoples* (London: Faber & Faber, 1991); Richard Fletcher, *The Cross and the Crescent: Christianity and Islam from Muhammad to the Reformation* (New York: Viking, 2003); Bernard Lewis, *The Arabs in History* (Oxford: Oxford University Press, 1993).

2. The account of the various embassies is in Muhammad Ibn Ishaq, *The Life of Muhammad: A Translation of Ishaq's Sirat Rasul Allah,* introduction and notes by A. Guillaume (Karachi: Oxford University Press, 1955), 652–59.

3. *Res Gestae,* XIV. 4.

4. Ibn Ishaq, *Life of Muhammad,* 181–87.

5. Similar claims are made elsewhere, e.g. 41:3, "A Book of which the verses are made plain, an Arabic Qur'an for a people who know"; 43:3,

"And before it the Book of Musa [Moses] was a guide and a mercy; and this is a Book verifying [it] in the Arabic language that it may warn those who are unjust and as good news for the doers of good." And cf. 13:37; 16:103; 20:113; 39:28; 42:7.

6. The first day of the first month (Muharram) corresponds to July 15 or 16 in the Julian calendar.

7. Cook, *Muhammad,* 41.

8. Patricia Crone, *God's Rule: Government and Islam* (New York: Columbia University Press, 2004), 13.

9. See Michael Cook, *Forbidding Wrong in Islam* (Cambridge: Cambridge University Press, 2003).

10. Bernard Lewis, "Politics and War," in *The Legacy of Islam,* ed. Joseph Schnact and C. E. Bosworth (Oxford: Oxford University Press, 1979), 156.

11. Quoted in Bernard Lewis, *The Crisis of Islam: Holy War and Holy Terror* (New York: Random House, 2003), 34. See the same author's *The Political Language of Islam* (Chicago: The University of Chicago Press, 1988), 71–90.

12. This is often known as "positive" tolerance. "Negative" tolerance, which is akin to indifference, derives from a general skepticism about the validity of any belief.

13. Patricia Crone and Marin Hinds, *God's Caliph: Religious Authority in the First Centuries of Islam* (Cambridge: Cambridge University Press, 1986), 19.

14. Quoted in Hourani, *History of the Arab Peoples,* 19.

15. Ibn Khaldûn, *The Muqaddimah: An Introduction to History,* trans. Franz Rosenthal (Princeton, N.J.: Princeton University Press, 1967), 330.

16. According to the Persian historian Ahmad ibn Yahya al-Baladhuri in *Kitâb Futûh al-Buldân* (*The Origins of the Islamic State*), trans. Philip Hitti (New York: Columbia University Press, 1916), 187.

17. Quoted in John Tolan, *Saracens: Islam in the European Medieval Imagination* (New York: Columbia University Press, 2002), 40.

18. Patricia Crone, *Medieval Islamic Political Thought* (Edinburgh: Edinburgh University Press, 2004), 334.

19. Lewis, *Political Language of Islam,* 75.

20. Henri Pirenne, *Mohammed and Charlemagne* (Mineola, N.Y.: Dover Publications, 2001), 152–53.

21. Colin Smith, ed., *Spanish Ballads* (Oxford: Pergamon Press, 1964), 55.

22. Quoted in Bartolomé de las Casas, *A Short Account of the Destruction of the Indies,* trans. Nigel Griffin (London: Penguin Press, 1992), xxxviii.

23. Quoted in Derek W. Lomax, *The Reconquest of Spain* (London: Longman, 1978), 26.

24. See, in general, Richard W. Bulliet, *Conversion to Islam in the Medieval Period: An Essay in Quantitative History* (Cambridge, Mass.: Harvard University Press, 1979), 114–27.

25. Quoted in Lewis, *Arabs in History*, 134.

26. Leo of Rozmital, *The Travels of Leo of Rozmital*, trans. and ed. Malcolm Letts (Cambridge: Cambridge University Press, 1957), 91–92.

27. Fletcher, *Cross and the Crescent*, 22–23.

28. Colin Smith, ed. *Christians and Moors in Spain* (Warminster, U.K.: Aris & Philips, 1988), 1:65–67; also see Jessica A. Coope, *The Martyrs of Córdoba: Community and Family Conflict in an Age of Mass Conversion* (Lincoln: University of Nebraska Press, 1995), 67–69.

29. Quoted in Bernard Lewis, *The Muslim Discovery of Europe* (New York: W. W. Norton, 1982), 19; and Lewis, "Europe and Islam: Muslim Perceptions and Experiences," in *From Babel to Dragomans: Interpreting the Middle East* (Oxford: Oxford University Press, 2004), 124, where he compares it to "the reverse suffered by some scouting party from nineteenth-century British India caught by tribesmen in the wilds of Afghanistan."

30. Gibbon, *Decline and Fall of the Roman Empire*, 52.

31. S. D. Goiten, "The Origin of the Vizierate and Its True Character," in *Studies in Islamic History and Institutions* (Leiden: Brill, 1966), 168–96.

32. Quoted in Richard Hodges and David Whitehouse, *Mohammed, Charlemagne and the Origins of Europe* (Ithaca, N.Y.: Cornell University Press, 1983), 126–27.

33. Quoted in Lewis, *Arabs in History*, 131.

34. G. Levi Della Vida, "La corrispondeza di Berta di Toscano col Califfo Muktafi," *Rivista Storica Italiana* 66 (1954): 21–38.

35. Lewis, *Muslim Discovery of Europe*, 76.

36. Quoted in Sanjay Subrahmanyam, "Taking Stock of the Franks: South Asian Views of Europeans and Europe, 1500–1800," *The Indian Economic and Social History Review* 42 (2005): 6–100; 69.

37. Crone, *Medieval Islamic Political Thought*, 171–72.

38. Fletcher, *Cross and the Crescent*, 50.

39. H. Fradkin, "The Political Thought of Ibn Tufayl," in *The Political Aspects of Islamic Philosophy*, ed. E. Butterworth (Cambridge, Mass.: Harvard University Press, 1992), 234–61.

40. Dante, *Inferno*, canto 4, ll. 36–39.

41. *Fasl al-Maqâl,* para. 5, in Averroès, *Discours décisif,* trans. Marc Geoffroy (Paris: Flammarion, 1996), 107.
42. *Fasl al-Maqâl,* para. 7, in *Discours décisif,* 109.
43. Ernest Renan, "L'Islamisme et la Science," in *Oeuvres complètes de Ernest Renan,* ed. Henriette Psichari (Paris: Calmann-Lévy, 1947), 954–956.
44. Ibid., 947–49.
45. Ernest Renan, "Averroès et l'Averroïsme" [1852], in *Oeuvres complètes de Ernest Renan,* ed. Psichari, 3:23.
46. Quoted in Franco Cardini, *Europa e Islam: storia di un malinteso* (Rome: Laterza, 2002), 130.
47. John Lamoreaux, "Early Eastern Christian Responses to Islam," in *Medieval Christian Perceptions of Islam: A Book of Essays,* ed. John Tolan (New York: Garland Press, 1996), 14–15.
48. *Adversus haereses,* I, xxvi 3.
49. James Kritzeck, *Peter the Venerable and Islam* (Princeton, N.J.: Princeton University Press, 1964), 17–18.
50. Pope Gregory VII, in *Patrologia Latina,* ed. Migne, CXLVIII, 450–52. Also see the comments of Albert Hourani in *Islam in European Thought,* 9.
51. *De Haeresibus,* in Daniel J. Sahas, *John of Damascus on Islam: The "Heresy of the Ishmaelites"* (Leiden: E. J. Brill, 1972), 133.
52. Ibid., 139.
53. Peter the Venerable, in *Patrologia Latina,* ed. Migne, CLXXXXI, 671.
54. Quoted in Kritzeck, *Peter the Venerable and Islam,* 142–43.
55. Quoted in Norman Daniel, *Islam and the West: The Making of an Image* (Edinburgh: Edinburgh University Press, 1960), 68.
56. Ibid., 102.
57. *De Haeresibus,* 139, and Daniel, *Islam and the West,* 96–100.
58. *De Haeresibus,* 32–48.
59. *Chanson de Roland,* v. 3164.
60. Machiavelli, *The Prince,* 55.
61. Quoted in Anthony Pagden, *European Encounters with the New World* (New Haven, Conn.: Yale University Press, 1993), 36.

## Chapter 6

1. Or so contemporary accounts and the reaction of his listeners suggest. Unfortunately, no full account of what he actually said survives.
2. Quoted in Jonathan Riley-Smith, *The First Crusaders, 1095–1131* (Cambridge: Cambridge University Press, 1997), 61.

3. Ibid., 20.

4. Jonathan Riley-Smith, *The First Crusade and the Idea of Crusading* (London: Athlone Press, 1986), 26.

5. Matthew 16:24.

6. Rosalind Hill, ed., *Gesta Francorum et aliorum Hierosolimitanorum (The Deeds of the Franks and the Other Pilgrims to Jerusalem)* (London: Thomas Nelson, 1962), 17.

7. Augustine, *Epist.* 189.6 [to Bonifatius], in *Patrologia Latina*, ed. Migne, XXXIII, 856.

8. From *La Chanson d'Antioche,* in Louise and Jonathan Riley-Smith, *The Crusades: Idea and Reality, 1095–1272* (London: Edward Arnold, 1981), 72.

9. From Baldric of Bourgueil, *Historia Jerosolimitana,* quoted in Riley-Smith, *The First Crusade and the Idea of Crusading,* 48–49.

10. Peter Partner, *God of Battles: Holy Wars in Christianity and Islam* (London: HarperCollins, 1997), 82.

11. From *Historia,* in Louise and Jonathan Riley-Smith, *The Crusades,* 54.

12. Carole Hillenbrand, *The Crusades: Islamic Perspectives* (Edinburgh: Edinburgh University Press, 1999), 295–96.

13. Riley-Smith, *The First Crusade and the Idea of Crusading,* 92.

14. Ibn al-Qalânisî, *The Damascus Chronicle of the Crusades,* extracted and translated by H.A.R. Gibb (London: Luzac and Co., 1932), 48.

15. Sir Steven Runciman, "The First Crusade: Antioch to Ascalon," in *A History of the Crusades,* ed. Kenneth M. Setton (Madison: University of Wisconsin Press, 1969), 1:308–42.

16. *Gesta Francorum et aliorum Hierosolimitanorum,* 92.

17. Quoted in Riley-Smith, *The First Crusaders,* 77.

18. T. E. Lawrence, *Crusader Castles,* a new edition with introduction and notes by Denys Pringle (Oxford: Oxford University Press, 1988), 77.

19. Usamah Ibn Munqidh, *An Arab-Syrian Gentleman and Warrior in the Period of the Crusades* (New York: Columbia University Press, 1929), 29.

20. Ibn al-Qalânisî, *Damascus Chronicle of the Crusades,* 269.

21. From *De consideratione,* in Louise and Jonathan Riley-Smith, *The Crusades: Idea and Reality,* 62.

22. Partner, *God of Battles,* 93.

23. Quoted in Sir Hamilton A. R. Gibb, "The Rise of Saladin, 1169–1189," in *A History of the Crusades,* ed. Kenneth M. Setton, 1:567.

24. Quoted in Marshall W. Baldwin, "The Decline and Fall of Jerusalem, 1174–1189," in *A History of the Crusades,* ed. Kenneth M. Setton, 1:612.

25. Quoted in Hillenbrand, *Crusades,* 180.

26. Geoffrey Hindley, *Saladin: A Biography* (London: Constable, 1976), 49.

27. Francesco Gabrieli, ed., *Storici arabi delle crociate* (Turin: Einaudi, 1957), 86–87.

28. Quoted in Karsh, *Islamic Imperialism,* 83.

29. For a more critical view of Saladin, see M. C. Lyons and D.E.P. Jackson, *Saladin: The Politics of Holy War* (Cambridge: Cambridge University Press, 1982).

30. Voltaire, *Essai sur les moeurs,* ed. R. Pomeau, 2 vols. (Paris: Bordas, 1990) 1:581. See p. 344–45.

31. Gibbon, *Decline and Fall of the Roman Empire,* LIX.

32. *The Talisman,* Cap. VI.

33. Hillenbrand, *Crusades,* 593.

34. Elizabeth Siberry, *The New Crusaders: Images of the Crusades in the Nineteenth and Early Twentieth Centuries* (Aldershot, U.K.: Ashgate, 2000), 67–68.

35. Described in Hillenbrand, *Crusades,* 594–601.

36. Gibbon, *Decline and Fall of the Roman Empire,* LX.

37. Chabrol, *Essai sur les moeurs,* vol. 1, 585.

38. Lord Macauley, *History of England,* Bk. I, Cap. XIV.

39. Gibbon, *Decline and Fall of the Roman Empire,* LXI.

40. Johann Gottfried von Herder, "This Too: A Philosophy of History for the Formation of Humanity" (1774), in *Herder: Philosophical Writings,* ed. Michael N. Forster (Cambridge: Cambridge University Press, 2002), 306.

41. François-René, vicomte de Chateaubriand, *Itinéraire de Paris à Jérusalem,* ed. Jean-Claude Berchet (Paris: Gallimard, 2003), 445–46.

42. Siberry, *The New Crusaders,* 67.

43. See 367–68.

44. Quoted in Elizabeth Siberry, "Images of the Crusades in the Nineteenth and Twentieth Centuries," in *The Oxford Illustrated History of the Crusades,* ed. Jonathan Riley-Smith (Oxford: Oxford University Press, 1997), pp. 565–85.

45. Sayyid Qutb, "Social Justice in Islam (Al-'adalat al-ijtima'iyya fi'l-Islam)," in William E. Shepard, *Sayyid Qutb and Islamic Activism: A Translation and Critical Analysis of "Social Justice in Islam"* (Leiden: E. J. Brill, 1996), 286–87. For a brilliantly vivid picure of Qutb, see Lawrence Wright, *The Looming Tower: Al-Qaeda and the Road to 9/11* (New York: Alfred A. Knopf, 2006), 7–31.

46. Todd S. Purdum, "Bush Warns of a Wrathful, Shadowy and Inventive War," *The New York Times,* September 17, 2001, A2.

47. Gilles Kepel, *The War for Muslim Minds: Islam and the West* (Cambridge, Mass.: Belknap Press, 2004), 117.

48. Ofra Bengio, *Saddam's Word: Political Discourse in Iraq* (New York: Oxford University Press, 1998), 82–84.

49. From an anonymous pamphlet entitled "Nationalist Documents to Confront the Crusader Attack on the Arab Homeland," quoted in Hillenbrand, *Crusades,* 609–10.

## Chapter 7

1. For the early history of the Ottomans, see Caroline Finkel, *Osman's Dream: The Story of the Ottoman Empire, 1300–1923* (New York: Basic Books, 2006), 1–47; Heath W. Lowry, *The Nature of the Early Ottoman State* (Albany: State University of New York Press, 2003); Paul Wittek, *The Rise of the Ottoman Empire* (New York: B. Franklin, 1971); Halil Inalcik, "The Question of the Emergence of the Ottoman State," *International Journal of Turkish Studies* 2 (1980): 71–79.

2. Muhammad Ibn Battûta, *The Travels of Ibn Battûta,* trans H.A.R. Gibb (Cambridge: Cambridge University Press, 1962), 2:453, and quoted in Finkel, *Osman's Dream,* 13–24.

3. Quoted in Steven Runciman, *The Fall of Constantinople, 1453* (Cambridge: Cambridge University Press, 1965), 21. This is still the most evocative and compelling account of the fall of the city.

4. Michael A. Sells, *The Bridge Betrayed: Religion and Genocide in Bosnia* (Berkeley: University of California Press, 1996), 38–45.

5. Quoted in Runciman, *Fall of Constantinople,* 1.

6. Ibid., 10.

7. Kristovoulos, *History of Mehmed the Conqueror,* trans. Charles T. Riggs (Princeton, N.J.: Princeton University Press, 1954), 29.

8. Ibid., 58–59.

9. Heers, *Chute et mort de Constantinople,* 247.

10. Kristovoulos, *History of Mehmed the Conqueror,* 60–61.

11. Michele Ducas, "Historia turco-byzantina," in Agostino Pertusi, *La caduta di Constantinopoli. L'eco nel mondo* (Milan: Mondadori, 1976), 167.

12. Quoted in Finkel, *Osman's Dream,* 52.

13. Kristovoulos, *History of Mehmed the Conqueror,* 72–73.

14. Quoted in Gustave Schlumberger, *Le Siège, la prise et le sac de Constantinople en 1453* (Paris: Plon, 1935), 330.

15. Kristovoulos, *History of Mehmed the Conqueror,* 76. He also says that there were 50,000 prisoners, which is certainly an exaggeration.

16. Isidore of Kiev to Cardinal Bessarion, Candia, July 6, 1453, in Agostino Pertusi, *La caduta di Constantinopoli,* vol. 1: *Le testimonianze dei contemporanei* (Milan: Mondadori, 1976), 76.

17. Quoted in Runciman, *The Fall of Constantinople,* 149. Afrasiab was the legendary hero-king of Turan.

18. Quoted in James Hankins, "Renaissance Crusaders: Humanist Crusade Literature in the Age of Mehmed II," *Dumbarton Oaks Papers* 49 (1995): 111–207, 122.

19. Agostino Pertusi, *Testi inediti e poco noti sulla caduta di Constantinopoli* (Bologna: Patron, 1983), 74, 76.

20. During the reigns of Abdülhamid I (1774–1789) and his successor, Selim III (1789–1807), and again during the reign of Mahmud II (1808–1839) the name "Constantinople" was replaced on the coinage by "Islambol." Finkel, *Osman's Dream,* 383.

21. Ibid., 53.

22. This is probably a fantasy. But Mehmed certainly had Greek books in his library and knew some Greek. See J. Raby, "Mehmed the Conqueror's Greek Scriptorium," *Dumbarton Oaks Papers* 37 (1983): 15–34.

23. Kristovoulos, *History of Mehmed the Conqueror,* 181–82.

24. Hankins, "Renaissance Crusaders," 139.

25. Runciman, *The Great Church in Captivity,* 182–85.

26. Quoted in Heers, *Chute et mort de Constantinople,* 263.

27. The affair is described in Runciman, *Fall of Constantinople,* 166–67.

28. Ibid., 166.

29. Aeneas Silvius Piccolomini (Pope Pius II), *The Memories of a Renaissance Pope: The Commentaries of Pius II,* trans. F. A. Gragg (New York: Capricorn Books, 1962), 237.

30. Aeneas Silvius Piccolomini (Pope Pius II), *Lettera a Maometto II* (*Epistola ad Mahumetem*), ed. Giuseppe Tofanin (Naples: R. Pironti, 1953).

31. Finkel, *Osman's Dream,* 72–73.

32. For this and other possible interpretations of the picture, see M.-P. Oedani-Fabris, "Simbologia ottomana nell'opera di Gentile Bellini," *Atti dell'Istituto veneto di scienze, lettere ed arti,* 155 (1996–97): 1–29.

33. Isidore of Kiev to the "Faithful of Christ," Candia, July 8, 1453, in Agostino Pertusi, *La caduta di Constantinopoli,* 84, 82.

34. Anonymous, *The Policy of the Turkish Empire* (London, 1597), f. A3v.

35. Quoted in Bernard Lewis, *Islam and the West* (Oxford: Oxford University Press, 1993), 72.

36. Norman Housley, *The Later Crusades, 1272–1580* (Oxford: Oxford University Press, 1991).

37. Paolo Giovio, *Commentario delle cose dei Turchi* (Venice, 1538), f. diiir.

38. Cornell H. Fleischer, "The Lawgiver as Messiah: The Making of the Imperial Image in the Reign of Süleymân," in *Soliman le Magnifique et son temps,* ed. Gilles Veinstein (Paris: École du Louvre, 1992), 159–78. The term "caliph" had been applied by others to Mehmed I as early as 1421. But Süleymân was the first to be described as caliph in an official document, in this case the "Law Book of Buda" from the 1540s. See Colin Imber, "Süleymân as Caliph of the Muslims: Ebû's-Su'ûd's Formulation of Ottoman Dynastic Ideology," in *Soliman le Magnifique et son temps,* ed. Gilles Veinstein, 176–84.

39. Quoted in Finkel, *Osman's Dream,* 115.

40. Ogier Ghiselin de Busbecq, *The Turkish Letters of Ogier Ghiselin de Busbecq,* trans. Edward Forster (Oxford: Clarendon Press, 1927), 112.

41. For the military and strategic details of the battle, see Hanson, *Carnage and Culture,* 233–39.

42. François de la Noue, *Discours politiques et militaires,* ed. F. E. Sutcliffe (Geneva: Droz, 1967), 439.

43. The march on Vienna has been vividly described in John Stoye, *The Siege of Vienna* (New York: Holt, Rinehart and Winston, 1964), 15–23.

44. Ibid., 52.

45. Quoted in Bernard Lewis, *What Went Wrong? The Clash Between Islam and Modernity in the Middle East* (London: Weidenfeld and Nicolson, 2002), 16.

46. Quoted in Runciman, *Fall of Constantinople,* 178.

47. Mark Mazower, *The Balkans: A Short History* (New York: Modern Library, 2002), 69.

48. Montesquieu, *L'Esprit des lois,* III, 14.

49. Lucien Febvre, *L'Europe: Genèse d'une civilisation* (Paris: Perrin, 1999), 176.

## Chapter 8

1. According to the historian and propagandist Florimond de Raemond. Quoted in Donald Kelley, *The Beginning of Ideology: Consciousness and So-*

*ciety in the French Reformation* (Cambridge: Cambridge University Press, 1981), 28.

2. Quoted in Euan Cameron, *The European Reformation* (Oxford: Clarendon Press, 1991), 1.

3. John Locke, "First Tract on Government," in *Political Essays,* ed. Mark Goldie, (Cambridge: Cambridge University Press, 1997), 48–49.

4. Quoted in Strobe Talbott, *A Gathering of Tribes: The Story of a Big Idea* (New York: Simon & Schuster, 2008).

5. Quoted in Theodore K. Rabb, *The Struggle for Stability in Early-Modern Europe* (New York: Oxford University Press, 1975), 81.

6. John Stuart Mill, "On Liberty" in *On Liberty and Other Writings,* ed. Stefan Collini (Cambridge: Cambridge University Press, 1989), 11.

7. Thomas Hobbes, *Leviathan,* ed. Richard Tuck (Cambridge: Cambridge University Press, 1991) I, 3, 24.

8. Quoted in Anthony Pagden, *The Fall of Natural Man: The American Indian and the Origins of Comparative Ethnology* (Cambridge: Cambridge University Press, 1982), 67.

9. Thomas Hobbes, *The Elements of Law, Natural and Political,* 2nd ed., ed. Ferdinand Tönnies (London: Frank Cass & Co., 1969), 188–89, 2. 10. 8.

10. John Donne, *An Anatomie of the World: The First Anniversary,* lines 205 8, 213–18.

11. Descartes, *Second Meditation,* 7:25.

12. John Locke, *An Essay Concerning Human Understanding,* I, III, 2.

13. Ibid., II, XXVIII.

14. Letter to Michael Ainsworth in 1709, in *Life, Unpublished Letters, and Philosophical Regimen of Anthony, Earl of Shaftesbury,* ed. Benjamin Rand (London: S. Sonnenschein & Co., 1900), 403–5.

15. Thomas Hobbes, *On the Citizen (De Cive),* trans. and ed. Richard Tuck and Michael Silverthorne (Cambridge: Cambridge University Press, 1998), 27.

16. See Richard Tuck, "The 'Modern' Theory of Natural Law," in *The Languages of Political Theory in Early-Modern Europe,* ed. Anthony Pagden (Cambridge: Cambridge University Press, 1987), 99–119.

17. James Boswell, *Journal of a Tour to the Hebrides with Samuel Johnson, 1733,* ed. Frederick A. Pootle and Charles H. Bennet (New Haven, Conn.: Yale University Press, 1961), 189.

18. Paul Henri Dietrich, baron d'Holbach, "Système de la Nature," in *Oeuvres philosophiques complètes,* ed. Jean-Pierre Jackson (2 vols.) (Paris: Editions Alive, 1999), 2:165.

19. Immanuel Kant, "An Answer to the Question: What Is Enlightenment?" in *Political Writings*, ed. Hans Reiss, trans. H. B. Nisbet (Cambridge: Cambridge University Press, 1991), 56.

20. Kant, *Critique of Pure Reason*, A. xii.

21. Kant, "An Answer to the Question: What Is Enlightenment?" 57.

22. Marie-Jean-Antoine-Nicolas Caritat, Marquis de Condorcet, *Esquisse d'un tableau historique des progrès de l'esprit humain*, ed. Alain Pons (Paris: Flammarion, 1988), 74.

23. Ibid., 208, 266.

24. Samuel Johnson, *The Prince of Abyssinia: A Tale* (London, 1759), 47, 116; and see Jack Goody, *The East in the West* (Cambridge: Cambridge University Press, 1992), 2–4.

25. See Ian Buruma and Avishi Margalit's brief and pointed account of these anti-Western stereotypes in *Occidentalism: The West in the Eyes of Its Enemies* (New York: Penguin Press, 2004).

26. Friedrich Nietzsche, *Daybreak: Thoughts on the Prejudices of Morality*, trans. R. J. Hollingdale (Cambridge: Cambridge University Press, 1993), 118.

## Chapter 9

1. Subrahmanyam, "Taking Stock of the Franks: South Asian Views of Europeans and Europe, 1500–1800," 88.

2. Jonathan Spence, *The Question of Hu* (New York: Vintage Books, 1989).

3. Charles-Louis de Secondat, baron de Montesquieu, "Some Reflections on the Persian Letters," in *Persian Letters*, trans. C. J. Betts (New York: Viking Penguin, 1973), 283.

4. Ibid., Letter 30, 83.

5. Ibid., Letter 59, 124.

6. Montesquieu, *L'Esprit des lois*, XX, 1.

7. Sir William Jones, *A Grammar of the Persian Language*, in *The Collected Works of Sir William Jones*, [1807] facsimile edition, 13 vols. (New York: New York University Press, 1993), 5:165.

8. Edward Said, *Orientalism* (New York: Vintage Books, 1979), 7.

9. For an account of just how crude, see Bernard Lewis, "The Question of Orientalism," in *Islam and the West* (Oxford: Oxford University Press, 1993), 99–118.

10. Jones, "The Third Anniversary Discourse, delivered 2 February, 1786" [to the Asiatic Society of Calcutta] in *Collected Works*.

11. Kapil Raj, *Relocating Modern Science: Circulation and the Construction of*

*Scientific Knowledge in South Asia and Europe, Seventeenth to Nineteenth Centuries* (Delhi: Permanent Black, 2006), 95–138.

12. Sir William Jones, *Dissertation sur la littérature orientale* (London, 1771), 10–11.

13. Jones, "The Fourth Anniversary Discourse, delivered 15 February, 1787" [to the Asiatic Society of Calcutta] in *Collected Works,* 3:50.

14. Jones, *Letters,* in *Collected Works,* 2:652.

15. Jones, *A Grammar of the Persian Language,* 167.

16. Jones, *Dissertation sur la littérature orientale,* 50.

17. Jones, "The Best Practicable System of Judicature for India," in *Collected Works,* 1:133.

18. Quoted in Bernard Cohn, "The Command of Language and the Language of Command," in *Subaltern Studies,* ed. Ranajit Guha (Delhi, 1985), 4:295.

19. See S. N. Muherjee, *Sir William Jones: A Study in Eighteenth-Century British Attitudes to India* (Cambridge: Cambridge University Press, 1968).

20. James Boswell, *The Life of Samuel Johnson* (Oxford: Oxford University Press, 1983), 159.

21. Jones, "The Fourth Anniversary Discourse," 36.

22. Jones, *Dissertation sur la littérature orientale,* 52.

23. Max Müller, *Lectures on the Science of Language* (London: Longman, 1864), 219–20.

24. Max Müller, *Theosophy or Psychological Religion,* in *Collected Works of the Right Hon. F. Max Müller,* 18 vols. (London: Longman, 1898), 4:73.

25. Arnaldo Momigliano, "Preludio settecentesco a Gibbon" (1977), in *Fondamenti della storia antica* (Turin: Einaudi, 1984), 312–27.

26. Quoted in Girolamo Imbruglia, "Tra Anquetil-Duperron e *L'Histoire de deux indies:* libertà, dispotismo e feudalismo," *Rivista Storica Italiana* 106 (1994): 141.

27. Abraham-Hyacinthe Anquetil-Duperron, "Discours préliminaire ou introduction au Zend-Avesta," in *Voyage en Inde, 1754–1762,* ed. Jean Deloche, Manonmani Filliozat, and Pierre-Sylvain Filliozat (Paris: École Française d'Extrême-Orient, 1997), 64.

28. Abraham-Hyacinthe Anquetil-Duperron, *Considérations philosophiques et géographiques sur les deux mondes (1780–1804),* ed. Guido Abattista (Pisa: Scuola Normale Superiore, 1993).

29. Abraham-Hyacinthe Anquetil-Duperron, *Législation orientale* (Amsterdam, 1778), 181.

30. The only biography of Anquetil-Duperron is Raymond Schwab, *Vie*

*d'Anquetil-Duperron* (Paris: Librairie Ernest Leroux, 1934), but it adds relatively little to Anquetil-Duperron's own account of his life between 1754 and 1762 in *Discours préliminaire ou introduction au Zend-Avesta*.

31. Anquetil-Duperron, *Discours préliminaire ou introduction au Zend-Avesta*, 74–75.

32. For more on Zoroaster and the religion he created, see pp. 153–55.

33. Anquetil-Duperron, *Discours préliminaire ou introduction au Zend-Avesta*, 95.

34. Ibid., 255.

35. Ibid., 342.

36. Quoted in Schwab, *Vie d'Anquetil-Duperron*, 85.

37. Anquetil-Duperron, *Discours préliminaire ou introduction au Zend-Avesta*, 449.

38. Ibid., 462.

39. Quoted in Schwab, *Vie d'Anquetil-Duperron*, 98–99.

40. Jones, "Letter to the University of Oxford," in *Collected Works*, 1:367; Jones, "Lettre à Monsieur A*** Du P*** dans laquelle est compris l'examen de sa traduction des livres attribués à Zoroastre" [November 23, 1771], in *Collected Works*, 10:410–13.

41. Jones, "Lettre à Monsieur A*** Du P***," in *Collected Works*, 10:417.

42. Anquetil-Duperron, *Discours préliminaire ou introduction au Zend-Avesta*, 74.

43. Jones, "Lettre à Monsieur A*** Du P***," in *Collected Works*, 10:408–409.

44. Ibid., 438.

45. John Richardson, *A Dissertation on the Languages, Literature and Manners of the East* (Oxford, 1777), 126.

46. Quoted in Schwab, *Vie d'Anquetil-Duperron*, 99.

47. See Garland Cannon, *The Life and Mind of Oriental Jones: Sir William Jones, the Father of Modern Linguistics* (Cambridge: Cambridge University Press, 1990), 44.

48. Max Müller, *The Sacred Books of the East* [1887] (Delhi: Motilal Banarsidass, 1992), 4:1, 16.

49. Voltaire, "Le Pyrrhonisme de l'histoire," in *Oeuvres completes de Voltaire*, 52 vols. (Paris: Garnier Freres, 1877–85), 27:237.

50. Voltaire, *Essai sur les moeurs,* ed. René Pomeau (Paris: Classiques Garnier, 1990), 1:268.

51. Montesquieu, *L'Esprit des lois*, XVII, 3-2.

52. David Hume, "Of National Characters," in *Essays: Moral, Political and Literary,* ed. Eugene F. Miller (Indianapolis: Liberty Classics, 1985), 204.

53. On the history of the term, see the classic articles by Franco Venturi, "Oriental Despotism," *Journal of the History of Ideas* 24 (1963): 133–42, and Richard Kroebner, "Despot and Despotism: Vicissitudes of a Political Term," *Journal of the Warburg and Courtauld Institutes* 14 (1951): 275–302, and, more recently, Joan-Pau Rubiés, "Oriental Despotism and European Orientalism: Botero to Montesquieu," *Journal of Early-Modern History* 9, no. 2. (2005): 109–80.

54. See Jürgen Osterhammel, *Die Entzauberung asiens: Europa und die asiatischen Reiche im 18. Jahrhundert* (Munich: C. H. Beck, 1988), 284–306.

55. François Bernier, *Événemens particuliers, ou ce qui s'est passé de plus considerable après la guerre pendant cinq ans . . . dans les états du Grand Mongol* (Paris, 1670), 1:256–57.

56. Montesquieu, *L'Esprit des lois,* I, 14.

57. Montesquieu, *Persian Letters,* 234, letter 131. Montesquieu's celebrated distinction between the driving forces (*principes*) behind the three kinds of government—monarchy, republicanism, and despotism—is in *The Spirit of the Laws,* III.

58. Montesquieu, *L'esprit des lois,* I, 3.

59. *Le fanatisme, ou Mahomet le prophète, tragédie en cinq actes* (first performed in Lille in April 1741), Act I, scene 5.

60. Condorcet, *Esquisse d'un tableau historique des progrès de l'esprit humain,* 172.

61. Ibn Khaldûn, *The Muqaddimah,* 120–21.

62. Voltaire, *Essai sur les moeurs,* 1:821–22.

63. Anquetil-Duperron, *Législation orientale,* 32.

64. Voltaire, *Essai sur les moeurs,* 1:832.

65. Ibid., 1:835.

66. Ibid., 2:415–16.

67. Ibid., 2:773.

68. Montesquieu, *L'Esprit des lois,* X, 14.

69. Voltaire, *Essai sur les moeurs,* 2:767.

70. Ibid., 1:231.

71. Ibid., 2:772.

72. Ibid., 1:271.

73. Johann Gottfried von Herder, "Reflections on the Philosophy of History," in *On World History: An Anthology,* ed. Hans Adler and Ernst A. Menze (Armonk, N.Y.: M. E. Sharpe, 1997), 247.

74. W. Schluchter, *The Rise of Western Rationalism: Max Weber's Developmental History* (Berkeley: University of California Press, 1981), 61–67.

75. Henry Sumner Maine, *Village Communities in the East and the West* (London: John Murray, 1881), 7.

76. Hegel, *Philosophy of History*, 99.

77. Ibid., 142–43.

78. Constantin-François Volney, *Voyage en Syrie et en Égypte* [1787–1799], in *Oeuvres*, ed. Anne Deneys-Tunney and Henry Deneys (Paris: Fayard, 1998), 3:15–16.

79. Ibid., 3:161–62.

80. Ibid., 3:194.

81. Volney's account of Palmyra is in *Voyage en Syrie et en Égypte*, 474–80. His reflections, however, occur in *Les ruines ou méditation sur les révolutions des empires* in *Oeuvres*, 1:171–173, 232–234. This work, translated into English as *The Ruins: or a Survey of the Revolutions of Empires*, argued that empires—by which Volney intended all civilizations—flourished only when what today would be called "liberal democracy" prospered and religions of all kinds were banished from the public sphere. It became immensely popular throughout Europe and the new United States, where it might possibly have been translated into English by Thomas Jefferson.

82. Volney, *Les ruines ou méditation sur les révolutions des empires*, 1:245–46.

83. *Le Moniteur*, 5 brumaire an VIII.

## Chapter 10

1. Volney, *Les ruines ou méditation sur les révolutions des empires*, 1:245–56.

2. Bernard Lewis, *The Emergence of Modern Turkey* (London: Oxford University Press, 1961), 317.

3. Jacob M. Landau, *The Politics of Pan-Islam: Ideology and Organization* (Oxford: Clarendon Press, 1990), 10–11.

4. Volney, "Considerations sur la guerre des Turks en 1788," in *Voyage en Syrie et en Égypte*, 3:641–43. See Lewis, *What Went Wrong?*, 21.

5. Text in Frédéric Masson and Guido Biagi, *Napoléon: manuscrits inédits, 1789–1791*, 8 vols. (Paris: P. Ollendorf, 1907), 3:17–19.

6. Quoted in Henry Laurens, *Les origines intellectuelles de l'expédition d'Égypte: l'orientalisme islamisant en France (1698–1798)* (Istanbul: Éditions Isis, 1987), 190–92.

7. Quoted in Yves Laissus, *L'Égypte, une aventure savante, 1798–1801* (Paris: Fayard, 1998), 18.

8. Quoted in Georges Lacour-Gayet, *Talleyrand* (Paris: Éditions Payot, 1990), 321.

9. Charles-Maurice-Camille, prince de Talleyrand-Périgord, *Mémoires du Prince de Talleyrand,* 5 vols. (Paris: C. Lévy, 1891–92), 1:77–78.

10. Quoted in Lacour-Gayet, *Talleyrand,* 321.

11. Quoted in Edward Ingram, *Commitment to Empire: Prophecies of the Great Game in Asia, 1797–1800* (Oxford: Oxford University Press, 1981), 52.

12. Gilbert-Joseph Volvic de Chabrol, *Description de l'Égypte, ou recueil des observations et des recherches qui ont été faites en Égypte pendant l'expédition de l'armée française, publié par les ordres de Sa Majesté l'empereur Napoléon le Grand,* 2nd ed. (24 vols.) (Paris: C. L.-F. Pankoucke, 1821–30), 1:cxlii–cxliii.

13. *Réimpression de l'ancien Moniteur depuis la réunion des Étas-Généraux jusq'au Consulat* (May 1789–November 1799), (Paris, 1893), 29:501.

14. Claire-Élisabeth-Jeanne Gravier de Vergennes, Comtesse de Rémusant, *Mémoires de Madame de Rémusant,* 3 vols. (Paris: Calmann Lévy, 1880), 1:274.

15. A complete list is provided in Laissus, *L'Égypte, une aventure savante,* 524–25.

16. Volney, *Voyage en Syrie et en Égypte,* 11–12.

17. Quoted in Niqula ibn Yusuf al-Turk, *Histoire de l'expédition des Français en Égypte par Nakoula-El-Turk, publiée et traduite par Desgranges aîné* (Paris: Imprimerie Royale, 1839), 19.

18. Quoted in Laissus, *L'Égypte, une aventure savante,* 75–76.

19. Ibid., 76.

20. Henry Laurens, *L'Expedition d'Égypte, 1798–1801* (Paris: Éditions de Seuil, 1997), 131.

21. This is taken from the Arabic version in Abd al-Rahman al-Jabarti, *Al-Jabarti's Chronicle of the First Seven Months of the French Occupation of Egypt,* ed. and trans. S. Morhe (Leiden: Brill, 1975), 41.

22. The word used by Napoleon was *huqûq,* which is closer to "claim" than "right" and is generally applied to God, as in "God's right" *huqûq allâh,* although it can also designate the claims an individual has on the community.

23. al-Jabarti, *Chronicle,* 41. Victor Hugo, *Orientales,* XL.

24. Sir Walter Scott, *The Life of Napoleon Buonaparte,* 9 vols. (Edinburgh: Cadell and Co., 1827), 4:83.

25. Quoted in Laurens, *L'Expédition d'Égypte,* 158.

26. Jean-Jacques Rousseau, *Du Contrat social,* in *Oeuvres complètes,* ed.

Bernard Gagnebin and Marcel Raymond (Paris: Bibliothèque de la Pléiade, 1964), 3:462–463.

27. Quoted in C. A. Bayly, *The Birth of the Modern World, 1780–1914* (Malden, U.K.: Blackwell, 2004), 108.

28. Al-Jabarti, *Chronicle,* 20–21.

29. Quoted in Lewis, *The Muslim Discovery of Europe,* 183.

30. Laurens, *Origines intellectuelles de l'expédition d'Égypte,* 184.

31. Elie Kedourie, *Afghani and 'Abduh: An Essay on Religious Unbelief and Political Activism in Modern Islam* (London: Frank Cass, 1966).

32. Dominique-Vivant Denon, *Voyage dans la Basse et la Haute Égypte, pendant les campagnes du général Bonaparte* (Paris: P. Dinot l'Ainé, 1802), 174.

33. Quoted in Hillenbrand, *The Crusade,* 226.

34. François Bernoyer, *Avec Bonaparte en Égypte et en Syrie, 1798–1800,* ed. Christian Tortel (Paris: Éditions Curandera, 1981), 46–47.

35. Quoted in Clément de la Jonquière, *L'Expedition d'Égypte, 1798–1801,* 5 vols. (Paris: H. Charles Lavauzelle, 1899–1907), 1:462.

36. Michael Haag, *Alexandria: City of Memory* (New Haven, Conn.: Yale University Press, 2005).

37. Quoted in Laurens, *Origines intellectuelles de l'expédition d'Égypte,* 96.

38. Chateaubriand, *Itinéraire de Paris à Jérusalem,* 459–60.

39. David Cannadine, *Ornamentalism: How the British Saw Their Empire* (London: Penguin Press, 2002), 78–79.

40. Chabrol, *Essai sur les moeurs des habitants modernes de l'Égypte,* in *Description de l'Égypte,* 18:26.

41. Bernoyer, *Avec Bonaparte en Égypte et en Syrie,* 48.

42. Quoted in Laurens, *L'expédition d'Égypte,* 172.

43. Ibid., 124.

44. Ibid., 128.

45. Ibid., 163.

46. *Réimpression de l'ancien Moniteur,* 29:654–55.

47. Quoted in Laissus, *L'Égypte, une aventure savante,* 129.

48. Marie-Joseph Las Cases, *Mémorial de Sainte-Hélène,* ed. Gérard Walter, 2 vols. (Paris: Bibliothèque de la Pléiade, 1956), 1:504.

49. Quoted in Maya Jasanoff, *Edge of Empire: Conquest and Collecting in the East, 1750–1850* (London: Fourth Estate, 2005), 201.

50. Ali Bahgat, "Acte de marriage du General Abdallah Menou avec la dame Zobaidah," *Bulletin de l'Institut Égyptien,* 9 (1899): 221–35.

51. Laissus, *L'Égypte, une aventure savante,* 350–51.

52. Alexis de Tocqueville, "Travail sur l'Algérie," in *Tocqueville sur l'Algérie* [1847], ed. Seloua Luste Boulbina (Paris: Flammarion, 2003), 112.

53. Laissus, *L'Égypte, une aventure savante,* 106–11.

54. Quoted in Laurens, *L'Expédition d'Égypte,* 160.

55. Ibid., 74.

56. Bernoyer, *Avec Bonaparte en Égypte et en Syrie,* 131.

57. Al-Turk, *Histoire de l'expédition des Français en Égypte par Nakoula-El-Turk,* 52.

58. Al-Jabarti, *Chronicle of the First Seven Months,* 112, and Laurens, *L'Expédition d'Égypte,* 235.

59. Bernoyer, *Avec Bonaparte en Égypte et en Syrie,* 71.

60. Laurens, *L'Expedition d'Égypte,* 40.

61. Denon, *Voyage dans la Basse et la Haute Égypte,* 39.

62. Laurens, *L'Éxpedition d'Égypte,* 165.

63. Ibid., 196.

64. *Réimpression de l'ancien Moniteur,* 29:492–93, 497–98.

65. Volney, *Voyage en Syrie et en Égypte,* 3:109.

66. See pp. 330–31.

67. Quoted in Laissus, *L'Égypte, une aventure savante,* 195.

68. Denon, *Voyage dans la Basse et la Haute Égypte,* 64–65.

69. Ibid., 64.

70. Quoted in Laissus, *L'Égypte, une aventure savante,* 82.

71. Chabrol, *Essai sur les moeurs des habitans modernes de l'Égypte,* 31.

72. Bernoyer, *Avec Bonaparte en Égypte et en Syrie,* 86.

73. Denon, *Voyage dans la Basse et la Haute Égypte,* 168.

74. Chabrol, *Essai sur les moeurs des habitans modernes de l'Égypte,* 68.

75. Ibid., 31–34.

76. Mill, "On Liberty," 70–71.

77. Chabrol, *Essai sur les moeurs des habitans modernes de l'Égypte,* 32.

78. Edward Lane, *The Manners and Customs of the Modern Egyptians* (New York: Dutton, 1966), 291.

79. Chabrol, *Essai sur les moeurs des habitans modernes de l'Égypte,* 213–14.

80. Bernoyer, *Avec Bonaparte en Égypte et en Syrie,* 95.

81. Benjamin Frossard, *Observations sur l'abolition de la traite des nègres présentées a la Convention Nationale* (N.P., 1793), 125. Slavery, however, never troubled Napoleon, provided it was safely beyond the frontiers of Europe. In 1802, at the prompting of his wife, Josephine, the daughter of a sugar planter from Martinique, he attempted to reestablish slavery on the French island of Saint Dominique, with disastrous and bloody consequences.

82. Bernoyer, *Avec Bonaparte en Égypte et en Syrie,* 99.

83. Al-Jabarti, *Chronicle,* 43–47.

84. Mill, "The Subjection of Women," in *On Liberty and Other Writings,* 119.

85. Chabrol, *Essai sur les moeurs des habitants modernes de l'Égypte,* 95–96.

86. Volney, *Voyage en Syrie et en Égypte,* 16.

87. Cf. Lane: "It has been asserted by many Christians that the Muslims believe women to have no souls" and thus are unfit for the afterlife. *The Manners and Customs of the Modern Egyptians,* 67–68. In fact, the Qur'an promises Paradise to all believers regardless of gender.

88. Chabrol, *Essai sur les moeurs des habitants modernes de l'Égypte,* 90, 95–96, 117.

89. "Revolutionary Proclamation for Law and Fatherland," in *The Movement for Greek Independence, 1770–1821: A Collection of Documents,* ed. Richard Clogg (London: Macmillan, 1976), 149.

90. Quoted in Henry Laurens, "Le mythe de l'expédition d'Égypte en France et en Égypte aux XIXe et XXe siècles," in *L'Égyptologie et les Champollion,* ed. Michel Dewachter and Alain Fouchard (Grenoble: Presses Universitaires de Grenoble, 1994), 321–30.

91. Albert Hourani, *Arabic Thought in the Liberal Age, 1798–1939* (Cambridge: Cambridge University Press, 1962), 67–102; Laurens, *L'Expédition d'Égypte,* 471.

92. *Congrès National des Forces Populaires: La Charte* (Cairo: Administration de l'Information, 1962), 24. Quoted in Henry Laurens, "Bonaparte a-t-il colonisé l'Égypte?," *L'Histoire* 216 (1997): 46–49.

93. Fouad Ajami, *The Arab Predicament: Arab Political Thought and Practice Since 1967* (Cambridge: Cambridge University Press, 1992), 92–94.

94. *Réimpression de l'ancien Moniteur,* 29:681. Another markedly different version, which appears nowhere in the *Moniteur,* is provided in Nahum Sokolow, *History of Zionism* (London: Longmans, 1919), 1:63–79, and 2:220–25, and repeated in Franz Kobler, *Napoleon and the Jews* (New York: Schocken Books, 1975), 72. This reads: *"Politics, Turkey, Constantinople,* 28 Germinal [April 17] Bonaparte caused to publish a proclamation in which he invites all the Jews of Asia and of Africa to come and range themselves under his banners in order to re-establish ancient Jerusalem. A great number have armed themselves already and their battalions threaten Aleppo."

95. Henry Laurens, who also cites the Sokolow/Kobler version of the notice in the *Moniteur,* believes that this might be Lucien Bonaparte. "Le projet de'état juif en Palestine, attribué à Bonaparte," *Orientales* (Paris: CNRS Éditions, 2004) 1:123–43.

96. Ibid.
97. Jacques Derogy and Hesi Carmel, *Bonaparte en Terre Sainte* (Paris: Fayard, 1992), 25. Laurens, "Le projet de'état juif en Palestine, attribué à Bonaparte."
98. Quoted in John Darwin, *Britain, Egypt and the Middle East: Imperial Policy in the Aftermath of War, 1918–1922* (London: Macmillan, 1981), 171.

## Chapter 11

1. Quoted in Bernard Lewis, "The 'Sick Man' of Today Coughs Closer to Home," in *From Babel to Dragomans: Interpreting the Middle East* (Oxford: Oxford University Press, 2004), 364.
2. See pp. 172–73.
3. Quoted in Karsh, *Islamic Imperialism*, 104–5.
4. Nur Bilge Criss, *Istanbul Under Allied Occupation, 1918–1923* (Leiden: Brill, 1999), 7.
5. Quoted in Mazower, *The Balkans*, 51.
6. Sir William Eton, "A Survey of the Turkish Empire," in *The Movement for Greek Independence*, ed. Clogg, 46–47.
7. "The Paternal Exhortation of Patriarch Anthimos of Jerusalem," in *The Movement for Greek Independence*, ed. Clogg, 59–60.
8. Sir William Eton, "Greece Under Ottoman Rule" [1791], in *The Movement for Greek Independence*, ed. Clogg, 3.
9. Ioannis Pringos, "The Journal of Ioannis Pringos of Amsterdam," in *The Movement for Greek Independence*, ed. Clogg, 42–43.
10. Quoted in David Brewer, *The Flame of Freedom: The Greek War of Independence, 1821–1833* (London: John Murray, 2001), 20.
11. "The Holy Synod Anathematises the *Philiki Etairia*," in *The Movement for Greek Independence*, ed. Clogg, 203.
12. Robert Walsh, "A Residence in Constantinople," in *The Movement for Greek Independence*, ed. Clogg, 207–8.
13. Thomas Smart Hughes, *Travels in Greece and Albania*, 2 vols. (London: H. Colburn and R. Bentley, 1830), 2:81, 97.
14. "Fight for Faith and Motherland," in *The Movement for Greek Independence*, ed. Clogg, 203, 201.
15. Reported in Thomas Gordon, *History of the Greek Revolution*, 2 vols. (London: T. Cadell, 1832), 1:183, and see *A Short History of Modern Greece*, Richard Clogg, (Cambridge, U.K.: Cambridge University Press, 1979), 47–49.

16. Quoted in William St. Clair, *That Greece Might Still Be Free: The Philhellenes in the War of Independence* (London: Oxford University Press, 1972), 59.

17. Ibid., 60.

18. William St. Clair offers a powerful account of the massacres in ibid., 1–2.

19. Thomas Smart Hughes, *An Address to the People of England in the Cause of the Greeks, Occasioned by the Late Inhuman Massacres on the Isle of Scio* (London: Simpkin and Marshall, 1822).

20. From the provisional constitution of 1821, quoted in Richard Clogg, *A Short History of Modern Greece,* 58.

21. Brewer, *Flame of Freedom,* 139.

22. St. Clair, *That Greece Might Still Be Free,* 177.

23. Quoted in Douglas Dakin, *The Greek Struggle for Independence, 1821–1833* (Berkeley: University of California Press, 1973), 107.

24. Quoted in Brewer, *Flame of Freedom,* 198.

25. St. Clair, *That Greece Might Still Be Free,* 174–75.

26. Quoted in C. M. Woodhouse, *The Philhellenes* (London: Hodder and Stoughton, 1969), 116.

27. Quoted in Dakin, *Greek Struggle for Independence,* 186.

28. Ibid., 202–3.

29. Quoted in Niall Ferguson, *Colossus: The Price of America's Empire* (London: Penguin Press, 2006), 221.

30. Sir Ronald Storrs, *The Memoirs of Sir Ronald Storrs* (New York: G. P. Putnam and Sons, 1937), 206.

31. Quoted in Peter Mansfield, *A History of the Middle East* (New York: Penguin Press, 2003), 99.

32. Quoted in Lewis, *Emergence of Modern Turkey.*

33. Ibid., 222.

34. In 1717, Lady Mary Wortley Montagu, whose husband was ambassador to the Sublime Porte, witnessed inoculation in Istanbul and had the embassy surgeon inoculate her five-year-old son, and later her four-year-old daughter. The practice seems, however, to have originated in China.

35. Lewis, *Emergence of Modern Turkey,* 53.

36. Ibid, 46.

37. Lewis, *Muslim Discovery of Europe,* 222–23. Despite the opposition of the Ulema, however, clocks and watches had been produced in limited numbers in the Galata district of Istanbul in the seventeenth century.

38. Quoted in Finkel, *Osman's Dream*, 376.

39. Lewis, *Emergence of Modern Turkey*, 57.

40. Virginia Aksan, "Ottoman Political Writing, 1768–1808," *Journal of Middle-Eastern Studies* 25 (1993): 53–69.

41. Lewis, *Emergence of Modern Turkey*, 71–72.

42. Paul Dumon, "La période des *Tanzimât* (1839–1878)," in *Histoire de l'empire ottoman*, ed. Robert Mantran (Paris: Fayard, 1989), 459–522.

43. Lewis, *Emergence of Modern Turkey*, 106.

44. Adolphus Slade, *Record of Travels in Turkey and Greece etc. and of a Cruise in the Black Sea with the Capitan Pasha, in the Years 1829, 1830, and 1831*, 2 vols. (Philadelphia: E. L. Carey, 1833), 1:275–56.

45. Ibid., 277.

46. Ibid., 271.

47. Elie Kedourie, *Arabic Political Memoirs and Other Studies* (London: Frank Cass, 1974), 2.

48. See Feroz Ahmed, *The Young Turks: The Committee of Union and Progress in Turkish Politics, 1908–1914* (Oxford: Clarendon Press, 1969).

49. Quoted in Mansfield, *History of the Middle East*, 126, and see Finkel, *Osman's Dream*, 510–18.

50. Kedourie, *Arabic Political Memoirs and Other Studies*, 260.

51. Mazower, *The Balkans*, 101.

52. Quoted in Karsh, *Islamic Imperialism*, 101–2. It is true, however, that once the sultan had declared war, hasty agreements were made between the Allies as to who was to get what piece of the Ottoman Empire once the war was over. These concluded with the Constantinople Agreement of March–April 1915 and the Treaty of London the following month.

53. Karsh, *Islamic Imperialism*, 103, quoting Arnold Toynbee, *Turkey: A Past and a Future* (New York: George H. Dorn, 1917), 28–29.

54. S. Tufan Buzpinar, "The Hijaz, Abdülhamid II and Amir Hussein's secret dealings with the British, 1877–80," *Middle Eastern Studies* 31 (1995): 99–123.

55. Elie Kedourie, *The Anglo-Arab Labyrinth: The McMahon-Husayn Correspondence and Its Interpretations, 1914–1939* (Cambridge: Cambridge University Press, 1976), 5.

56. Storrs, *Memoirs*, 168.

57. Ibid., 192.

58. Quoted in Efraim Karsh and Inari Karsh, *Empires of the Sand: The Struggle for Mastery in the Middle East, 1789–1923* (Cambridge, Mass.: Harvard University Press, 1999), 173.

59. Kedourie, *The Anglo-Arab Labyrinth*, 113–16.

60. Quoted in Karsh, *Islamic Imperialism*, 129.

61. Storrs, *Memoirs*, 168; David Fromkin, *A Peace to End All Peace: The Fall of the Ottoman Empire and the Creation of the Modern Middle East* (New York: Henry Holt, 1989), 221.

62. Quoted in Karsh, *Islamic Imperialism*, 180.

63. Ibid., 187–88.

64. Quoted in Karsh and Karsh, *Empires of the Sand*, 197.

65. October 7, 1916. Quoted in Fromkin, *A Peace to End All Peace*, 221.

66. Quoted from the *Arab Bulletin*, the official publication of the Arab Bureau, in Fromkin, *Peace to End All Peace*, 222.

67. Storrs, *Memoirs*, 202, 238.

68. Cannadine, *Ornamentalism*, 72.

69. Quoted in Elie Kedourie, *England and the Middle East: The Destruction of the Ottoman Empire, 1914–1921* (Hassock, Sussex: Harvester Press, 1978), 118.

70. Ibid., 101.

71. T. E. Lawrence, "Nationalism Amongst the Tribesmen," *Arab Bulletin* 26 (November 1916), in *Secret Despatches from Arabia by T. E. Lawrence* (Cambridge: The Golden Cockerel Press, n.d.), 38–39.

72. T. E. Lawrence, *The Seven Pillars of Wisdom: The Complete 1922 Text* (Fordingbridge, U.K.: Castle Hill Press, 1997), i.

73. "Personal Notes on the Sherifial Family," *Arab Bulletin* 26 (November 1916), quoted in Lawrence, *Secret Despatches from Arabia*, 35.

74. Lawrence, *Seven Pillars of Wisdom*, 1:239.

75. Fromkin, *Peace to End All Peace*, 312.

76. Quoted in Briton Cooper Busch, *Britain, India and the Arabs, 1914–1921* (Berkeley: University of California Press, 1971), 137, n. 57.

77. Ibid., 137–38.

78. Ibid., 139–40. It was Sir Arnold Wilson, the civil commissioner of Mesopotamia, who had dubbed Sykes an "ebullient Orientalist."

79. Vivian Gilbert, *The Romance of the Last Crusade: With Allenby to Jerusalem* (New York: D. Appleton and Co., 1925), 176–79.

80. The circumstances surrounding the seizure of Damascus, and of Lawrence's self-inflated role in it, are very murky. See Elie Kedourie, "The Capture of Damascus 1 October, 1918," in *The Chatham House Version and Other Middle Eastern Studies* (New York: Praeger, 1970), 48–51.

81. Kedourie, *England and the Middle East*, 97.

82. T. E. Lawrence, *The Letters of T. E. Lawrence,* ed. David Garnett (London and Toronto: Jonathan Cape, 1938), 291.

83. Fromkin, *Peace to End All Peace,* 343.

84. Woodrow Wilson, *The Public Papers of Woodrow Wilson,* ed. Ray Stannard Baker and William E. Dodd, 6 vols. (New York and London: Harper and Bros., 1925–27), 5:159–61.

85. Quoted from *England in Egypt,* in Kedourie, in *England and the Middle East,* 25–26.

86. Salvador de Madariaga, *Portrait of Europe* [*Bosquejo de Europa*] (New York: Roy Publishers, 1955), 23.

87. Quoted in Finkel, *Osman's Dream,* 488.

88. Leonard S. Woolf, *The Future of Constantinople* (London: George Allen and Unwin, 1917).

89. Darwin, *Britain, Egypt and the Middle East,* 171–72.

90. Criss, *Istanbul Under Allied Occupation,* 7–9.

91. Christopher M. Andrew and A. S. Knaya-Forstner, *The Climax of French Imperial Expansion, 1914–1924* (Stanford, Calif.: Stanford University Press, 1981), 180.

92. Quoted as the epithet to David Fromkin, *A Peace to End All Peace.* It of course provides the title of this superb book.

93. Andrew and Knaya-Forstner, *Climax of French Imperial Expansion,* 189.

94. Ibid., 203.

95. Quoted in Fromkin, *Peace to End All Peace,* 401.

96. Jean-Baptiste Tavernier, *Travels in India,* 2 vols., trans. V. Ball (London: Macmillan and Co., 1889) 1:381–84.

97. The phrase was in fact first used by Colonel Charles Stoddart, who died playing the game in 1842. See Peter Hopkirk, *The Great Game: The Struggle for Empire in Central Asia* (New York: Kodansha International, 1994).

98. See pp. 364–67.

99. George Nathaniel Curzon, *Persia and the Persian Question,* 2 vols. (London: Longmans, 1892), 1:480.

100. Nikki R. Keddie, *Religion and Rebellion in Iran: The Tobacco Protest of 1891–1892* (London: Cass, 1966).

101. Viscount Grey of Fallodon, *Twenty-Five Years, 1892–1916* (London: Hodder and Stoughton, 1925), 153.

102. Ibid., 165–166.

103. Quoted in Karsh, *Islamic Imperialism,* 125–26.

104. Quoted in Cannadine, *Ornamentalism*, 77.

105. Quoted in Fromkin, *Peace to End All Peace*, 452.

106. Mansfield, *History of the Middle East*, 228.

107. Quoted in Karsh and Karsh, *Empires of the Sand*, 288.

108. Benjamin Disraeli, *Tancred; or the New Crusade* (London: Longman Green, 1894), 309.

109. Theodor Herzl, *The Jew's State: An Attempt at a Modern Solution to the Issue of the Jews*, trans. Henk Overberg (Northvale, N.J.: Jason Aronson, 1991), 134, 145.

110. Ibid., 196.

111. Ibid., 148.

112. Quoted in Ronald Hyam, *Britain's Declining Empire: The Road to Decolonisation, 1918–1968* (Cambridge: Cambridge University Press, 2006), 51.

113. Quoted in Isaiah Friedman, *The Question of Palestine: British-Jewish-Arab Relations, 1914–1918* (New Brunswick, N.J.: Transaction Publishers, 1992), 285–86.

114. See p. 417.

115. Herzl, *The Jew's State*, 147.

116. Quoted in Hyam, *Britain's Declining Empire*, 55.

117. Quoted in Friedman, *Question of Palestine*, 126–27.

118. Hyam, in *Britain's Declining Empire*, 53–54, comments that "it would be hard to find a more shocking illustration of the extent and depth of the West's complacency in the early twentieth century about the supposedly inevitable decline of Islam, of the West's contempt for Muslim interests."

119. Walter Laquer, *The Israeli-Arab Reader* (London: Penguin Press, 1970), 37.

120. Michael Cohen, *Palestine and the Great Powers, 1945–1948* (Princeton, N.J.: Princeton University Press).

121. David Armitage, "Declaration of the Establishment of the State of Israel," in *The Declaration of Independence: A Global History* (Cambridge, Mass.: Harvard University Press, 2007), 240.

122. Quoted in Karsh, *Islamic Imperialism*, 190.

123. See pp. 257–58.

124. John Darwin, *After Tamerlane: The Global History of Empire* (London: Penguin Press, 2007), 457.

125. Gamal Abdel Nasser, *Egypt's Liberation: The Philosophy of Revolution* (Washington, D.C.: PublicAffairs Press, 1955), 108.

126. See, in general, Keith Kyle, *Suez* (London: Weidenfeld and Nicolson, 1991).

127. Quoted in Karsh, *Islamic Imperialism,* 145.

128. Quoted in C. Ernest Dawn, "The Origins of Arab Nationalism," in *The Origins of Arab Nationalism,* ed. Rashid Khalidi et al., (New York: Columbia University Press, 1991), 5.

129. Leonard Binder, *The Ideological Revolution in the Middle East* (New York: John Wiley and Sons, 1964), 154–97.

130. An independent report prepared by the Institute of International Affairs in London shortly before the creation of the United Arab Republic compared the "Arab speaking world" to the "German-speaking world . . . before Bismarck" and predicted that regional loyalties and rivalries would be "too strong for the development of a unitary Arab state." *British Interests in the Mediterranean and Middle East: A Report by a Chatham House Study Group* (London: Oxford University Press, 1958), 54.

131. Amatzia Baram, "Mesopotamian Identity in Ba'thi Iraq," in *Middle Eastern Studies* 19 (1983): 39–57, and "A Case of Imported Identity: The Modernizing Secular Ruling Elites of Iraq and the Concept of Mesopotamian-inspired Territorial Nationalism," *Poetics Today* 15 (1994): 279–319.

132. Thomas Erskine May, *Democracy in Europe* (London: Longman Green, 1877), 29.

133. The phrase "clash of civilizations" derives, of course, from Samuel Huntingdon's much-discussed book, *The Clash of Civilizations and the Remaking of World Order* (New York: Simon and Schuster, 1996).

134. Both cited in Kepel, *War for Muslim Minds,* 123–24.

135. Gilles Kepel, *Jihad: The Trial of Political Islam,* trans. Anthony F. Roberts (Cambridge, Mass.: Belknap Press, 2002), 63.

136. Kedourie, *Afghani and 'Abduh,* and see Hourani, " 'Abduh's Disciples: Islam and Modern Civilization," in *Arabic Thought in the Liberal Age, 1798–1939,* 161–92.

137. Quoted in Elie Kedourie, *Islam in the Modern World and Other Studies* (London: Mansell, 1980), 26. Al-Afghânî was replying to Ernest Renan's account of Islam as incompatible with science. See pp. 204–6.

138. Ajami, *Arab Predicament,* 50–56, describing the views of Sami al-Jundi, one of the founders of the original Ba'th party.

139. Quoted in ibid., 63–64.

140. Quoted in Wright, *Looming Tower,* 25.

141. Nikki R. Keddie, *An Islamic Response to Imperialism: Political and Religious*

*Writings of Sayyid Jamâl ad-Dîn "al-Afghânî"* (Berkeley: University of California Press, 1968), 81.

142. Quoted in Karsh, *Islamic Imperialism,* 214.

143. Quoted in Yossef Bondansky, *Bin Laden: The Man Who Declared War on America* (Roseville, Calif.: Prima Publishing, 2001), 8–9.

144. Ibid., 12.

145. Kepel, *Jihad,* 147–148.

## Epilogue

1. Quoted in Bondansky, *Bin Laden,* 382.

2. See p. 175.

3. Quoted in Daniel Benjamin and Steven Simon, *The Age of Sacred Terror* (New York: Random House, 2002), 149.

4. Quoted in Wright, *Looming Tower,* 176.

5. Gilles Kepel, *Fitna: guerre au coeur de l'Islam* (Paris: Gallimard, 2004), 99–105.

6. Benjamin and Simon, *Age of Sacred Terror,* 167.

7. Quoted in ibid., 104–5.

8. The phrase comes from a communiqué from the Abu Hafs al-Misri Brigades, which claimed responsibility for the Madrid bombings, quoted in Kepel, *War for Muslim Minds,* 145.

9. See, e.g., Richard W. Bulliet, *The Case for Islamo-Christian Civilization* (New York: Columbia University Press, 2004): "According to the 'clash of civilizations' hypothesis, the (Judeo-Christian) West has always been and always will be at odds with Islam. According to the Islamo-Christian civilization model, Islam and the West are historical twins whose resemblance did not cease when their paths parted."

10. Mawlana Abdullah Mawdudi, "Political Theory of Islam," in *Islam in Transition: Muslim Perspectives,* eds. John Donohue and John L. Esposito (Oxford: Oxford University Press, 1982), 252.

11. Quoted in Nader A. Hashemi, "Change from Within," in *Islam and the Challenge of Democracy,* ed. Khaled Abou El Fadl (Princeton, N.J.: Princeton University Press, 2004), 53.

12. Seyyed Vali Reza Nasr, *Mawdudi and the Making of Islamic Revivalism* (New York: Oxford University Press, 1996), 84, 88.

13. Olivier Carré, *Mysticism and Politics: A Critical Reading of Fi Zilal al Qur'an by Sayyid Qutb (1906–1966),* trans. Carol Artigues (Leiden: Brill, 2003), 153.

14. Sayyid Qutb, *Social Justice in Islam* (Leiden: Brill, 1996), 4.

15. Sayyid Qutb, *Milestones* (Damascus: Dar al-Ilm, n.d.), 93.

16. Quoted in Wright, *Looming Tower,* 15.

17. Qutb, *Milestones,* 21.

18. James P. Piscatori, *Islam in a World of Nation-States* (Cambridge: Cambridge University Press, 1986), 22.

19. Kepel, *Jihad,* 25–26.

20. Qutb, *Social Justice in Islam,* 24–25.

21. Quoted in Nazih N. Ayubi, *Political Islam: Religion and Politics in the Arab World* (London: Routledge, 1991), 140.

22. Quoted in Wright, *Looming Tower,* 31.

23. The pope's divisive speech in Regensburg on September 12, 2006, ends with a plea for a reconciliation between Christianity and modern science—although it is clear that for the pope all of what might broadly be called ethics and morals belongs to theology and not to what he calls "human science." Furthermore the Christianity he has in mind is the rationalized Hellenism—his term—of Saint Thomas Aquinas, not the "literalism" of the Protestant churches, much less the crass fundamentalism of the modern evangelists. *Faith, Reason and the University.*

24. Kepel, *Jihad,* 72.

25. Emmanuel Sivan, *Radical Islam: Medieval Theology and Modern Politics* (New Haven, Conn.: Yale University Press, 1985), 97.

26. An excellent, succinct account of Ibn Taymîyah and his teachings is given in Benjamin and Simon, *Age of Sacred Terror,* 38–94.

27. Quoted in ibid., 42.

28. Ajami, *Arab Predicament,* 63–65.

29. Albert Hirschman, "Exit, Voice and the Fate of the German Democratic Republic," in *A Propensity to Self-Subversion* (Cambridge, Mass.: Harvard University Press, 1995), 9–44.

30. See pp. 16–17.

31. *The New York Times,* November 28, 2006, A12.

32. Quoted in Lewis, *Crisis of Islam,* 159.

33. Qutb, *Milestones,* 7, 11–12.

34. www.washingtonpost.com/wrp-srv/world/documtns/ahmadinejad0509.pdf.

35. Mill, "On Liberty," 70.

36. Gary Wills, "A Country Ruled by Faith," *The New York Review of Books,* Novmber 16, 2006, 8.

37. George Hourani, *Reason and Tradition in Islamic Ethics* (Cambridge: Cambridge University Press, 1985), 210.

38. John L. Esposito, *The Islamic Threat: Myth or Reality?* (New York: Oxford University Press, 1999), 55.

39. Clifford Geertz, *Islam Observed* (New Haven, Conn.: Yale University Press, 1968), 64.

# BIBLIOGRAPHY

This is a list of the works cited in the endnotes. It does not, however, include classical sources or most works that exist in multiple editions.

Abou el Fadl, Khaled. *Islam and the Challenge of Democracy*. Ed. Joshua Cohen and Deborah Chasman. Princeton, N.J.: Princeton University Press, 2004.

Adkins, A.W.H. *Moral Values and Political Behaviour in Ancient Greece*. New York: W. W. Norton, 1972.

Aeschylus. *The Persians*. Trans. Janet Lembke and C. J. Herington. New York: Oxford University Press, 1981.

Ahmed, Feroz. *The Young Turks: The Committee of Union and Progress in Turkish Politics, 1908–1914*. Oxford: Clarendon Press, 1969.

Ajami, Fouad. *The Arab Predicament: Arab Political Thought and Practice Since 1967*. Cambridge: Cambridge University Press, 1992.

Akçam, Taner. *A Shameful Act: The Armenian Genocide and the Question of Turkish Responsibility*. Trans. Paul Bessemer. New York: Metropolitan Books, 2007.

Aksan, Virginia. "Ottoman Political Writing, 1768–1808." *Journal of Middle-Eastern Studies* 25 (1993): 53–69.

Ando, Clifford. *Imperial Ideology and Provincial Loyalty in the Roman Empire*. Berkeley: University of California Press, 2000.

Andrew, Christopher M., and A. S. Knaya-Forstner. *The Climax of French Imperial Expansion, 1914–1924.* Stanford, Calif.: Stanford University Press, 1981.

Anquetil-Duperron, Abraham-Hyacinthe. *Considérations philosophiques et géographiques sur les deux mondes* (1780–1804). Ed. Guido Abattista. Pisa: Scuola Normale Superiore, 1993.

———. *Législation orientale.* Amsterdam, 1778.

———. *Voyage en Inde, 1754–1762.* Eds. Jean Deloche, Manonmani Filliozat, and Pierre-Sylvain Filliozat. Paris: École Française d'Extrême-Orient, 1997.

Aristides, Aelius. "The Roman Oration." In *The Ruling Power: A Study of the Roman Empire in the Second Century After Christ Through the Roman Oration of Aelius Aristides,* ed. James H. Oliver. Transactions of the American Philosophical Society New Series, 23 (1953).

Armitage, David. *The Declaration of Independence: A Global History.* Cambridge, Mass.: Harvard University Press, 2007.

Averroès [Abû al-Walîd Muhammad ibn Rushd]. *Discours décisif.* Trans. Marc Geoffroy. Paris: Flammarion, 1996.

Ayubi, Nazih N. *Political Islam: Religion and Politics in the Arab World.* London: Routledge, 1991.

Bahgat, Ali. "Acte de marriage du General Abdallah Menou avec la dame Zobaidah." *Bulletin de l'Institut Egyptien* 9 (1899): 221–35.

Bailey, Cyril, ed. *The Legacy of Rome.* Oxford: Oxford University Press, 1923.

al-Baladhuri, Ahmad ibn Yahya. *Kitâb Futûh al-Buldân (The Origins of the Islamic State).* Trans. Philip Hitti. New York: Columbia University Press, 1916.

Balsdon, F.P.V.D. *Romans and Aliens.* London: Duckworth, 1979.

Baram, Amatzia. "A Case of Imported Identity: The Modernizing Secular Ruling Elites of Iraq and the Concept of Mesopotamian-Inspired Territorial Nationalism." *Poetics Today* 15 (1994): 279–319.

———. "Mesopotamian Identity in Ba'thi Iraq." *Middle Eastern Studies* 19 (1983).

Barnes, Jonathan. "Ciceron et la guerre juste." *Bulletin de la Société Française de Philosophie* 80 (1986): 41–80.

Bayly, C. A. *The Birth of the Modern World, 1780–1914.* Malden, U.K.: Blackwell, 2004.

Benedict XVI (pope). *Faith, Reason and the University: Memories and Reflections.* Vatican City: Libreria Editrice Vaticana, 2006.

Bengio, Ofra. *Saddam's Word: Political Discourse in Iraq.* New York: Oxford University Press, 1998.

Benjamin, Daniel, and Steven Simon. *The Age of Sacred Terror*. New York: Random House, 2002.

Bernier, François. *Événemens particuliers, ou ce qui s'est passé de plus considerable après la guerre pendant cinq ans . . . dans les états du Grand Mongol*. Paris, 1670.

Bernoyer, François. *Avec Bonaparte en Égypte et en Syrie, 1798–1800*. Ed. Christian Tortel. Paris: Éditions Curandera, 1981.

Binder, Leonard. *The Ideological Revolution in the Middle East*. New York: John Wiley and Sons, 1964.

Boardman, John, Jasper Griffin, and Oswyn Murray, eds. *The Oxford History of the Classical World*. Oxford: Oxford University Press, 1986.

Bondansky, Yossef. *Bin Laden: The Man Who Declared War on America*. Roseville, Calif.: Prima Publishing, 2001.

Boswell, James. *Journal of a Tour to the Hebrides with Samuel Johnson, 1733*. Ed. Frederick A. Pootle and Charles H. Bennet. New Haven, Conn.: Yale University Press, 1961.

———. *The Life of Samuel Johnson*. Oxford: Oxford University Press, 1983.

Bosworth, A. B. *Conquest and Empire: The Reign of Alexander the Great*. Cambridge: Cambridge University Press, 1988.

Brewer, David. *The Flame of Freedom: The Greek War of Independence, 1821–1833*. London: John Murray, 2001.

Briant, Pierre. "La date des révoltes babyloniennes contra Xersès." *Studia Iranica* 21 (1992): 12–13.

———. *Histoire de l'Empire Perse: De Cyrus à Alexandre*. Paris: Fayard, 1996.

———. "La vengeance comme explication historique dans l'oeuvre d'Hérodote." *Revue des Études Grecques* 84 (1971): 319–35.

*British Interests in the Mediterranean and Middle East: A Report by a Chatham House Study Group*. London: Oxford University Press, 1958.

Brown, Peter. *Augustine of Hippo: A Biography*. London: Faber and Faber, 1976.

———. *The World of Late Antiquity*. New York: W. W. Norton, 1989.

Bulliet, Richard W. *The Case for Islamo-Christian Civilization*. New York: Columbia University Press, 2004.

———. *Conversion to Islam in the Medieval Period: An Essay in Quantitative History*. Cambridge, Mass.: Harvard University Press, 1979.

Buruma, Ian, and Avishi Margalit. *Occidentalism: The West in the Eyes of Its Enemies*. New York: Penguin, 2004.

Busbecq, Ogier Ghiselin de. *The Turkish Letters of Ogier Ghiselin de Busbecq*. Trans. Edward Forster. Oxford: Clarendon Press, 1927.

Busch, Briton Cooper. *Britain, India and the Arabs, 1914–1921*. Berkeley: University of California Press, 1971.

Buzpinar, S. Tufan. "The Hijaz, Abdülhamid II and Amir Hussein's Secret Dealings with the British, 1877–80." *Middle Eastern Studies* 31 (1995): 99–123.

Calasso, Roberto. *The Marriage of Cadmus and Harmony.* New York: Vintage Books, 1993.

Cameron, Euan. *The European Reformation.* Oxford: Clarendon Press, 1991.

Cannadine, David. *Ornamentalism: How the British Saw Their Empire.* London: Penguin, 2002.

Cannon, Garland. *The Life and Mind of Oriental Jones: Sir William Jones, the Father of Modern Linguistics.* Cambridge, Mass.: Cambridge University Press, 1990.

Cardini, Franco. *Europa e Islam: storia di un malinteso.* Rome: Laterza, 2002.

Carré, Olivier. *Mysticism and Politics: A Critical Reading of Fi Zilal al Qur'an by Sayyid Qutb (1906–1966).* Trans. Carol Artigues. Leiden: Brill, 2003.

Cartledge, Paul. *Alexander the Great: The Hunt for a New Past.* London: Macmillan, 2004.

——. *Thermopylae: The Battle That Changed the World.* London: Pan Books, 2006.

Chabrol, Gilbert-Joseph Volvic de. "Essai sur les moeurs des habitans modernes de l'Égypte." *Description de l'Égypte, ou recueil des observations et des recherches qui ont été faites en Égypte pendant l'expédition de l'armée française, publié par les ordres de Sa Majesté l'Empereur Napoléon le Grand,* 2nd ed., vol. 18. Paris: C. L.-F. Pankoucke, 1821–30.

Chateaubriand, François-René, Vicomte de. *Itinéraire de Paris à Jérusalem.* Ed. Jean-Claude Berchet. Paris: Gallimard, 2003.

Clogg, Richard, ed. *The Movement for Greek Independence, 1770–1821: A Collection of Documents.* London: Macmillan, 1976.

——. *A Short History of Modern Greece.* Cambridge: Cambridge University Press, 1979.

Cohen, Michael. *Palestine and the Great Powers, 1945–1948.* Princeton, N.J.: Princeton University Press, 1982.

Cohn, Bernard. "The Command of Language and the Language of Command." In *Subaltern Studies,* vol. 4, ed. Ranajit Guha. Delhi: N.P., 1985, 286–301.

Condorcet, Marie-Jean-Antoine-Nicolas Caritat, marquis de. *Esquisse d'un tableau historique des progrès de l'esprit humain.* Ed. Alain Pons. Paris: Flammarion, 1988.

*Congrès National des Forces Populaires: La Charte.* Cairo: Administration de l'Information, 1962.

Cook, Michael. *Forbidding Wrong in Islam.* Cambridge: Cambridge University Press, 2003.

———. *Muhammad.* Oxford: Oxford University Press, 1983.

Coope, Jessica A. *The Martyrs of Córdoba: Community and Family Conflict in an Age of Mass Conversion.* Lincoln: University of Nebraska Press, 1995.

Criss, Nur Bilge. *Istanbul under Allied Occupation, 1918–1923.* Leiden: Brill, 1999.

Crone, Patricia. *God's Rule: Government and Islam.* New York: Columbia University Press, 2004.

———. *Medieval Islamic Political Thought.* Edinburgh: Edinburgh University Press, 2004.

———. *Slaves on Horses: The Evolution of the Islamic Polity.* Cambridge: Cambridge University Press, 1980.

Crone, Patricia, and Marin Hinds. *God's Caliph: Religious Authority in the First Centuries of Islam.* Cambridge: Cambridge University Press, 1986.

Curzon, George Nathaniel. *Persia and the Persian Question.* London: Longmans, 1892.

Dakin, Douglas. *The Greek Struggle for Independence, 1821–1833.* Berkeley: University of California Press, 1973.

Daniel, Norman. *Islam and the West: The Making of an Image.* Edinburgh: Edinburgh University Press, 1960.

Darwin, John. *After Tamerlane: The Global History of Empire.* London: Penguin, 2007.

———. *Britain, Egypt and the Middle East: Imperial Policy in the Aftermath of War, 1918–1922.* London: Macmillan, 1981.

Dauge, Yves Albert. *Le Barbare: recherches sur la conception de la barbarie et de la civilisation.* Bruxelles: Latomus Revue d'Études Latines, 1981.

Davies, Norman. *Europe: A History.* Oxford: Oxford University Press, 1997.

Della Vida, G. Levi. "La Corrispondeza di Berta di Toscana col Califfo Muktafi." *Rivista Storica Italiana* 66 (1954): 21–38.

Denon, Dominique-Vivant. *Voyage dans la Basse et la Haute Égypte, pendant les campagnes du général Bonaparte.* Paris: P. Dinot l'Ainé, 1802.

Derogy, Jacques, and Hesi Carmel. *Bonaparte en Terre Sainte.* Paris: Fayard, 1992.

Disraeli, Benjamin. *Tancred; or the New Crusade.* London: Longman Green, 1894.

Donohue, John, and John L. Esposito. *Islam in Transition: Muslim Perspectives.* Oxford: Oxford University Press, 1982.

Droysen, Johann Gustav. *Geschichte Alexanders des Grossen,* vol. 1 of *Geschichte des Hellenismus.* Basel: Schwabe, 1952.

Dumézil, Georges. *Idées romaines.* Paris: Gallimard, 1969.

Erdmann, Carl. *The Origin of the Idea of Crusade.* Trans. Marshall W. Baldwin and Walter Goffart. Princeton, N.J.: Princeton University Press, 1977.

Esposito, John L. *The Islamic Threat: Myth or Reality?* New York: Oxford University Press, 1999.

Euben, Peter. "Political Equality and the Greek Polis." In M. J. Gargas McGrath, *Liberalism and Modern Polity.* New York: Marcel Dekker, 1959, 207–29.

Eusebius of Caesarea. *Eusebius: Life of Constantine.* Trans. A. Cameron and S. Hall. Oxford: Oxford University Press, 1999.

Evans, J.A.S., "Father of History or Father of Lies? The Reputation of Herodotus." *Classical Journal* 64 (1968): 11–17.

Febvre, Lucien. *L'Europe: Genèse d'une civilisation,* Paris: Perrin, 1999.

Ferguson, Niall. *Colossus: The Price of America's Empire.* London: Penguin, 2006.

Ferrary, Jean-Louis. *Philhellénisme et imperialisme: Aspects idééologiques de la conquête du monde hellénistique.* Rome: Bibliothèque des Écoles Françaises d'Athènes et de Rome, 1988.

Finkel, Caroline. *Osman's Dream: The Story of the Ottoman Empire, 1300–1923.* New York: Basic Books, 2006.

Finley, M. I. *Ancient Slavery and Modern Ideology.* Harmondsworth, U.K.: Penguin Books, 1983.

Fletcher, Richard. *The Cross and the Crescent: Christianity and Islam from Muhammad to the Reformation.* New York: Viking, 2003.

Fox, Robin Lane. *The Search for Alexander.* Boston: Little, Brown, 1980.

Fradkin, H. "The Political Thought of Ibn Tufayl." In *The Political Aspects of Islamic Philosophy,* ed. E. Butterworth. Cambridge, Mass.: Harvard University Press, 1992, 234–61.

Freeman, Charles. *The Closing of the Western Mind: The Rise of Faith and the Fall of Reason.* New York: Alfred A. Knopf, 2003.

Frend, W.H.C. *Martyrdom and Persecution in the Early Church.* Oxford: Oxford University Press, 1965.

Friedman, Isaiah. *The Question of Palestine: British-Jewish-Arab Relations, 1914–1918.* New Brunswick, N.J.: Transaction Publishers, 1992.

Fromkin, David. *A Peace to End All Peace: The Fall of the Ottoman Empire and the Creation of the Modern Middle East.* New York: Henry Holt, 1989.

Frossard, Benjamin. *Observations sur l'abolition de la traite des nègres présentées a la Convention Nationale.* N.P., 1793.

Frye, Richard N. *The History of Ancient Iran.* Munich: C. H. Beck'sche Verlagsbuchhandlung, 1984.

Gabrieli, Francesco, ed. *Storici arabi delle crociate.* Turin: Einaudi, 1957.

Garcin, J.-C., ed. *États, sociétés et cultures du monde musulman médiéval, Xe.–XVe. siècle.* 3 vols. Paris: PUF, 1995–2000.

Garnsey, P. A., and C. R. Whittaker, eds. *Imperialism in the Ancient World.* Cambridge: Cambridge University Press, 1978.

Garnsey, Peter. *Ideas of Slavery from Aristotle to Augustine.* Cambridge: Cambridge University Press, 1996.

Gauthier, R. A. *Magnanimité, l'idéal de la grandeur dans la philosophie païenne et dans la théologie chrétienne.* Paris: Vrin, 1951.

Geertz, Clifford. *Islam Observed.* New Haven, Conn.: Yale University Press, 1968.

*Gesta Francorum et aliorum Hierosolimitanorum* [*The Deeds of the Franks and the Other Pilgrims to Jerusalem*]. Ed. Rosalind Hill. London: Thomas Nelson, 1962.

Gilbert, Vivian. *The Romance of the Last Crusade: With Allenby to Jerusalem.* New York: D. Appleton and Co., 1925.

Giovio, Paolo. *Commentario delle cose dei Turchi.* Venice, 1538.

Goiten, S. D. *Studies in Islamic History and Institutions.* Leiden: Brill, 1966.

Goody, Jack. *The East in the West.* Cambridge: Cambridge University Press, 1992.

Gordon, Mary. "The Nationality of Slaves Under the Early Roman Empire." In *Slavery in Classical Antiquity,* ed. M. I. Finley. Cambridge: W. Heffer and Sons, 1960, 171–189.

Gordon, Thomas. *History of the Greek Revolution.* 2 vols. London: T. Cadell, 1832.

Grant, Michael. *The World of Rome.* London: Weidenfeld and Nicolson, 1960.

Green, Peter. *Alexander of Macedon, 356–323 B.C.: A Historical Biography.* Berkeley: University of California Press, 1991.

Gress, David. *From Plato to NATO: The Idea of the West and its Opponents.* New York: Free Press, 1998.

Grey, Edward (Viscount Grey of Fallodon). *Twenty-five Years, 1892–1916.* London: Hodder and Stoughton, 1925.

Haag, Michael. *Alexandria: City of Memory.* New Haven, Conn.: Yale University Press, 2005.

Hammond, N.G.L. "The Kingdom of Asia and the Persian Throne." In *Alexander the Great: A Reader,* ed. Ian Worthington. London: Routledge, 2003.

Hankins, James. "Renaissance Crusaders: Humanist Crusade Literature in the Age of Mehmed II." *Dumbarton Oaks Papers* 49 (1995): 111–207.

Hanson, Victor Davis. *Carnage and Culture: Landmark Battles in the Rise of Western Power.* New York: Anchor Books, 2001.

———. "Take Me to My Leader," *The Times Literary Supplement,* Oct. 2, 2004: 11–27.

Heers, Jacques. *Chute et mort de Constantinople, 1204–1453.* Paris: Perrin, 2005.

Hegel, Georg Friedrich. *The Philosophy of History.* Trans. J. Sibree. New York: Dover Publications, 1956.

Herder, Johann Gottfried von. *Herder: Philosophical Writings.* Ed. Michael N. Forster. Cambridge: Cambridge University Press, 2002.

———. *On World History: An Anthology.* Ed. Hans Adler and Ernst A. Menze. Armonk, N.Y.: M. E. Sharpe, 1997.

Herodotus. *The Histories.* Trans. Aubrey de Sélincourt, rev. by John Marincola. London: Penguin, 1996.

Herzl, Theodor. *The Jew's State: An Attempt at a Modern Solution to the Issue of the Jews.* Trans. Henk Overberg. Northvale, N.J.: Jason Aronson, 1991.

Hillenbrand, Carole. *The Crusades: Islamic Perspectives.* Edinburgh: Edinburgh University Press, 1999.

Hindley, Geoffrey. *Saladin: A Biography.* London: Constable, 1976.

Hirschman, Albert. *A Propensity to Self-Subversion.* Cambridge, Mass.: Harvard University Press, 1995.

Hobbes, Thomas. *The Elements of Law, Natural and Political,* 2d ed. Ed. Ferdinand Tönnies. London: Frank Cass & Co., 1969.

———. *Leviathan.* Ed. Richard Tuck. Cambridge: Cambridge University Press, 1991.

———. *On the Citizen.* Trans. and ed. Richard Tuck and Michael Silverthorne. Cambridge: Cambridge University Press, 1998.

Hodges, Richard, and David Whitehouse. *Mohammed, Charlemagne and the Origins of Europe.* Ithaca, N.Y.: Cornell University Press, 1983.

Holbach, Paul Henri Dietrich, baron de. *Oeuvres philosophiques complètes.* 2 vols. Ed. Jean-Pierre Jackson. Paris: Éditions Alive, 1999.

Hopkirk, Peter. *The Great Game: The Struggle for Empire in Central Asia.* New York: Kodansha International, 1994.

Hourani, Albert. *Arabic Thought in the Liberal Age, 1798–1939.* Cambridge: Cambridge University Press, 1962.

———. *A History of the Arab Peoples.* London: Faber & Faber, 1991.

———. *Islam in Western Thought.* Cambridge: Cambridge University Press, 1991.

Hourani, George. *Reason and Tradition in Islamic Ethics.* Cambridge: Cambridge University Press, 1985.

Housley, Norman. *The Later Crusades, 1272–1580.* Oxford: Oxford University Press, 1991.

Hughes, Lindsey. *Russia in the Age of Peter the Great.* New Haven and London: Yale University Press, 1998.

Hughes, Thomas Smart. *An Address to the People of England in the Cause of the Greeks, Occasioned by the Late Inhuman Massacres on the Isle of Scio.* London: Simpkin and Marshall, 1822.

——. *Travels in Greece and Albania.* 2 vols. London: H. Colburn and R. Bentley, 1830.

Hume, David. *Essays: Moral, Political and Literary.* Ed. Eugene F. Miller. Indianapolis: Liberty Classics, 1985.

Huntingdon, Samuel. *The Clash of Civilizations and the Remaking of World Order.* New York: Simon and Schuster, 1996.

Hyam, Ronald. *Britain's Declining Empire: The Road to Decolonisation, 1918–1968.* Cambridge: Cambridge University Press, 2006.

Ibn al-Qalânisî. *The Damascus Chronicle of the Crusades.* Extracted and trans. H.A.R. Gibb. London: Luzac and Co., 1932.

Ibn Battuta, Muhammad. *The Travels of Ibn Battuta.* 2 vols. Trans H.A.R. Gibb. Cambridge: Cambridge University Press, 1962.

Ibn Ishaq, Muhammad. *The Life of Muhammad: A Translation of Ishaq's* Sirat Rasul Allah. Introduction and notes by A. Guillaume. Karachi and Oxford: Oxford University Press, 1955.

Ibn Khaldûn, Abd Al-Rahman, Ibn Muhammad. *The Muqaddimah: An Introduction to History.* Trans. Franz Rosenthal. Princeton, N.J.: Princeton University Press, 1967.

Ibn Munquid, Usamah. *An Arab-Syrian Gentleman and Warrior in the Period of the Crusades.* New York: Columbia University Press, 1929.

Imbruglia, Girolamo. "Tra Anquetil-Duperron e *L'histoire des deux Indies:* libertà, dispotismo e feudalismo." *Rivista storica italiana* 56 (1994), 140–193.

Inalcik, Halil. "The Question of the Emergence of the Ottoman State." *International Journal of Turkish Studies* 2 (1980): 71–79.

Ingram, Edward. *Commitment to Empire: Prophecies of the Great Game in Asia, 1797–1800.* Oxford: Oxford University Press, 1981.

Isaac, Benjamin. *The Invention of Racism in Classical Antiquity.* Princeton, N.J.: Princeton University Press, 2004.

al-Jabarti, Abd al-Rahman, *Al-Jabarti's Chronicle of the First Seven Months of the French Occupation of Egypt.* Ed. and trans. S. Morhe. Leiden: Brill, 1975.

Jasanoff, Maya. *Edge of Empire: Conquest and Collecting in the East, 1750–1850.* London: Fourth Estate, 2005.

Johnson, Samuel. *The Prince of Abyssinia: A Tale.* London, 1759.

Jolivet, J., et al., eds. *Multiple Averroès.* Paris: les Belles Lettres, 1973.

Jones, A.H.M. *Constantine and the Conversion of Europe.* London: Hodder and Stoughton, 1948.

Jones, Sir William. *The Collected Works of Sir William Jones* [1807] facsimile edition. 13 vols. New York: New York University Press, 1993.

———. *Dissertation sur la litérature orientale.* London, 1771.

Kant, Immanuel. *Lectures on Ethics.* Eds. Peter Heath and J. B. Schneewind. Cambridge: Cambridge University Press, 1997.

———. *Political Writings.* Ed. Hans Reiss, trans. H. B. Nisbet. Cambridge: Cambridge University Press, 1991.

Karsh, Efraim. *Islamic Imperialism: A History.* New Haven, Conn.: Yale University Press, 2006.

Karsh, Efraim, and Inari Karsh. *Empires of the Sand: The Struggle for Mastery in the Middle East, 1789–1923.* Cambridge, Mass.: Harvard University Press, 1999.

Keddie, Nikki R. *An Islamic Response to Imperialism: Political and Religious Writings of Sayyid Jamâl ad-Dîn "al-Afghâni."* Berkeley: University of California Press, 1968.

———. *Religion and Rebellion in Iran: The Tobacco Protest of 1891–1892.* London: Cass, 1966.

Kedourie, Elie. *Afghani and 'Abduh: An Essay on Religious Unbelief and Political Activism in Modern Islam.* London: Frank Cass, 1966.

———. *The Anglo-Arab Labyrinth: The McMahon-Husayn Correspondence and Its Interpretations, 1914–1939.* Cambridge: Cambridge University Press, 1976.

———. *Arabic Political Memoirs and Other Studies.* London: Frank Cass, 1974.

———. "The Capture of Damascus 1 October, 1918." In *The Chatham House Version and Other Middle Eastern Studies.* New York: Praeger, 1970, 48–51.

———. *England and the Middle East: The Destruction of the Ottoman Empire, 1914–1921.* Hassock, Sussex: The Harvester Press, 1978.

———. *Islam in the Modern World and Other Studies.* London: Mansell, 1980.

Kelley, Donald R. *The Beginning of Ideology: Consciousness and Society in the French Reformation.* Cambridge: Cambridge University Press, 1981.

———. *Historians and the Law in Postrevolutionary France.* Princeton, N.J.: Princeton University Press, 1984.

Kepel, Gilles. *Fitna: Guerre au coeur de l'Islam.* Paris: Gallimard, 2004.

———. *Jihad: The Trial of Political Islam.* Trans. Anthony F. Roberts. Cambridge, Mass.: Belknap Press, 2002.

———. *The War for Muslim Minds: Islam and the West.* Cambridge, Mass.: Belknap Press, 2004.

Khalidi, Rashid, et al., eds. *The Origins of Arab Nationalism.* New York: Columbia University Press, 1991.

Kobler, Franz. *Napoleon and the Jews.* New York: Schocken Books, 1975.

Kristovoulos. *History of Mehmed the Conqueror.* Trans. Charles T. Riggs. Princeton, N.J.: Princeton University Press, 1954.

Kritzeck, James. *Peter the Venerable and Islam.* Princeton, N.J.: Princeton University Press, 1964.

Kroebner, Richard. "Despot and Despotism: Vicissitudes of a Political Term." *Journal of the Warburg and Courtauld Institutes* 14 (1951): 275–302.

Kuhrt, A. "The Cyrus Cylinder and Achaemenid Imperial Policy." *Journal for the Study of the Old Testament* 25 (1983): 83–94.

Kupperman, Karen Ordahl. *Settling with the Indians: The Meeting of English and Indian Cultures in America, 1580–1640.* Totowa, N.J.: Rowman and Littlefield, 1980.

Kyle, Keith. *Suez.* London: Weidenfeld and Nicolson, 1991.

Lacour-Gayet, Georges. *Talleyrand.* Paris: Éditions Payot, 1990.

Laissus, Yves. *L'Égypte, une aventure savante, 1798–1801.* Paris: Fayard, 1998.

La Jonquière, Clément de. *L'Expedition d'Égypte, 1798–1801.* 5 vols. Paris: H. Charles-Lavauzelle, 1899–1907.

Landau, Jacob M. *The Politics of Pan-Islam: Ideology and Organization.* Oxford: Clarendon Press, 1990.

Lane, Edward. *The Manners and Customs of the Modern Egyptians.* New York: Dutton, 1966.

La Noue, François de. *Discours politiques et militaires.* Ed. F. E. Sutcliffe. Geneva: Droz, 1967.

Laquer, Walter. *The Israeli-Arab Reader.* London: Penguin, 1970.

Las Casas, Bartolomé de. *A Short Account of the Destruction of the Indies.* Trans. Nigel Griffin. London: Penguin, 1992.

Las Cases, Marie-Joseph-Emmanuel-Auguste-Dieudonné. *Mémorial de Sainte-Hélène.* 2 vols. Ed. Gérard Walter. Paris: Bibliothèque de la Pléiade, 1956.

Laurens, Henry. "Bonaparte a-t-il colonisé l'Égypte?" *L'Histoire* 216 (1997): 46–49.

———. *L'expédition d'Égypte, 1798–1801.* Paris: Éditions du Seuil, 1997.

———. "Le mythe de l'expédition d'Égypte en France et en Égypte aux XIXe et XXe siècles." In *L'égyptologie et les Champollion,* ed. Michel Dewachter and Alain Fouchard. Grenoble: Presses Universitaires de Grenoble, 1994, 321–30.

———. *Les origines intellectuelles de l'expédition d'Égypte: l'orientalisme islamisant en France (1698–1798).* Istanbul: Éditions Isis, 1987.

———. "Le projet de'état juif en Palestine, attribué à Bonaparte." In *Orientales*. Paris: CNRS Éditions, 2004, 1:123–43.

Lawrence, Thomas Edward. *Crusader Castles*. A new edition with introduction and notes by Denys Pringle. Oxford: Oxford University Press, 1988.

———. *The Letters of T. E. Lawrence*. Ed. David Garnett. London: Jonathan Cape, 1938.

———. *Secret Despatches from Arabia by T. E. Lawrence*. Cambridge, England.: The Golden Cockerel Press, n.d.

———. *The Seven Pillars of Wisdom: The Complete 1922 Text*. Fordingbridge, England: The Castle Hill Press, 1997.

Levene, Mark. *Genocide in the Age of the Nation-State*. 2 vols. London: I. B. Tauris, 2005.

Lévi-Strauss, Claude. *The Elementary Structures of Kinship*. Trans. James Hare Bell. London: Eyre and Spottiswoode, 1968.

Levy, Reuben. *A Baghdad Chronicle*. Cambridge: Cambridge University Press, 1929.

Lewis, Bernard. *The Arabs in History*. Oxford: Oxford University Press, 1993.

———. *The Crisis of Islam: Holy War and Unholy Terror*. New York: Random House, 2003.

———. *The Emergence of Modern Turkey*. London: Oxford University Press, 1961.

———. *From Babel to Dragomans: Interpreting the Middle East*. Oxford: Oxford University Press, 2004.

———. *Islam and the West*. Oxford: Oxford University Press, 1993.

———. *The Muslim Discovery of Europe*. New York: W. W. Norton, 1982.

———. *The Political Language of Islam*. Chicago: The University of Chicago Press, 1988.

———. "Politics and War." In *The Legacy of Islam*, ed. Joseph Schnact and C. E. Bosworth. Oxford: Oxford University Press, 1979.

———. *What Went Wrong? The Clash Between Islam and Modernity in the Middle East*. London: Weidenfeld and Nicolson, 2002.

Lintott, Andrew. "What Was the *Imperium Romanum*?" *Greece and Rome* 28 (1981): 53–67.

Locke, John. "First Tract on Government." In *Political Essays,* ed. Mark Goldie. Cambridge: Cambridge University Press, 1997.

Lomax, Derek W. *The Reconquest of Spain*. London: Longman, 1978.

Loraux, Nicole. *The Invention of Athens: The Funeral Oration in the Classical City*. Cambridge, Mass.: Harvard University Press, 1986.

Lowry, Heath W. *The Nature of the Early Ottoman State.* Albany: State University of New York Press, 2003.

Lyons, M. C., and D.E.P. Jackson. *Saladin: The Politics of Holy War.* Cambridge: Cambridge University Press, 1982.

Machiavelli, Niccolò. *The Prince.* Ed. David Wootton. Indianapolis: Hackett, 1995.

MacMullen, Ramsay. *Christianizing the Roman Empire, A.D. 100–400.* New Haven, Conn.: Yale University Press, 1984.

———. *Romanization in the Time of Augustus.* New Haven, Conn.: Yale University Press, 2000.

Madariaga, Salvador de. *Portrait of Europe [Bosquejo de Europa].* New York: Roy Publishers, 1955.

Maine, Henry Sumner. *Village Communities in the East and the West.* London: John Murray, 1881.

Mansfield, Peter. *A History of the Middle East.* New York: Penguin, 2003.

Mantran, Robert, ed. *Histoire de l'empire ottoman.* Paris: Fayard, 1989.

Masson, Frédéric, and Guido Biagi. *Napoléon: manuscrits inédits, 1789–1791.* 8 vols. Paris: P. Ollendorf, 1907.

May, Thomas Erskine. *Democracy in Europe.* London: Longman Green, 1877.

Mazower, Mark. *The Balkans: A Short History.* New York: Modern Library, 2002.

Mazza, F. "The Phoenicians as Seen by the Ancient World." In *The Phoenicians,* ed. Sabatino Moscati. London: I. B. Tauris, 2001, 548–567.

Mill, John Stuart. *The Basic Writings of John Stuart Mill.* New York: Modern Library, 2002.

———. *On Liberty and Other Writings.* Ed. Stefan Collini. Cambridge: Cambridge University Press, 1989.

Millar, Fergus. *Rome, the Greek World and the East.* Vol. 1: *The Roman Republic and the Augustan Revolution.* Hannah M. Cotton and Guy M. Rodgers, ed. Chapel Hill: The University of North Carolina Press, 2002.

Momigliano, Arnaldo. *Alien Wisdom: The Limits of Hellenization.* Cambridge: Cambridge University Press, 1975.

———. *Fondamenti della storia antica.* Turin: Einaudi, 1984.

Montesquieu, Charles-Louis de Secondat, Baron de. *Persian Letters.* Trans. C. J. Betts. New York: Viking Penguin, 1973.

Muherjee, S. N. *Sir William Jones: A Study in Eighteenth-Century British Attitudes to India.* Cambridge: Cambridge University Press, 1968.

Müller, Max. *Collected Works of the Right Hon. F. Max Müller.* 18 vols. London: Longman, 1898.

———. *Lectures on the Science of Language.* London: Longman, 1864.

———. *The Sacred Books of the East* [1887]. Delhi: Motilal Banarsidass, 1992.

Nasr, Seyyed Vali Reza. *Mawdudi and the Making of Islamic Revivalism.* New York: Oxford University Press, 1996.

Nasser, Gamel Abdel. *Egypt's Liberation: The Philosophy of Revolution.* Washington, D.C.: PublicAffairs Press, 1955.

Nicolet, Claude. *The World of the Citizen in Republican Rome.* Trans. P. S. Falla. Berkeley: University of California Press, 1980.

Nietzsche, Friedrich. *Daybreak: Thoughts on the Prejudices of Morality.* Trans. R. J. Hollingdale. Cambridge: Cambridge University Press, 1993.

———. "On the Uses and Disadvantages of History for Life." In *Untimely Meditations,* trans. R. J. Hollingdale. Cambridge: Cambridge University Press, 1983.

Oedani-Fabris, M.-P. "Simbologia ottomana nell'opera di Gentile Bellini." *Atti dell'Istituto veneto di scienze, lettere ed arti* 155 (1996–97): 1–29.

Olmstead, Albert Ten Eyck. *History of the Persian Empire.* Chicago: University of Chicago Press, 1948.

Osterhammel, Jürgen. *Die Entzauberung Asiens. Europa und die asiatischen Reiche im 18. Jahrhundert.* Munich: C. H. Beck, 1988.

Pagden, Anthony. *European Encounters with the New World.* New Haven, Conn.: Yale University Press, 1993.

———. *The Fall of Natural Man: The American Indian and the Origins of Comparative Ethnology.* Cambridge: Cambridge University Press, 1982.

———, ed. *The Languages of Political Theory in Early-Modern Europe.* Cambridge: Cambridge University Press, 1987.

Parker, R. A., and W. Dubberstein. *Babylonian Chronology.* Princeton, N.J.: Princeton University Press, 1956.

Partner, Peter. *God of Battles: Holy Wars in Christianity and Islam.* London: HarperCollins, 1997.

Passerini, Luisa. *Il mito d'Europa: radici antiche per nuovi simboli.* Florence: Giunti, 2002.

Pertusi, Agostini. *La caduta di Constantinopoli. L'eco nel mondo.* Milan: Mondadori, 1976.

———. *La caduta di Constantinopoli.* Vol. 1: *Le testimonianze dei contemporanei.* Milan: Mondadori, 1976.

———. *Testi inediti e poco noti sulla caduta di Constantinopoli.* Bologna: Patron, 1983.

Peters, F. E. *The Monotheists: Jews, Christians, and Muslims in Conflict and Competition.* 2 vols. Princeton, N.J.: Princeton University Press, 2003.

Piccolomini, Aeneas Silvius (Pope Pius II). *Lettera a Maometto II* (*Epistola ad Mahumetem*). Ed. Giuseppe Tofanin. Naples: R. Pironti, 1953.

——. *The Memories of a Renaissance Pope: The Commentaries of Pius II*. Trans. F. A. Gragg. New York: Capricorn Books, 1962.

Pirenne, Henri. *Mohammed and Charlemagne*. Mineola, N.Y.: Dover Publications, 2001.

Piscatori, James P. *Islam in a World of Nation-States*. Cambridge: Cambridge University Press, 1986.

Pocock, J.G.A. "Some Europes in Their History." In *The Idea of Europe: From Antiquity to the European Union*. Ed. Anthony Pagden. Cambridge: Cambridge University Press, 2002, 55–71.

*The Policy of the Turkish Empire*. London, 1597.

Pomeroy, Sarah B., Stanley Burstein, Walter Donolan, and Jennifer Tolbert Roberts. *Ancient Greece: A Political, Social and Cultural History*. New York: Oxford University Press, 1999.

Prosdocimi, Luigi. "Ex facto oritur ius: breve nota di diritti medievale." *Studi senesi*, 1954–55: 66–67, 808–19.

Purchas, Samuel. *Hakluytus Posthumus or Purchas His Pilgrimes, Contayning a History of the World, in Sea Voyages and Lande-Travells by Englishmen & Others*. 5 vols. London, 1625.

Purdum, Todd S. "Bush Warns of a Wrathful, Shadowy and Inventive War." *The New York Times*, September 17, 2001.

Qutb, Sayyid. *Milestones*. Damascus: Dar al-Ilm, n.d.

——. *Social Justice in Islam* [*Al-'adalat al-ijtima'iyya fi'l-Islam*]. In William E. Shepard, *Sayyid Qutb and Islamic Activism: A Translation and Critical Analysis of Social Justice in Islam*. Leiden: Brill, 1996.

Rabb, Theodore K. *The Struggle for Stability in Early-Modern Europe*. New York: Oxford University Press, 1975.

Raby, J. "Mehmed the Conqueror's Greek Scriptorium." *Dumbarton Oaks Papers* 37 (1983): 15–34.

Raj, Kapil. *Relocating Modern Science: Circulation and the Construction of Scientific Knowledge in South Asia and Europe, Seventeenth to Nineteenth Centuries*. Delhi: Permanent Black, 2006.

*Réimpression de l'ancien Moniteur depuis la réunion des États-Généraux jusq'au consulat* (May 1789–November 1799). Paris, 1893.

Rémusant, Claire-Élisabeth-Jeanne Gravier de Vergennes, Comtesse de. *Mémoires de Madame de Rémusant*. 3 vols. Paris: Calmann Lévy, 1880.

Renan, Ernest. *Oeuvres complètes de Ernest Renan*. 4 vols. Ed. Henriette Psichari. Paris: Calmann-Lévy, 1947.

Richardson, John. *A Dissertation on the Languages, Literature and Manners of the East.* Oxford, 1777.

Riley-Smith, Jonathan. *The First Crusade and the Idea of Crusading.* London: The Athlone Press, 1986.

——. *The First Crusaders, 1095–1131.* Cambridge: Cambridge University Press, 1997.

——. *The Oxford Illustrated History of the Crusades.* Oxford: Oxford University Press, 1997.

Riley-Smith, Louise and Jonathan. *The Crusades: Idea and Reality, 1095–1272.* London: Edward Arnold, 1981.

Roseman, L. J. "The construction of Xerxes' Bridge over the Hellespont." *Journal of Hellenic Studies* 116 (1996): 88–108.

Rousseau, Jean-Jacques. *Du Contrat social.* In *Oeuvres complètes,* ed. Bernard Gagnebin and Marcel Raymond. Paris: Bibliothèque de la Pléiade, 1964, 3:279–470.

Rubiés, Joan-Pau. "Oriental Despotism and European Orientalism: Botero to Montesquieu." *Journal of Early-Modern History* 9, no. 2. (2005): 109–80.

Runciman, Steven. *The Fall of Constantinople, 1453.* Cambridge: Cambridge University Press, 1965.

——. *The Great Church in Captivity: A Study of the Patriarchate of Constantinople from the Eve of the Turkish Conquest to the Greek War of Independence.* Cambridge: Cambridge University Press, 1968.

Russell, Peter. *Prince Henry "The Navigator": A Life.* New Haven, Conn.: Yale University Press, 2000.

Sahas, Daniel J. *John of Damascus on Islam: The "Heresy of the Ishmaelites."* Leiden: Brill, 1972.

Said, Edward. *Orientalism.* New York: Vintage Books, 1979.

Schiavone, Aldo. *The End of the Past: Ancient Rome and the Modern West.* Trans. Margaret J. Schneider. Cambridge: Harvard University Press, 2000.

Schluchter, W. *The Rise of Western Rationalism: Max Weber's Developmental History.* Berkeley: University of California Press, 1981.

Schlumberger, Gustave. *Le Siège, la prise et le sac de Constantinople en 1453.* Paris: Plon, 1935.

Schwab, Raymond. *Vie d'Anquetil-Duperron.* Paris: Libraire Ernest Leroux, 1934.

Scott, Sir Walter. *The Life of Napoleon Buonaparte.* 9 vols. Edinburgh: Cadell and Co., 1827.

Sells, Michael A. *The Bridge Betrayed: Religion and Genocide in Bosnia* Berkeley: University of California Press, 1996.

Setton, Kenneth M., ed. *A History of the Crusades.* Madison: University of Wisconsin Press, 1969.

Shaftesbury, Anthony Ashley Cooper, Third Earl of. *Life, Unpublished Letters, and Philosophical Regimen of Anthony, Earl of Shaftesbury.* Ed. Benjamin Rand. London: S. Sonnenschein & Co., 1900.

Shaw, S. J. *Ottoman Egypt in the Age of the French Revolution.* Cambridge, Mass.: Harvard University Press, 1966.

Siberry, Elizabeth. *The New Crusaders: Images of the Crusades in the Nineteenth and Early Twentieth Centuries.* Aldershot, U.K.: Ashgate, 2000.

Sissa, Giulia. "The Irony of Travel: Democracy and Ethnocentrism in Herodotus." Forthcoming in *Metis.*

Sivan, Emmanuel. *Radical Islam: Medieval Theology and Modern Politics.* New Haven, Conn.: Yale University Press, 1985.

Slade, Adolphus. *Record of Travels in Turkey and Greece etc. and of a Cruise in the Black Sea with the Capitan Pasha, in the years 1829, 1830, and 1831.* 2 vols. Philadelphia: E. L. Carey, 1833.

Smith, Colin, ed. *Christians and Moors in Spain.* Warminster, U.K.: Aris & Philips, 1988.

———. *Spanish Ballads.* Oxford: Pergamon Press, 1964.

Sokolow, Nahum. *History of Zionism.* 2 vols. London: Longmans, 1919.

Spence, Jonathan. *The Question of Hu.* New York: Vintage Books, 1989.

St. Clair, William. *That Greece Might Still Be Free: The Philhellenes in the War of Independence.* London: Oxford University Press, 1972.

Storrs, Sir Ronald. *The Memoirs of Sir Ronald Storrs.* New York: G. P. Putnam and Sons, 1937.

Stoye, John. *The Siege of Vienna.* New York: Holt, Rinehart and Winston, 1964.

Subrahmanyam, Sanjay. "Taking Stock of the Franks: South Asian Views of Europeans and Europe 1500–1800." *The Indian Economic and Social History Review* 42 (2005): 6–100.

Suetonius (Gaius Suetonius Tranquilus). *Lives of the Twelve Caesars.* Trans. Robert Graves. New York: Welcome Rain, 2001.

Talbott, Strobe. *A Gathering of Tribes: The Story of a Big Idea.* New York: Simon & Schuster, 2008.

Talleyrand-Périgord, Charles-Maurice-Camille, prince de. *Mémoiries du Prince de Talleyrand.* 5 vols. Paris: Calmann Lévy, 1891–92.

Tarn, W. W. *Alexander the Great.* Cambridge: Cambridge University Press, 1948.

Tavernier, Jean-Baptiste. *Travels in India.* 2 vols. Trans. V. Ball. London: Macmillan and Co., 1889.

Thomas, Rosalind. *Herodotus in Context: Ethnography, Science and the Art of Persuasion.* Cambridge: Cambridge University Press, 2002.

Thompson, Norma. *Herodotus and the Origins of the Political Community.* New Haven, Conn.: Yale University Press, 1996.

Tocqueville, Alexis de. *Tocqueville sur l'Algérie, 1847.* Ed. Seloua Luste Boulbina. Paris: Flammarion, 2003.

Tolan, John. *Saracens: Islam in the European Medieval Imagination.* New York: Columbia University Press, 2002.

———, ed. *Medieval Christian Perceptions of Islam: A Book of Essays.* New York: Garland Press, 1996.

Toynbee, Arnold. *Turkey: A Past and a Future.* New York: George H. Dorn, 1917.

al-Turk, Niqula ibn Yusuf. *Histoire de l'expédition des Français en Égypte par Nakoula-El-Turk, publiée et traduite par Desgranges aîné.* Paris: Imprimerie Royale, 1839.

Valéry, Paul. *Variété: Essais quasi politiques.* Paris: Gallimard, 1957.

Veinstein, Gilles, ed. *Soliman le Magnifique et son temps.* Paris: École du Louvre, 1992.

Venturi, Franco. "Oriental Despotism." *Journal of the History of Ideas* 24 (1963): 133–42.

Veyne, Paul. *L'Empire gréco-romain.* Paris: Seuil, 2005.

Vlastos, Gregory. *Platonic Studies.* Princeton, N.J.: Princeton University Press, 1981.

Volney (Constantin-François Chasseboeuf). *Les Ruines ou méditation sur les révolutions des empires.* In *Oeuvres,* ed. Anne Deneys-Tunney and Henry Deneys. Paris: Fayard, 1989, vols. 4–5.

———. *Voyage en Syrie et en Égypte (1787–1799)* In *Oeuvres,* ed. Anne Deneys-Tunney and Henry Deneys. Paris: Fayard, 1998, vol. 3.

Voltaire (François-Marie Arouet). *Essai sur les moeurs.* Ed. R. Pomeau. 2 vols. Paris: Bordas, 1990.

———. *Oeuvres complètes de Voltaire.* 52 vols. Paris: Garnier Frères, 1877–85.

White, Sherwin A. N. *The Roman Citizenship.* Oxford: Oxford University Press, 1973.

Wills, Gary. "A Country Ruled by Faith," *The New York Review of Books* 53, no. 18 (November 16, 2006).

Wilson, James. *The Works of James Wilson.* 2 vols. Ed. Robert Green McCloskey. Cambridge, Mass.: Harvard University Press, 1967.

Wilson, Woodrow. *The Public Papers of Woodrow Wilson.* 6 vols. Eds. Ray Stannard Baker and William E. Dodd. New York: Harper and Bros., 1925–27.

Wittek, Paul. *The Rise of the Ottoman Empire.* New York: B. Franklin, 1971.

Woodhouse, C. M. *The Philhellenes.* London: Hodder and Stoughton, 1969.

Woolf, Leonard S. *The Future of Constantinople*. London: George Allen and Unwin, 1917.

Worthington, Ian, ed. *Alexander the Great: A Reader*. London: Routledge, 2003.

Wright, Lawrence. *The Looming Tower: Al-Qaeda and the Road to 9/11*. New York: Alfred A. Knopf, 2006.

Zurara, Gomes Eanes de. *Crónica dos feitos na conquista de Guiné*. 2 vols. Ed. Torquato de Sousa Soares. Lisbon: Academia Portuguesa da Historia, 1961.

Abbasids, 173, 194, 206–8, 237, 251, 252–53, 278, 365, 416
'Abduh, Muhammad, 504, 505, 521, 536
Abdûlhamid I, 442–43, 445–46, 450
Abdûlhamid II, 450–52, 454, 456, 477, 488, 499
Aboukir, battle at, 395, 396, 397
Abû Bakr, 175, 178–79, 180, 202
Achaemenid dynasty: and Alexander, 32, 52, 60, 61, 62, 63; Aristides's views about, 101; decline of, 46, 47, 52, 66, 350; and divine kingship, 64–65; and Egypt, 26, 46, 47, 364; founding of, 58; and Greeks, 350; heirs of, 123, 124, 352; Islamic overthrow of, 66; lands of, 61; and Ottoman Empire, 353; and Pahlavi rulers, 57–58, 508; and Parthians, 350; power of, 21; rise of, 6; Sassanid restoration of legacy of, 175–76; succession in, 54; Susa as capital of, 20; as threat to West, 156; unification under, 62; *vanitas* of, 84; and Zoroastrianism, 13, 20–21, 154. *See also* Cyrus the Great; Darius I
Achilles, 50, 51, 59, 64, 258, 270, 332
Acre, 233, 235, 238, 242, 243, 250, 389, 396, 399–400, 418
Acropolis, 34, 49, 53, 435–36
Actium, battle at, 91–93, 95, 282
ad-Din, Nasir, 477, 478–79
Aeneas, 72, 73, 143
*Aeneid* (Virgil), 72, 73, 92, 244
Aeschylus, 8–10, 14, 21, 35, 44

Afghanistan, 59, 61, 249, 509–10, 512, 514, 516
Africa, 129, 151, 183, 323, 324, 348, 471
Agamemnon, 5, 29, 49, 51, 332
Ahmadinejad, Mahmoud, 534–35
Ahura Mazda "Light," 20, 50, 56, 154, 176
Ain Jalut, battle of, 252, 516
al-Afghâni, Jamâl ad-Dîn, 382, 504, 506, 521
al-Andalus, 186, 189, 190, 191, 192, 196, 197, 198, 219
al-Aqsâ Mosque (Jerusalem), 232, 233, 234, 239
al-Fârâbi, 203, 208, 209, 521
al-Jabarti, 'Abd al-Rahman, 377–79, 381, 389, 391, 394–95, 410
al-Jazzar, Ahmad Pasha, 396, 398, 399
al-Kindî, 201, 203, 208, 209
al-Ma'mûn, 196, 200, 207
al-Mansûr, 187, 195–96, 200, 207
al-Qadisiyya, battle at, 177, 178–79, 464
al-Qaeda, 502, 522, 523
al-Wahhâb, Muhammad ibn 'Abd, 421, 514, 522, 523, 524
al-Zawahiri, Ayman, 512, 516
Alaric, 106, 126, 129, 145, 267
Albania, 257, 277, 285, 434
Albigensians, 156
Aleppo, 176, 235, 460
Alexander the Great: accomplishments of, 70; ambitions of, 27, 51, 103, 357; Aristotle as tutor for, 42–43, 50, 53, 62, 63, 64, 67; and Babylon, 11; and barbarians, 42–43;

Alexander the Great (cont'd)
concept of civilization of, 138; death of, 54, 60; and decline of Macedon, 84; and democracy, 528; divine ancestry of, 63; and Egypt, 364, 366, 369; empire-building of, 54–55; and gods, 50; heirs of, 277, 352; and hellenization of Asia, 76; image of, 68; as "Iskandar," 66; justification of conquests by, 117; legacy of, 61–68, 175; magnanimity of, 53–54; marriages of, 58–59, 60, 62; Mehmet's identification with, 270, 275; mummy of, 93; mutiny against, 59, 62; myths about, 65–66; and need for common culture, 102; Persepolis burning by, 10, 55–58; and Persia, 32, 51–60, 61, 62, 73–74, 350; reconciliation policy of, 52; and rise of Roman Empire, 73–74; role in history of, 416; as role model, 61, 66–67, 94, 137, 366, 367, 369, 374, 398; Roman images of, 76; and Rome as greatest of all civilizations, 101; as ruler of all Asia, 53; Seneca's view about, 90; and Stoicism, 121; world view of, 62, 103

Alexandria, Egypt: Alexander's body in, 61, 93; and Antony and Cleopatra, 89, 90, 91, 94; Bernoyer's views about, 383–84; and Bonaparte's campaign, 373–74, 386, 390, 400; British in, 368, 373, 395; burning of library at, 206; and Byzantine Empire, 142; Caesar's capture of, 88; as "capital" of Roman Empire, 90, 91; modern, 384–85; Octavian in, 93; Persian capture of, 176; Volney trip to, 357, 358, 412

Algeria, 211, 276, 372, 391, 413, 469, 504, 517, 530, 537

'Ali, Muhammad, 414–15, 416, 435, 436, 437, 484, 495

Anabaptists, 304

Anatolia, 79, 192, 252, 253, 254, 260, 266, 274, 455

Ankara, Turkey, 254, 259, 287, 494

Anquetil-Duperron, Abraham-Hyacinthe, 335–43, 344, 350–51, 371

anti-Semitism, 486, 487, 488

Antichrist, 65, 164, 267, 272, 275, 308

Antioch, 142, 176, 231, 232, 233, 235, 236, 239

Antoninus Pius, 38, 69, 71, 100, 111, 148

Antony, Marc, 59, 61, 88–95

apostates, jihad against, 523, 524

Aqaba, 464–65

Aquinas, Thomas, 150, 203, 209, 313

"Arab Renaissance," 193–209

Arabia/Arabian Peninsula, 158, 174, 421–22, 455–66. See also Arabs; Saudi Arabia/Saudis

Arabs: and Balfour Declaration, 489–90;

Bonaparte's views of, 382; and Charlemagne, 192; Christian, 158–59, 168, 169; and Christian views of rise of Islam, 210; and Crusades, 219; cultural identity of, 174; and fall of Istanbul, 471–72; image of, 157, 404; and Islam, 158, 168, 169, 175, 195, 198, 508–9; and Israel, 488, 501–3, 508–9; and Jews as outpost of civilization against barbarism, 491; Lawrence's views about, 469; military abilities of, 158; and Mozarabs, 189–91; muslim roots of, 498; Nasser as hero to, 497; nationalism of, 416, 463, 497–98, 519, 520; and Ottoman Turks, 253, 455–66, 504; and Palestine, 484–93; and pan-Islamicism, 363; and Paris Peace Conference, 474–75; rebellions/uprisings of, 455–66, 482–83, 493; and self-determination, 470, 474; and Sykes declaration, 466–67; and Sykes-Picot agreement, 459–60; and terrorism, 493; unification of, 198, 495–96, 497, 498, 503. See also Arabia/Arabian Peninsula; Saudi Arabia/Saudis; specific person or nation/empire

Arafat, Yasser, 460

Aristides, Aelius: and beneficence/glories of Roman Empire, 29, 69–71, 74, 101–3, 107, 109–12, 121–23, 140, 360; and citizenship in Christianity, 130; and decline of Roman Empire, 122–23; and diversity in empires, 103; and "end of history," 128, 131, 531; and Rome as legitimate ruler of Greeks, 85; and survival of empires, 103; and unification of Roman Empire, 111; and universalism, 100, 104, 121–22; views about Persians of, 102–3; and Xerxes, 29, 38

Aristotle: as Alexander's tutor, 42–43, 50, 53, 62, 63, 64, 67; canonical status of, 295; and climate theory, 345; and Enlightenment, 319; ethics of, 129; exile of, 61; and expansion of Islam, 208; and Mehmed's academy, 350; praise for Romans by, 81; and psyche, 42; and Scholasticism, 313; translations of, 200, 201, 202, 203, 208, 209; and universalism, 118

Armenia, 98, 156, 181, 284, 422

Artaxerxes "Great King," 46, 47, 60

Ascalon, 232, 238, 242

Asia and Europe/West: accommodation between, 538; Aeschylus's views about, 8–10, 14; and Alexander-Persian war, 54–55; and Alexander's legacy, 61, 62, 63, 66; and association of cultures, 333–35; and Athenian Empire, 37–38; balance of power between, 356–57; and Bonaparte's "new Byzantine Empire," 398; and Byzantine Empire, 260–61; and church–state relations, 141; and civilization, 332, 344–55; climate theory about, 345–46; and colonialism, 330–33; and "constitutional debate," 19; and decline and fall of empires, 358–60; differences between, 6–7, 343, 381–82, 524–25; and economic power, 32; enmity between, 51; and European interest in Asia, 330; Europeans' travels to, 326–27; and expansion of Ottoman Turks, 254–55; Greeks as middle ground between, 75; and Hannibal, 82; images of, 44; and Islam, 174; and Jews as outpost of civilization against barbarism, 491; link between, 4; "marriage" of, 59; and Persian Wars, 5–10, 21, 25–26, 29, 41, 44; polarization of, 530; and reform of Islam, 505–6; and Roman civil wars, 88; and Roman Empire, 72–73, 111, 112, 137; in seventeenth century, 326; and superiority–inferiority status, 291–92, 330–31; travels between, 326, 327–28; unification of, 21, 54, 62, 63, 112, 137; and Xerxes interest in Europe, 26; and Xerxes–Demaratus encounter, 31–33; and Zoroastrianism, 343

Asia/East: and Alexander the Great, 42, 53; as barbaric, 84; and Byzantine Empire, 143; and Enlightenment, 323, 324; first Panhellenic invasion of, 49; Greek conquest of, 51; Hegel's views about, 356; hellenization of, 76; images of, 18, 44, 79, 81, 85, 329, 344–55, 401, 405–6, 413, 485; as land of *vanitas*, 78; and Oriental despotism, 344–55; origins of Christianity in, 128–29, 136, 153; Ottoman Turks as masters of, 39; paganism in, 135; and Peace of Paris, 494; Roman conquests in, 79, 80, 83, 85, 95; Russia as part of, 289; Scipio's dream about, 82; in seventeenth century, 326; as source of civilization, 333; as source for redemption, 485; vices of, 81; wealth of, 32. See also Asia and Europe/West; Oriental despotism; Orientalism; Orientalist(s); *specific person, nation, or topic*

at-Tahtâwî, Rifa'ah Rafi', 415, 416, 497
Atatürk (aka Mustafa Kemal), 494–95, 499, 500
atheism, 379, 380–81, 444
Athena (mythological character), 29, 34, 49, 50–51, 52
Athens/Athenians: and Alexander's death, 60; Antony's visit to, 89; consolidation of empire of, 46; constitution of, 23; Council of Five Hundred in, 23; decline of, 47; and

Delian League, 45; democracy in, 7, 15, 16–17, 23, 67, 527, 528; distinctions between Asians and, 7; Duchy of, 276; as empire, 86; Goths' sacking of, 126; and Greek War of Independence, 435–36; Hippias's expulsion from, 23; and Macedon, 48; as naval power, 37; paganism in, 135; Periclean, 68; and Persian Wars, 21–22, 24–25, 27, 28, 33, 34–37, 45–46, 525; power of, 46; preeminence of, 37; re-creation of, 68, 430; and selection of Persian king, 19; wealth of, 56. See also Greece/Greeks; Peloponnesian Wars

Augustus (emperor), 61, 72, 74, 88, 89, 91–95, 96, 97, 101, 128, 130, 150
Aurelius, Marcus, 69, 71, 120, 197
Austria, 147, 279, 288, 300, 301, 362, 364, 367, 432
Averroës, 202, 203, 204, 209
Avesta, 336–37, 340, 341, 343
Avicenna, 201–2, 203, 208, 209
ayatollahs, 170

Babylon, 11, 55, 59, 61
Bactria, 11, 59, 61. See also Afghanistan
Badr, battle at, 167–68
Baghdad: and Arab Revolt, 458; as capital of Islamic Middle East, 195–96, 197, 206–7; Christians in, 207; and creation of Iraq, 483; and Crusades, 236; and expansion of Islam, 195–96, 197, 206–7, 208; fall of, 11, 466, 526; and Gulf War, 514; and Iraqi War, 526; Jews in, 467; as mixed society, 207; Mongols sacking of, 251; rule of, 231; Shiite as predominant in, 467; and Sykes–Picot agreement, 459; translators in, 200, 208–9
Balfour Declaration, 489–90, 491, 492
Balkans, 156, 230, 257, 274, 362, 413, 423, 452, 453, 455
balloons, hot-air, 394–95, 404
barbarians: and Alexander the Great, 42–43; Asiatic, 84; categories of, 77; and East as land of *vanitas*, 78; Europeans/Westerners as, 77, 143, 193; and Greeks, 42–44, 45, 77, 102, 115; Jews as outpost of civilization against, 491; and Macedon, 48; meaning of term, 42; and Romans, 74, 77–78, 82, 104, 106–7, 127; and universalism of Christianity, 139; as wicked, 67
Basra, 458, 459, 467, 483
Ba'th party, 497–98, 513, 521
Battle of the Pyramids, 387, 393, 419
Bedouins, 206, 385, 386, 396, 397, 404, 460, 462, 464, 465, 466

Berbers, 181, 183, 206

Bernoyer, François, 383–84, 386, 392, 395, 404, 408–9, 411

Bible: Amiroutzes' views about, 271; Avesta compared with, 343; and Gnostic Gospels, 294–95; interpretation of, 295; Islamic interest in, 198; Judaic influence on, 294; King James, 302; Luther's translation of, 302, 303; New Testament of, 129, 155, 161, 211, 212, 294, 295, 302; Old Testament of, 155, 161, 211, 212, 294; parts of, 294; Qur'an compared with, 294, 295; Qur'an as parody of, 214; and Scholasticism, 313; and science, 313; Vulgate, 295, 300, 303

bin Laden, Osama, 501, 502, 503, 512, 513, 514, 515, 519, 523, 524, 532

Bismarck, Otto von, 454, 487

Bonaparte, Napoleon: accomplishments of, 396–97; Alexander as model for, 61, 366, 367, 369, 374, 398; ambitions/mission of, 368–69, 398, 417–19; and Britain, 367, 368, 372–73, 399; and church-state relations, 418; Egyptian campaign of, 93–94, 246, 366–70, 372–400, 401–2, 409–10, 413–14, 415, 416, 417, 419, 420, 444, 495, 503, 518; and fall of Holy Roman Empire, 147; and Greece, 425; image of, 389; and India, 367; and Islam, 374, 375–76, 377, 378, 379–80, 387, 388–89, 393, 413–14, 417–18; and Italy, 377; and Judaism, 417–19, 484–85, 486; as "Muhammad of the West," 375–76, 398; and "new Byzantine Empire," 397; as new Mahdi, 388; as Oriental despot, 390; as "Orientalist," 365–66, 388–89, 413–14; and Ottoman Empire, 367, 368, 413, 417, 420, 441; and Palestine, 484–85, 486; and Persia, 477; and religion, 369–70, 376–77, 417–18; Russian invasion by, 421; self-transformation of, 401; and slavery, 408; and Volney, 360

Britain: and Arabs, 455–67, 468, 493; and Bonaparte, 367, 368, 372–73, 399; Caesar's conquest of, 287; church-state relations in, 299, 307; and colonialism, 330; and contract theory of government, 318; and Damascus, 469; and East-West relations, 327; and Egypt, 368, 373, 374, 389, 395, 400, 414, 416, 437–38, 455–58, 471, 475, 484, 495, 496–97; emergence of Protestant, 308; and France, 338, 340, 367, 368, 372–73, 389, 395–96, 399, 400, 469–70, 475, 485; "Great Game" of, 477; and Greek War of Independence, 432, 436–37; and India, 347, 356, 367, 368, 417, 456, 477; and Iran, 479, 480, 500;

and Iraq, 459, 467, 475, 483–84, 513; and Israel, 493; Istanbul occupation by, 291; and Jordan, 459, 475, 482; legal system in, 449; mandates of, 481–82; medicine in, 319; and Mesopotamia, 482; Middle East policies of, 482–83; Muslim immigrants in, 537; and Ottoman Empire, 259, 272, 291, 437, 439, 454–66, 469, 470; and Ottoman-Byzantine war, 262; and Ottoman-Russian relations, 290; and Palestine, 417, 419, 464, 468, 475, 482, 485–93; and Paris Peace Conference, 473–75; parliamentary system in, 528; and Persia, 26, 475, 478, 500; Roman conquest of, 105, 106; Saladin tomb flag in, 241; and self-determination, 470, 471, 475; and Shiites and Sunnis, 482–83; in Suez, 368; and Sykes-Picot agreement, 459, 469, 470; and Syria, 469, 475; and Treaty of Westphalia, 307; and World War I, 454–66, 468–69

Buda, battle at, 288

Bulgaria, 257, 361, 453

Bush, George W., 248, 368, 465, 534–35

Byron, George Gordon, Lord, 432–34, 435

Byzantine Empire: Arab conquest of, 248; and Arabs, 158–59, 192; and Asia, 143; attempts to restore, 397, 425, 442; and "barbarian" West, 143; beginning of, 142; Bonaparte's "new," 397; Charlemagne recognized by, 146; and Christianity, 143; and church-state relations, 143, 144–45, 255–56, 260–61; and coming of Islam, 157; and Crusades, 230, 231, 241–42, 248, 350; decline and fall of, 12, 252, 254–55, 260–69, 350; and despotism, 352; diarchy of, 145; and division of Middle East, 175; and East-West relations, 260–61; emperors of, 144; and expansion of Islam, 158, 159, 191–92, 195, 248; geography of, 143; and Greeks, 142, 143; hellenization of, 142–43, 175; images of, 143; influence on Islam of, 182; and Mongols, 259–60; and Ottoman Turks, 142, 254–55, 260–69, 276–77, 353, 424; and papacy-patriarchy relations, 255–56; and Persia, 143; and Russia, 288; Saladin treaty with, 239; and Sassanids, 176. *See also* Constantinople

Caesar, Julius, 61, 65, 86–87, 88, 103, 190, 287

Caesarea, 132, 133, 134

Cairo, Egypt: Bonaparte's address to people of, 388–89; and Bonaparte's Egyptian campaign, 386–91, 396, 399, 400; British in, 437; Ramses II statue in, 495; Shiite

Caliphate in, 181, 231; Volney's trip to, 357–58

Caligula (emperor), 80, 96, 98–99, 117

caliphates: dissolution of, 494; Fatimid, 231, 232; Hegel's views about, 356; Istanbul as seat of, 472, 473; as monarchs, 182; as Muhammad's successors, 164; Ottoman Empire as replacement for, 270; popes as analogous to, 363, 456; and religion, 207; role in history of, 416; and "sultan-caliphs," 362–63; Sunnis as, 279; and translations of ancient work, 200; Umayyad, 179, 180, 181, 194, 195, 196, 416; universalism of, 497. *See also* Abbasids; *specific caliph*

Calvinism, 303, 304, 440

Campo Formio, Treaty of, 367

Cannae, battle at, 82

Caracalla, Edict of, 110

Carlowitz, Treaty of, 290–91

Carolingian Empire, 147, 197, 228

Carthage/Carthaginians, 80–84, 348

Castile, 189, 213, 219, 220

Cathars, 156, 296

Catherine (empress of Russia), 442–43

Catholics/Catholic Church: as Anticrist, 308; and expansion of Ottoman Turks, 254–55; founding of, 138; and impact of Reformation, 308; and Luther, 297–304; origins of, 138; persecutions of Protestants by, 226; priests of, 303; and Reconquest, 220; in Spain, 186; and theology, 254–55; and Treaty of Westphalia, 306–8; and Wars of Religion, 304–6; "works" doctrine of, 298, 302. *See also* papacy; *specific person*

Chabrol, Gilbert-Joseph Volvic de, 385, 402, 403, 404, 405, 407, 411–12

Chaeronea, battle of, 7, 48

Chardin, Jean, 326, 328, 329–30, 342, 353, 354–55

Charlemagne, 146–47, 185, 197, 207, 216, 217, 254, 273

Charles V, 277, 278, 280, 281, 301, 302, 304

Chasseboeuf, Constantin-François. *See* Volney, Constantin-François

Chateaubriand, François-René, vicomte de, 245–46, 385

Chernomen (Cirmen), battle of, 257

China: and Asia-Europe relations, 356; and democracy, 525, 526–27; and despotism, 346, 353; as empire, 86; European images of, 346, 353, 355; and expansion of Islam, 195, 196, 198, 199; and geographical exploration, 309–10; Hegel's views about, 356; and Manichaeism, 155–56; Mongols from, 182, 352; Qing emperors of, 356;

and Soviet Union, 524; and Spain, 151; Western impact on, 506, 538

choice, human, as basic to West, 533

Christ, Jesus: and Arians, 137; Asian origins of, 129; and Bible, 294; birth of, 128; call for followers by, 225; and church-state relations, 150, 165, 167, 424; and Constantine, 133; and divisiveness within Christianity, 295–96; and end of time, 164; and ethics, 152; and Europe, 128–29; Graindor's poem about, 227; influence on Bush of, 535; Islamic views of, 211, 212; kingdoms of, 131–32, 140, 150; and Last Supper, 255; Lazar represented as, 257; as messenger of God, 155; miracles of, 214; Muhammad's meeting with, 160; and origins of Christianity, 128; as prophet, 168, 379; Qur'an compared with teachings of, 299–300; Second Coming of, 128, 140, 141, 151, 294; and Stoicism, 120; and universalism, 138, 164; wise men pay homage to, 13. *See also* Bible; Gospels

Christianity/Christians: as apex of civilization, 128; breakaway churches of, 296; and Byzantine Empire, 143; and changing world, 161; characteristics of, 130–31, 186; and Christians as victims of Rome, 97; and church-state relations, 139–41, 145–51, 152, 165, 167, 299, 532–33; and citizenship, 110, 113–15, 130; and City of God, 141; *civitas* of, 128, 141; clergy in, 170; as community of believers, 139; and Constantine, 132–38; conversions in, 171; and differences between East and West, 381–82; divisiveness within, 141–42, 146–47, 243, 276, 294, 295–96, 297–308; and Edict of Milan, 133–34; and "end of history," 532; and expansion of Islam, 159, 164, 168–69, 172–74, 176, 183, 184–85, 186–94, 219; and free will, 152; fundamentalism in, 535–36; good-evil in, 153–54; and Greeks, 141; Heavenly City for, 130; as "Holy Republic," 136; and image of Arabs, 157; internal divisiveness within, 153–54; in Iraq, 143; irreconcilable differences between West and, 524; Islam as similar to, 211–12; Islamic contribution to, 208; and Judaism/Jews, 128, 136, 174; as "the Law," 135; and Manichaeism, 155, 156; martyrdom of, 130; in Mediterranean world, 183; Montesquieu's portrayal of, 328; as mystery cult, 131; as one true religion, 279; origins of, 128–29, 136, 153; and Ottoman Empire, 422, 440; paganism's influence on, 152–53; as "peoples of the Book," 172–73; persecution of, 135–36,

Christianity/Christians (*cont'd*)
173–74; and Reconquest of Europe,
210–22; as "republic of all the world," 118;
and rise of Islam, 210–22; and Roman
Empire, 97, 110, 113–15, 127–38, 143–44,
145; and salvation, 201; and science, 205,
208; and slavery, 139; spread of, 79–80,
130–31; as state religion, 129; and
Stoicism, 138; strengths of, 148; survival
of, 296–97; theology of, 200, 206, 211,
212, 295–96; transformation of, 291–92;
and translations of ancient works, 209;
and Trinity, 137; unity within, 294;
universalism of, 124, 128, 130, 134, 138,
152, 164, 165; virtues of, 76–77, 130. *See
also specific person or topic*
Church, role in world of, 296
"Church Fathers," 295, 296, 313
church–state relations: and Asia–Europe
relations, 141; and Bonaparte's Egyptian
campaign, 378–79; Bonaparte's views
about, 418; and Byzantine Empire, 143,
144–45, 255–56, 260–61; and Christianity,
139–41, 145–51, 152, 165, 167, 299,
532–33; and *civitas*, 141; and Constantine,
165–66, 300; and democracy, 536; and
"Dictate of the Pope" (1075), 148–50; and
Enlightenment, 318; in Europe, 148–50,
354; formula for, 307; in Germany, 299;
and individual freedom, 151–52; and
"Investiture Controversy," 149–50; and
Islam, 164, 165, 167, 299–300, 536; and
Islam–Christianity differences, 532–33;
and legal system, 150; and Oriental
despotism, 354; in Ottoman Empire,
271, 424, 445, 446–47; and paganism,
141; and papacy, 141, 147–51, 299; and
papacy–patriarchy relations, 254–56,
260–61; and Reformation, 304; and
Roman Empire, 137–38, 139–41, 143,
147–51; and secularization, 533–34; in
Spain, 190–91; and Treaty of Westphalia,
307; in United States, 535–36; and values,
141; and war, 307; and Zoroastrianism,
167
Churchill, Winston, 455, 472, 480, 482, 483,
484, 490, 491, 529
Cicero, Marcus Tullius: and barbarians, 77;
and beneficence of Roman Empire, 111;
and Carthage, 80; and *civitas*, 104; and
Greek character, 75, 80; on Herodotus, 5;
as pagan, 202; and Parthians, 85;
"republic of all the world" of, 136; and
Roman Empire as world empire, 101, 102;
slave of, 108; universal human republic of,
115, 118–19; on warfare, 117, 118
citizenship, 110–15, 130, 139. *See also* civitas

civilization(s): Asia/East as source of human,
333, 334; at-Tahtâwî's defense of, 415;
Christianity as apex of, 128; "clash" of,
538; and East–West relations, 332; and
Enlightenment, 322; Greece as "cradle" of
Western, 428–29; and Jews as outpost of
civilization against barbarism, 491; Jones'
views about, 332; rise and fall of, 358–60,
361–62; and Western values, 138
*civitas*, 104–6, 109, 110–18, 121, 123–24, 128,
141, 151, 153, 292
classical writings, 161, 198–204, 208–9, 313.
*See also specific source*
Claudius (emperor), 70, 74, 77, 99, 110–11, 126
Clavijo, battle of, 186
Cleisthenes, 16–17, 23, 28
Cleopatra (queen of Egypt), 59, 88–94
climate theory, 345–46
clocks, 439, 440
Cluny (Benedictine abbey), 187–88, 213
Code Napoléon, 397, 415
colonialism, 330–33, 366, 401, 460, 470, 471,
496, 517
Columbus, Christopher, 221, 310
Condorcet, Marquis de, 322–23, 324, 349, 383,
527
Congress of Mantua (1459), 273
conscientious objectors, 150–51
Constantine the Great: and Charlemagne's
coronation, 146; and Christianity,
132–38; and church–state relations, 138,
143, 147, 165–66, 255, 300; *civitas* of,
123–24; death of, 275; and Delphi's
serpent monument, 38–39; and divine
kingship, 144; and division of Roman
Empire, 132; and founding of
Constantinople, 137; and Hagia Sophia,
269; invasion of Italy by, 132; at Malvian
Bridge, 132–33, 136; and paganism,
132–33, 135; reunification of Roman
Empire by, 136; Rome captured by,
132–33
Constantine IX (emperor), 261, 262, 265–66
Constantinople: as capital of Ottoman
Empire, 39; as capital of Roman Empire,
261; and Charlemagne's coronation, 146;
Christian concerns about fall of, 272–73;
Christians in, 173–74, 207; and Crusades,
230–31, 242–43, 261, 267, 276; culture of,
207; excommunication of patriarch of,
255; and expansion of Islam, 179; fall of,
179, 242–43, 272–73; founding of, 137; as
"Golden Apple," 261–62, 267, 270, 271;
holy relics of, 243; Kingdom of, 243; as
"New Rome," 137, 142; Ottoman
conquest of, 253, 258, 260, 262–73, 274,
288, 350, 472; Rome compared with, 142;

serpent monument in, 38–39; Süleyman's attack on, 191–92. *See also* Istanbul
"Constitution of Medina," 163–64, 165
contract theory of government, 318–19
Copts, 66, 396, 397
Córdoba, Spain, 190, 196, 197, 198, 208, 219
cosmopolitanism, 112, 120, 138
Council of Florence, 260
Crimea, 362, 363, 364, 420, 442
Croesus (king of Lydia), 10, 21, 22
Crusader States of the Levant ("Outremer"), 233–35, 236, 239, 242, 502, 514, 522
Crusades: as acting on God's behalf, 226–27; and Arab destruction of Israel, 501–2; battle cry of, 224; and Bonaparte's Egyptian campaign, 375; and breakaway Christian sects, 276; and collapse of Byzantine Empire, 350; and Constantinople, 261, 267; and Egypt, 364; end of, 243–44; financing of, 276; First, 217, 223–35, 244; Fourth, 242–43; and Hattin battle, 238, 248–49; historical significance of, 244–50; as holy war, 235; incompetence of, 367; inspiration/ motivation for, 146, 367; and Jews, 229–30, 233, 234; Knights of Saint John as relics of, 372; Muslim views of, 247–50; Nicholas's call for, 272; objectives of, 224, 225, 228; and Ottoman Empire, 272, 276; Reconquest as, 219, 222; as religious fanaticism, 244–45; and resurgence of Islamic fundamentalism, 508; role in history of, 416; romantic views of, 245; Second, 213; *The Song of Roland* as text of, 217; terrorism as phase of war dating back to, 516; Third, 242; three, 503; Urban's call for, 224–28; as war of terror, 227; of words and ideas, 503–4
Ctesiphon (Persia), 159, 176, 178
Cultural Revolution, 524
culture: of Islam, 196, 197–204, of Sassanid Empire, 206–7; in Spain, 188–89, 197–98, 202–3; and survival of empires, 102
Curzon, Nathaniel, Lord, 470, 472, 478, 481–82, 490, 491–92
Cyprus, 46, 90, 247, 280, 281, 282
Cyrus "the Great," 8, 10–12, 17, 21, 55, 56, 57, 65, 490
Czech Republic, 147, 300, 301, 305

Damascene, John, 212, 214
Damascus: Arab capture of, 176; and Britain, 469, 471; and Crusades, 235, 236, 237, 241, 244; Faisal exiled from, 475; opposition to caliphate in, 180; political significance of, 469; Saladin statue/ tomb in, 241; splendor of, 197;

Volney's trip to, 358; Wilhelm II visit to, 241
Dante, 202–3, 217
*dar al–Islam*: and Abbasids, 196; and Arab–Israeli conflict, 501, 509; and Buda, 288; and Crusades, 230, 233, 504; and divided Middle East societies, 537; and Egypt, 364, 497; and fall of Istanbul, 472; and Gulf War, 514; and Islamic domination of Rome, 271; Islamic Iberia as part of, 186; and *jihad*, 172; and Mozarabs, 189; official languages of, 198; and Ottoman capture of Constantinople, 268; and Treaty of Carlowitz, 291
Darius I (king): accomplishments of, 20–21, 26; in Aeschylus's play, 44, 477, 528; and Alexander–Persian war, 51–52, 53–54, 55; death of, 26, 68; as empire builder and lawmaker, 5, 20, 21; family of, 53–54, 60; image of, 20; motivation of, 62; Pahlavi as heir of, 57; and Persian "constitutional debate," 14, 16, 18; and Persian Wars, 8, 21–26, 29; selection as king of, 20; and Zoroastrianism, 154
Declaration of the Rights of Man, 388, 393, 518
"Defenestration of Prague," 305
Demaratus, 31–33, 115
democracy: and Alexander's legacy, 67; basic assumption of, 531; and characteristics of modern West, 517; and church–state relations, 536; and contract theory of government, 319; and despotism, 352; and Enlightenment values, 318–19, 535; Greek/Athenian, 15, 16–17, 23, 67, 527; and imperialism, 527; Indo–European experiment in, 335; "introduction" of, 529–30; in Iraq, 525; and Islam as religion of submission, 518–19; Islamic resistance to, 382; and Islamic terrorism, 516; as *isonomia*, 14–15; and "loyal opposition," 528; as major distinction between East and West, 525; and Muslims, 525–31; and nationalism, 499; and Ottoman Empire, 449, 450; Persian "constitutional debate" about, 13–17; and Persian Wars, 25, 39, 44; principles/characteristics of, 15, 19; and reform of Islam, 505; and regime change, 525–31; and revolution, 529; and Roman Empire, 71; and Salamis as Greek victory, 37; and self–determination, 471; in Turkey, 494; as unnatural, 528; as weak and corrupt, 516
Deng Xiaoping, 526–27
Denon, Dominique–Vivant, 382–83, 396, 401, 403, 405, 406
Descartes, René, 313, 314, 440

despotism: and Egypt, 404–6; and images of
East, 413; Oriental, 38, 346–55, 383–84,
390, 452, 457; of Ottoman Empire, 413;
and women, 407, 411
dialogue, West-Islam, 517–18
"Dictate of the Pope" (1075), 148–50
Diet of Worms, 301
Diocletian (emperor), 65, 105, 106, 123, 132,
146, 155
diplomacy, and *jihad,* 291
distinctiveness, origin of European sense of,
6–7
divine kingship, 63–64, 144
divorce, 215, 299
*diwan,* 388, 390, 393
Donne, John, 312, 313, 314
Dragatsani, battle of, 429
Dreyfus, Alfred, 486
dualist religions, 153–56

East India Company, 331, 356
East. *See* Asia and Europe/West; Asia/East
Edirne, 254, 263, 271, 285
Egypt/Egyptians: and Abbasids, 237; and
Achaemenid dynasty, 26, 46, 47, 364; and
Alexander, 55, 63, 64, 364, 366, 369; and
Antony and Cleopatra, 88–95; and "Arab
Renaissance," 209; and Arab–Israeli
conflict, 502; and Arabs, 364, 495–96, 497;
as barbarians, 77; and Ba'th party, 498;
Bonaparte's campaign in, 93–94, 246,
366–70, 372–400, 401–2, 409–10, 413–17,
419, 420, 444, 495, 503, 518; and Britain,
368, 373, 374, 389, 395, 400, 414, 416,
437–38, 455–58, 471, 475, 484, 495,
496–97; Caesar in, 88; and church-state
relations, 418; and climate theory, 346;
Coptic Church of, 66; and Crusades, 232,
243, 364; and *dar al–Islam,* 364; decline of,
361; democracy in, 530; and despotism,
404–5; and France, 365–400, 401, 413,
414–16, 495, 496–97; Fû'ad I as king of,
484; fusion of Greek culture and, 94; gods
of, 89, 94; images of, 401, 402, 403–4; and
Islam, 179, 180, 181, 206, 209, 504, 505,
517; and Israel, 496–97; July Revolution
in, 495; Kingdom of, 484; and Mamluks,
252; modernization of, 413–16, 495;
Muslim Brotherhood in, 247, 505;
nationalism in, 415–16, 484, 495, 497;
and Ottoman Empire, 285, 364–65, 368,
374, 400, 416, 436–38; and Persians, 11,
12, 26, 38, 46, 47, 176; in post–World War
I years, 495–97; Ptolemy capture of, 61;
and Qutb, 521; religion in, 12, 181, 406–7;
and Roman Empire, 364; "savants" in,
370–71, 391–95; as source of ideas for

Greeks, 402; as source of Persian and
Greek science and philosophy, 344; and
Soviet invasion of Afghanistan, 510; and
Soviet Unions, 496; and UAR, 498; as
under suzerainty of Baghdad Caliphate,
237; and U.S., 496, 497; Volney's trip to,
357–58, 371–72, 384, 412
Egyptomania, 364
Eighty Years' War, 304–5
elections, and democracy, 529–30
Eliot, George (pseud.), 485–86
emperors: and church-state relations, 141;
divine right of, 144; as gods, 144; as God's
representatives on Earth, 144; kings
distinguished from, 97; titles of, 144. *See
also specific person or nation*
empires: building of, 46, 54–55; and culture,
102; decline of, 85; diversity in, 103;
emergence of Islamic, 182; Livy's views
about, 103; as monarchies, 86, 95; rise and
fall of, 358–60, 361–62; Rome as last of
world, 101–2; survival of, 102, 103. *See also
specific ruler or empire*
"end of history," 128, 131, 164, 531–35
England. *See* Britain
Enlightenment: at-Tahtâwi's reconciliation of
Islam with, 415; and Bonaparte's
Egyptian campaign, 391, 393; and
Crusades, 245; and democracy, 318–19,
535; enduring values of, 535; impact of,
320–25; and Judaism, 417; and morality,
316–17, 322; Orientalists of, 333–35; and
philosophy, 315–18, 322, 325; pre-
domination of values of, 535; and
rationalism, 317–18, 320, 322, 324, 325;
and reason/rationalism, 317–18, 320, 322,
324, 325; and reform efforts, 443; rise of
secular, 329; Rousseau's views of, 404; and
science, 315–20, 322, 323, 325; values of,
321–22, 325, 535; Western images of
Orient during, 335–43; and women, 411
Enver Bey, 451, 452, 453
Erasmus, Desiderius, 299, 310
ethics, 15, 152
"ethnic cleansing," 61
ethnocentrism, 19, 43
Euclid, 208, 209
Europa (mythological character), rape of, 3–4,
5, 27
"Europe of Nations," creation of, 147
Europe/West: balance of power in, 429; as
barbaric, 77, 143, 193; and Byzantine
Empire, 143; chaos in, 146; characteristics
of modern, 517; and Christ as European,
129; church-state relations in, 141, 144,
148–50, 354, 533; as "civilization," 292;
constitutions of, 71; Darius's advances

into, 23–24; despotism in, 384; dialogue between Islam and, 517–18; discussions of best forms of government in modern, 13; expansion of, 11, 291; failure of, 519–20; feudalism in, 351; freedom/choice in, 39, 533; and Greek War of Independence, 431–32; as heir to Roman Empire, 350; identity of, 292; images of, 44, 75; individualism of, 147; integration of Muslims in, 537; irreconcilable differences between Islam/Muslims and, 524, 538; Islam indifference to culture of, 199–200; kingship in, 97; legal system in, 115, 116, 117, 167; meaning of term, 323; medieval, 216; mythological stories about founding of, 3–5; naming of, 3, 6; new conception of, 292; Orient as inferior to, 330–31; Oriental studies in, 198; origins of civilization in, 74, 76, 343, 371; poverty of, 32; and Roman Empire, 76, 79, 103; science/medicine in, 202, 208; secularization of, 244; superiority of, 404; terrorism's role in Islam's struggle with, 516–17; and translations of ancient works, 209; and treaties for lasting international peace, 306–7; and universalism, 124; values/virtues of, 76, 88, 138, world view of, 292; Xerxes interest in, 26, 27. See also Asia and Europe/West; specific person, nation, or topic
European Union, 247, 448, 530
Eusebius, 132, 133, 134, 136
excommunication, 145, 149, 301
exploration, geographical, 309–10

Faisal I (emir/king), 460, 463, 464–65, 469, 474, 475, 483, 484, 492
Farouk, 416, 495, 502
fatalism, 406–7
Feraios, Rigas. See Velestinlis, Rigas
Ferdinand II, archduke, 277, 279, 280
Ferdinand, 186, 220–21, 439
Field of Blackbirds. See Kosovo Polje
France: and Algeria, 391, 413; and Arabs, 193–94, 459; Asian's travels to, 327–28, 329; and Balfour Declaration, 490; Bonaparte's return to, 395–400; and Britain, 338, 340, 367, 368, 372–73, 389, 395–96, 399, 400, 469–70, 475; Cathars in, 156; and church-state relations, 307; colonization by, 330; constitution of, 425, 431, 434; and contract theory of government, 318; and Crusades, 224, 225, 229, 245, 246; and democracy, 527; and East-West relations, 327; and Egypt, 365–400, 401, 413, 414–16, 495, 496–97; "Great Game" of, 477; and Greece, 425,

431; and Greek War of Independence, 432, 436, 437; in India, 340; and Italy, 399; Jacobins in, 527; and Lebanon, 459, 484; and Libya, 413; mandates of, 481–82; Montesquieu's views about, 327–28, 329; Napoleonic, 366–400; nationalism in, 474; and Ottoman Empire, 272, 284, 362, 380, 396, 397, 439, 440, 441, 444, 447, 454; and Ottoman-Byzantine war, 262; overseas empire of, 151; and Palestine, 246, 413, 489, 490, 492; and Paris Peace Conference, 473–75; as part of Holy Roman Empire, 301; and Persia, 477; power of king in, 228; reform efforts in, 445; Republic of, 367, 378, 384, 393; rise of kingdom of, 147; and self-determination, 470; and Sykes–Picot agreement, 459–60, 469, 470; and Syria, 246–47, 413, 459, 469, 475, 482, 484; and Treaty of Westphalia, 307; Vendée in, 399–400; and Wars of Religion, 304, 449; and women, 475; and World War I, 455, 459. See also Franks; French Revolution; specific person
Franco, Francisco, 150, 187
Franks, 185, 192–93, 194, 198, 199–200, 219, 439, 444; and Crusades, 230, 231, 233, 234, 235, 241. See also Charlemagne
French Revolution, 318, 322, 369, 374, 376, 377, 378, 382, 383, 393, 399, 425, 452, 529

Galen, 199, 202, 208, 209, 319
Galileo Galilei, 313, 316, 319, 440
Gaul, 88, 98, 103, 104, 110, 123
Gaza Strip, 398, 496, 502
German Democratic Republic (GDR), 526
Germany/Germans: Asian, 500, 507; Carolingian Empire in, 147; church-state relations in, 299, 300, 304; creation of nation of, 487; and Crusades, 229–30, 245, 246; emergence of Protestant, 308; and fall of Rome, 145; "Great Game" of, 477; as Holy Roman Empire, 300–301; kingship traditions in, 300–301; legal system of, 115; and Luther, 297, 299, 303; and Ottoman Empire, 241, 453–54, 455, 457, 461, 463; and Palestine, 490; Peasants' War in, 303–4; politics in, 299; rise of kingdom of, 147; Roman defeats by, 131; and Treaty of Westphalia, 306; and Turkey, 453–54, 457; and Wars of Religion, 306; and World War I, 453–54, 455, 457, 461, 463
Gibbon, Edward, 71–72, 116, 123, 193, 194, 210, 240, 243, 245, 349, 358
Gladstone, William, 454, 472
Godfrey of Bouillon, 230, 244, 246, 248

gods/goddesses: and Alexander the Great, 50, 63; Egyptian, 94; emperors as, 144; Greek, 73, 135; and Persian Wars, 44; as punishing Rome, 126; role in human affairs of, 44; Roman, 73, 80, 127, 130, 131, 132–33, 135; and Xerxes, 30. *See also* paganism; *specific god/goddess*

Golden Age: as end of history, 131; of Latin literature, 96; of Roman Empire, 71–72, 96–97, 131

good–evil dualism, 153–55, 217

"Gordian knot," 52–53

Gospels, 120, 126, 128, 134, 155, 161, 162, 168, 202, 294–95, 296, 300, 380, 423, 535

Goths: and image of Romans, 107; romanization of, 182; sacking of Rome by, 106, 126–27, 129, 131, 145, 267, 270; in Spain, 183, 184, 185, 186, 197, 219

government: contract theory of, 318–19; as major distinction between East–West, 524–25; Persian "constitutional debate" about, 13–17. *See also* politics

Granada, 186, 197, 219–20, 221–22

Granicus, battle at, 51, 52, 61

Great Britain. *See* Britain

Great Mosque (Istanbul), 278–69, 421

Greece/Greeks: accomplishments of, 353; and Alexander, 61, 62–63, 67–68; ancient objects in, 371; and Asia–Europe relations, 174; Asians distinguished from, 7; Augustinian views of, 140; as Balkan League member, 453; and barbarians, 42–44, 45, 77, 102, 115; beginning of empire of, 350; and Bonaparte, 413, 425; and Byzantine Empire, 142, 143; and Christianity, 139, 141; as Christians, 173; city-states of, 8, 11, 46, 61, 201; and climate theory, 346; conquest of Asia by, 51; constitution of, 425, 431, 434; as "cradle of Western civilization," 428–29; and Crusades, 242; decline of, 292, 361, 413; and Delian League, 45, 46; and despotism, 350; discussions about best form of government among, 13; divisiveness among, 47; and Egypt, 94, 402; and Enlightenment, 322; as ethnocentrists, 43; and fall of Constantinople, 263; and France, 365; and free will, 152; and fusion of Greek and Roman culture, 74–77; gods of, 73, 135; and growth of civilization, 128; images of, 27–28, 75–76, 79, 81, 350; images of Persians among, 13–14, 17, 64, 65; Jones's views about, 332; king of, 437; legal system of, 32–33, 67, 115, 117; Marathon as turning point in history of, 25–26; and Mehmed the Conqueror, 427;

as middle ground between Asia and Europe, 75; orientalization of, 76; origins of, 6, 337, 343; and Ottoman Turks, 253, 423–24, 425–37, 494; and Persia, 5, 7–10, 11, 13, 19, 41, 44, 48–49, 59, 102, 115; Phanariot, 423, 424; religion in, 63; and Renaissance, 309; as republic, 423, 431; and Roman Empire, 70, 73, 85; and Sassanids, 176; and slavery, 108, 109; as Stoics, 119; strength of, 32–33; translations of ancient, 200–203, 204, 209; and Troy, 72; Turkish relations with, 439; values of, 66–67, 80; War of Independence of, 425–37; world view of, 7–8. *See also* Athens/Athenians; Persia; Sparta/Spartans; *specific person*

"Greek fire," 259

Greek Orthodox Church, 239, 254–56, 260, 271, 288, 422, 424, 427–28, 432. *See also* patriarchy, Greek

"Greekness," 6–7, 31

Gregory V (patriarch), 427, 429

Gregory VII (pope), 136, 148–50, 151, 211–12

Gulf War (1991), 248, 249, 512–14, 516

Hadith ("traditions"), 161–62, 164, 166, 172, 179, 181, 269–70, 295, 299, 300, 446, 501, 521

Hadrian (emperor), 71, 111, 197

Hagia Sophia (Constantinople/Istanbul), 144, 243, 267, 268, 269, 270, 271, 442

Hamas, 509

Hanafis/Hanafism, 166–67, 173

Hannibal, 81–83, 86, 105

Hapsburg Empire, 277, 284–87, 288, 301, 304–5, 307

Hârûn ar-Rashîd, 196, 197, 200, 207

Hastings, battle of, 25

Hattin, battle of, 238, 248–49, 252, 516

Hegel, Georg Friedrich, 38, 356

Helen of Troy, 4, 27, 51

Heraclius (emperor), 157–58, 176, 239

Hercules, 50, 63, 64

Herodotus: and Asia–Europe relations, 51; and Athenian democracy, 23; and Cyrus–Tomyris battle, 12; and Darius's image, 20; and decline of empires, 85; and distinctions between Europe and Asia, 6–7; and Egyptian–Persian battles, 26; as "Father of History," 5; and freedom of Otanes's family, 16; and government as major distinction between East–West, 525; and Greek democracy (*isonomia*), 19, 67; and Greek Empire, 292; and Greek–Persian hatred, 5; "Greekness" views of, 6–7; *Histories* of, 6, 9, 17, 21; and images of Greeks, 28; and images of

Orientals, 78; and images of Persians, 6, 15, 17, 18, 79, 102, 103; and "inhabited world," 71; and Persian "constitutional debate," 13, 14, 17, 19, 71; and Persian virtues, 154; and Persian Wars, 21, 22, 24, 25, 30, 35–36, 41, 44, 58; and politics and ethics, 15; and Rape of Europa, 4; and Roman law, 115; and selection of Persian king, 19; and strength of Persian army, 29; Thersander's conversation with, 18; and vulnerability of Persians, 19; and Xerxes, 31–33, 41

Herzl, Theodor, 418, 486–87, 488–89, 491, 492, 493

Hindus, 173, 202, 333

Hitler, Adolf, 33, 499, 530

Hobbes, Thomas, 308, 312, 313, 316–17, 318

Holland. *See* Netherlands

Holocaust, 230, 493

Holy Land, 231, 236, 242, 243, 249, 250, 485, 503. *See also* Palestine

Holy League, 288, 290

Holy Roman Empire, 147, 277, 279, 300–301, 305

Homer, 48, 50, 54, 72, 80, 273, 428

Horace, 74, 96, 332

House of War. See *jihad*

humanism, 309

Hume, David, 245, 345, 346, 349

Hungary: Carolingian Empire in, 147; church–state relations in, 300; and decline of empires, 361; and Greek War of Independence, 432; and Ottoman Empire, 256, 277, 282, 283, 284; as part of Holy Roman Empire, 301; Soviet invasion of, 496; and Treaty of Carlowitz, 290; and Treaty of Passarowitz, 440

Hussein ibn Ali, 456–61, 466, 467, 468, 482

Hussein, Saddam, 177, 178, 248, 249, 498–99, 512–14, 531

Ibn Khaldûn, 175, 349, 383

ibn Taymîyah, Taqi ad-Din, 522–23, 524

*Iliad* (Homer), 50, 54, 72, 80, 270, 332

imams, 170, 181–82, 201

immigrants, Muslims as; 537–38

imperialism, 11, 247–48, 416, 460, 469, 472, 497, 503, 527, 536–37

India: and accommodation with West, 538; Alexander the Great in, 59–60; Anquetil-Duperron in, 338–40, 341, 342; and Asia–Europe relations, 356; Bernier's writings about, 346, 347; and Bonaparte's ambitions, 367, 368, 417; and Bonaparte's Egyptian campaign, 369, 397, 400; and Britain, 347, 356, 367, 368, 417, 456, 477; colonialism in, 331, 332–33; and

despotism, 347, 350, 353, 355; and expansion of Islam, 173, 183, 196, 201; and geographical exploration, 310; Hegel's views about, 356; immigrants from, 537; Jones' views about, 332, 333; legal system in, 333; Manichaeism in, 154; Mughals of, 210, 326, 346, 347, 350, 356, 407, 473, 476; and pan Islamicism, 363; and Persian Empire, 476; politics in, 340;"protected peoples" in, 173; reform in, 504; as source of civilization, 334, 371; Zoroastrianism in, 154

indulgences, 298, 302

Inquisition, 245

Institut d'Égypte, 369, 370–71, 382–83, 385–86, 391–95, 400, 401, 415

"Investiture Controversy," 149

Io (mythological character), 4, 5, 27

Ionia, 10, 17, 21, 22, 23, 45

Iran: Alexander's invasion of, 59; and American invasion of Iraq, 34; and Asia–Europe relations, 356; as authentically Islamic, 525; and Britain, 479, 480, 500; clergy in, 170, 527; Constitution Reform in, 480; corruption in, 500, 521; and democracy/elections, 519, 527, 529, 531; and expansion of Islam, 181, 194; Hasharid dynasty of, 356; and Iraq, 177–78, 513, 531; Islamic fundamentalism in, 507–8, 509; as Islamic republic, 58, 479–81, 508, 519, 527; Khomeini uprisings in, 58, 507–8, 509; legal system in, 500; modernization/ westernization of, 249, 500, 521; naming of, 500; national epic of, 178; oil in, 479–80, 500, 508; and Ottoman Empire, 362; overthrow of monarchy in (1979), 58, 476; Pahlavi dynasty in, 249; reform in, 479–81; religious resurgence in, 507–8, 509; restoration of Xerxes' Great Hall in, 57–58; role in modern world of, 500; and Russia/Soviet Union, 479, 480, 500; Safavid dynasty of, 66, 326; secularization of, 500; Shiites in, 181; and terrorism, 493; titles of rulers of, 11; and U.S., 479; White Revolution in, 507; women in, 500; Zoroastrianism in, 154. See also Persia

Iraq: ancient objects in, 371; and Arabs, 483, 499; Arabs in, 483; attempt to unify, 498–99; Ba'th party in, 498, 513; and Britain, 459, 467, 475, 483–84, 513; Christianity in, 143; corruption in, 513; creation of state of, 483; decline of, 361; democracy/elections in, 525, 531; divisiveness within, 483; and expansion of Islam, 179, 180, 182; Faisal as king of, 483, 484, 498; Hashemite kingdom of,

Iraq (cont'd)
    484; and Iran, 177–78, 513, 531; and
    Kurds, 526; and Mamluks, 252;
    modernization of, 498–99; Mongol
    conquest of, 251–52; nationalism in, 498;
    oil in, 514; and Ottoman Empire, 468;
    pre-Islamic past of, 499; and resurgence
    of Islam, 509; Seljuq Turk rule over, 230;
    Shiites in, 467, 498–99, 526, 531; and
    Soviet Union, 498; Sunnis and Shiites in,
    467, 498–99, 526, 531; Wahhâbism in,
    421. See also Baghdad; Gulf War; Hussein,
    Saddam; Iraqi war
Iraqi war (2003), 34, 248, 249, 368, 382, 468,
    514, 516–17, 526, 531
Isis, cult of, 79
Islam: and Alexander's legacy, 66; as Antichrist,
    308; and Arab Revolt, 456–66; and
    Asia–Europe relations, 174; as based on
    divine decree, 533; and British control of
    Iraq and Palestine, 468–69; change in
    relationship between Christianity and,
    287–91; characteristics of, 196;
    Christianity and rise of, 210–22; and
    church-state relations, 164, 165, 167,
    299–300, 532–33, 536; clergy in, 170; as
    compatible with collective rule, 527;
    contribution to Christianity to, 208;
    conversions to, 171–72, 182, 183, 188,
    207, 352, 389, 441, 523; culture of, 196,
    197–204; dealing with changing world in,
    161–62; decline of, 204–5, 219–20, 221;
    dialogue between West and, 517–18; and
    differences between East and West, 381–82;
    divisiveness within, 179–82; domination
    of, 205–6; and "end of history," 532; and
    Enlightenment, 415; expansion of, 157–58,
    162–210; extremists in, 517; fatalism in,
    406–7; five pillars of, 171; as form of
    Manichaeism, 156; fundamentalism in,
    455, 507–8, 509, 532–33; as heresy,
    213–14; holy places of, 469; and
    immigrants, 537–38; impact of, 407; and
    international law, 290–91; irreconcilable
    differences between West and, 538; legal
    system in, 116, 166–67, 172, 173, 347,
    422; meaning of, 354; military of, 175,
    179–80, 182, 195; modernization of,
    205–6, 415, 416, 504; and Muhammad's
    successor, 174–75; and nationalism, 415,
    419; and need for new Qur'anic generation,
    521; need for return to roots of, 521, 522;
    as one true religion, 279; and Oriental
    despotism, 354; origins of, 158–62;
    "philosophers" of, 521; predictions about
    end of, 288–89; progressive, 520;
    purification of, 505; rebirth of, 505;
    reform of, 505–6; and Reformation, 308;
    religion as enemy of, 517; as religion of
    submission, 518–19; resistance to
    imperialism by, 536–37; resurgence of,
    498, 506–10, 521; sects in, 170; and sex,
    407–12; and sultan's spiritual jurisdiction
    over all Muslims, 363; symbols of, 232;
    tawhid doctrine of, 421; and terrorism's
    role in struggle against West, 516–17;
    theology of, 170–74, 200, 203–4, 206, 211,
    212, 295; three crusades against, 503–4;
    toleration in, 172–73, 517–18; unification
    of, 237, 453; as unifying force among
    Arabs, 198; universalism of, 164, 195, 380,
    436, 444, 508–9, 513, 524; wealth of, 180;
    and women, 407–12; and World War I,
    455; and Zoroastrianism, 159, 173. See also
    jihad; Mecca; Medina; Muhammad;
    Muslims; Shari'a; Shiites; Sunnis;
    terrorism; Ulema; Umma; specific person or
    nation/empire
Islamic Salvation Party, 530
Isocrates, 31, 32, 47, 48–49, 65, 69
isonomia, 14, 19, 28, 66–67, 525, 528
Israel: Arab focus on destruction of, 501–3,
    513; and Britain, 493; creation of state of,
    249, 484, 486, 493; and distinctions
    between Israelis and Jews, 501; and Egypt,
    496–97; military of (Haganah), 487, 492;
    and nationalism, 419; and Six-Day War
    (1967), 502–3; and Soviet invasion of
    Afghanistan, 510; and Soviet Union, 493;
    and U.S., 493; and War of Independence
    (1948), 502
Istanbul: and Ankara as capital of Turkey, 494;
    British occupation of, 291, 471–72; as
    capital of Ottoman Turks, 39;
    Constantinople renamed, 269; French in,
    440–41; Great Mosque in, 278–79;
    importance of, 472–73; Mehmed
    academy in, 350; proposed international
    administration for, 423, 472–73; and
    Russia–Ottoman wars, 362; as seat
    of caliph-sultan, 472, 473; slavery in,
    372
Italy: and Balfour Declaration, 490; and
    Charlemagne, 147; church-state relations
    in, 299; Constantine's invasion of,
    132–34; and crusade against Ottoman
    Turks, 273; and expansion of Islam, 194,
    196, 216; geographical exploration by,
    309; and Greek War of Independence,
    432; Metternich's views of, 429; and
    Napoleonic wars, 377, 399; and Palestine,
    489; as part of Holy Roman Empire, 277,
    301; Pius II tour of, 273. See also
    Rome/Romans

Jaffa, 242, 398–99

Janissary Corps, 254, 266, 268, 352, 393, 444, 445

Japan, 473, 506, 538

Jassy, Treaty of, 443

Jerusalem: Bonaparte in, 396; Bonaparte's views about, 417, 418, 485; and Britain, 419, 468, 471; and Charlemagne's extension of Roman Empire, 146, 147; Christians in, 239; Church of the Holy Sepulcher in, 134; and creation of Jewish state, 487; and Crusades, 222, 224, 227, 229, 231–33, 234–35, 237, 238–39, 241, 242; and expansion of Islam, 157; fall of, 252; as freed from Ottoman rule, 241; and Gulf War, 513; holy sites in, 239; and Islam-Christian relations, 233; Jesus's birth in, 13; Jews in, 233, 239, 486, 488, 492; Kingdom of, 221, 233, 246; Muhammad's visit to, 160; Persian attack on, 176; political significance of, 469; St. James in, 186; St. John in, 216; St. Paul in, 113, 114; and Six-Day War, 503; and terrorism, 516; Volney's trip to, 358; Wilhelm's visit to, 246

Jews/Judaism: Amiroutzes' views about, 271; and anti-Semitism, 486, 487, 488; and "Arab Renaissance," 209; Babylonian exile of, 11; in Baghdad, 467; and Balfour Declaration, 489–90; Bonaparte's views about, 417–19, 484–85, 486; Christianity as hellenized, 136; Christianity as heresy of, 128; and Crusades, 229–30, 233, 234, 239; and "end of history," 532; and Enlightenment, 417; and expansion of Islam, 168–69, 172–73, 174, 189, 208, 209; and Greek War of Independence, 427; immigration of, 492; influence on Bible of, 294; as instruments of West, 501; Islam's struggle with, 248; Israelis distinguished from, 501; in Jerusalem, 233, 239; Justinian's plans to convert, 135; in Mainz, 229, 230; and Muhammad, 501; and Orthodox-Catholic relations, 254; and Ottoman Empire, 417, 422, 427; as outpost of civilization against barbarism, 491; and Palestine, 417–19, 484–93; as "peoples of the Book," 172–73; persecution of, 135, 150, 183, 226; and St. Paul, 114; in Spain, 183, 221; and Syria, 492; and universalism, 139, 164. See also Israel; Palestine; Zionism

jihad: and apostates, 523, 524; and Arab–Israeli conflict, 508–9; and Bonaparte's Egyptian campaign, 390, 396; and diplomacy, 291; and Islamic treaties with non-muslim states, 290; and Libya–U.S. relations, 249;

meaning of, 171–72; and modern terrorism, 517; 9/11 attack as renewal of, 516; and resurgence of Islam, 508; and Treaty of Carlowitz, 288; World War I as, 454, 455, 456, 463. See also Crusades

Jones, Sir William, 330, 331, 332–34, 335, 341–42, 343, 371

Jordan, 158, 234, 459, 475, 482, 503, 517

Judea, 94, 486, 488, 489

justice, and politics, 15

Justinian (emperor), 116–17, 135, 144, 333, 449

Kant, Immanuel, 43, 321

Khomeini, Ruhollah, Ayatollah, 58, 249, 507, 508, 519, 529

kings/kingship, 65, 97, 228, 300–301, 318. See also specific person

Kitchener, Lord, 437, 455–56, 457

Kléber, Jean-Baptiste, 389, 400

Knights of Saint John "Hospitalers," 233, 234, 238, 280, 372

Knights Templar, 233, 234–35, 238

Koran, 213–14, 215–16, 348, 384, 404

Kosovo Polje "Field of Blackbirds," 257, 258

Kristovoulos, 263, 265, 266, 267, 270

Küçük Kaynarca, Treaty of, 362–64, 420, 442

Kurds, 467, 498, 499, 526, 530

Kuwait, 460, 512–14

Lactantius, 4, 134, 136

Latin League, 73

"Law Against Demons," 336–37

Law Code of 438, 144

Lawrence, T. E. "Lawrence of Arabia," 234, 241, 457–66, 469–70, 482, 484

lead poisoning, 97–98

League of Nations, 62, 470, 481, 482

"learned ignorance," 218

Lebanon, 252, 459, 469, 484, 504

legal system: and changes in law, 167; and church-state relations, 150; and emperor as source of law, 144; European, 115, 116, 117; of Germanic tribes, 115; Greek, 32–33, 67, 115, 117; in India, 333; international, 117, 290–91; in Iran, 500; Islamic, 116, 166–67, 172, 173, 422; Justinian, 116–17; and Oriental despotism, 347, 348, 354; in Ottoman Empire, 446–47, 449–50; in Roman Empire, 115–19, 449; of Theodosius II, 144; for Turkey, 494; and Twelve Tables, 115–16; for warfare, 117–18; in West, 167; and Xerxes' army, 19

Leo III (pope), 146, 151, 273

León, 186, 187, 213, 219

Lepanto, battle of, 281–82, 284, 362

"Letter to America" (Internet letter), 532–33
"Letter to Mehmed," 273–74
Libya, 90, 179, 413
literature, Golden Age of Roman, 96–97
"living together" (*convivencia*), 188–91, 221, 234, 253
Livy, 76, 82, 96, 103, 270
Lloyd George, David, 418–19, 473, 474, 475, 492
Locke, John, 299, 313, 315, 316, 317, 318
Lollards, 296
Lomellino, Angelo Giovanni, 271, 272
London, Treaty of, 436
Louis XIV, 327, 328, 445, 499
loyalty, in Islam, 164
Lucan, 79, 94, 112
Luther, Martin, 297–99, 301–2, 303–4, 308, 421, 440

Macedon, 27, 47–48, 49, 52, 58, 63, 83–84
Machiavelli, 150, 220, 221
Magi, 13, 14, 15, 80
Mahmud II, 254, 363, 435, 445, 448–49, 450
Maine, Henry Sumner, 331, 335, 355
majority rule, 16–17
Malikism, 167, 173
Malta, 280, 284, 288, 372, 376
Mamluks, 195, 252, 364, 365, 368, 375, 383, 387, 390, 391, 396, 398, 404, 406, 414
mandates, 481–82
Manichaeism, 20, 154–56, 171, 212, 508
Mantua, Congress of (1459), 273
Marathon, battle at (490 B.C.E.), 21, 24–26, 28, 37, 39, 44, 82, 279, 350, 428
Marcellinus, Ammianus, 95–96, 158
Mardonius, 18, 26, 27, 28, 38
Maronite Christians, 504
marriage(s): of Alexander, 58–59, 60, 62; of Antony and Cleopatra, 94; of Claudius, 99; in Islam, 215–16; of Nero, 101; political, 59; of Theodora and Orhan, 253
Martel, Charles "Charles the Hammer," 192, 193
martyrs, 130, 191, 225–26, 521
Marx, Karl, 347, 505, 531
Marxism, 505, 506, 507, 524
*The Masked Prophet* (Bonaparte), 365, 388–89
materialism, 379, 410
mathematics, 402–3, 441
Mawdudi, Mawlana Abdullah, 518, 519, 522, 524, 527
Maxentius, 132, 133, 135, 137
Mecca: attack on Medina by, 169; and British–Ottoman relations, 469; as holy place of Islam, 169, 174; holy sites in, 239; Hussein's capture of, 460; Ka'ba in, 163, 169, 178, 179, 421; and Muhammad, 167,

175; Muslim occupation of, 169; oligarchy of, 163; paganism in, 162; pilgrimages to, 163, 169, 171; polytheism in, 162; rejection of Muhammad in, 162–63; and Saudi dynasty, 422; U.S. forces near, 514; and Wahhâbinism, 421
Median Empire, 10, 11, 12, 13
medicine, 202, 319, 440
Medina, 163–65, 167–69, 174, 179, 180, 239, 421, 422, 464, 469, 514
Medjelle (Ottoman law code), 446, 447, 449
Mehmed II, 256, 262–72, 273–75, 277, 279, 285–87, 290–91, 350, 417, 441
Mehmed IV, 284–87
Mehmed Said, 327, 440, 441
Mehmed V, 452, 454
Menou, Jacques, 389–90, 400
merchants, 331, 390, 424–25, 428
Mesopotamia, 459, 469, 482–83, 531
Metternich, Prince Klemens von, 429, 435
Middle Ages, 188, 226, 245, 449, 521
Middle East: Baghdad as capital of Islamic, 195–96, 197, 206–7; benefits of democracy in, 526; British policies in, 482–83; and British–French relations, 485; Byzantine–Sassanid division of, 175; Crusades in, 226, 246; divisions within, 537; Gulf War as plot to dominate, 513; and Paris Peace Conference, 474. *See also specific location or nation*
Milan, Edict of, 133–34
Mill, John Stuart, 25, 307–8, 347, 405–6, 411, 535
*millet* system, 173, 422–24, 446
Mithras/Mithraism, 80, 131
modernization, 205–6, 247, 499, 504, 505–6. *See also specific nation*
Mohács, battle of, 277
monarchy: and caliphs, 182; and despotism, 352–53; divine, 63–64; as empire, 86, 95; and Islam, 195; Macedon as, 48; Ottoman Empire as, 443, 450; and Persian "constitutional debate," 13–17; and Rome, 71, 87, 95, 97; weaknesses of, 28–29
Mongols: in Baghdad and Syria, 251–52; and Byzantine Empire, 259–60, 416, 516; in China, 182; and Christian views of rise of Islam, 210; as converts to Islam, 523; cultural integration of, 352; and expansion of Islam, 196; Ibn Taymîyah fatwa against, 523; methods of, 251; Mughals as, 59–60; and Ottoman Empire, 494; and Russia, 288
Monophysites, 173, 296
Montecassino (Benedictine monastery), 194, 216
Montesquieu, Charles-Louis, baron of, 67–68,

289, 327–29, 345–46, 348, 369, 383, 407, 415
Moors, 183, 185–86, 187, 192, 197, 275
morality, 316–17, 322, 416–17
Morocco, 187, 372, 504, 524
Moscow, 288, 397, 420
Moses, 152, 160, 209, 214
Mozarabs, 188–91
Mughals. *See* India: Mughals of
Muhammad: and aethism and materialism, 379; alcohol ban by, 158–59; as Arab, 168, 483, 498; biography of, 159–60; and Bonaparte's Egyptian campaign, 374, 375, 378; Christian views about, 212–13; and church-state relations, 165–66; and "Constitution of Medina," 163–64; and creation of Islam, 521; and cultural identity for Arabs, 174; death of, 174, 294; as deity, 210–11; as despot, 348–49, 354; European images of, 348–50, 354; and expansion of Islam, 157; and fatalism, 406; Gabriel's messages to, 160, 162, 217, 538; and irreconcilable differences between Islam and West, 538; and Islamic domination of Rome, 271; and Jews, 501; as *mahdi*, 182; marriages of, 215; in Medina, 163–64, 168; as Messenger of God, 160, 161, 162, 165; 171, 215; and Muslim betrayal of old ways, 506; and Ottoman capture of Constantinople, 269; persecution of, 163; as precursor of Antichrist, 267; predictions of, 179; relics of, 454; and salvation, 201; sex life of, 214–15; and Shari'a, 166; stories about, 210–11; successor to, 174–75; titles of, 155; and universalism of Islam, 181, 436; vision of, 160–61; as war leader, 167, 172; and wars among Muslims, 512–13; and women's status, 412. *See also* Hadith; Qur'an
Muhammad (sultan). *See* Mehmed II
Müller, Max, 334, 335, 343
Mulvian Bridge (Ponte Milvio), 132, 136
Murad I, 254, 257, 258, 262
Muscovy, Duchy of, 274, 284
Muslim Brotherhood, 247, 501, 505
Muslims: as apostates, 524; and church-state relations, 533; and democracy, 525–31; and expansion of Islam, 163; and failure of Muslim societies, 506; images of, 329, 401, 411, 413; as immigrants, 537–38; integration into Europe of, 537; irreconcilable differences between West and, 524; and nationalism, 497; non-Arab, 181; persecution of, 135, 226; as roots of Arabs, 498; terrorism as means for mobilizing, 516; tolerance of/by, 135,

136; Turkey as inspiration for, 494–95; unification of, 514. *See also* Islam; *specific Muslim*
Muwashashah poetry, 188
Mycale, battle of, 38

Nafpaktos, battle of, 281
Nairobi, 489, 515
Nasrids, 186, 219, 220
Nasser, Gamal Abdel, 416, 484, 495–97, 498, 499, 503, 520
Nasserites, 498, 499, 519
nationalism: Arab, 416, 455, 463, 497–98, 519, 520; and Arab-Israeli conflict, 501–3; and Ba'th party, 513; and Bonaparte's Egyptian campaign, 419; and democracy, 499; as failure, 509; and Gulf War, 513–14; and Islam, 415, 419, 521; and Israel, 419, 487, 488; and Marxist-Leninism, 505; and Muslims, 497; and religion, 504, 505–6; and Soviet invasion of Afghanistan, 509; and Turkey as inspiration for Muslim states, 494–95. *See also specific nation*
natural law, 310–12, 317, 318
Navarino Bay, battle of, 436–37
Nazis, 150, 335
Nelson, Horatio, 372–73, 395, 396
Nero (emperor), 80, 99–100
Nestorians, 173
Netherlands, 147, 277, 290, 301, 304–5, 307, 308, 399
new world order, and Paris Peace Conference, 473–74
Nicaea: Council of, 138, 165–66, 214; and Crusades, 230, 231, 232; Ottoman capture of, 253; siege of, 226
Nicene Creed, 138
"noble savages," 386
Normans, 150, 194, 219
North Africa, 106, 127, 137, 181, 183, 212, 219, 276, 372

Octavian. *See* Augustus
oil, 479–80, 500, 508, 514
oligarchy, 13–17, 23, 97, 163
Omdurman, battle of, 455
Orient/Orientals: and "Arab Renaissance," 206; and Bonaparte's Egyptian campaign, 374; and images of Greeks, 75; images of, 78, 79, 88, 89, 123, 259, 326, 329, 330–31, 335–55, 401, 405–6, 409; as inferior to Europe, 330–31, 332; and Manichaeism, 155; Phoenicians as, 80; and Roman Empire, 123; stereotypes of, 6; virtues of, 78
Oriental despotism, 38, 346–55, 383–84, 390, 452, 457

Oriental studies, 198
Orientalism: of Antony, 93; of Greeks, 76; and
    Orthodox-Catholic relations, 255; and
    Roman Empire, 94, 95; of Russia, 289
Orientalist(s), 333–35, 341, 365–66, 371, 374,
    375, 378, 388–89, 413–14
original sin, 153
Otanes, 13, 14–15, 16, 17, 19, 23, 28, 66, 528
Ottoman Empire: absolutism of, 457; and
    Achaemenid dynasty, 353; ancient objects
    in, 371; and Arab Revolt, 455–66; and
    Asia-Europe relations, 356; and Balkan
    League, 453; and Balkans, 423, 452, 455;
    and Bonaparte, 367, 368, 413, 414, 417,
    420, 441; Bursa as capital of, 253; call for
    crusade against, 272; and Christian views
    of rise of Islam, 210; and Christianity,
    253, 271–75, 422; church-state relations
    in, 271, 424, 445, 446–47; consolidation
    of, 259–60, 274; Constantinople as capital
    of, 39; Constantinople captured by, 253,
    258, 260–72; constitution of, 363, 450–51;
    and Crimea, 442; and crusades, 276;
    decline and fall of, 287–91, 345, 359, 364,
    401, 420–25, 429, 437, 454–66, 471–72,
    476, 490, 494; and democracy, 449, 450;
    despotism in, 347, 350, 351–52, 353–54,
    413, 499; dismemberment of, 473–76,
    481, 489, 504; diversity within, 361, 422;
    divisiveness within, 283, 421–22, 444, 451,
    455; and East-West relations, 327; as
    Eastern empire, 326; education in, 450;
    and European balance of power, 429; and
    expansion of Islam, 195; expansion of,
    192, 253, 256–58, 263–66, 274, 276–77,
    279–81, 284, 353; image of, 259, 268–69,
    407; and Islam, 170, 270, 421, 436, 450,
    455; and *jihad*, 172; and Judaism, 417; and
    Knights of Saint John, 372; legal system
    in, 446–47, 449–50; as masters of Asia, 39;
    military of, 279, 285, 286, 353, 362, 439,
    441, 442, 443–44, 445, 453, 455; *millet*
    system in, 173, 422–24, 446; monarchy in,
    443, 450, 499; and mongols, 494; as
    Muslims, 253; "New Order" in, 443, 444,
    445–46; "Noble Rescript of the Rose
    Chamber" in, 446; non-Muslims in,
    422–25, 446; and papacy, 268, 273–74,
    276, 279–80, 281, 288; and patriarchy,
    271; in post-World War I years, 249;
    power of, 270; reform/modernization of,
    438–53, 454, 504; reputation of, 253,
    275–76; rise of, 252–54; and Rome, 278,
    279; and science/technology, 439, 443–44;
    as "The Sick Man of Europe," 420, 421,
    454; strengths of, 352; "sultan-caliphs"
    in, 362–63; and sultan's spiritual
    jurisdiction over all Muslims, 363;
    Tanzimat in, 445–47, 448, 451, 477–78; as
    terror of the world, 276; trade/commerce
    in, 424–25; and World War I, 363, 453–66,
    468–69, 504. *See also specific ruler or
    nation/empire*
Outremer. *See* Crusader States of the Levant

paganism, 130, 131–33, 135–36, 141, 152–53,
    162, 172, 177, 207. *See also* gods/goddesses
Pahlavi dynasty, 57–58, 64, 249, 500
Pakistan, 510, 517, 530, 537
Palestine: Ain Jalut battle in, 252; and
    Arab-Israeli conflict, 501–3; Arabs in,
    484–93; and Balfour Declaration, 489–90;
    Bonaparte in, 398–99, 418; Bonaparte's
    views about, 417–19, 484–85, 486; and
    Britain, 417, 419, 464, 468, 475, 482,
    485–93; creation of, 484–93; and
    Crusades, 240; democracy/elections in,
    529–30; and dismemberment of Ottoman
    Empire, 504; flag in, 468; and France, 246,
    413, 489, 490, 492; as future Jewish state,
    417–19; and Germany, 490; and
    imperialism, 460; and Islam-Christian
    relations, 287; and Italy, 489; Jews in,
    484–93; and Mamluks, 252; mandate for,
    482; Mongols in, 252; Muslim
    Reconquest of, 235; and papacy, 489; and
    Paris Peace Conference, 474; and
    resurgence of Islamic fundamentism,
    508–9; and Russia, 490; Saladin's desire to
    drive Christians out of, 240; and
    self-determination, 474, 492; Seljuq Turk
    rule over, 230; and Soviet invasion of
    Afghanistan, 510; and U.S., 489, 490; and
    Zionism, 485–87. *See also* Aqaba
Palmyra, 358–59, 361
pan-Arabism, 497, 503, 509, 513
pan-Islamicism, 363, 451, 488
papacy: and Bonaparte's Egyptian campaign,
    379; and caliphs as analogous to popes,
    363, 456; and church-state relations, 141,
    147–51, 299; domination by, 205–6; and
    fall of Constantinople, 268; and Holy
    League, 288; as international power,
    145–46; and Luther, 298–99, 301, 308;
    and Ottoman Empire, 268, 273–74, 276,
    279–80, 281, 288; and Palestine, 489;
    patriarchy relations with, 254–56, 260–61;
    power of, 147, 148. *See also*
    Catholics/Catholic Church
Paris Peace Conference (1919), 246–47,
    473–75, 481, 484, 494
Parsees "Parsis," 20, 154, 207, 336, 339, 341
Parthians, 77, 84–85, 87, 94, 98, 156, 175, 176,
    350, 359

Pascal, Blaise, 89, 311–12
Pasha, Kara Mustafa, 284, 286, 287
Passarowitz, Treaty of, 440
patriarchy, Greek, 145, 254–56, 260–61, 271, 424
patriotism, 497
peace: treaties for lasting international, 306–7. *See also* Paris Peace Conference; *specific treaty*
Peace of Westphalia, 306
Peacock Throne, 476
Peasants' War, 303–4
Peloponnesian Wars, 46, 47, 70
Persepolis (Persia), 10, 21, 55–58, 59
Persia: Aeschylus play about, 8–10, 44; and al-Qadisiyya battle, 177, 178; and Alexander the Great, 32, 51–60, 62, 73–74, 76, 101, 350; and Alexander's capture of women, 53–54, 60; and Alexander's legacy, 63, 66–68; and Antonine emperors, 128; and Arabs, 176–77; Aristides' views of, 102–3; and Athens/Athenians, 21–22, 24–25, 27, 28, 34–37, 45–46; as barbarians, 77; Bernier's writings about, 347; and Britain, 475, 478, 500; and Byzantium, 143; changed to Iran, 500; characteristics of, 41; clergy in, 477, 478–79; commerce/trade in, 478; "constitutional debate" in, 13–17, 19, 23, 71; Constitutional Revolution in, 479; corruption of, 478; and culture of Islam, 202; Cyrus's control of, 11–12; and Darius, 21–26, 29; decline and fall of, 38–39, 41, 176, 476–81; and Delian League, 45; democracy (*isonomia*) in, 66; and despotism, 347, 350, 352; destruction of army of, 12; divisiveness within, 11–12, 478–79; and East-West relations, 41, 44, 327; and Egypt, 26, 46, 47; ethnocentrism of, 19; expansion of, 27; and expansion of Islam, 176, 177, 181, 194–95, 206, 207, 248; and France, 477; and Greece/Greeks, 5, 7–10, 11, 13, 19, 41, 44, 48–49, 59, 65, 102, 115, 525; and growth of civilization, 128; Herodotus's story of, 5, 6, 9, 17, 21, 44; images of, 6, 13–14, 17, 18, 64, 65, 76, 79, 123; impact of, 41–42, 44; imperialism of, 11; and India, 476; influence on Islam of, 182; and irreconcilable differences between Islam and West, 538; Isfahan as capital of, 283; Jones' views about, 332, 333; King's Peace in, 46; and Macedone, 48; and Manichaeism, 155; military in, 28, 29, 30–31, 477, 478; and military power versus debate, 17–18; as monarchy, 18–19, 528; Montesquieu's views about, 327–28, 329; origins of, 10; and Ottoman Empire,

274, 283–84, 353, 468, 477; partitioning of, 480; Persepolis as capital of, 21; Philip's war against, 49; in post-World War I years, 500; Qajar dynasty in, 477, 500; reform in, 479–81; religion in, 20–21, 63; and Russia, 476, 477; Safavid dynasty in, 66, 210, 278, 283–84, 326, 350, 352, 354, 476; selection of king in, 19–21; Shiites in, 283–84, 477; as struggle between Asia and Europe, 5–10; Sunnis in, 477; Timurids in, 269; titles for rulers of, 11; and translations of ancient work, 200–201; upheavals within, 10–13, 21–22; vulnerabilities of, 19; wealth of, 55–56; world view of, 7–8; and Xerxes, 26–33, 38, 39, 45, 47, 58. *See also* Achaemenid dynasty; Iran; Sassanid Empire
*The Persian Letters* (Montesquieu), 327–28, 329, 348
*The Persians* (Aeschylus), 8–10, 35, 44
Peter the Great (czar), 289, 290, 476, 499
Phanariot Greeks, 423, 424
Pharsalus, battle of, 86
Philhellenes "Greece Lovers," 432, 434
Philip II (emperor), 7, 32, 48, 49, 67, 73, 281, 282, 304, 445
philosophy, 204, 207, 208, 209, 308, 309, 312, 313, 315–18, 322, 325, 344. *See also specific philosopher or type of philosophy*
Phoenicia, 54, 80–81, 359. *See also* Carthage
Pius II (pope), 268, 272–73
Plassey, battle of, 356
Plataea, battle of, 18, 38, 275
Plato, 37, 42, 43, 69, 70, 89, 118, 129, 200, 201, 202, 273, 332, 345, 531
Pliny the Elder, 102, 129
Plutarch, 42–43, 50–51, 66, 67, 68, 69, 87, 88, 89, 90, 92, 94, 120–21
Poitiers, battle of, 192–93, 194
Poland, 257, 274, 288, 306, 362, 432
politics: and contract theory of government, 318–19; and Enlightenment, 318; and ethics, 15; as "the good life," 15; and justice, 15; as major distinction between East and West, 525; as man-made, 532; and public celebrations as means of propagating political messages, 393–94; and religion, 306–8, 318, 367. *See also* government
Pompey, 61, 86–87
popes. *See* papacy; *specific pope*
Portugal, 151, 186, 284, 309, 432
Prague, defenestration of, 305
Priam, 51, 271
printing, 439, 440–41
"Proclamation to the Egyptians" (Bonaparte), 374–76

Prophet. *See* Muhammad
protected people, 422, 446
Protestants, 226, 297–309
Ptolemies, 56, 61, 93, 94, 208, 209, 310, 416
public celebrations, 393–94
Punic Wars, 81, 82

Qur'an: Alexander as figure in, 66; Amiroutzes' views about, 271; Avesta compared with, 343; Bible compared with, 161, 162, 294, 295, 299–300; and Bonaparte's Egyptian campaign, 369, 370, 374, 375, 376, 378, 380, 388, 393, 413; centrality to Islam of, 382; Christian views about, 214, 218; and church–state relations, 299–300, 536; as divine and immutable, 161; interpretations of, 181, 295, 354; Jesus as prophet in, 168; and Jews, 168, 501; Latin translations of, 213; and Muhammad as god, 211; and Muhammad and Seal of the Prophets, 155; and need for return to original sources of Islam, 521; and unbelievers, 171–73; organization of, 161; and philosophy, 203, 204, 208; printing of, 440, 441; and reform of Ottoman Empire, 440; and Saladin as hero, 239; Shari'a and Hadith as derived from, 166; and signs of divine favor, 168; and terrorism, 515; translations of, 188; and Uthmân, 179, 180; women in, 412; writing of, 160–61
Qutb, Sayyid, 247–48, 519 22, 523, 533

*Rasselas* (Johnson), 323–24
Raymond of Saint-Gilles, Count of Toulouse, 230, 232, 233
reason/rationalism, 317–18, 320, 322, 324, 325, 356, 381–82, 412, 444
"Reconquest," 186, 187, 220, 235
Reformation, 302–9
reformers: of late nineteenth century, 504. *See also specific person*
regime change, democratic, 525–31
Reginald of Châtillon, 237–38, 241, 248
relativism, 317–18, 325
religion: and Alexander's legacy, 63; and Bonaparte's Egyptian campaign, 369–70, 376–77, 378; Bonaparte's views about, 368, 417–18; and decline and fall of empires, 359; distinction between society and, 44; domination by, 205–6; as enemy of Islam, 517; and Enlightenment, 325; and failure of West, 519–20; fundamentalism in, 325, 532; as God-made, 532; minimization of, 533; and nationalism, 504, 505–6; and Oriental despotism, 348, 354; and

philosophy, 203; and politics, 306–8, 318, 367; and science, 205–8, 315–16. *See also specific religion or nation/empire*
Rémusat, Claire de, 369, 376
Renaissance, 13, 53, 193–209, 309–12
Renan, Ernest, 205–6, 207–8
Roman Empire: accomplishments of, 353; and Actium, 91–93, 95; and Alexander's legacy, 62, 65, 68; Antonine emperors of, 71, 100–101, 118, 122, 124; and Antony and Cleopatra, 88–95; Aristides views about, 531; and Asia, 72–73, 79, 80, 83, 112; and Asia–Europe/West, 72–73, 111, 112, 137; and barbarians, 127; beginnings of, 74; beneficence of, 70, 71, 111–12, 121; boundaries of, 70–71; characteristics of, 85–95; and Christianity, 97, 110, 113–15, 127–28, 129, 130, 132–38, 143–44, 145; and church–state relations, 137–38, 139–41, 143, 147–51; civil wars in, 86–88, 90–95, 131; *civitas*/citizenship in, 104–6, 109, 110–18, 121, 123–24, 139; Constantinople as capital of, 261; constitution of, 71, 85, 531; and Crusades, 242; decline and fall of, 12, 106–7, 122–24, 126, 131–32, 145, 179, 196–97, 242, 288, 322, 359; and decline of paganism, 135–36; democracy in, 531; divisiveness within, 90, 106, 132, 142; as "Earthly City," 140–41; Eastern Empire of, 135, 148; emperors of, 144; and "end of history," 531; and Enlightenment, 322; European colonization by, 103; expansion of, 70–71, 73, 77, 79, 81, 83, 117, 145–46; Golden Age of, 71–72, 96–97; and growth of civilization, 128; Holy Roman Empire as successor to, 300; internal divisions in, 100; Julio-Claudian dynasty of, 84, 97, 98–100, 122; and kingship as theocracy, 65; legal system in, 115–19, 449; as legitimate ruler of Greeks, 85; military strength of, 29, 86, 101, 103, 131; as monarchy, 95, 97; Octavian as master of, 93; and oligarchy, 97; oppression of, 111; and Orientalism, 94, 95; and Orientals, 123; Pax Romana of, 97; population of, 71; power of, 107–10; as principate, 95, 96, 107; rebirth of, 146; reunification of, 136; rise of, 73–74; and Seleucid Empire, 62, 63; symbols of, 85–86, 101, 111; technology of, 101; toleration in, 135–36; unification of, 111; and unity of mankind, 65; universalism of, 118–19, 120–22, 124, 128; and *vanitas*, 95; as worldwide civilization, 101–2, 350. *See also* Rome/Romans; *specific person or nation/empire*

Roman Republic, 367

Romania, Empire of. *See* Constantinople: Kingdom of

"Romanness," 104–6

Romantics, 325, 462, 485

Rome/Romans: and Alexander the Great, 70, 76; Arch of Constantine in, 135; Aristides' views about, 360; Asian civilization as equal to, 344; and barbarians, 74, 77–78, 82, 104, 106–7; beginning of historical, 73–74; burning of, 100; and Byzantine Empire, 261; *civitas* of, 104–6, 109, 110–18, 121, 123–24; Constantinople as "new," 137, 142; constitution of, 74; cosmopolitanism of, 103; and *dar al–Islam*, 270–71; discussions about best form of government among, 13; domination of Greeks by, 70; as eternal city, 127, 129; founding of, 143, as fusion of Asia and Europe, 72–73; and fusion of Greek and Roman culture, 74–77; gods of, 73, 80, 127, 130, 131, 132–33, 135; Goths' sacking of, 106, 126–27, 129, 131, 145, 267, 270; as greatest of all civilizations, 101; Henry IV attack on, 149–50; and humanism, 309; images of, 81, 97, 101, 103, 111; and images of Persians, 79; Jones' views about, 332; lead poisoning in, 97–98; Luther's visit to, 297–98; as monarchy, 87; mythological origins of, 72, 73; and Ottoman Empire, 271–75, 278, 279; and pagan influence on Christianity, 152–53; perception of East by, 83–85; and Renaissance, 309; as republic, 85–87, 95, 96, 97, 112; role in history of, 416; and "Romanness," 104–6; and Sassanids, 176; and universalism of Christianity, 139; and U.S. as Rome of the West, 107; values of, 76, 78, 79; virtue of, 84, 102; wall around, 279; and West, 292; world view of, 69–70, 101. *See also* papacy; Roman Empire; *specific person*

Rousseau, Jean-Jacques, 349, 369, 376–77, 383, 386, 404, 415, 441, 531

Roxana, 58–59, 60

rule of law. *See* legal system

Russia: Bolshevik Revolution in, 495; Bonaparte invasion of, 421; and Bonaparte's Egyptian campaign, 397; and Bonaparte's "new Byzantine Empire," 397; and Byzantine Empire, 288; and Crimea, 442; European views of, 289; expansion of, 290; "Great Game" of, 477; and Greek War of Independence, 428, 436, 437; and Iran, 479, 480; military of, 290; modernization of, 289, 290; and Ottoman Empire, 288–90, 362–64, 387, 420–21, 428, 429, 436, 442–43; and Palestine, 490; as part of Asia, 289; and Persia, 476, 477; religion in, 288, 362; and Wars of Religion, 306. *See also* Soviet Union

Rycaut, Paul, 326, 329–30

sacraments, 144, 255, 256

Sâdât, Anwar, 510, 522

Said, Edward, 330, 423

St. Augustine, 126–28, 129–30, 139, 140–42, 152, 226, 298, 310

St. Irenaeus, 211

St. James, 186–87

St. Jerome, 137, 295, 300, 303

St. John, 120, 211, 212–13, 216

St. Martin de Tours, 192

St. Paul, 113–14, 115, 126, 129, 134, 138–39, 148, 153, 165, 294, 296, 297, 299

St. Peter's church (Rome), 134, 194, 298

Saladin, 236–41, 242, 246, 248, 416, 503, 513

*salafism*, 504, 521

Salamis, battle at (480 B.C.E.), 8–10, 21, 25, 34–38, 39, 44, 45, 95, 282, 350

Saracens, 189, 191, 210, 216, 217, 244, 261

Sardis (kingdom of Lydia), 10, 22, 24, 29, 52

Sassanid Empire: and Alexander's legacy, 175; and Arabs, 464; as Archaemenid heirs, 124; and Byzantium, 176; culture of, 206–7; and division of Middle East, 175; divisiveness within, 176; end of, 177, 206; and expansion of Islam, 158, 159, 176–77, 195; influence on Arabs of, 353; and Mithraism, 131; and Parsees, 336; and Parthians, 85, 350; and Romans, 143; as threat to West, 156; Umar as caliph of, 179; Zoroastrianism in, 155, 175–76, 207

Saudi Arabia/Saudis, 421–22, 457, 461, 503, 510, 512, 514, 515, 517, 522, 523

"savants," 370–71, 391–95. *See also* Institut d'Égypte

Schmalkaldic League, 304

Scholasticism, 313

science: ancient Egypt as source of, 402; beneficent nature of, 324; and Bonaparte's Egyptian campaign, 369, 379, 392, 394, 416; and Christianity, 205; and creation of new sciences, 320; Egypt as source of Persian and Greek, 344; emergence of modern, 312–20; and Enlightenment, 315–20, 322, 323, 325; and images of Orientals, 355; and Islam, 205; as "new philosophy," 313; observation and experiment in, 315; and Oriental despotism, 354; and Ottoman Empire, 439, 444; and paganism, 207; and Qutb's views of West, 520; and religion, 208, 315–16; revolution in, 315–16; and wealth, 320

Scipio Aemilianus, 70, 83

Scipio Africanus "Scipio the Elder," 83, 86, 275

Scipio, Publius Cornelius, 82, 267

"Seal of the Prophets," 155, 161, 168, 174, 214, 349

secularism: beginning of, 322; and Bonaparte's Egyptian campaign, 384, 413; and characteristics of modern West, 517; and church-state relations, 533–34; and differences between East and West, 381–82; and education, 447, 450; and failure of Muslim societies, 506; and Islam, 499, 518–19; and Qutb's teachings, 522; and science, 354; and Tanzimat, 446. See also church-state relations; specific nation

Seleucid Empire, 62, 63, 83, 84, 86

self-determination, 470–71, 474, 492

Selim III, 396, 414, 422, 443, 444–45, 446, 448, 499

Seneca, 61, 77, 84–85, 90, 197, 202

September 11, 2001, 515–16, 523

Serbia, 256, 257, 258, 263, 277, 361, 420–21, 453

Sermon on the Mount, 300

Sèvres Agreement, 496

sexual issues, 214–15, 327, 328, 407–12, 520

Shari'a: and Bonaparte's Egyptian campaign, 377, 380, 388; and conversions to Islam, 171, 523; interpretation of, 170; and Iranian revolution, 500, 508; and Islam's rejection of democracy, 519; as laws of Islam, 116, 166–67; and "Letter to America," 533; and Oriental despotism, 347; and Ottoman Empire, 449–50; and reform of Islam, 505; and reform of Ottoman Empire, 442; and resurgence of Islam, 508; and separation of church and state, 533; and Treaty of Carlowitz, 290; and Turkey, 494

Shiites: and Abbasids, 194; basic beliefs of, 181, 182; and Ba'th party, 513; and British, 482–83; and divisions within Islam, 170, 180–81, 231, 482–83; in Egypt, 181; hierarchy of, 477; in Iran, 181; in Iraq, 467, 498–99, 526, 531; and jihad, 171; and non-believers, 181; in Persia, 283–84, 477; and Saladin, 237; and Wahhâbism, 421

Sibylline Books, 87

Sicily, 73, 80, 81, 194, 219, 281

Six-Day War (1967), 502–3, 507

Skepticism, 313–14, 318

slaves/slavery: in Algeria, 276; and Bonaparte's Egyptian campaign, 372, 376; and Christianity, 139; and church-state

relations, 151; and Crusades, 238, 239; and expansion of Islam, 194, 200; Franks as source of, 200; Greek, 108, 109; and Greek War of Independence, 430, 431; and images of Orientals, 79; Mamluks as, 195; Muslim views of, 403; Muslims as, 173, 195; and Oriental despotism, 347–48, 383; and Ottoman Empire, 276; and overseas empires, 151; "political," 348; and Roman Empire, 107–8, 109; and women, 408

Social Contract (Rousseau), 376

Socrates, 61, 202

The Song of Roland (French poem), 216–17, 218

Soviet Union: Afghanistan invasion by, 509–10, 512; and Arabs, 497; and China, 524; and democracy, 526; and Egypt, 496; as empire, 86; Hungarian invasion by, 496; and Iran, 500; and Iraq, 498; Israel recognized by, 493; Marxist-Leninism of, 505; and warfare, 118. See also Russia

Spain: Carthaginians in, 81; Catholics in, 186; and China, 151; Christians in, 184–85, 186, 187, 188, 190–91, 213, 219; church-state relations in, 190–91, 307; culture in, 188–89, 197–98, 202–3, 206; despotism in, 282; and expansion of Islam, 183–91, 194, 196, 197, 206, 216; Franco regime in, 150; as "Gallic Empire," 123; geographical exploration by, 309; Goths in, 183, 184, 185, 186, 197, 219; and Greek War of Independence, 432; Inquisition in, 206; Jews in, 183, 221; Moors in, 183, 185–86, 197; Mozarabs in, 189–91; Netherlands as independent from, 307; and Ottoman Empire, 281; overseas empire of, 151; as part of Holy Roman Empire, 277, 301; Reconquest of, 219–20, 221–22; reform efforts in, 443, 445; and Roman Empire, 98; and Treaty of Westphalia, 307; and Wars of Religion, 305

Sparta/Spartans, 17, 23, 24–25, 32, 33, 46, 47. See also Peloponnesian Wars

Spring-Rice, Cecil, 480–81, 506

Stoics, 119–21, 122, 138, 151

Storrs, Ronald, 437, 457, 459, 461, 466

Sudan, 414, 437, 455, 456, 517

Suetonius, 93, 98, 99–101

Suez, 368, 465–66, 496–97

Sufis, 241

Süleyman (caliph), 191–92

Süleyman I "the Magnificent," 277–80, 291, 353, 362, 440

sultans, 362–63, 449. See also specific person

Sunnis: basic beliefs of, 180–81, 182; and British, 482–83; as caliphs, 279; and

divisions within Islam, 170, 180–81, 231, 482–83; in Iraq, 467, 498–99, 526; in Persia, 477; and resurgence of Islam, 508; Shiite wars against, 231, 482–83; Süleyman as upholder of beliefs of, 278
Susa (Persia), 21, 38, 54, 60, 63
Sweden, 305, 306, 308, 432
Switzerland, 304, 307, 357, 432
Sykes, Mark, 459, 460, 466–67, 468, 470, 491
Sykes-Picot agreement, 459–60, 469, 470
Syria/Syrians: and Abbasids, 195; ancient objects in, 371; and Antony and Cleopatra, 90; and Arab Revolt, 458; Arabs in, 158; as barbarians, 77; Ba'th party in, 497–98; Bonaparte in, 398, 420; and Britain, 469, 475; and Byzantine Empire, 192; Christians in, 234, 239; and church-state relations, 418; and Crusades, 230, 231, 234, 236, 237, 238, 240, 246, 247; decline of, 361; and expansion of Islam, 176, 179, 182; Faisal as king of, 483, 492; and France, 246–47, 413, 459, 469, 475, 482, 484; images of, 79; and Jews, 492; and Mamluks, 252; mandate for, 482; Mongol conquest of, 251–52; and Ottoman Empire, 455, 504; paganism in, 207; and Palestine as future Jewish state, 417; and Paris Peace Conference, 474, 475; in post–World War I years, 246–47, 504; reform in, 504; as republic, 484; and Roman Empire, 79, 98; Saladin as ruler of, 237; and self-determination, 474; Seljuq Turk rule over, 230; and Six-Day War, 503; and slavery, 372; and Sykes-Picot agreement, 459, 469, 470; and UAR, 498; Volney's trip to, 357, 358, 371–72, 398; Wilhelm II's visit to, 240–41; Willibald's arrest in, 189; and World War I, 455, 458

Tacitus, 99, 100, 110
Taliban, 510, 516, 522, 524
Talleyrand-Périgord, Charles-Maurice de, 365, 367–68
Tamerlane, 259–60, 477, 494
Tancred, 232, 233, 244, 485
Tanzimat, 445–47, 448, 451, 477–78
Tarn, W. W., 62, 65, 68
Tartars, 206, 285, 286
Tasso, Torquato, 244, 392
Tavernier, Jean-Baptiste, 326, 328, 329–30, 476
tawhid doctrine, 421
technology, 101, 319–20, 353, 379, 404, 439, 443–44, 506
Ten Commandments, 151–52
terrorism, 227, 248, 493, 502, 510, 514–17
Thebes, 18, 47, 48

Themistocles, 34, 35, 37, 428, 436
theocracy, kingship as, 65
theology, 209, 312, 313, 316. See also specific religion
Thermopylae, battle at, 30–31, 33–34, 39, 428
Thirty Years' War, 305
Thucydides, 37, 44, 45
Tocqueville, Alexis de, 391
Toledo, Spain, 183, 184, 197, 219
Topkapi Palace (Istanbul), 269, 280, 454
Tott, Baron de, 365, 441
trade/commerce, 131, 326, 329–30, 424–25, 428, 478
Trajan (emperor), 61, 70, 71, 99, 100, 105, 152, 197
treaties, 306–7. See also specific treaty
Treaty of Campo Formio, 367
Treaty of Carlowitz, 290–91, 362, 420, 440
Treaty of Jassy (1792), 443
Treaty of Küçük Kaynarca, 362–64, 420, 442
Treaty of London (1827), 436
Treaty of Passarowitz (1718), 440
Treaty of Westphalia, 306–8
Tripoli, 233, 234, 235, 239, 280, 372
Trojan War, 4–5, 48, 50, 51, 73, 270
Troy, 4–5, 29, 49–50, 52, 72, 76, 99, 270
Truman, Harry S., 493
Tunisia, 243, 372, 504, 524
Turkey: Ankara as capital of, 494; Bernier's writings about, 347; and Bonaparte, 366; and Britain, 459; constitution for, 494; and Crusades, 234; democracy in, 530; and European Union, 448, 530; formation of modern, 454; and Germany, 453–54, 457; as inspiration for Muslim states, 494–95; legal system for, 494; military of, 494; nationalism in, 494–95; and Ottoman Empire, 254; post–World War I years in, 494–95; reform/modernism in, 477, 524; Republic of, 454
"Turkification," 452–53, 455
Turks: and Crusades, 230–31; Greek relations with, 439; as identified with Islam, 530–31; as Mamluks, 195; and national identity in Turkey, 494; number of, in Ottoman Empire, 423; Seljuq, 230–31, 353. See also Ottoman Empire; Uyghur Turks
Twelve Tables, 115–16
Tyre, 235, 239, 242

Ukraine, 306, 362, 442
Ulema (community of scholars), 166, 170, 382, 388, 393, 414, 447, 477, 494, 499, 507, 524
Umayyad Caliphate, 178–81, 192, 194–96, 206, 268, 363, 416

Umma (Islamic community), 163, 164, 167,
174, 195, 201, 497, 521, 522, 538
United Arab Republic (UAR), 498
United Nations, 62, 482, 496, 514
United States: and Alexander's legacy, 62; and
Balfour Declaration, 490; church-state
relations in, 532, 535–36; and Egypt, 496,
497; as empire, 86; and Enlightenment,
322–23; and Greek War of Independence,
429, 430–31, 432; and Gulf War, 512–14;
imperialism of, 460; and Iran, 479; and
Israel, 493; and Jordan, 482; as
manifestation of crusader, 508; and
military technology, 320; and Palestine,
489, 490; and Paris Peace Conference, 470,
473; and Qutb's views of Islam, 520;
Qutb's visit to, 526; religious
fundamentalism in, 519; as Rome of the
West, 107; and Soviet invasion of
Afghanistan, 510; and Taliban, 516;
terrorism against, 514–17; and warfare,
118. See also Gulf War; Iraqi War; Wilson,
Woodrow
universalism: and Alexander the Great, 60, 62;
and Arabs, 497; and Aristides, 121–22; as
of Asian origin, 11; of caliphate, 497; as
central to European expansion, 11; of
Charlemagne, 147; and Christianity, 124,
128, 130, 134, 138, 152, 164, 165; of
Cicero, 118–19; continuing dream of, 124;
and differences between East and West,
381–82; and Edict of Milan, 134; and
Europe, 124; of Islam, 164, 195, 380, 436,
444, 508–9, 513, 524; and Judaism, 164; of
Muhammad, 181; and Roman civitas, 151;
of Roman Empire, 118–19, 120–22, 124,
128; of St. Paul, 114, 138–39; and
self-determination, 471; and Stoics, 151;
and values, 128
Urban II (pope), 224–25, 226, 227, 233
USS Cole, 515

values: and church-state relations, 141; and
democracy, 19; and differences between
East and West, 381–82; Enlightenment,
321–22, 325, 535; of Greeks, 66–67, 80;
Qutb's views of West, 520–21; of Rome,
76, 79; and universalism, 128; of West,
138
vanitas, 78–79, 95
Vatican. See papacy
Velestinlis, Rigas, 413, 425–26, 434, 435
Vendîdâd, 336, 339
Venice, 242, 243, 262, 280, 281, 282, 284, 288
Venture de Paradis, 374, 378
Vienna, Austria, 279, 283, 286–87, 288, 420,
473

Vienne, Council of, 217
Virgil, 72, 73, 76, 92, 96, 102, 143
virtues (virtus), 76–77, 78, 84, 88, 102, 130, 141,
154, 316, 322
Visigoths. See Goths
Volney, Constantin-François, 357–63, 366,
367, 371–72, 374, 383, 384, 396–98, 412,
417
Voltaire, 147, 240, 243, 342, 344–45, 346,
348–55, 357, 369, 383, 401, 415, 428, 444

Wahhâbism, 421, 510
Waldesians, 296
war, 117–18, 147, 226, 228, 307. See also
Crusades; jihad; specific war
Wars of Religion, 304–9
Weizmann, Chaim, 490, 492
West. See Asia and Europe/West; Europe/West
Westoxification "gharbzadegi," 521
Westphalia, Treaty of, 306–8
"White Revolution," 500, 507
Wilhelm II (kaiser), 240–41, 246, 418, 454
Wilson, Woodrow, 470, 471, 474, 475
"Winnie the Pooh" cartoon, 531
women, 407–12, 474–75, 494, 500
World Trade Center, terrorist attacks on, 248,
515–16
world view, 7–8, 62, 69–70, 101, 292
World War I: and Britain, 454–66, 468–69;
and dismemberment of Ottoman
Empire, 468–69, 473–76; and Egypt,
437, 495; and fall of Istanbul, 471–72;
and France, 455, 459; and Germany,
453–54, 455, 457, 461, 463; and Islam,
455; as jihad, 454, 455, 456, 463; and
Ottoman Empire, 363, 453–66, 468–69,
504; and Paris Peace Conference, 246,
473–75; and Sykes-Picot agreement,
459–60, 469, 470; and Zionism,
489
World War II, 33, 471, 492, 493
Worms, Diet of, 149
Worms, Edict of, 302

Xerxes (king): and Alexander-Persian war, 50,
57; as character in Aeschylus play, 8–9;
death of, 58; Demaratus's encounter with,
31–33; and gods, 30; heirs of, 352, 353;
Herodotus's views of, 41; Hundred-
Columned Hall of, 56–59; and Macedon-
Persian War, 49; motivation of, 62;
Pahlavi as heir of, 57; and Persian Wars,
26–31, 38, 39, 45, 47, 58; and rule of
law, 115; at Salamis, 8–9, 34–37

Young Ottomans, 450, 451
Young Turks, 451, 452–53, 461, 494, 504

Yousef, Ramzi, 514–15
Ypsilantis, Alexandros, 428–29, 434, 435

Zangwill, Israel, 418–19
Zeno of Citium, 67, 119, 120, 121
Zenobia (queen), 123, 358
Zeus (mythological character), 44, 53, 64, 100–101, 121
Zionism, 418, 419, 485–93, 513
Zoroastrianism: and Achaemenid dynasty, 13, 20–21, 154; and Alexander the Great, 64; and Anquetil–Duperron's works, 337, 341–43, 344; and Asian roots of Greeks, 337; basic beliefs of, 154; calendar of, 58; and church-state relations, 167; and dualism, 153–54; and East–West relations, 343; and Islam, 159, 173; and Manichaeism, 155; and Mithraism, 131; in Sassanid Empire, 155, 175–76, 207; spread of, 154. *See also* Avesta

Anthony Pagden is distinguished professor of political science and history at the University of California, Los Angeles. He was educated in Chile, Spain, and France, and at Oxford. He has been the reader in intellectual history at Cambridge, a fellow of King's College, a visiting professor at Harvard, and Harry C. Black Professor of History at Johns Hopkins University. He is the author of many prize-winning books, including *Peoples and Empires: A Short History of European Migration, Exploration, and Conquest, from Greece to the Present* and *European Encounters with the New World: From Renaissance to Romanticism.* Pagden contributes regularly to such publications as *The New York Times, Los Angeles Times,* and *The New Republic.*